ASTROPHYSICAL MASERS:
UNLOCKING THE MYSTERIES OF THE UNIVERSE

IAU SYMPOSIUM 336

COVER ILLUSTRATION: SA GENTI ARRUBIA

Pink flamingos, or *sa genti arrubia* (the red people), as they are called in Sardinia, populate the ponds and salt marshes surrounding Cagliari. In particular, in the Ponds of Santa Gilla and the Molentargius Regional Park it is possible to admire hundreds of them, together with several other species of aquatic birds, wintering and nesting during their migration between Europe and Africa.

They are often seen with their heads under the shallow water in search of tiny crustaceans on which they feed, or, as shown in this photograph, during their pleasant and graceful flight, with their beautiful shades of colors offered by their spread wings.

It is particularly amazing, and likely the only place in the world, that these birds, usually so shy, stop and nest in Cagliari, a modern city, so highly urbanized and built up. This has led to the flamingos becoming a symbol of the town.

Photo credit: Gian Paolo Vargiu, INAF-Osservatorio Astronomico di Cagliari

IAU SYMPOSIUM PROCEEDINGS SERIES

Chief Editor
PIERO BENVENUTI, IAU General Secretary
IAU-UAI Secretariat
98-bis Blvd Arago
F-75014 Paris
France
iau-general.secretary@iap.fr

Editor
MARIA TERESA LAGO, IAU Assistant General Secretary
Universidade do Porto
Centro de Astrofísica
Rua das Estrelas
4150-762 Porto
Portugal
mtlago@astro.up.pt

INTERNATIONAL ASTRONOMICAL UNION

UNION ASTRONOMIQUE INTERNATIONALE

ASTROPHYSICAL MASERS: UNLOCKING THE MYSTERIES OF THE UNIVERSE

PROCEEDINGS OF THE 336th SYMPOSIUM OF THE INTERNATIONAL ASTRONOMICAL UNION HELD IN CAGLIARI, ITALY SEPTEMBER 4–8, 2017

Edited by

ANDREA TARCHI
INAF-Osservatorio Astronomico di Cagliari, Italy

MARK J. REID
Harvard-Smithsonian Center for Astrophysics, USA

and

PAOLA CASTANGIA
INAF-Osservatorio Astronomico di Cagliari, Italy

CAMBRIDGE
UNIVERSITY PRESS

CAMBRIDGE UNIVERSITY PRESS

University Printing House, Cambridge CB2 8BS, United Kingdom

1 Liberty Plaza, Floor 20, New York, NY 10006, USA

10 Stamford Road, Oakleigh, Melbourne 3166, Australia

First published 2018

Printed in the UK by Bell & Bain, Glasgow, UK

Typeset in System LaTeX 2_{ε}

A catalogue record for this book is available from the British Library Library of Congress Cataloguing in Publication data

This journal issue has been printed on FSC^{TM}-certified paper and cover board. FSC is an independent, non-governmental, not-for-profit organization established to promote the responsible management of the world's forests. Please see www.fsc.org for information.

ISBN 9781107192454 hardback
ISSN 1743-9213

Table of Contents

Dedication of the Symposium

Theory of masers and maser sources
Chairs: M.J. Reid, T. Robishaw, S. Goedhart

Galaxies and Supermassive Black Holes
Chairs: P. Castangia, J. Moran, W. Baan

The Structure of the Milky Way
Chairs: Y. Xu, S. Ellingsen

Star Formation
Chairs: J.-M. Torrelles, K.-T. Kim, Z. Abraham, C. Goddi

Evolved Stars
Chairs: W.H.T. Vlemmings, A. Richards, L. Humphreys

New facilities
Chairs: H.-J. van Langevelde, S. Breen

Preface

The IAU Symposium No. 336, titled "Astrophysical Masers: Unlocking the Mysteries of the Universe", was held from the 4th to 8th of September 2017 in Cagliari (Italy), an enjoyable city with a variety of attractions for visitors, located in the southern part of Sardinia that hosts among the most beautiful seashores in the Mediterranean. The Symposium was attended by over 150 participants from 25 countries.

This Symposium was the fifth in a series of conferences dedicated to astrophysical masers. The previous conferences were held in the United States (Arlington, Virginia in 1992), Brazil (Angra dos Reis in 2001), Australia (Alice Springs in 2007), and South Africa (Stellenbosch in 2012). The Symposium was opened by the IAU General Secretary, Prof. Piero Benvenuti, who introduced the main aims and tasks of the IAU, underlined the relevance of the IAU Symposia, welcomed the participants, and officially started the Symposium.

Astronomical masers touch on a very broad range of astrophysical phenomena, from those taking place in massive star forming regions and evolved stars, to those in accretion disks around super-massive black holes. Thus, masers provide invaluable tools to estimate fundamental astronomical quantities, such as gas motions, distances, and black hole masses. Since the last conference there has been an explosion of work on masers, especially related to the cosmic distance scale, the structure of the Milky Way, and the masses of (AGN) black holes.

Indeed, the important results presented in over 140 contributions (oral and posters), all of very high quality, and arranged in six oral sessions and two poster sessions, have once again demonstrated that molecular masers are fundamental tools to provide answers to many of the most important and puzzling mysteries of modern astronomy. In addition, the Symposium testified to the dramatic growth of the field due to improvements in astrometry with VERA and the VLBA, and the coming online of major new facilities, like ALMA, eMERLIN, and the KVN, and the promising opportunities offered by SKA and its pathfinders, and the Next Generation VLA.

In the framework of the intense programme spiced up with stimulating discussions, the participants had the chance to visit the site of the 64-m Sardinia Radio Telescope, located in the Gerrei area. This facility, as illustrated during the Symposium in some scientific talks, is indeed an excellent instrument to perform forefront observational programmes in maser science, both in Galactic and extragalactic environments, as a single-dish and as part of VLBI arrays.

In order to leave a significant legacy in the community after the Symposium, educational and outreach activities took place in parallel with the scientific programme.

The editors would like to thank our sponsors and all the personnel of the Osservatorio Astronomico di Cagliari for their support. In addition, we would like to thank the SOC and LOC members. We are particularly grateful to Alberto Sanna, Ciriaco Goddi, Gabriele Surcis, and Luca Moscadelli who conceived of holding this Symposium in Sardinia and supported it throughout its realization. Last, but not least, we thank the staff of the Hotel Regina Margherita in Cagliari, the venue of the conference, for their professionalism and kindness.

Andrea Tarchi, Mark Reid & Paola Castangia

THE ORGANIZING COMMITTEE

Scientific

Z. Abraham (Brazil)
A. Bartkiewicz (Poland)
P.J. Diamond(UK)
S. Ellingsen (Australia)
G. Garay (Chile)
S. Goedhart (South Africa)
M. Honma (Japan)
L. Loinard (Mexico)

N. McClure-Griffiths (Australia)
K.M. Menten (Germany)
L. Moscadelli (Italy)
Y. Pihlstrom (USA)
M.J. Reid (co-Chair, USA)
A. Sanna (Germany)
A. Tarchi (Chair, Italy)
X.-W. Zheng (China)

Local

P. Castangia
S. Casu (Chair)
D. Coero Borga
T. Coiana
G. Melis

S. Poppi
P. Soletta
G. Surcis
A. Tarchi (co-Chair)

Acknowledgements

The symposium is sponsored and supported by the IAU Divisions H (Interstellar Matter and Local Universe), C (Education, Outreach and Heritage), G (Stars and Stellar Physics), J (Galaxies and Cosmology); and by the IAU Commissions A1 (Astrometry) and B4 (Radio Astronomy).

Funding and support by the
International Astronomical Union,
Istituto Nazionale di Astrofisica,
Associazione Culturale Cefalú and Astronomy,
Comune di Cagliari,
European Union,
Regione Autonoma della Sardegna,
and
Unione dei Comuni del Gerrei
are gratefully acknowledged.

CONFERENCE PHOTOGRAPH

Photo credit: Gianni Alvito, INAF-Osservatorio Astronomico di Cagliari

Participants

Artis **Aberfelds**, Ventspils International Radio Astronomy Center, Latvia — artis.aberfelds@venta.lv
Zulema **Abraham**, Astronomy Department University of Sao Paulo, Brazil — zulema.abraham@iag.usp.br
Aleksei **Alakoz**, Astro Space Center of the Lebedev Physical Institute, Russia — alexey.alakoz@gmail.com
Megan **Argo**, University of Central Lancashire, UK — margo@uclan.ac.uk
Yoshiharu **Asaki**, National Astronomical Observatory of Japan, Tokyo, Japan — yoshiharu.asaki@nao.ac.jp
Kitiyanee **Asanok**, National Astronomical Research Institute of Thailand, Thailand — kitiyanee@narit.or.th
Willem A. **Baan**, ASTRON Netherlands Institute for Radio Astronomy, The Netherlands — baan@astron.nl
Alejandro **Báez Rubio**, Universidad Nacional Autónoma de México, Mexico — abaez@astro.unam.mx
Anna **Bartkiewicz**, Centre for Astronomy, Nicolaus Copernicus University, Poland — annan@astro.umk.pl
Olga **Bayandina**, Astro Space Center of the Lebedev Physical Institute, Russia — bayandina@asc.rssi.ru
Maite **Beltrán**, INAF-Osservatorio Astrofisico di Arcetri, Firenze, Italy — mbeltran@arcetri.astro.it
Piero **Benvenuti**, International Astronomical Union, France — iau-general.secretary@iap.fr
Kārlis **Bērziņš**, Ventspils International Radio Astronomy Centre, Latvia — karlis.berzins@venta.lv
Anna **Bonaldi**, SKA Organization, Jodrell Bank, Macclesfield, UK — a.bonaldi@skatelescope.org
Jim **Braatz**, National Radio Astronomy Observatory, Charlottesville, USA — jbraatz@nrao.edu
Jan **Brand**, INAF-Istituto di Radioastronomia, Bologna, Italy — brand@ira.inaf.it
Shari **Breen**, The University of Sydney, Australia — shari.breen@sydney.edu.au
Crystal Lee **Brogan**, National Radio Astronomy Observatory, Charlottesville, USA — cbrogan@nrao.edu
Ross Alexander **Burns**, Joint institute for VLBI ERIC, The Netherlands — Burns@jive.eu
Paola **Castangia**, INAF-Osservatorio Astronomico di Cagliari, Italy — pcastang@oa-cagliari.inaf.it
Silvia **Casu**, INAF-Osservatorio Astronomico di Cagliari, Italy — silvia@oa-cagliari.inaf.it
Thanapol **Chanapote**, Department of Physics, Khon Kaen University, Thailand — t.chanapote@gmail.com
Xi **Chen**, Center for Astrophysics, Guangzhou University, China — chenxi@shao.ac.cn
James Okwe **Chibueze**, SKA South Africa, South Africa — jchibueze@ska.ac.za
Se-Hyung **Cho**, Korea Astronomy and Space Science Institute, Republic of Korea — cho@kasi.re.kr
Mark J. **Claussen**, National Radio Astronomy Observatory, Socorro, USA — mclausse@nrao.edu
Tiziana **Coiana**, INAF-Osservatorio Astronomico di Cagliari, Italy — tcoiana@oa-cagliari.inaf.it
Francisco **Colomer**, Joint institute for VLBI ERIC, The Netherlands — f.colomer@oan.es
Timea **Csengeri**, Max Planck Institute für Radioastronomy, Germany — csengeri@mpifr-bonn.mpg.de
Nichol **Cunningham**, Green Bank Observatory, USA — ncunning@nrao.edu
Claudia **Cyganowski**, School of Physics and Astronomy, University of St. Andrews, UK — cc243@st-andrews.ac.uk
Daria **Dall'Olio**, Chalmers University of Technology, Onsala, Sweden — daria.dallolio@chalmers.se
Jean-François **Desmurs**, Observatorio Astronómico Nacional, Spain — jf.desmurs@oan.es
Philip J. **Diamond**, SKA Organisation, Jodrell Bank, Macclesfield, UK — p.diamond@skatelescope.org
Engels **Dieter**, Hamburger Sternwarte, Universität Hamburg, Germany — dengels@hs.uni-hamburg.de
Richard **Dodson**, University of Western Australia, Australia — richard.dodson@icrar.org
Simon **Ellingsen**, School of Physical Sciences, University of Tasmania, Australia — Simon.Ellingsen@utas.edu.au
Sandra **Etoka**, Hamburger Sternwarte, Universität Hamburg, Germany — Sandra.Etoka@googlemail.com
Elena **Fedorova**, National Taras Shevchenko University of Kyiv, Ukraine — efedorova@ukr.net
Yasuo **Fukui**, Department of Physics, Nagoya University, Japan — fukui@a.phys.nagoya-u.ac.jp
Ortwin E. **Gerhard**, Max Planck Institute für Extraterrestrishe Physics, Germany — gerhard@mpe.mpg.de
Ciriaco **Goddi**, Radboud University; ALLEGRO/Leiden Observatory, The Netherlands — c.goddi@astro.ru.nl
Sharmila **Goedhart**, SKA South Africa, Cape Town, South Africa — sharmila@ska.ac.za
Jose-Francisco **Gomez**, Instituto de Astrofisica de Andalucia, Spain — jfg@iaa.es
Malcolm **Gray**, Jodrell Bank Centre for Astrophysics, Univ. of Manchester, UK — Malcolm.Gray@manchester.ac.uk
Jenny E. **Greene**, Princeton University, Dept of Astrophysics, USA — jgreene@astro.princeton.edu
Yuxin **He**, Xinjiang Astronomical Observatory, Chinese Academy of Sciences, China — heyuxin@xao.ac.cn
Christian **Henkel**, Max-Planck Institut für Radioastronomie, Germany — chenkel@mpifr-bonn.mpg.de
Tomoya **Hirota**, National Astronomical Observatory of Japan, Japan — tomoya.hirota@nao.ac.jp
Mareki **Honma**, Mizusawa VLBI Observatory, NAOJ, Japan — mareki.honma@nao.ac.jp
Bo **Hu**, Nanjing University, China — hubonju@gmail.com
Liz **Humphreys**, European Southern Observatory, Germany — ehumphre@eso.org
Todd R. **Hunter**, National Radio Astronomy Observatory, Charlottesville, USA — thunter@nrao.edu
Eodam **Hwang**, Korea Astronomy and Space Science Institute, Republic of Korea — ziueo@kasi.re.kr
Lucas J. **Hyland**, University of Tasmania, Australia — Lucas.Hyland@utas.edu.au
Hiroshi **Imai**, Department of Physics and Astronomy, Kagoshima University, Japan — hiroimai@sci.kagoshima-u.ac.jp
Katharina **Immer**, Joint Institute for VLBI ERIC, The Netherlands — ksimmer09@gmail.com
Sergey **Kalenskii**, Astro Space Center of the Lebedev Physical Institute, Russia — kalensky@asc.rssi.ru
Fateme **Kamali**, Max Planck Institute für Radio Astronomy, Germany — fkamali@mpifr-bonn.mpg.de
Ji-hyun **Kang**, Korea Astronomy and Space Science Institute, Republic of Korea — jkang@kasi.re.kr
Athol **Kemball**, Department of Astronomy, Urbana, USA — akemball@illinois.edu
Jeoung Sook **Kim**, Mizusawa VLBI Observatory, NAOJ, Japan — evony08@gmail.com
Soon-Wook **Kim**, Korea Astronomy and Space Science Institute, Republic of Korea — xrnovae@gmail.com
Kee-Tae **Kim**, Korea Astronomy and Space Science Institute, Republic of Korea — ktkim@kasi.re.kr
Jaeheon **Kim**, Shanghai Astronomical Observatory, Shanghai, China — jhkim@shao.ac.cn
Jungha **Kim**, SOKENDAI National University; NAOJ, Japan — jungha.kim@nao.ac.jp
Yuta **Kojima**, Yamaguchi University, Japan — g010vb@yamaguchi-u.ac.jp
Liudmyla **Kozak**, National Taras Shevchenko University of Kyiv, Ukraine — gutovska@ukr.net
Busaba Hutawarakorn **Kramer**, MPIfR, Germany; NARIT, Thailand — bkramer@mpifr-bonn.mpg.de
Vasaant **Krishnan**, INAF-Osservatorio Astrofisico di Arcetri, Italy — vasaantk@arcetri.astro.it
Elisabetta **Ladu**, Università degli Studi di Cagliari, Italy — ladueli91@gmail.com
Boy **Lankhaar**, Chalmers University of Technology, Onsala, Sweden — lankhaar@chalmers.se
Katharina **Leiter**, University of Würzburg Lehrstuhl für Astronomie, Germany — kleiter@astro.uni-wuerzburg.de
Silvia **Leurini**, INAF-Osservatorio Astronomico di Cagliari, Italy — sleurini@oa-cagliari.inaf.it
Jingjing **Li**, Purple Mountain Observatory, China — jjli@pmo.ac.cn
Dalei **Li**, Xinjiang Astronomical Observatory, China — lidalei@xao.ac.cn
Ivan **Litovchenko**, Astro Space Center of the Lebedev Physical Institute, Russia — grosh@asc.rssi.ru

Gordon **MacLeod**, Hartebeesthoek Radio Astronomy Observatory, South Africa   gord@hartrao.ac.za
Alberto **Masini**, INAF-Osservatorio Astronomico di Bologna, Italy   alberto.masini4@unibo.it
Fabrizio **Massi**, INAF-Osservatorio Astrofisico di Arcetri, Italy   fmassi@arcetri.astro.it
Jabulani P. **Maswanganye**, North-West University Centre for Space Research, South Africa   pop7paul@gmail.com
Tiege P. **McCarthy**, University of Tasmania, Australia   tiegem@utas.edu.au
Giuseppe **Melis**, INAF-Osservatorio Astronomico di Cagliari, Italy   melis@oa-cagliari.inaf.it
Karl M. **Menten**, Max-Planck-Institut für Radioastronomie, Germany   kmenten@mpifr.de
Francois **Mignard**, Observatoire de la Côte d'Azur, France   francois.mignard@oca.eu
James M. **Moran**, Harvard-Smithsonian Center for Astrophysics, USA   jmoran@cfa.harvard.edu
Luca **Moscadelli**, INAF-Osservatorio Astrofisico di Arcetri, Italy   mosca@arcetri.astro.it
Kazuhito **Motogi**, Yamaguchi University, Japan   kmotogi@yamaguchi-u.ac.jp
Eric J. **Murphy**, National Radio Astronomy Observatory, Charlottesville, USA   emurphy@nrao.edu
Akiharu **Nakagawa**, Faculty of Science, Kagoshima University, Japan   nakagawa@sci.kagoshima-u.ac.jp
Jun-ichi **Nakashima**, Ural Federal University, Russia   nakashima.junichi@gmail.com
Mateusz **Olech**, Centre for Astronomy, Nicolaus Copernicus University, Poland   olech@astro.umk.pl
Luca **Olmi**, INAF-Osservatorio Astrofisico di Arcetri, Italy   olmi.luca@gmail.com
Gabor **Orosz**, Kagoshima University, Japan   gabor.orosz@gmail.com
Juergen **Ott**, National Radio Astronomy Observatory, Socorro, USA   jott@nrao.edu
Tomoaki **Oyama**, National Institutes of Natural Sciences, NAOJ, Japan   t.oyama@nao.ac.jp
Francesca **Panessa**, INAF-Istituto di Astrofisica e Planetologia Spaziali, Italy   francesca.panessa@iaps.inaf.it
Sonu Tabitha **Paulson**, Indian Institute of Space Science and Technology, India   sonutabitha@gmail.com
Alison **Peck**, Gemini Observatory, USA   apeck@gemini.edu
Andres Felipe **Perez-Sanchez**, European Southern Observatory, Chile   aperezsa@eso.org
Guy **Perrin**, CNRS Institut des Sciences de l'Univers, France   guy.perrin@obspm.fr
Dominic **Pesce**, University of Virginia, USA   dpesce@virginia.edu
Ylva **Pihlstrom**, University of New Mexico, USA   ylva@unm.edu
Sergio **Poppi**, INAF-Osservatorio Astronomico di Cagliari, Italy   spoppi@oa-cagliari.inaf.it
Hai-Hua **Qiao**, Shanghai Astronomical Observatory, China; Curtin University, Australia   qiaohh@shao.ac.cn
Luis Henry **Quiroga Nuñez**, Leiden Observatory; JIVE, The Netherlands   quiroganunez@strw.leidenuniv.nl
Mark J. **Reid**, Harvard-Smithsonian Center for Astrophysics, USA   reid@cfa.harvard.edu
Anita M. S. **Richards**, Jodrell Bank Centre for Astrophysics, University of Manchester, UK   amsr@jb.man.ac.uk
Maria **Rioja**, International Center for Radio Astronomy Research, Australia; OAN, Spain   maria.rioja@icrar.org
Tim **Robishaw**, Dominion Radio Astrophysical Observatory, Canada   tim.robishaw+drao@gmail.com
Carolina B. **Rodríguez-Garza**, Inst. de Radioastronomia y Astrofisica, UNAM, Mexico   ca.rodriguez@crya.unam.mx
Nobuyuki **Sakai**, National Astronomical Observatory of Japan, Japan   nobuyuki.sakai@nao.ac.jp
Daisuke **Sakai**, The University of Tokyo, Japan   sakai.daisuke@nao.ac.jp
Alberto **Sanna**, Max-Planck-Institut für Radioastronomie, Germany   asanna@mpifr-bonn.mpg.de
Hidetoshi **Sano**, Institute for Advanced Research, Nagoya University, Japan   sano@a.phys.nagoya-u.ac.jp
Rafał **Sarniak**, Centre for Astronomy, Nicolaus Copernicus University, Poland   kain@astro.umk.pl
Till **Sawala**, University of Helsinki, Department of Physics, Finland   till.sawala@helsinki.fi
Evgeni **Semkov**, Institute of Astronomy and NAO, Bulgaria   esemkov@astro.bas.bg
Nadezhda N. **Shakhvorostova**, Astro Space Center, Lebedev Physical Institute, Russia   nadya.shakh@gmail.com
Hiroko **Shinnaga**, Dept. of Physics and Astronomy, Kagoshima University, Japan   shinnaga@sci.kagoshima-u.ac.jp
Ivar **Shmeld**, Ventspils International Radio Astronomy Center, Latvia   ivarss@venta.lv
Lorant **Sjouwerman**, National Radio Astronomy Observatory, Socorro, USA   lsjouwer@nrao.edu
Andrey M. **Sobolev**, Astronomical Observatory Ural Federal University, Russia   Andrej.Sobolev@urfu.ru
Paolo **Soletta**, INAF-Osservatorio Astronomico di Cagliari, Italy   psoletta@oa-cagliari.inaf.it
Bringfried **Stecklum**, Thüringer Landessternwarte Tautenburg Sternwarte, Germany   stecklum@tls-tautenburg.de
Angelica Erica **Strack**, Western Illinois University, USA   ae-strack@wiu.edu
Anton Atanasov **Strigachev**, Institute of Astronomy and NAO, Bulgaria   anton@astro.bas.bg
Michael **Stroh**, University of New Mexico, USA   mstroh@unm.edu
Koichiro **Sugiyama**, Mizusawa VLBI Observatory, NAOJ, Japan   koichiro.sugiyama@nao.ac.jp
Yan **Sun**, Purple Mountain Observatory, China   yansun@pmo.ac.cn
Kazuyoshi **Sunada**, Mizusawa VLBI Observatory, NAOJ, Japan   kazu.sunada@nao.ac.jp
Gabriele **Surcis**, INAF-Osservatorio Astronomico di Cagliari, Italy   surcis@oa-cagliari.inaf.it
Sherry **Suyu**, MPI for Astrophysics; Technical University of Munich, Germany   suyu@mpa-garching.mpg.de
Daniel **Tafoya**, Onsala Space Observatory, Chalmers University of Technology, Sweden   daniel.tafoya@chalmers.se
Kazuhiro **Takefuji**, National Institute of Information and Comm. Technology, Japan   takefuji@nict.go.jp
Andrea **Tarchi**, INAF-Osservatorio Astronomico di Cagliari, Italy   atarchi@oa-cagliari.inaf.it
Taylor **Tobin**, Department of Astronomy University of Illinois, USA   tltobin2@illinois.edu
Jose Maria **Torrelles**, Institut de Ci'encies de l'Espai, Spain   chema.torrelles@ice.cat
Miguel Angel **Trinidad**, Department of Astronomy University of Guanajuato, Mexico   trinidad@astro.ugto.mx
Lucero **Uscanga**, University of Guanajuato, Mexico   luscag@gmail.com
Stefanus Petrus **van den Heever**, Hartebeeshoek Radio Astronomy Obs., South Africa   sp.vandenheever@gmail.com
Johan **van der Walt**, North West University, South Africa   vanderwalt.dj@gmail.com
Huib J. **van Langevelde**, Joint institute for VLBI ERIC, The Netherlands   langevelde@jive.eu
Ruby **Van Rooyen**, North West University; SKA South Africa, South Africa   ruby@ska.ac.za
Wouter **Vlemmings**, Chalmers University of Technology, Sweden   wouter.vlemmings@chalmers.se
Maxim **Voronkov**, CSIRO Astronomy And Space Science, Australia   Maxim.Voronkov@csiro.au
Marion E. **West**, Hartebeesthoek Radio Astronomy Observatory, South Africa   marion@hartrao.ac.za
Pawel **Wolak**, Centre for Astronomy, Nicolaus Copernicus University, Poland   wolak@astro.umk.pl
Gang **Wu**, Xinjiang Astronomical Observatory, China   wug@xao.ac.cn
Ye **Xu**, Purple Mountain Observatory, China   xuye@pmo.ac.cn
Jiaerken **Yeshengbieke**, Xinjiang Astronomical Observatory, China   jarken@xao.ac.cn
Dong-Hwan **Yoon**, Korea Astronomy and Space science Institute, Republic of Korea   dhyoon83@kasi.re.kr
Jinghua **Yuan**, National Astronomical Observatories, China   jhyuan@nao.cas.cn
Youngjoo **Yun**, Korea Astronomy and Space Science Institute, Republic of Korea   yjyun@kasi.re.kr
Jiang-Shui **Zhang**, Center for Astrophysics, Guangzhou University, China   jszhang@gzhu.edu.cn
Jianjun **Zhou**, Xinjiang Astronomical Observatory, China   zhoujj@xao.ac.cn

Dedication of the Symposium:
This symposium is dedicated to the memory of
Charles Malcolm Walmsley

Astrophysical Masers:
Unlocking the Mysteries of the Universe
Proceedings IAU Symposium No. 336, 2017
A. Tarchi, M.J. Reid & P. Castangia, eds.

© International Astronomical Union 2018
doi:10.1017/S1743921318000923

Malcolm Walmsley's Maser Science

Karl M. Menten

Max-Planck-Institut für Radioastronomie, Auf dem Hügel 69, D-53121 Bonn, Germany
email: kmenten@mpifr.de

Charles Malcolm Walmsley passed away on 1 May 2017. Over a long and highly productive career, Malcolm made numerous and fundamental contributions to the science of the interstellar medium and star formation. These have recently been summarized elsewhere (Menten & Cesaroni 2017). Here I would like to describe some of his work related to masers.

Malcolm became strongly engaged in astronomical maser research in the mid-1980s, when a relatively strong and narrow spectral line from the methanol molecule (CH_3OH) at 23.1 GHz was serendipitously discovered with the Effelsberg 100 meter radio telescope of the Max-Planck-Institut für Radioastronomie (MPIfR). At that time, the only known methanol *maser* lines were the series of $J_{k=2} - J_{k=1}$ transitions of E-type CH_3OH ($J = 2, 3, 4, \ldots$) near 25 GHz that had *only* been found in the Orion Kleinmann-Low nebula in 1971 by Barrett *et al.* (1971) and nowhere else, despite extensive searches.. It became clear that toward Orion-KL, the newly detected 23.1 GHz line was *not* masing, although it was detected there. The title of the article, by Wilson *et al.* (1984), reporting this line, *Detection of a new type of methanol maser*, gave the first hint at the class I/class II methanol maser dichotomy – the 23.1 GHz $9_2 - 10_1$ A^+ being the first class II line, to which soon the even stronger 19.9 GHz $2_1 - 3_0$ E line was added (Wilson *et al.* 1985). In contrast, the 25 GHz lines belong to class I.

These discoveries triggered Malcolm's interest in methanol masers, which became a major portion of my dissertation that he had started supervising. With the great sensitivity of Effelsberg, more 25 GHz maser sources were detected in regions much farther away than Orion (Menten *et al.* 1986). Malcolm also became curious about the excitation of methanol masers. At that time, collisional rate coefficients didn't exist for CH_3OH, so he used what published (experimental) information he could find in the literature and conducted statistical equilibrium calculations that indeed predicted maser action in the class I maser lines newly discovered by Morimoto *et al.* (1985) with the Nobeyama 45 meter telescope. Most interestingly, these calculations also predicted enhanced absorption against the cosmic microwave background ("over-cooling") in the 12.1 GHz $2_0 - 3_{-1}$ E transition toward cold dark clouds. This he indeed discovered toward two objects, a result he particularly cherished (Walmsley *et al.* 1988). Toward high-mass star forming regions with strong far infrared (FIR) radiation he and his collaborators had found this line to be the (then) strongest class II methanol maser line (Batrla *et al.* 1987).

At that time, Malcolm also was studying hyperfine structure (hfs) lines from rotationally excited levels of the hydroxyl radical (OH) at multiple radio frequencies, also with the Effelsberg telescope and with the VLA and VLBI (e.g., Walmsley *et al.* 1986). A conclusion of this work is a comprehensive study of OH excitation (with R. Cesaroni) that consisted of innovative statistical equilibrium calculations (considering FIR line overlaps; Cesaroni & Walmsley 1991). It was aimed at reproducing the observed pattern of maser emission in some hfs lines and absorption in others and constrained the physical conditions in the OH-bearing regions, i.e, the warm, dense molecular envelopes of ultracompact HII regions, which also harbor class II CH_3OH masers.

Figure 1. The MPIfR 100 meter radio telescope at Effelsberg, Germany. It started operations in 1972 and was extensively used by Malcolm Walmsley, his collaborators and students for a wide range of studies on interstellar radio recombinations lines and molecules. The picture of Malcolm was taken in 2005 in Acireale on Sicily at the IAU Symposium 227 *Massive star birth: A Crossroads of Astrophysics*, a conference he co-organized. Credits: MPIfR/N. Junkes and Cambridge University Press

The above work was done while Malcolm was a staff member of the MPIfR in Bonn, where he worked from 1969–1994. During the second part of his career, at Arcetri Observatory, he collaborated in numerous projects on high mass star forming regions, many of which involved masers. Finally, in one of his last papers, he returned to the topic of class I methanol masers and made major contributions to a comprehensive study of their excitation. It was published 30 years after his first paper on the subject (Leurini *et al.* 2016, see also the contribution by Leurini & Menten to these proceedings).

Malcolm Walmsley was invited to give the summary talk at this conference, a task he sadly could not perform. This meeting has been dedicated to his memory.

References

Barrett, A. H., Schwartz, P. R., & Waters, J. W. 1971, *ApJ*, 168, L101
Batrla, W., Matthews, H. E., Menten, K. M., & Walmsley, C. M. 1987, *Nature*, 326, 49
Cesaroni, R. & Walmsley, C. M. 1991, *A&A*, 241, 537
Leurini, S., Menten, K. M., & Walmsley, C. M. 2016, *A&A*, 592, A31
Menten, K. M., Walmsley, C. M., Henkel, C., & Wilson, T. L. 1986, *A&A*, 157, 318
Menten, K. & Cesaroni, R. 2017, *Nature Astronomy*, 1, 0173
Morimoto, M., Kanzawa, T., & Ohishi, M. 1985, *ApJ*, 288, L11
Walmsley, C. M., Baudry, A., Guilloteau, S., & Winnberg, A. 1986, *A&A*, 167, 151
Walmsley, C. M., Batrla, W., Matthews, H. E., & Menten, K. M. 1988, *A&A*, 197, 271
Wilson, T. L., Walmsley, C. M., Jewell, P. R., & Snyder, L. E. 1984, *A&A*, 134, L7
Wilson, T. L., Walmsley, C. M., Menten, K. M., & Hermsen, W. 1985, *A&A*, 147, L19

Theory of masers and maser sources

Astrophysical Masers:
Unlocking the Mysteries of the Universe
Proceedings IAU Symposium No. 336, 2017
A. Tarchi, M.J. Reid & P. Castangia, eds.

© International Astronomical Union 2018
doi:10.1017/S1743921317011358

Maser Theory: Old Problems and New Insights

M. D. Gray

Jodrell Bank Centre for Astrophysics, School of Physics and Astronomy,
Alan Turing Building, University of Manchester,
M13 9PL, Manchester, UK
email: Malcolm.Gray@manchester.ac.uk

Abstract. Maser theory continues to be driven by advances in observational techniques. Here, I consider the responses to VLBI with space-Earth baselines and cross-correlation spectroscopy (a re-consideration of coherence properties), routine observation in full-Stokes polarization (a re-casting of the polarization transfer equations), and long-term variability monitoring (3-D modelling of irregular domains).

Keywords. masers, molecular processes, radiative transfer, ISM: molecules, radio lines: general

1. Introduction

I consider in detail three areas of active maser research that may benefit from some new, or re-discovered, theoretical insight in addition to the available observational data. These areas are the coherence properties of astrophysical masers, the modelling of maser polarization, and the analysis of maser variability with 3-D radiative transfer and saturation models.

2. Coherence Properties

Two types of coherence are discussed in Elitzur (1992): phase coherence and bandwidth coherence. Phase coherence is maintained between two rays from some point on a source, diverging through some beam angle θ, and propagated over a distance L, provided that the phase difference $\Delta\phi = L\pi\theta^2/\lambda$ is much smaller than 1 radian. Large values of L, of order 10^{12} m in astrophysical masers, drove the conclusion that astrophysical masers are not phase coherent. This probably remains a valid conclusion, but the recent detection, with an interferometer including the RadioAstron instrument, of a 22-GHz $(\lambda = 0.0136\,\mathrm{m})\mathrm{H_2O}$ maser in Cep A that is unresolved at an angular scale of $10\,\mu\mathrm{as}$, improves the condition by several orders of magnitude. At the distance of Cep A, this angular scale corresponds to a linear distance of $\sim 10^9$ m. Using this distance as L, but keeping the original estimate of $\pi\theta^2 = 3.14\times 10^{-4}$, we obtain $\Delta\phi = 2.42\times 10^6$, so the conclusion of no phase coherence still appears safe, even with the smallest maser features so far discovered.

The second type of coherence discussed in Elitzur (1992) is bandwidth coherence. For a source to be bandwidth coherent, the essential requirement is that the homogeneous response profile of the transition is broader than the inhomogeneous profile. The homogeneous profile broadens as a maser saturates due to power, or saturation, broadening. This arises from Stark effect distortion of the molecule by the electric field of the maser radiation. According to Elitzur (1992), power broadening becomes sufficient to enforce bandwidth coherence for a brightness temperature of $T_b > 3 \times 10^{14}(4\pi/\Omega)\,\mathrm{K}$ for the

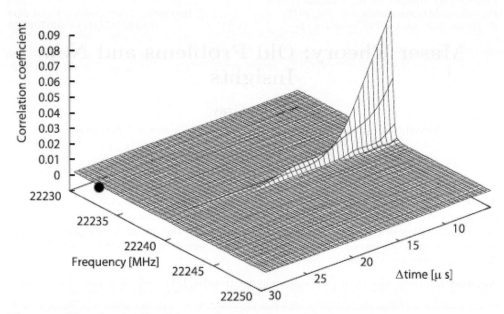

Figure 1. The coherence function as a function of time delay and of frequency across the line in an observation of the star-forming region W3(H₂O) by Takefugi *et al.* (2016)

22-GHz transition of H_2O. For reasonable maser beam angles, see above, the brightest maser sources, particularly flares, could be marginally bandwidth coherent. For example water-maser flares have been observed in Orion with $T_b > 10^{16}$ K (Kobayashi *et al.* 2000), and with $T_b \simeq 8 \times 10^{17}$ K (Strelnitskii 1982).

Recent evidence from observationally measured coherence functions (Takefugi *et al.* 2016) strongly suggests that very bright astrophysical masers are indeed bandwidth coherent. The key point is the shape of the decay of the coherence function (Fig. 1), which has a long tail towards longer time delays that is inconsistent with the Gaussian behaviour expected if the function were controlled by the inhomogeneous profile. Note that the relation between the coherence function and the line shape is a Fourier transform: a Gaussian inhomogeneous profile transforms to a Gaussian decay in the coherence function; a Lorentzian homogeneous profile transforms to an exponential, with a much slower decay. The observed coherence time in this example is $18\,\mu$s, suggesting a homogeneous width of 56 kHz. This compares to an observed inhomogeneous line profile width (full width at half power) of 96 kHz.

The Stark broadening effect by the electric field of the maser radiation imposes an ultimate limit on the brightness temperature, above which the line will be split into a distinctive pattern (Slysh 2003). The fact that a pattern of this type has not been observed suggests that maser beam angles for the brightest sources are of order 10^{-3}, somewhat smaller than the estimates in Elitzur (1992). The measured coherence times in Takefugi *et al.* (2016) indicate that the brightness temperatures of some H_2O maser features in W3(H_2O) and W49 reach 8.5×10^{18} and 6.2×10^{22} K respectively.

Another old, but excellent, reference that covers many aspects of Doppler-broadened maser propagation, including the statistics of the observed radiation, which is related to the work described above, is Menegozzi & Lamb (1978).

3. Polarization

Perhaps no aspect of theoretical maser research has been more controversial. In particular, the case of Zeeman splittings small compared to the inhomogeneous line width has historically been very problematic. Another major source of difficulty is the use of different conventions to describe left- and right-handed polarization, and in the definitions of the Stokes parameters, particularly Stokes-V. Definitions conforming to the IEEE definition of right- and left-handed polarization and the IAU convention for Stokes V are set out in Green *et al.* (2014).

One of the reasons that the polarization problem is so difficult is that four, coupled, simultaneous radiative transfer equations must be solved to recover the Stokes parameters. A piece of old insight into this problem was provided by Landi Degl'Innocenti (1987), who developed a method of writing the Stokes problem in the style of the integrating factor method that is used to write a formal solution of the scalar radiative transfer equation. The result is a Stokes vector solution that may be frequency and/or angle averaged as necessary, allowing the polarization problem to be cast into a Milne-Schwarzschild integral equation form, see, for example, King & Florance (1964), by eliminating the radiation averages from the density matrix (DM) equations.

In detail, Landi Degl'Innocenti solves the full polarization radiative transfer equation, $d\mathbf{i}/d\tau = \eta\mathbf{i}$, in this case assuming negligible spontaneous emission, as $\mathbf{i}(\tau) = \mathrm{O}(\tau, \tau')\mathbf{i}(\tau')$, where τ is the optical depth and $\mathrm{O}(\tau, \tau')$ is a 4×4 matrix operator, the propagation operator that relates the dimensionless Stokes vector, $\mathbf{i}(\tau)$, at two different depths. In the limit where $\tau' \to 0$, this operator multiplies the Stokes vector for a ray just entering the maser medium, and can be shown to be equal to (Landi Degl'Innocenti 1987),

$$\mathrm{O}(\tau, \tau') = \mathrm{I} + \sum_{n=1}^{\infty} \frac{1}{n!} \int_{\tau'}^{\tau} d\tau_1 \int_{\tau'}^{\tau} d\tau_2 ... \int_{\tau'}^{\tau} d\tau_n P \left\{ \eta(\tau_1)\eta(\tau_2)...\eta(\tau_n) \right\} \quad (3.1)$$

In eq. 3.1, I is the identity matrix, and P is an operator that orders the Stokes matrices, η by depth, with the greatest depth on the left. A general Stokes matrix has seven unique elements, and may be written,

$$\eta = \begin{bmatrix} \eta_i & \eta_q & \eta_u & \eta_v \\ \eta_q & \eta_i & \rho_v & -\rho_u \\ \eta_u & -\rho_v & \eta_i & \rho_q \\ \eta_v & \rho_u & \rho_q & \eta_i \end{bmatrix} \quad (3.2)$$

where many models assume $\eta_u = \rho_v = \rho_q = \rho_u = 0$. The remaining terms, η_i, η_q and η_v are functions of the molecular populations or inversions and trigonometrical functions of the angle between the magnetic field, at a certain depth, and the ray direction, possibly with an additional continuum component.

Radiation averages derived from eq. 3.1 can now be used to eliminate these quantities from the DM equations. Such averages, denoted by an over-bar may be grouped into the convenient functions,

$$\begin{aligned} f^{0,\pm} &= \frac{1}{4\pi} \oint 2(\bar{i}^{0,\pm} - \bar{q}^{0,\pm}) \sin^2 \theta d\Omega \\ h^{0,\pm} &= \frac{1}{4\pi} \oint 2\bar{v}^{0,\pm} \cos \theta d\Omega \\ g^{0,\pm} &= \frac{1}{4\pi} \oint \left[(1 + \cos^2 \theta)\bar{i}^{0,\pm} + \bar{q}^{0,\pm} \sin^2 \theta \right] d\Omega, \end{aligned} \quad (3.3)$$

where the superscripts denote the transition type of the line shape function that has been used for the frequency integration. The DM equations that the results from eq. 3.3 need to be inserted into are somewhat complicated, but for a 3D model under the complete redistribution approximation, they are

$$
\begin{aligned}
\Delta^+ \;=\; & \frac{\alpha^+}{D}\left\{(1+2f^0)\left[1-\frac{\alpha^-}{\alpha^+}(g^++h^+)+2(g^-+h^-)\right]+(\frac{\alpha^-}{\alpha^+}f^+-f^-)(g^0+h^0)\right. \\
& \left. +\;\frac{\alpha^0}{\alpha^+}\left[f^-(g^++h^+)-f^+(1+2g^-+2h^-)\right]\right\},
\end{aligned} \tag{3.4}
$$

where Δ^+ is the inversion in the σ^+ transition of a $J=1-0$ Zeeman group at a specific node or depth of the model. The $\alpha^{\pm,0}$ constants are the ratio of the pump and loss rates for the transition denoted by the superscript. The expression for Δ^- has a similar structure to eq. 3.4, whilst the π-inversion is given by

$$
\begin{aligned}
\Delta^0 \;=\; & \frac{\alpha^0}{D}\left\{1+2(g^-+h^-)+2(g^+-h^+)+3(g^+g^--h^+h^-)+5(g^+h^--g^-h^+)\right. \\
& -\;\frac{\alpha^+}{\alpha^0}\left[g^0(1+g^-+3h^-)-h^0(1+3g^-+h^-)\right] \\
& \left. -\;\frac{\alpha^-}{\alpha^0}\left[g^0(1+g^+-3h^+)-h^0(1+3g^-+h^-)\right]\right\}. \tag{3.5}
\end{aligned}
$$

The denominator, D, common to eq. 3.4 and eq. 3.5 is

$$
\begin{aligned}
D \;=\; & (1+2f^0)[1+2(g^-+h^-)+2(g+-h^+)+3(g^+g^--h^+h^-)+5(g^+h^--g^-h^+)] \\
& +\;f^+[h^0(1+h^-+3g^-)-g^0(1+g^-+3h^-)] \\
& -\;f^-[h^0(1-h^++3g^+)+g^0(1+g^+-3h^+)]. \tag{3.6}
\end{aligned}
$$

These equations now form a set of non-linear algebraic equations in the $\Delta^{\pm,0}$ at all positions or depths of the model.

4. Nature of Masing Objects

When the physical objects that support astrophysical masers are discussed, we often talk of them as 'clouds', but our knowledge of what they actually are is rather limited. An exception is a group of masers found in our own Solar System, known to originate from the comae of comets. The fact that we can reliably use masers for trigonometric parallax distance measurement strongly suggests that at least the masing objects are more like 'bullet' clouds than a transient wave-like phenomenon. Clues to the nature of maser objects include the geometry and frequency dependence of their emission, and likely also their variability, since the brightness of maser rays is strongly dependent on the available gain length in the line-of-sight. The considerations above call for a fully 3-D maser model with irregular geometry.

A good way to construct a general 3-D maser model is to use the finite-element method of discretization. Elements can in turn be generated from the Delaunay triangulation of an irregular distribution of points (nodes), see for example Zienkiewicz & Taylor (2000). A triangulated domain of this type is shown in Fig. 2. The radiation solution for all nodes of the domain is carried out via a finite-element implementation of the CEP method (Elitzur & Asensio Ramos 2006,Gray 2012) under a quasi-two-level CVR approximation at present. The solution for a certain optical thickness is straightforwardly used as a starting guess for the next, thicker, model, allowing highly saturated solutions to be

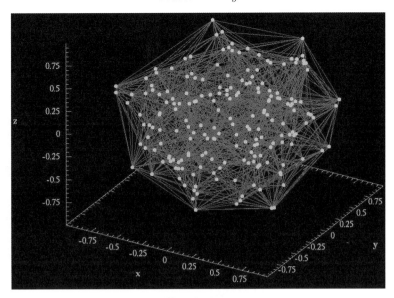

Figure 2. A triangulated domain of 1177 tetrahedral (simplex) elements with 202 nodes

obtained. If the model has J nodes and Q ray paths, then the inversion Δ_i, at the ith node, is given by

$$\Delta_i - \left[1 + \frac{i_{BG}}{4\pi} \sum_{q=1}^{Q} \frac{a_q}{s_q^2} \sum_{n=0}^{\infty} \frac{1}{n!(n+1)^{1/2}} \left(\sum_{j=1}^{J} \Phi_{j,q}\Delta_j \right)^n \right]^{-1} = 0, \qquad (4.1)$$

where the functions $\Phi_{j,q}$ are constructed from the shape-functions of the elements of the nodes j found along ray path q to the target (node i). Solutions to eq. 4.1 show the expected concentration of the most saturated nodes to the outside of the model, and the least saturated nodes towards the centre. The qth ray path has a facet area a_q at a distance s_q from the target.

Once nodal solutions have been computed, formal solutions can be calculated for the brightnesses of rays along lines of sight to an observer's position. These can be converted to maps that show the expected decrease in the observed size of the cloud with increasing amplification and saturation, for example Alcock & Ross (1985).

It is possible to investigate maser variability by simulating rotation of the cloud depicted in Fig. 2. The simulation was performed by computing the radiation transfer solution in the frame of the cloud, which can have no internal velocity gradients assuming that it rotates as a solid body. An external observer then sees the cloud with a Doppler correction that appears mostly as broadening with a small shift in the peak frequency of the maser line. The plane perpendicular to the rotation axis was chosen so as to present the longest axis of the domain periodically to the observer.

Variability of Class 2 methanol masers has been investigated in detail by Goedhart, Gaylard & van der Walt (2004), where light curves of several years duration are plotted for a variety of sources that show patterns that include monotonic growth and decay and periodic and aperiodic low-amplitude flaring. We show our light curve in Fig. 3 (left panel), while the observer's view of the cloud at ten different times (a simulated VLBI map) is shown in the right-hand panel. The model maser has a light curve, in terms of the brightest specific intensity, that varies by a factor of $\sim$3 over one rotation. The

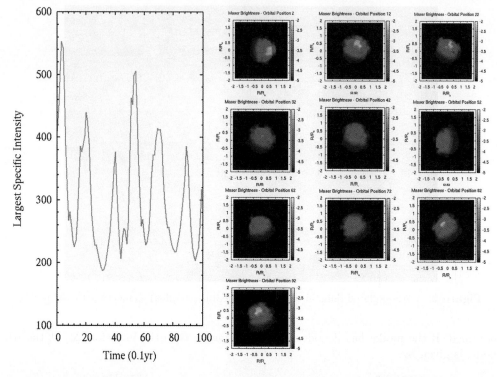

Figure 3. Left panel: The light curve of the rotating model maser in terms of the peak specific intensity (in multiples of the saturation intensity). Right panel: Snap-shots of the brightness map at various times during the cycle. A lower-depth model has been used so that the whole cloud is easily visible.

simulated map (right-hand panel) shows that the position of brightest emission also varies considerably with time.

References

Alcock, C. & Ross, R. R. 1985 *ApJ*, 290, 433

Elitzur, M. 1992, *Astronomical Masers* (Dordrecht: Kluwer)

Elitzur, M. & Asensio Ramos, A. 2006 *MNRAS*, 365, 779

Goedhart, S., Gaylard, M. J. & van der Walt, D. J. 2004 *MNRAS*, 355, 553

Gray, M. D. 2012, *Maser Sources in Astrophysics* (New York: CUP)

Green, J. A., Gray, M. D., Robishaw, T., Caswell, J. A. & McClure-Griffiths, N. M. 2014 *MNRAS*, 440, 2988

Kobayashi, H., Shimoikura, T., Omodaka, T., & Diamond, P. J. 2000, in: H. Hirabayashi, P. G. Edwards & D. W. Murphy (eds.), *Astrophysical Phenomena Revealed by Space VLBI* (Institute of Space and Astronautical Science), p. 109-112

King, J. I.F. & Florance, E. T. 1964 *ApJ*, 139, 397

Landi Degl'Innocenti, E. 1988, in: W. Kalkofen (ed.), *Numerical Radiative Transfer* (Cambridge, CUP), p. 265

Menegozzi, L. N. & Lamb Jr., W. E. 1978 *Phys. Rev. A*, 17, 701

Slysh, V. I. 2003, in: J. A. Zensus, M. H. Cohen & E. Ros (eds.), *Radio Astronomy at the Fringe* (ASP Conference Series, Vol. 300), p. 239

Strelnitskii, V. S. 1982 *Soviet Astron. Lett.*, 8, 86

Takefuji, K, Imai, H. & Sekido, M. 2016 *PASJ*, 68, 86

Zienkiewicz, O. & Taylor, R. 2000, *The Finite Element Method: The basis* (Oxford: Butterworth-Heinemann)

Astrophysical Masers:
Unlocking the Mysteries of the Universe
Proceedings IAU Symposium No. 336, 2017
A. Tarchi, M.J. Reid & P. Castangia, eds.

© International Astronomical Union 2018
doi:10.1017/S1743921317011371

MASERS: A Python package for statistical equilibrium calculations applied to masers

R. van Rooyen[1] and D. J. van der Walt[2]

[1]SKA South Africa,
3rd Floor, The Park, Park Road, Pinelands, 7405, Western Cape, South Africa
email: `ruby@ska.ac.za`

[2]Department of Space Physics, North-West University,
Potchefstroom Campus, 11 Hoffman Street, Potchefstroom, 2531
email: `johan.vanderwalt@nwu.ac.za`

Abstract. The study of astrophysical maser formation provides a useful probe of the chemical composition and physical conditions of the sources they are observed in. This exploration requires continuously solving the SE equations for the populations of the energy levels in search of conditions that will produce an inversion. After evaluation of available implementations applying the Escape Probability approximation, the MASERS solver was developed to provide an efficient and robust matrix inversion calculation. This open source package is hosted at `https://bitbucket.org/ruby_van_rooyen/masers`.

Keywords. masers, molecules, numerical modelling, instabilities

1. Introduction

In order to understand the pumping mechanisms of masers, suitable models must solve the rate equations for non-LTE statistical equilibrium (SE). A realistic model for a line-emitting system must contain a sufficiently large number of levels and take into account all processes describing population exchange (Sobolev & Gray 2012). These SE equations must be solved simultaneously over all levels using robust iterative solvers.

The Escape Probability approximation simplifies the coupling between radiative transfer and the SE equations by pre-applying some assumption based on photon propagation through various mediums. In addition, it also uses adjustable variable selection to computationally approximate the physical environment, which allows easy manipulation of parameter space under investigation to inspect pumping mechanisms. The main disadvantage with the use of adjustable, empirical parameters, is that with a departure from their optimum values, very slow convergence and even divergence may occur.

The most efficient way of solving for a large number of level populations is to express the rate equations as a matrix of coefficients acting on a vector of populations. In this form, the level populations can be obtained by a number of standard numerical methods, with the only prerequisite for a successful solution being a reasonable initial guess. Matrix computation requires memory to store big matrices for the rate equations of molecules with a large number of levels. Calculations of pseudo-inversion can lead to destructive numerical instabilities if the implementation does not properly represent the mathematical nature of the equations and variables. Plus, numerical procedures are strongly convergent if the starting solution is not too far from the final solution, but fails to converge in typical non-LTE conditions.

The MASERS package is developed in Python, provides a reasonably fast, stable algorithm that deals with the solution method's inherent numerical sensitivities; allows

13

different maser geometries for calculation; includes the contribution of interacting background radiation fields, as well as other sources of opacity such as line overlap.

2. Non-equilibrium inversion

Applying the escape probability to the rate equations, the level populations calculation can be expressed as presented in Equation 2.7.1 from Elitzur (1992).

$$
\begin{aligned}
\frac{dn_i}{dt} = & -\sum_{j<i}\left\{A_{ij}\beta_{ij}\left[n_i + W\aleph_{ij}(n_i - n_j)\right] + C_{ij}\left[n_i - n_j\exp\left(\frac{-h\nu_{ij}}{kT}\right)\right]\right\} \\
& + \sum_{j>i}\frac{g_j}{g_i}\left\{A_{ji}\beta_{ji}\left[n_j + W\aleph_{ji}(n_j - n_i)\right] + C_{ji}\left[n_j - n_i\exp\left(\frac{-h\nu_{ji}}{kT}\right)\right]\right\}
\end{aligned}
\tag{2.1}
$$

As a parametric model Equation 2.1 capture all its information within the following parameters: the rate coefficients A_{ij} and C_{ji}; the number density of level i, $N_i = g_i n_i$, where g_i is the statistical weight; the information related to the maser environment such as the dilution factor W and $\aleph_{ij}$, the photon occupancy number at transition frequency ν_{ij}; as well as, geometry and kinematics of the masing region in β_{ij}.

Expressed as a matrix equation it becomes

$$\mathbf{Qx} = \mathbf{b}$$

where $\mathbf{b} = [0, \cdots, 0, 1]^T$, $x_i = \frac{n_i}{n_{mol}}$ is the normalised fractional population density, with n_{mol} the total population density of the molecule and $\mathbf{Q} = \mathbf{R} + \widetilde{\mathbf{C}}$.

$$
R_{ij} = \begin{cases} A_{ji}\beta_{ji}(1+X_{ji}) & i<j \\ A_{ij}\beta_{ij}\left(\frac{g_i}{g_j}\right)X_{ij} & j>i \\ -\sum_{i\neq j}R_{ji} & j=i \end{cases}
\quad\text{and}\quad
\widetilde{C}_{ij} = \begin{cases} C_{ji} & j>i \\ -\sum_{i\neq j}C_{ij} & j=i \end{cases}
$$

where R is the radiative and $\widetilde{C}$ the collisional components, and $X = \aleph_{bb} + [\aleph(T_{\ell d}) + w_d\aleph(T_{xd})](1 - e^{-\tau_d}) + w_{HII}\aleph(T_e)(1 - e^{-\tau_{HII}})$ is the radiative contribution of the background.

Numerical limitation and instabilities. Iterative methods are sensitive to divergence and oscillation (multiple valid solutions) since the initial estimate for the next iteration is simply the solution of the previous iteration and the process is repeated until calculated solution satisfies some convergence criteria.

Badly chosen convergence criteria can also contribute to numerical instabilities. The natural choice of excitation temperature calculation as convergence criteria showed sensitivity to divergence, caused by catastrophic cancellation due to the difference in small numbers in the denominator of the excitation temperature calculation, $T_{ex} \propto [\ln(x_l g_u) - \ln(x_u g_l)]^{-1}$. More reliable convergence is obtained by directly comparing level populations with x_n and x_{n-1}, the level populations calculated during the current and previous evaluation.

Oscillating behaviour is only pronounced at lower transition levels and the solutions were found to become more stable if the next iteration is given some "memory" of the previous solutions. This was done using a running average calculation, $x_n = c_1 \times x_n + c_2 \times x_{n-1}$ where $c_1 \leqslant 1$ and $c_2 \leqslant 1$ are some weighting coefficients with $c_1 + c_2 = 1$. The larger the coefficient for the previous solution the "longer" the memory of the next solution. For MASERS this "memory" was found to be fairly large with $x_n = 0.05 \times x_n + 0.95 \times x_{n-1}$. It should be noted that a "long memory", $c_2 \to 1$, requires a more stringent convergence limit, $|x_n - x_{n-1}|/x_{n-1} < 10^{-7}$.

Lastly, all numerical implementations must carry some awareness of precision limits and computer number formats. To take extremely small rate coefficients into account a larger number representation and small number calculation libraries must be used. Such specialised libraries use more memory and computation takes longer. In order ensure consistent calculations, the MASERS calculator limits A-coefficients to be $\geqslant 10^{-13}$. This limit can be imposed safely since the Einstein-A coefficients describe the transition probability per unit time that an atom currently in the upper level will go to the lower level, which is unlikely if the Einstein A coefficient is extremely small.

Results. The MASERS software was used to investigate the pumping of the H_2CO maser in G37.55+0.20, using the formaldehyde molecular data, Wiesenfeld & Faure (2013), and comparing the output with results published in Van der Walt (2014). Implementing the environments described in Section 3 of Van der Walt (2014), MASERS successfully recreates stimulated inversions of the 1_{10}–1_{11} (4.8 GHz) transition. Good agreement was obtained with the published inversion results for collisionally, as well as radiatively excited dust and free-free continuum emission.

False positives. A parametric model captures all its information within its parameters with no reference to actual data, making these methods very reliant on intelligent input by the user to accurately describe the physical environment. A proper understanding of the model inputs, implementation and assumptions are essential. Even then the user should always evaluate the outcome against expectation to guard against false positives.

This behaviour can be simulated for formaldehyde masers with the following environment parameters: Calculate the optical depth over the molecular column density range $N_{H2CO} = 10^{13} - 10^{18}\, cm^{-2}$ for total density $n_{H2} = 10^3,\ 10^4,\ 10^5\, cm^{-3}$ and kinetic temperature $T_{kin} = 10, 20, 30\, K$. To investigate radiative pumping by thermal dust emission, apply a dust background radiation contribution of $50\, K$. If the dust temperature is applied as a flat temperature, $T_{bb} = 50\, K$, the model will calculate the background continuum contribution for a black body emitter, which will result in inversions over the column density range. However, when applying it as a dust temperature, $T_d = 50\, K$ there will be no inversions, as expected, since the dust grey body continuum contribution model is used.

3. Line overlap

A detailed model of the escape probability representation of the rate equations with line overlap is given in Elitzur & Netzer (1985), which can be generalised to obtain:

$$
\begin{aligned}
\frac{dN_i^a}{dt} =\ & -N_i^a \left[\sum_{j<i} A_{ij}^a \hat{\beta}_a (1 + X_{ij}^a) + \sum_{j>i} A_{ji}^a \hat{\beta}_a \left(\frac{g_j}{g_i} \right) X_{ji}^a \right] \\
& + \left[\sum_{j<i} N_j^a A_{ij}^a \hat{\beta}_a \left(\frac{g_i}{g_j} \right) X_{ij}^a + \sum_{j>i} N_j^a A_{ji}^a \hat{\beta}_a (1 + X_{ji}^a) \right] \\
& + N_i^a \left[\sum_{j<i} A_{ij}^a f_a (1 - \beta_T)(x_a + X_{ij}^a) + \sum_{j>i} A_{ji}^a f_a (1 - \beta_T) \left(\frac{g_j}{g_i} \right) X_{ji}^a \right] \quad (3.1) \\
& - \left[\sum_{j<i} N_j^a A_{ij}^a f_a (1 - \beta_T) \left(\frac{g_i}{g_j} \right) X_{ij}^a + \sum_{j>i} N_j^a A_{ji}^a f_a (1 - \beta_T)(x_a + X_{ji}^a) \right] \\
& + \sum_{j<i} (1 - \beta_T) x_a \left(\sum_{\alpha}^{\alpha \neq a} f_\alpha A_{ul}^\alpha N_u^\alpha \right)_{ij} - \sum_{j>i} (1 - \beta_T) x_a \left(\sum_{\alpha}^{\alpha \neq a} f_\alpha A_{ul}^\alpha N_u^\alpha \right)_{ji}
\end{aligned}
$$

where *line a* is the transition line being evaluated and $\hat{\beta}_a = [1 - (1 - f_a)(1 - \beta_a)]$.

Note, only the radiative emission is affected by line overlap, thus Equation 2.1 is rewritten separating non-overlap and overlap, without explicitly including the collisional contributions in Equation 3.1. As with Equation 2.1 this grouping can be represented with a matrix equation $(\mathbf{R} + \tilde{\mathbf{C}})\mathbf{n} = \mathbf{0}$, where the radiative matrix now has three components

$$\mathbf{R} = \mathbf{R_a} - \mathbf{R_T} + \mathbf{O},$$

with $\mathbf{R_a}$ the non-overlap radiation matrix of *line a*, $\mathbf{R_T}$ the overlap region of *line a* and $\mathbf{O}$ all contribution of overlapping lines to *line a* matrices as follows:

$$R_{ij}^a = \begin{cases} A_{ji}\hat{\beta}_a(1 + X_{ji}) & i < j \\ A_{ij}\hat{\beta}_a\left(\frac{g_i}{g_j}\right)X_{ij} & i > j \\ -\sum_{i \neq j} R_{ji} & i = j \end{cases} \quad \text{and} \quad R_{ij}^T = \begin{cases} A_{ji}\hat{\beta}_T(x_a + X_{ji}) & i < j \\ A_{ij}\hat{\beta}_T\left(\frac{g_i}{g_j}\right)X_{ij} & i > j \\ -\sum_{i \neq j} R_{ji} & i = j \end{cases}$$

with $\hat{\beta}_T = f_a(1 - \beta_T)$.

$$O_{ij} = \begin{cases} -(1 - \beta_T)x_a\alpha_{\mathbf{ji}} & i < j \\ (1 - \beta_T)x_a\alpha_{\mathbf{ij}} & i > j \\ 0 & i = j \end{cases} \quad \text{and} \quad \alpha_{ij}[a_u] = \sum_{\substack{\alpha_{ul} \\ \alpha \neq a}} f_{\alpha_{ul}} A_{\alpha_{ul}}$$

where $\alpha_{\mathbf{ij}}$ is a row vector with elements for the upper, u, and lower, l, indices of all lines overlapping *line a*.

4. Summary

Fundamental to the study of astrophysical masers is the calculation of level populations for various physical conditions under which a population inversion can occur. The well-known escape probability method provides a powerful numerical algorithm for solving these non-linear multi-level problems within reasonable computational time.

Given the ease of implementation of the escape probability method, it is the de-facto choice of many level population calculator codes available, such as RADEX (Van der Tak *et al.*2007). However, care should be taken since the conditions under which population inversion occurs often causes instabilities during the calculation of the inevitable matrix inversion, making the results unreliable.

For this reason the MASERS Python package was developed to provide stable solutions for the investigation of maser pumping in particular and level population calculations in general. This paper described the development and implementation details of this open source package, as well as some preliminary results for formaldehyde maser pumping mechanisms.

References

Sobolev, A. M. & Gray, M. D. 2012, *IAU Symposium*, 287, 13

Elitzur, M., 1992, *Kluwer Academic Publishers*, 0-7923-1217-1 PB

Elitzur, M. & Netzer, H., 1985, *ApJ*, 291, 464

Wiesenfeld, L. & Faure, A., 2013, *MNRAS*, 432, 2573

van der Tak, F. F. S., Black, J. H., Schöier, F. L., Jansen, D. J., & van Dishoeck, E. F., 2007, *A&A*, 468, 627

Van der Walt, D. J., 2014, *A&A*, 562, A68

Astrophysical Masers:
Unlocking the Mysteries of the Universe
Proceedings IAU Symposium No. 336, 2017
A. Tarchi, M.J. Reid & P. Castangia, eds.

© International Astronomical Union 2018
doi:10.1017/S1743921317010705

Physical properties of Class I methanol masers

Silvia Leurini[1] and Karl M. Menten[2]

[1] INAF - Osservatorio Astronomico di Cagliari
Via della Scienza 5, 09047, Selargius (CA), Italy
email: sleurini@oa-cagliari.inaf.it

[2] Max-Planck-Institut für Radioastronomie
Auf Dem Hügel 69, 53121, Bonn, Germany
email: kmenten@mpifr.de

Abstract. As first realised in the late 1980s, methanol masers come in two varieties, termed Class I and Class II. While Class II masers had observationally been extensively studied in the past, until recently relatively little attention was paid to Class I methanol masers due to their low luminosities compared to other maser transitions. In this review, we will focus on the recent progress in our understanding of Class I methanol masers both from an observational and from a theoretical point of view.

Keywords. masers, ISM: molecules

1. Introduction – the two classes of methanol masers

Methanol masers are common phenomena in regions of massive star formation. Early studies empirically revealed two different classes based on observational properties (Wilson *et al.* 1985; Batrla *et al.* 1987). These were termed Class I and Class II (Menten 1991a). Soon after, it was also recognised (Menten 1991b) that different pumping mechanisms produce inversion in Class I methanol masers (CIMMs), first detected in a series of lines near 25 GHz (Barrett *et al.* 1971) and, later, at 36 GHz and 44 GHz (Morimoto *et al.* 1985) on the one hand and in Class II methanol masers (CIIMMs) on the other. The most common CIIMM transition is by far the 6.7 GHz line, followed by the 12.18 GHz line (Menten 1991a; Batrla *et al.* 1987).

CIIMMs are found in the closest environment of massive young stellar objects (MYSOs) and are pumped by the intense mid-infrared radiation from these objects' dense dust shells. Their presence is actually a sufficient, if not a necessary, signpost for a MYSO (Urquhart *et al.* 2015). Early calculations (e.g., Cragg *et al.* 1992) showed that inversion population in CIIMMs happens when the brightness temperature of the external radiation is greater than the kinetic temperature of the gas. Sobolev & Deguchi (1994) and Sobolev *et al.* (1997) first suggested that CIIMMs are excited through the pumping of molecules from the torsional ground state to rotational energy levels within the torsionally excited $v_t = 1, 2$ states as consequence of absorption of mid-infrared photons followed by cascade back to the ground state. Later, Cragg *et al.* (2005) examined the effects of including higher energy levels, and found that this effect is significant only at dust temperatures above 300 K. One should notice, however, that up to now models of CIIMMs have only made use of limited sets of collisional rates and that the results of new calculations of Rabli & Flower (2010, 2011) have not been implemented yet. It is of paramount importance to revise models of CIIMMs in light of these new calculations and investigate in details the exact pumping cycle of CIIMMs.

CIMMs are, on the other hand, collisionally pumped and their occurrence can be explained from the basic properties of the CH_3OH molecule (Lees 1973). First statistical equilibrium studies on CIMMs made use of collisional rates based on experimental results by Lees & Oka (1969) and Lees & Haque (1974). These rates show a propensity for $\Delta k = 0$ collisions and a dependence upon ΔJ as $1/\Delta J$. The propensity for $\Delta k = 0$ over $\Delta k = 0, 1$ collisions causes molecules to preferentially de-excite down a k stack, causing overpopulation in the lower levels in the $k = -1$ stack for CH_3OH-E, leading to maser action in the $J_{-1} \rightarrow (J-1)_0$-$E$ series and to enhanced absorption ($\equiv$ anti-inversion) in the $2_0 \rightarrow 3_{-1}$-E line (Walmsley $et\ al.$ 1988). An analogous situation holds for the $K = 0$ ladder of CH_3OH-A, leading to the $J_0 \rightarrow (J-1)_1$-A^+ masers and also to the anti-inversion of the $5_1 \rightarrow 6_0$-A^+ transition†. However, while previously used collisional rates can reproduce the anti-inversion at 12.18 GHz and 6.7 GHz and inversion for many CIMM lines, they do not account for the $J_2 \rightarrow J_1$-E series of masers at 25 GHz. For this reason, ad hoc selection rules were adopted to reproduce inversion population in these transitions (e.g., Johnston $et\ al.$ 1992 and Cragg $et\ al.$ 1992). Recent models (McEwen $et\ al.$ 2014; Nesterenok 2016) have confirmed that CIMMs mase in different density ranges.

In this contribution, we review the observational and theoretical properties of CIMMs in view of the increasing interests that these masers gained over the last decades not only in the star formation community but also in other fields of astronomy.

2. Overview of observational properties of CIMMs

CIMMs are associated with astrophysical shocks. They are often observed in the swept-up interaction region of outflows with ambient material predominantly from MYSOs (Plambeck & Menten 1990), but also from low-mass YSOs, (e.g. Kalenskii $et\ al.$ 2010; Rodríguez-Garza $et\ al.$ 2017). They have also been detected in the interaction regions of supernova remnants with the molecular clouds (e.g., Szczepanski $et\ al.$ 1989; Haschick $et\ al.$ 1990; Pihlstrom $et\ al.$ 2011), in cloud-cloud collisions (e.g. Sobolev 1992), and in layers where expanding HII regions interact with the ambient molecular environment (Voronkov $et\ al.$ 2010). Finally, over the last years detections of CIMMs were reported also in external galaxies (e.g., in NGC 253 and NGC 4945 by Ellingsen $et\ al.$ 2014 and 2017, and McCarthy $et\ al.$ 2017) with emission reported to be up to 10^5 times more luminous than the corresponding Galactic CIMMs.

In the last decade, CIMMs were routinely used as signposts of star formation activity, especially in very early phases of massive star formation for which it is difficult to detect outflow activity in mid-infrared observations due to high extinction (e.g., Cyganowski $et\ al.$ 2009, Yanagida $et\ al.$ 2014, Towner $et\ al.$ 2017). The recent detection of circular polarization in CIMMs lines shows these masers' potential for probes of magnetic fields in interstellar shocks (Sarma & Momjian 2009, 2011). Finally, their association with supernova remnants make CIMMs interesting tracer of density which may be useful for modeling of cosmic ray particle acceleration (Frail 2011).

CIMMs in regions of massive star formation have typical luminosities of $10^{-5} - 10^{-6}\ L_\odot$, spot sizes are of the order of tens of AU based on VLA observations of the $6_2 \rightarrow 6_1$-E line at 25.018 GHz (Johnston $et\ al.$ 1997) and the recent first VLBI detection of the $7_0 \rightarrow 6_1$-A^+ maser at 44 GHz (Matsumoto $et\ al.$ 2014). In star-forming regions, the 44 GHz CIMM is usually stronger than the other transitions: Voronkov $et\ al.$ (2014) found

---

† We note that a strong IR field causes the $5_1 \rightarrow 6_0$-A^+ and the $2_0 \rightarrow 3_{-1}$-E line to become the most prominent Class II $maser$ transitions.

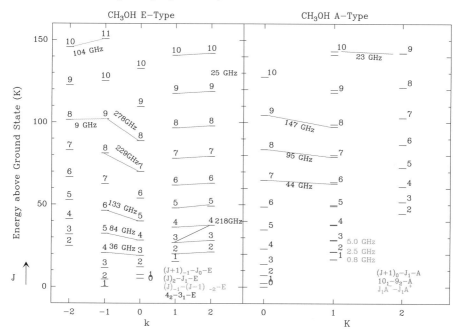

Figure 1. Partial rotational level diagram of E-type (right) and A-type (left) methanol (adapted from Leurini *et al.* 2016). Different colours (see the on-line version of the Figure) are used for different families of transitions: red indicates the $(J+1)_1 \rightarrow J_0$-E (left) and the $(J+1)_0 \rightarrow J_1$-A series (right), blue the $J_2 \rightarrow J_1$-E series (left) and the $10_1 \rightarrow 9_2$-A line (right), magenta the $(J-1) \rightarrow (J-1)_{-2}$-$E$ series (left) and the $J_1 A^- \rightarrow J_1 A^+$ transitions (right), black the $4_2 \rightarrow 3_1$-E maser (left).

that the 36 GHz line $(4_{-1} \rightarrow 3_0$-$E)$ is stronger than the 44 GHz maser only in 40 out of 292 maser spots in a recent survey of massive star-forming regions in both transitions. On the contrary, toward supernova remnants, the 36 GHz line is often stronger than the 44 GHz transition and in some cases the latter is completely absent (e.g, Pihlström *et al.* 2014). Recently, Ellingsen *et al.* (2017) suggested that extragalactic CIMMs (at least in NGC 253) have similar properties than those in supernova-molecular cloud interactions.

3. Statistical equilibrium calculations

In the following, we will review the recent results from Leurini *et al.* (2016) and complement their analysis with the discussion of other rarer CIMMs. In particular, we will report on the properties of the $9_{-1} \rightarrow 8_{-2}$-$E$ 9.9 GHz, $10_1 \rightarrow 9_2$-A^- 23.4 GHz, and $11_{-1} \rightarrow 10_{-2}$-$E$ 104 GHz lines; see Voronkov *et al.* (2012) for a detailed discussion of their observational properties. We also discuss predictions for the low-frequency $J_1 A^- \rightarrow J_1 A^+$ doublet transitions, the lowest of which ($J = 1$ at 834 MHz) is the first CH_3OH line discovered in the interstellar medium (Ball *et al.* 1970). Our main focus are CIMMs in regions of massive star formation. All lines discussed in this paper are shown in Fig. 1. For models of CIMMs in supernova remnants, we refer the reader to McEwen *et al.* (2014).

Following the formalism developed by Elitzur *et al.* (1989) and Hollenbach *et al.* (2013) for H_2O masers, we use the pump rate coefficient, q, the maser loss rate, Γ, and the inversion efficiency of the pumping scheme, η (see Eqs. 6–8 in Leurini *et al.* 2016) to define the photon production rate $\Phi_m = \frac{g_u\, g_l}{g_u + g_l} \times 2n^2 X(CH_3OH)\eta q$ to characterise the maser emission (g_u and g_l are the statistical weights for the upper and lower levels of a given maser system). Indeed, the photon production rate at line centre of a saturated maser transition is directly linked to its observed flux through the geometry of the maser. Therefore, under the assumption that different maser lines are emitted by the same volume of gas and that they are saturated, the ratio of the photon production rates of two lines is directly proportional to the ratio between the observed fluxes of those lines.

3.1. *The low-frequency $J_1\,A^- \to J_1\,A^+$ transitions*

Low-frequency transitions in the $J_1\,A^- \to J_1\,A^+$ series were detected in emission at 834 MHz and 5005 MHz towards bright centimeter continuum sources in the Central Molecular Zone (CMZ) surrounding the Galactic center (e.g., Ball *et al.* 1970, Robinson *et al.* 1974, Mezger & Smith 1976). Given that their flux densities are only a small fraction of that of continuum background radiation, these lines do clearly not show high gain maser action. High angular and velocity resolution observations are not available to directly confirm their maser nature. However, not only are they in emission against the CMZ's very strong radio continuum emission, they are also associated with absorption in the 12.18 GHz line toward at least two lines of sight (Peng & Whiteoak 1992). As explained in Section 1, *absorption* in this line, which is a cardinal CIIMM transition, is characteristic of CIMMs. This strongly suggests that the $J_1\,A^- \to J_1\,A^+$ lines are weak CIMMs. Indeed, previous calculations from Cragg *et al.* (1992) confirmed that these lines are inverted over a broad range of physical parameters. Their inversion under the special physical (and chemical) conditions of the CMZ follows a pattern observed for low radio frequency lines of several other molecules; see Menten (2004) for a discussion.

Our modelling predicts inversion in the low-frequency $J_1\,A^- \to J_1\,A^+$ transitions (and in all other lines for which maser emission is observed, see Fig. 1). The $J_1\,A^- \to J_1\,A^+$ series has low opacities ($|\tau| < 0.4$) and is inverted over a broad range of densities and temperatures. The brightest of the series is the $1_1\,A^- \to 1_1\,A^+$ line at 834 MHz.

3.2. *Modelling results*

We investigated the behaviour of the photon production rate Φ_m as function of the physics of the gas for the most common $(J+1)_{-1} - J_0$-*E*, $(J+1)_0 - J_1$-*A* and $J_2 - J_1$-*E* CIMMs and for the rarer 9.9 GHz, 23.4 GHz and 104 GHz lines. The computations show that for all these CIMMs, Φ_m grows with the emission measure ξ, the ratio of the product of the molecular hydrogen and methanol number densities over the velocity (see Eqs. 3 and 5 in Leurini *et al.* 2016). Therefore, bright CIMMs trace high methanol emission measures. These values are reached at high densities, high temperatures, and high methanol column densities. Moreover, all bright CIMMs (the $(J+1)_{-1} - J_0$-*E*, $(J+1)_0 - J_1$-*A* and $J_2 - J_1$-*E* series) quench at a constant specific column density of some $10^{17}\,\mathrm{cm}^{-2}\,\mathrm{km}^{-1}\,\mathrm{s}$ independently of volume density as long as it is in the range $10^5 - 10^7\,\mathrm{cm}^{-3}$, suggesting that there is a critical methanol column density at which quenching occurs. The rare CIMM lines at 9.9 GHz, 104 GHz and 23.4 GHz seem to trace more extreme conditions than the more common CIMM lines in terms of density, column density and temperature as already suggested by previous models (see Sobolev *et al.* 2005).

Leurini *et al.* (2016) showed that CIMMs can be divided in at least three families, depending on their behaviour as function of photon production rate: the $(J+1)_{-1} - J_0$-*E* type series, the $(J+1)_0 - J_1$-*A* type and the $J_2 - J_1$-*E* series at 25 GHz. They differ in the slope of the Φ_m dependence on ξ and on density. This is well illustrated in Fig. 2 (adapted from Figs. 10 and 12 of Leurini *et al.* 2016) where ratios between photon production rates of different masers are plotted as function of the emission measure: while ratios between Φ_m of lines in within the same family are relatively flat (left panel),

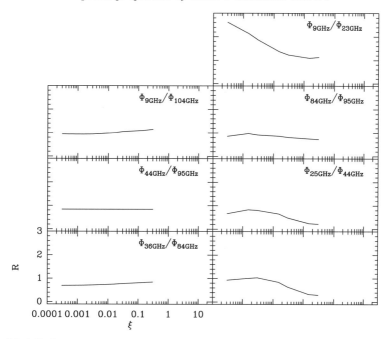

Figure 2. Modelled ratios between photon production rates of several CIMMs for $T = 200\,K$ and a volume density $n = 10^7\ \mathrm{cm}^{-3}$. In the left panel, we show ratios among transitions within same families, while in the right panel we plot ratios among CIMMs from different families.

ratios among different families have a strong dependence on ξ (right panel). The ratio of the $\Phi_{9.9\,\mathrm{GHz}}$ to $\Phi_{23.4\,\mathrm{GHz}}$ shows the strongest dependence on the physics of the gas. In addition, the $J_2 - J_1$-E lines have a much flatter ξ dependence than the other lines at high densities (Fig. 12 in Leurini *et al.* 2016).

4. Conclusions

CIMMs are powerful tools to investigate the physics of astrophysical shocks. Our calculations show that

• The production rates of all known CIMMs increase with ξ till thermalisation is reached. Therefore, bright CIMMs trace high methanol emission measure reached at high densities ($n(\mathrm{H_2}) \sim 10^7 - 10^8\ \mathrm{cm}^{-3}$), high temperatures ($> 100\,\mathrm{K}$) and high methanol column densities;

• CIMMs can reasonably be separated into different families depending on the behaviour of their photon production rate as function of ξ;

• The 25 GHz lines and the rare lines at 9.9 GHz, 23.4 GHz and 104 GHz trace higher densities and temperatures than the other lines.

This contribution, as this whole conference, is dedicated to the memory of Malcolm Walmsley.

References

Ball, J. A., Gottlieb, C. A., Lilley, A. E., & Radford, H. E. 1970, *Ap. Lett.*, 162, L203

Barrett, A. H., Schwartz, P. R., & Waters, J. W. 1971, *ApJ*, 168, L101

Batrla, W., Matthews, H. E., Menten, K. M., & Walmsley, C. M. 1987, *Nature*, 326, 49

Cragg, D. M., Johns, K. P., Godfrey, P. D., & Brown, R. D. 1992, *MNRAS*, 259, 203

Cragg, D. M., Sobolev, A. M., & Godfrey, P. D. 2005, *MNRAS*, 360, 533

Cyganowski, C. J., Brogan, C. L., Hunter, T. R., & Churchwell, E. 2009, *ApJ*, 702, 1615

Ellingsen, S. P., Chen, X., Qiao, H.-H., *et al.* 2014, *ApJ*, 790, L28

Ellingsen, S. P., Chen, X., Breen, S. L., & Qiao, H.-H. 2017, *MNRAS*, 472, 604

Elitzur, M., Hollenbach, D. J., & McKee, C. F. 1989, *ApJ*, 346, 983

Frail, D. A. 2011, *Mem. Soc. Astr. Italiana*, 82, 703

Haschick, A. D., Menten, K. M., & Baan, W. A. 1990, *ApJ*, 354, 556

Hollenbach, D., Elitzur, M., & McKee, C. F. 2013, *ApJ*, 773, 70

Johnston, K. J., Gaume, R., Stolovy, S., *et al.* 1992, *ApJ*, 385, 232

Johnston, K. J., Gaume, R. A., Wilson, T. L., Nguyen, H. A., & Nedoluha, G. E. 1997, *ApJ*, 490, 758

Kalenskii, S. V., Johansson, L. E. B., Bergman, P., *et al.* 2010, *MNRAS*, 405, 613

Lees, R. M. 1973, *ApJ*, 184, 763

Lees, R. M. & Oka, T. 1969, *J. Chem. Phys*, 51, 3027

Lees, R. M. & Haque, S. 1974, *Can. J. Phys.*, 52, 2250

Leurini, S., Menten, K. M., & Walmsley, C. M. 2016, *A&A*, 592, A31

McCarthy, T. P., Ellingsen, S. P., Chen, X., *et al.* 2017, *ApJ*, 846, 156

Matsumoto, N., Hirota, T., Sugiyama, K., *et al.* 2014, *Ap. Lett.*, 789, L1

McEwen, B. C., Pihlström, Y. M., & Sjouwerman, L. O. 2014, *ApJ*, 793, 133

Menten, K. M. 1991a, *ApJ*, 380, L75

Menten, K. 1991b, Atoms, Ions and Molecules: New Results in Spectral Line Astrophysics, 16, 119

Menten, K. M. 2004, in *The Dense Interstellar Medium in Galaxies*, 91, 69

Mezger, P. G. & Smith, L. F. 1976, *A&A*, 47, 143

Morimoto, M., Kanzawa, T., & Ohishi, M. 1985, *ApJ*, 288, L11

Nesterenok, A. V. 2016, *MNRAS*, 455, 3978

Peng, R. S. & Whiteoak, J. B. 1992, *MNRAS*, 254, 301

Pihlström, Y. M., Sjouwerman, L. O., & Fish, V. L. 2011, *Ap. Lett.*, 739, L21

Pihlström, Y. M., Sjouwerman, L. O., Frail, D. A., *et al.* 2014, *AJ*, 147, 73

Plambeck, R. L. & Menten, K. M. 1990, *ApJ*, 364, 555

Rabli, D. & Flower, D. R. 2010, *MNRAS*, 406, 95

Rabli, D. & Flower, D. R. 2011, *MNRAS*, 411, 2093

Robinson, B. J., Brooks, J. W., Godfrey, P. D., & Brown, R. D. 1974, *Australian Journal of Physics*, 27, 865

Rodríguez-Garza, C. B., Kurtz, S. E., Gómez-Ruiz, A. I., *et al.* 2017, arXiv:1709.09773

Sarma, A. P. & Momjian, E. 2009, *Ap. Lett.*, 705, L176

Sarma, A. P. & Momjian, E. 2011, *Ap. Lett.*, 730, L5

Sobolev, A. M. 1992, *Soviet Astron.*, 36, 590

Sobolev, A. M. & Deguchi, S. 1994, *A&A*, 291, 569

Sobolev, A. M., Cragg, D. M., & Godfrey, P. D. 1997, *A&A*, 324, 211

Sobolev, A. M., Ostrovskii, A. B., Kirsanova, M. S., *et al.* 2005, Massive Star Birth: A Crossroads of Astrophysics, 227, 174

Szczepanski, J. C., Ho, P. T. P., Haschick, A. D., & Baan, W. A. 1989, The Center of the Galaxy, 136, 383

Towner, A. P. M., Brogan, C. L., Hunter, T. R., *et al.* 2017, *ApJS*, 230, 22

Urquhart, J. S., Moore, T. J. T., Menten, K. M., *et al.* 2015, *MNRAS*, 446, 3461

Voronkov, M. A., Caswell, J. L., Ellingsen, S. P., & Sobolev, A. M. 2010, *MNRAS*, 405, 2471

Voronkov, M. A., Caswell, J. L., Ellingsen, S. P., *et al.* 2012, Cosmic Masers - from OH to H0, 287, 433

Voronkov, M. A., Caswell, J. L., Ellingsen, S. P., Green, J. A., & Breen, S. L. 2014, *MNRAS*, 439, 2584

Yanagida, T., Sakai, T., Hirota, T., *et al.* 2014, *Ap. Lett.*, 794, L10

Walmsley, C. M., Batrla, W., Matthews, H. E., & Menten, K. M. 1988, *A&A*, 197, 271

Wilson, T. L., Walmsley, C. M., Menten, K. M., & Hermsen, W. 1985, *A&A* 147, L19

Astrophysical Masers:
Unlocking the Mysteries of the Universe
Proceedings IAU Symposium No. 336, 2017
A. Tarchi, M.J. Reid & P. Castangia, eds.

© International Astronomical Union 2018
doi:10.1017/S1743921318000686

Quantum-Chemical calculations revealing the effects of magnetic fields on methanol masers

Boy Lankhaar[1], Wouter Vlemmings[2], Gabriele Surcis[3], Huib Jan van Langevelde[4,5], Gerrit C. Groenenboom[6] and Ad van der Avoird[6]

[1]Department of Space, Earth and Envoirment, Chalmers University of Technology,
Onsala Space Observatory, 439 92 Onsala, Sweden
email: boy.lankhaar@chalmers.se

[2]Department of Space, Earth and Envoirment, Chalmers University of Technology,
Onsala Space Observatory, 439 92 Onsala, Sweden

[3]INAF-Osservatorio Astronomico di Cagliari,
Via della Scienza 5, I-09047 Selargius, Italy

[4]Joint Institute for VLBI ERIC,
Postbus 2, 7990 AA Dwingeloo, The Netherlands

[5]Sterrewacht Leiden, Leiden University,
9513, 2330 RA Leiden, the Netherlands

[6]Theoretical Chemistry, Institute for Molecules and Materials, Radboud University,
Heyendaalseweg 135, 6525 AJ Nijmegen, The Netherlands

Abstract. Maser observations of both linearly and circularly polarized emission have provided unique information on the magnetic field in the densest parts of star forming regions, where non-maser magnetic field tracers are scarce. While linear polarization observations provide morphological constraints, magnetic field strengths are determined by measuring the Zeeman splitting in circularly polarized emission. Methanol is of special interest as it is one of the most abundant maser species and its different transitions probe unique areas around the protostar. However, its precise Zeeman-parameters are unknown. Experimental efforts to determine these Zeeman-parameters have failed. Here we present quantum-chemical calculations of the Zeeman-parameters of methanol, along with calculations of the hyperfine structure that are necessary to interpret the Zeeman effect in methanol. We use this model in re-analyzing methanol maser polarization observations. We discuss different mechanisms for hyperfine-state preference in the pumping of torsion-rotation transitions involved in the maser-action.

Keywords. magnetic fields, masers, molecular data, polarization

1. Introduction

The presence of a magnetic field within an astrophysical maser is known to produce partially polarized radiation. Linear polarization provides information on the magnetic field direction, and the magnetic field strength can be determined by comparing the field-induced frequency shifts between left- and right-circularly polarized emission. In OH, H_2O, SiO, and CH_3OH (methanol) masers, polarized radiation has been observed and analyzed for the information it contains on the magnetic field in the regions these masers probe. Extracting quantitative information on the magnetic field in these regions requires knowledge of the Zeeman parameters, describing the response of the maser molecule/atom to a magnetic field. These Zeeman parameters are known for all but methanol masers. Here we describe a quantitative theoretical model of the magnetic properties of methanol, including the complicated hyperfine structure that results from

23

its internal rotation (Lankhaar *et al.* 2016). With this model, we can determine the Zeeman splitting of the hyperfine states within all the known methanol maser transitions. We will use this model in (re-)interpreting methanol maser polarization observations.

Hyperfine interactions and Zeeman effects in methanol

The elucidation of methanol's hyperfine structure has been a challenging problem. The CH_3-group in methanol can easily rotate with respect to the OH-group, which leads to an extension of the usual rigid-rotor hyperfine Hamiltonian with nuclear spin-torsion interactions. In contrast with the nuclear spin-rotation coupling parameters, the torsional hyperfine coupling parameters cannot be obtained from quantum chemical calculations. Experiments probing the hyperfine structure of methanol have proven difficult to interpret, because the hyperfine transitions cannot be individually resolved. Lankhaar *et al.* (2016), revised the derivation of a Hamiltonian which includes the torsional hyperfine interactions and obtained the coupling parameters in this Hamiltonian from *ab initio* calculations and experimental data by Heuvel & Dymanus (1973) and Coudert *et al.* (2015). The hyperfine spectra of methanol calculated from this Hamiltonian agree well with the spectra observed for several torsion-rotation transitions of both *A*- and *E*-symmetry.

Zeeman interactions are governed by the same magnetic moments that determine the hyperfine structure, interacting with an external magnetic field. For a diamagnetic molecule as methanol three contributions to the molecules Zeeman effect are important:

• overall rotation. Rotational Zeeman effects are represented by the molecule-specific g-tensor, which for rigid non-paramagnetic molecules has been extensively studied experimentally for its valuable information on the electronic structure. Quantum chemical calculations are able to reproduce these experiments with high accuracy. We carried out quantum chemical calculations to obtain the rotational g-tensor for methanol.

• internal rotation or torsion. Torsional Zeeman interactions are represented by the molecule-specific b-vector. The calculation of the b-vector has not been implemented in the available quantum-chemical program packages. For nitromethane and methyl-boron-difluoride, the torsional Zeeman effect has been investigated experimentally by Engelbrecht *et al.* (1973). In order to estimate the torsional Zeeman effects in methanol, we have extrapolated the torsional *b*-vectors for these molecules, by comparing their internally rotating CH_3-groups to the CH_3-group of methanol.

• nuclear spins. The nuclear spin of methanol, CH_3OH, comes from the three protons in the CH_3 group and the proton in the OH group. The Zeeman effect of the protons scale with the proton g-factor: $g_p = 5.585$.

We combine these Zeeman interactions with the model of the hyperfine structure to determine the Zeeman splitting of the hyperfine states within all the known methanol maser transitions.

Zeeman interactions are usually described in a first-order approximation by the Landé g-factor. In methanol, each torsion-rotation transition is actually split into a number of transitions between individual hyperfine levels of the upper and lower torsion-rotation states (Figure 1). The Landé g-factors calculated for the different hyperfine transitions differ strongly which is important for the interpretation of the measured maser polarization effects (Lankhaar *et al.* 2017).

Polarization in methanol masers

Methanol maser circular polarization observations have been made for the transitions: 6.7 GHz (5_{15} A_2 → 6_{06} A_1) by, *e.g.*, Vlemmings *et al.* (2011), Surcis *et al.* (2012),

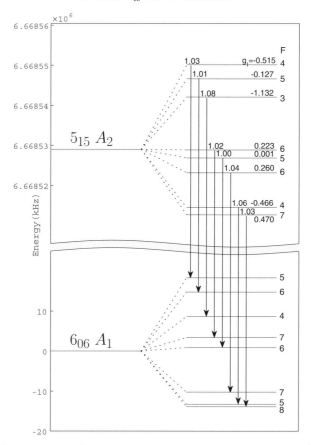

Figure 1. Hyperfine structure of the torsion-rotation levels in the 6.7 GHz (5_{15} $A_2 \to 6_{06}$ A_1) transition. The energy of the 6_{06} A_1 torsion-rotation level is set to zero. Arrows indicate the strongest hyperfine transitions with $\Delta F = \Delta J = 1$, with the Einstein A-coefficients (in 10^{-9} s^{-1}) indicated above. Landé g-factors of the transitions in a magnetic field of 10 mG are given at the righthand side of the upper energy levels. The rightmost numbers are the F quantum numbers of the hyperfine states.

44 GHz (7_{07} $A_2 \to 6_{16}$ A_1) by Momjian & Sarma (2016) and 36 GHz (4_{-1} $E \to 3_0$) E by Sarma & Momjian (2009). As the magnetic characteristics of methanol were not known, (hyperfine unspecific) estimates of the Zeeman parameters were used. In the following, we will re-analyze some of the observations using our calculated Zeeman parameters.

Hyperfine-specific effects in the maser action. The individual hyperfine lines are not spectrally resolved, but the maser action can favor specific hyperfine transitions by the following mechanisms:

• different radiative rates for stimulated emission

• kinematic effects, when there are two maser clouds along the line of sight with different velocities, such that a hyperfine transition in the foreground cloud amplifies emission from a different hyperfine transition in the background cloud

• population inversion of the levels involved in maser action is preceded by collisional and radiative de-excitation of higher torsion-rotation levels, with rate coefficients that are hyperfine-state specific.

Polarization observations of class II 6.7 GHz methanol masers. We assume that the transition with the largest Einstein coefficient for stimulated emission, the

$F = 3 \to 4$ transition, will be favored and that the maser action is limited to this transition. Then, the Zeeman-splitting coefficient is 10 times larger than the value currently used for magnetic field estimates. In the methanol maser regions probed by these class II masers, with an H_2 number density of $n_{H_2} \approx 10^8$ cm^{-3} , application of our new results to the large sample of maser observations reported in Vlemmings *et al.* (2011) indicates an average field strength of $\langle |B| \rangle \approx 12$ mG. This is in good agreement with OH-maser polarization observations by Wright *et al.* (2004), as well as with the extrapolated magnetic field vs. density relation, see Crutcher (1999).

Polarization observations of class I methanol 36 GHz and 44 GHz masers. We assume that the $F = 3 \to 2$ (36 GHz) and $F = 5 \to 4$ (44 GHz) hyperfine lines are favored and that the maser action is limited to these transitions. The observed class I methanol masers are expected to occur in shocked regions of the outflows at densities, an order of magnitude lower in comparison to class II masers. Using our analysis, the Zeeman splitting of the 36 GHz and 44 GHz lines would indicate magnetic field strengths of $20 - 75$ mG. Since, class I masers are shock excited, shock compression is expected to increase the magnetic field strength.

Oppositely polarized masers. Observations have shown reversals in the sign of polarization over areas of small angular extent in the sky. Such reversals have previously been interpreted as a change in field direction. However, reversals on au-scales would be surprising if one considers the agreement between the fields probed by methanol masers and dust emission. A more plausible explanation favored by our results is that in the masers with opposite signs of polarization, the masing process itself is due to the dominance of different hyperfine transitions. Such a mechanism is able to explain opposite circular polarization along the line of sight without assuming a change in magentic field direction, and to obtain magnetic fields comparable with the results from other masers that trace similar areas around the protostar.

Summary

We have presented a model for the Zeeman interactions in methanol, in combination with the hyperfine structure. In contrast to previous models of methanol's Zeeman effect, where a single effective g-factor was assumed, we show that each hyperfine transition in the maser line has its own unique Landé g-factor, and that these g-factors vary over a large range of values. We have applied our results to existing circular polarization measurements, which leads to substantially different conclusions, and confirms the presence of dynamically important magnetic fields around protostars.

References

Lankhaar, B., Groenenboom, G. C., & van der Avoird, A. 2016, *JCP*, 24, 145

Lankhaar, B., Vlemmings, W. H. T., Surcis, G., van Langevelde, H. J., Groenenboom, G. C., & van der Avoird, A. 2017, *Nature Astronomy*, 2, 145L

Coudert, L. H., Gutlé, C., Huet, T. R., Grabow, J.-U., & Levshakov, S. A. 2015, *JCP*, 4, 143

Heuvel, J. E. M. & Dymanus, A. 1973, *J Mol S*, 2, 45

Engelbrecht, L., Sutter, D., & Dreizier, H. 1973, *Z Nat A*, 5, 28

Vlemmings, W. H. T., Torres, R. M., & Dodson, R. 2011, *A&A*, 529, A95

Surcis, G., Vlemmings, W. H. T., van Langevelde, H. J., & Hutawarakorn Kramer, B. 2012, *A&A*, 541, A47

Momjian, E. & Sarma, A. P. 2016, *ApJ*, 2, 834

Sarma, A. P. & Momjian, E. 2009, *ApJ Lett*, 2, 705

Crutcher, R. M. 1999, *ApJ*, 2, 520

Wright, M. M., & Gray, M. D., & Diamond, P. J. 2004, *MNRAS*, 4, 350

Astrophysical Masers:
Unlocking the Mysteries of the Universe
Proceedings IAU Symposium No. 336, 2017
A. Tarchi, M.J. Reid & P. Castangia, eds.

© International Astronomical Union 2018
doi:10.1017/S1743921317011632

Maser Polarization

Gabriele Surcis[1], Wouter H. T. Vlemmings[2], Boy Lankhaar[2] and Huib Jan van Langevelde[3,4]

[1]INAF, Osservatorio Astronomico di Cagliari
Via della Scienza 5, I-09047, Selargius, Italy
email: `surcis@oa-cagliari.inaf.it`

[2]Department of Space, Earth and Environment, Chalmers University of Technology
Onsala Space Observatory, 439 92 Onsala, Sweden

[3]Joint Institute for VLBI ERIC
Postbus 2, 7990 AA Dwingeloo, The Netherlands

[4]Sterrewacht Leiden, Leiden University
Postbus 9513, 2330 RA Leiden, The Netherlands

Abstract. Through the observations and the analysis of maser polarization it is possible to measure the magnetic field in several astrophysical environments (e.g., star-forming regions, evolved stars). In particular from the linearly and circularly polarized emissions we can determine the orientation and the strength of the magnetic field, respectively. In these proceedings the implications, on observed data, of the new estimation of the Landé g-factors for the CH_3OH maser are presented. Furthermore, some example of the most recent results achieved in observing the polarized maser emission from several maser species will also be reported.

Keywords. masers, polarization, magnetic fields, radiative transfer

1. Introduction

Measuring magnetic fields in the proximity of astrophysical objects, like massive young stellar objects (YSOs) or evolved stars, has always been a strong desire. This tough challenge bothered the astronomers till few decades ago when the foundations of maser polarization theory were strengthened (e.g., Nedoluha & Watson 1992). Through interferometric observations and the analysis of the polarized emission of masers it is nowadays possible both to derive the morphology and to determine the strength of magnetic field at milliarcsecond resolution, which translates for close-by objects in astronomical unit (au) scale.

The main maser species for which the polarized emission is commonly detected are OH, CH_3OH, SiO, and H_2O masers. OH is a paramagnetic molecule, i.e. the molecule has a magnetic permeability greater or equal to unity, and the splitting (ΔV_Z) due to the Zeeman effect of its masering emission lines is larger than the linewidth (Δv_L) of the maser lines themselves. The behavior of non-paramagnetic molecules (CH_3OH, SiO, and H_2O) is much less pronounced, i.e. $\Delta V_Z < \Delta v_L$. This implies that from the OH maser emissions the direct measurement of the Zeeman-splitting, and consequently of the magnetic field strength (B), is straightforward while for the other three maser species it requires a more detailed analysis. The maser emission lines arise under different physical conditions (e.g., temperatures and densities) and consequently they trace the magnetic fields in different regions of the same astronomical object. The observations and analysis of all the four maser species are therefore fundamental.

In 1992 Nedoluha & Watson developed a full radiative transfer model for the polarized emission of 22-GHz H_2O maser. The transfer equations of this model are solved in the presence of a magnetic field that causes $\Delta V_Z < \Delta v_L$, and under the following conditions:

- the Zeeman frequency shift $g\Omega$, where g is the Landé g-factor and $\Omega = eB/m_ec$, is much larger than the rate of stimulated emission R;
- the Zeeman frequency shift is much larger than the decay rate Γ and the cross-relaxation between the magnetic substates Γ_ν.

Therefore the model is valid only for unsaturated H_2O masers. Although the model was developed for the 22-GHz H_2O maser, this is valid for all the non-paramagnetic maser species that meet the above conditions. This is the case both for the CH_3OH maser emissions (Vlemmings *et al.* 2010) and for the SiO maser emissions (Peréz-Sánchez & Vlemmings 2013). However, for the SiO masers it is necessary to include the anisotropic pumping mechanism that is not considered in Nedoluha & Watson's model.

2. CH_3OH maser

CH_3OH maser emission is divided into two classes: Class I (e.g., rest frequency = 36 GHz, 44 GHz) and Class II (e.g., 6.7 GHz, 12.2 GHz). All the maser lines originate from torsion-rotation transitions and only recently an accurate model of their hyperfine structure has been calculated (Lankhaar *et al.* 2016). The model shows that each single CH_3OH maser emission is composed of several hyperfine transitions (> 8) which are not spectrally resolved due to the typical "poor" spectral resolution of the observations ($\gtrsim 2$ kHz). Indeed the frequency separations of the hyperfine transitions of a maser emission are of the order of few kHz (Tables I-XIV of Lankhaar *et al.* 2016). Actually, it is still unknown how much a hyperfine transition contributes to its maser emission, therefore a detailed pumping model is absolutely fundamental to resolve this issue.

In addition, Lankhaar *et al.* (2017) investigated the split of the hyperfine transitions when the CH_3OH molecule is immersed in a magnetic field, providing the Landé g-factor for all the transitions. Although the main results obtained from Lankhaar *et al.* can be read in this book, we just underline here that the g-factors varies with the magnetic field strength and they can be considered constant for $B \lesssim 50$ mG.

The Full Radiative Transfer Method Code

In 2010 we adapted the Full Radiative Transfer Method (FRTM) code, developed by Vlemmings *et al.* (2006) for the H_2O maser and based on the model of Nedoluha & Watson (1992), for modeling the polarized emission of the 6.7-GHz CH_3OH maser. In the FRTM code the 6.7 GHz maser emission was assumed to be composed of only one transition (no hyperfine structure was considered), and later the assumed g-factor value was found to be innacurate (Vlemmings et sl. 2011). Note that the fact that the g-factor and the hyperfine structure are unknown does not influence the analysis of the linearly polarized emission, from which is possible to determine the emerging brightness temperature ($T_b\Delta\Omega$) and the intrinsic thermal linewidth (ΔV_i) of the maser line, and the θ angle (the angle between B and the maser propagation direction). On the contrary, the measurement of the Zeeman-splitting and the estimates of the magnetic field strength strongly depend on both the hyperfine structure and the g-factors. For this reason no magnetic field strength measured from the circularly polarized emission of CH_3OH maser has been provided (e.g., Surcis *et al.* 2015). Thanks to the work of Lankhaar *et al.* (2017) we were able to modify the FRTM code in order to model properly not only the polarized emission of the

6.7 GHz CH_3OH maser, but the polarized emission of all the torsion-rotation transitions. We assumed that eight hyperfine transitions contribute equally to each CH_3OH maser emission (see Lankhaar *et al.* in this book), with the g-factors and Einstein coefficients as tabled in Lankhaar *et al.* (2017). We also assumed that $B = 10$ mG and that the temperature of the incoming radiation is 25 K. Furthermore, we have implemented a new subroutine for calculating the Clebsch-Gordan coefficients.

Theoretical results

We have run the code, so far, for three of the CH_3OH maser emissions: 6.7 GHz, 36 GHz, and 44 GHz. Part of the results are plotted in Fig. 1. We run the code considering a linewidth of the maser (Δv_L) of 0.2 $km\,s^{-1}$ and an intrinsic thermal linewidth (ΔV_i) of 1.0 $km\,s^{-1}$.

For the 6.7-GHz maser transition (Class II) the rebroadening of the maser line, i.e. when the maser is entering the saturation state, happens when $T_b\Delta\Omega = 10^9 - 10^{10}$ K sr and consequently the expected linear polarization fraction for unsaturated maser is $P_L \lesssim 5\%$. If we assume that there exists one hyperfine transition dominating the maser emission, for instance that one with the largest Einstein coefficient for stimulated emission (A; Lankhaar *et al.* this book), we obtain similar results for P_L but the expected circular polarization fraction (P_V) increases considerably. The upper limit changes from $P_V < 0.2\%$ to $P_V < 0.7\%$, the letter matching the observations.

For the 36-GHz and 44-GHz maser transitions (Class I) the rebroadening is observed at lower brightness temperature than for the 6.7-GHz transition. The model predicts that this happens for $T_b\Delta\Omega = 5 \cdot 10^7 - 5 \cdot 10^8$ K sr and $T_b\Delta\Omega = 10^7 - 10^8$ K sr for the 36-GHz and the 44-GHz, respectively. Consequently for unsaturated masers we have $P_L^{36\,GHz} \lesssim 7\%$ and $P_L^{44\,GHz} \lesssim 4\%$. The upper limit of P_V increases, for both 36-GHz and 44-GHz maser emission, from a fraction of *per thousands* to fraction of *percent* if the hyperfine transition with the largest A coefficient is considered.

The Flux-Limited sample

Since 2008 we have observed 30 massive star-forming regions to detect the polarized emission of 6.7-GHz CH_3OH masers (the so-called Flux-Limited sample), 25 of which have been already analyzed and partially published (e.g., Surcis *et al.* 2015 and references therein). Most of the maser features were modeled by using the old FRTM code and so at the time of the pubblications no magnetic field strength was estimated due to the uncertainty of the g-factor. Thus we modeled again the maser features by using the new version of the code described above and for which the hyperfine transition F = 3$\longrightarrow$ 4 is assumed to dominate the maser emission (see Lankhaar *et al.* this book). We found that the obtained values of $T_b\Delta\Omega$, ΔV_i, and θ are the same within 1% to the previously measured ones and that the magnetic field strength ranges between 1 mG and 15 mG. Only in the case of NGC 7538 the magnetic field is particularly strong, $B \lesssim 50$ mG, needing a closer investigation with further observations.

3. SiO maser - the case of VY CMa

Since the last maser symposium in Stellenbosch (South Africa) several results have been achieved in observing the polarized emission of SiO masers (e.g., Assaf *et al.* 2013, Richter *et al.* 2016). One of the most recent results is the detection of the polarized emission of the SiO masers for the first time with the Atacama Large Millimeter/submillimeter Array (ALMA) around the red supergiant VY CMa (Vlemmings *et al.* 2017). Using Band 5 of ALMA they detected varying levels of P_L for ^{28}SiO (J=4-3, ν=0,1,2)

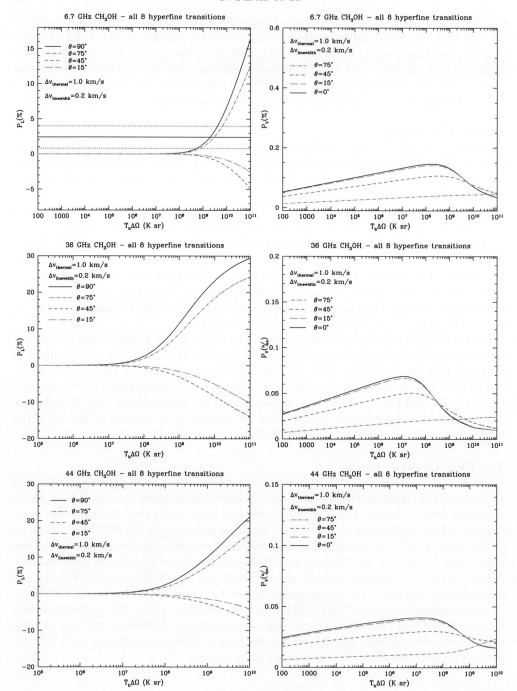

Figure 1. Outputs of the FRTM code for the 6.7 GHz, 36 GHz, and 44 GHz CH$_3$OH maser emissions. The plots are obtained by assuming that all the eight hyperfine transitions, which are listed in Lankhaar *et al.* (2017), contribute equally to the maser emission. Left panels: the fractional linear polarization vs. the emerging brightness temperature. The horizontal black dashed lines indicate the range of typical P_l measured towards 6.7 GHz CH$_3$OH masers. Right panels: the fractional circular polarization vs. the emerging brightness temperature.

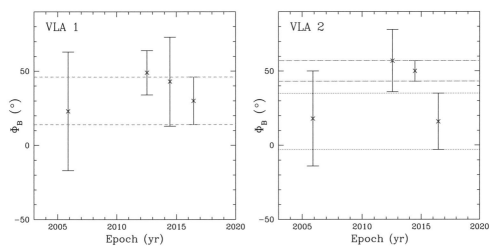

Figure 2. Multi-epoch comparison of magnetic field angles (Φ_B) for the massive YSOs VLA 1 (left panel) and VLA 2 (right panel) in W75N(B). The red dashed lines indicate the common range of angles among the four epochs (2005.89, 2012.54, 2014.46, and 2016.45). The black dotted lines indicate the common range of angles between epochs 2005.89 and 2016.46, and the blue dash-dotted ones between epochs 2012.54 and 2014.46.

and ^{29}SiO (J=4-3, ν=0,1) maser lines (rest frequency $\sim$ 170 GHz). In particular they observed a clear structure in the PA of the linear polarization vectors of ^{28}SiO (J=4-3, ν=1) similar to that observed by Herpin *et al.* (2006) at 86 GHz. The vectors rotate from $\sim$130° at blue-shifted velocities to $\sim$ −50° around the stellar velocity and back to $\sim$140° on the red-shifted side, suggesting the presence of a possible complex toroidal magnetic field morphology (Vlemmings *et al.* 2017).

4. H_2O maser - the case of W75N(B)

From 1999 to 2012 we have monitored the expansion of a 22-GHz H_2O maser shell around an unresolved continuum source excited by the massive YSO W75N-VLA2 with VLBI observations. We found that this shell is expanding at about 5 mas/yr and, more importantly, that it has evolved from an almost circular wind-driven shell to an elliptical morphology (see also Kim *et al.* this book). This suggests that we are observing in "real time" the transition from a non-collimated outflow event into a collimated outflow/jet structure during the first stages of evolution of a massive YSO. Moreover we have measured the magnetic field in two epochs separated by 7 years (in 2005 and in 2012). In this time interval the magnetic field changed its orientation following the rotation of the major-axis of the elliptical structure and decreases its strength. At 1200 au NW a more evolved YSO, named VLA 1, shows immutable H_2O maser distribution and magnetic field morphology. The presence of this nearby source reinforces the results of VLA 2 (Surcis *et al.* 2014).

In 2014 we started an European VLBI Network (EVN) monitoring project (four epochs separated by two years from one another) with the aim to follow both the expansion of the outflow/jet structure in VLA 2 and, more importantly, the variation of the magnetic field in the region. The first two epochs were observed in 2014 and 2016.

While the magnetic field around VLA 1 is still the same over time, with only an increment of the magnetic field strength of the order of 2.5 times, the magnetic field around VLA 2 changed again (Fig.2). In 2014 it was perfectly in agreement with the orientation measured in 2012, while in 2016 the magnetic field rotated back to its 2005 orientation (Fig.2). With respect to the conclusions of Surcis *et al.* (2014), who stated that the magnetic field around VLA 2 changed its orientation according to the new direction of the major-axis of the shell-like structure, this is unexpected. The next two EVN epochs (2018 and 2020) will hopefully help to clarify the phenomenon. In 2016 we also measured a magnetic field around VLA 2 15 times stronger than what was measured in 2012 by Surcis *et al.* (2014). This could be due to a further compression of the gas at the shock front caused by the encounter with a much denser medium than in the past, suggested also by the flaring of the H_2O maser features and to the stop of the maser expansion. More details of the recently achieved results could be read in Surcis *et al.* (*in prep.*).

5. The Future

The detection of polarized emission from several maser species is common nowadays. Besides the snapshots presented here there have been more interesting results that are highlighted in this book, both as talk contributions and as poster contributions. Although some scientific wishes that were expressed in 2012 in Stellenbosch (South Africa, IAUS 287) have been realized, like the determination of g-factors for CH_3OH maser emission or maser polarization observations with ALMA, many issues are still open or just popped up. Among these we would like to focus the attention on some of them:

• model the pumping mechanisms for Class I and Class II CH_3OH masers in order to determine the contributions of each single hyperfine transitions to the observed maser lines;

• confirm observationally the theoretical predictions made by the FRTM code;

• modify the FRTM code in order to include the anisotropic pumping mechanism of the SiO masers.

A bright future for maser polarization is beginning and the improvement of existing facilities and the construction of new facilities will help to make it even brighter.

References

Assaf, K. A., Diamond, P. J., Richards, A. M. S., & Gray, M. D. 2013, *MNRAS*, 431, 1077
Herpin, F., Baudry, A., Thum, C., Morris, D., & Wiesemeyer, H. 2006, *A&A*, 450, 667
Lankhaar, B., Groenenboom, G., & van der Avoird, A. 2016, *J. Chem. Phys.*, 145, 24
Lankhaar, B., Vlemmings, W. H. T., Surcis, G., van Langevelde, H. J., Groenenboom, G., & van der Avoird, A. 2017, *Nature Astronomy*, 2, 145L
Nedoluha, G. E. & Watson, W. D. 1992, *ApJ*, 384, 185
Peréz-Sánchez, A. F. & Vlemmings, W. H. T. 2013, *A&A*, 551A, 15
Richter, L., Kemball, A., & Jonas, J. 2016, *MNRAS*, 461, 2309
Surcis, G., Vlemmings, W. H. T., van Langevelde, H. J., Goddi, C., Torrelles, J. M., Cantó, J., Curiel, S., Kim, S.-W., & Kim, J.-S. 2014, *A&A*, 565, L8
Surcis, G., Vlemmings, W. H. T., van Langevelde, H. J., Hutawarakorn Kramer, B., Bartkiewicz, A., & Blasi, M. G. 2015, *A&A*, 578, 102
Vlemmings, W. H. T., Diamond, P. J., van Langevelde, H. J., & Torrelles, J. M. 2006, *A&A*, 448, 597
Vlemmings, W. H. T., Surcis, G., Torstensson, K. J. E., & van Langevelde, H. J. 2010, *MNRAS*, 404, 134
Vlemmings, W. H. T., Torres, R. M., & Dodson, R. 2011, *A&A*, 529, 95
Vlemmings, W. H. T., Khouri, T., Martí-Vidal, I., Tafoya, D., Baudry, A., Etoka, S., Humphreys, E. M. L., Jones, T. J., Kemball, A., O'Gorman, E., Peréz-Sánchez, A. F., & Richards, A. 2017, *A&A*, 603, 92

Astrophysical Masers:
Unlocking the Mysteries of the Universe
Proceedings IAU Symposium No. 336, 2017
A. Tarchi, M.J. Reid & P. Castangia, eds.

© International Astronomical Union 2018
doi:10.1017/S1743921317010961

Class I methanol masers in low-mass star formation regions

S. Kalenskii[1], S. Kurtz[2], P. Hofner[3], P. Bergman[4], C.M. Walmsley[5] and P. Golysheva[6]

[1]Lebedev Physical Institute, Astro Space Center,
84/32 Profsoyuznaya st., Moscow, GSP-7, 117997, Russia
email: kalensky@asc.rssi.ru

[2]Instituto de Radioastronomia y Astrofizika, Universidad Nacional Autonoma de Mexico,
Morelia, Michoacan, Mexico
email: s.kurtz@irya.unam.mx

[3]National Radio Astronomy Observatory, 1003 Lopezville Road, Socorro, NM 87801, USA
email: hofner_p@yahoo.com

[4]Onsala Space Observatory, Chalmers Univ. of Technology, 439 92 Onsala, Sweden
email: per.bergman@chalmers.se

[5]Dublin Institute for Advanced Studies, 31 Fitzwilliam Place, Dublin 2, Ireland (deceased)

[6]119992, Universitetski pr., 13, Sternberg Astronomical Institute, Moscow University, Moscow,
Russia
email: polina-golysheva@yandex.ru

Abstract. We present a review of the properties of Class I methanol masers detected in low-mass star forming regions (LMSFRs). These masers, henceforth called LMMIs, are associated with postshock gas in the lobes of chemically active outflows in LMSFRs NGC1333, NGC2023, HH25, and L1157. LMMIs share the main properties with powerful masers in regions of massive star formation and are a low-luminosity edge of the total Class I maser population. However, the exploration of just these objects may push forward the exploration of Class I masers, since many LMSFRs are located only 200–300 pc from the Sun, making it possible to study associated objects in detail. EVLA observations with a 0.2″ spatial resolution show that the maser images consist of unresolved or barely resolved spots with brightness temperatures up to 5×10^5 K. The results are "marginally" consistent with the turbulent model of maser emission.

Keywords. ISM: jets and outflows, masers, radio lines: ISM.

1. Introduction

Bright and narrow maser lines of methanol (CH_3OH) have been found towards many star-forming regions. Methanol masers can be divided into two classes, I and II (Menten, 1991b), with each class characterized by a certain set of transitions. Class I maser transitions are the $7_0 - 6_1 A^+$ transition at 44 GHz, $4_{-1} - 3_0 E$ transition at 36 GHz, $5_{-1} - 4_0 E$ transition at 84 GHz, $6_{-1} - 5_0 E$ at 132 GHz, $8_0 - 7_1 A^+$ transition at 95 GHz etc., while Class II transitions are the $5_1 - 6_0 A^+$ transition at 6.7 GHz, $2_0 - 3_{-1} E$ transition at 12 GHz, the series of $J_0 - J_{-1} E$ transitions at 157 GHz, etc. The strongest Class I masers (usually called MMIs) emit at 44 GHz and demonstrate flux densities up to 800 Jy (Haschick *et al.* 1990), while the strongest Class II masers (MMII) emit at 6.7 GHz and some of these achieve flux densities of ~ 4000 Jy (Menten, 1991a). Here we consider only MMIs; for a more thorough description of their main properties see the contribution by Leurini & Menten (this volume).

Until recently it was considered that methanol masers arise only in massive star formation regions (MSFRs). But in the past few years several MMIs have been found in

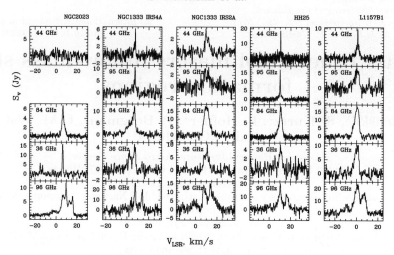

Figure 1. LMMI spectra at 44, 95, 84, and 36 GHz, taken at the Onsala Space Observatory. Purely thermal $2_K - 1_K$ methanol lines at 96 GHz are shown in the bottom row.

nearby low-mass star formation regions (LMSFRs) NGC1333, NGC2023, HH25, and L1157 (Kalenskii *et al.* 2006, Kalenskii *et al.* 2010a, Kang *et al.* 2013, Lyo *et al.* 2014). The masers were detected in the Class I lines at 36 GHz, 44 GHz, 95 GHz, and 132 GHz.

2. LMMI properties

MMIs in LMSFRs (hereafter called LMMIs) have been studied using the 20-m Onsala radio telescope and the KVN 21-m telescopes in a single-dish mode (Kalenskii *et al.* 2010a, Kang *et al.* 2013, Lyo *et al.* 2014). In addition, four objects have been observed at 44 GHz with the VLA in the D configuration (Kalenskii *et al.* 2010b, Kalenskii *et al.* 2013) with a spatial resolution of $\sim 1.5''$. The spectra of the lines observed at Onsala are shown in Fig. 1. The main LMMI properties are discussed in Kalenskii *et al.* (2013). They can be summarised as follows:

• All known LMMIs are associated with chemically active bipolar outflows, where the gas-phase abundance of methanol is enhanced due to grain mantle evaporation. VLA observations show that LMMIs are related to the shocked gas in the outflow lobes.

• The known LMMIs are associated with clouds where the column densities of methanol are no less than 10^{14} cm^{-2}. Kalenskii *et al.* (2010a) suggested that MMIs can arise only when methanol column density is above this value.

• Flux densities of LMMIs do not exceed 18 Jy at 44 GHz and are lower in the other Class I lines (see Fig 1). However, LMMI luminosities at 44 GHz match the relation between the protostar and maser luminosities $L_{\mathrm{CH_3OH}} = 1.71 \times 10^{-10} (L_{\mathrm{bol}})^{1.22}$, established for high- and intermediate-mass protostars by Bae *et al.* (2011).

• No variability at 44 GHz was detected in NGC1333I4A, HH25, or L1157 during the time period 2004–2011.

• Radial velocities of most LMMIs are close to the systemic velocities of associated regions. The only known exception is the maser detected at 36 GHz toward the blue lobe of the extra-high-velocity outflow in NGC 2023, whose radial velocity is 3.5 km s^{-1} lower than the systemic velocity.

Thus, one can see that the main properties of LMMIs are similar to those of HMMIs. LMMIs are likely to be a low luminosity edge of the overall MMI population. Therefore the question arises, why should we study these few weak objects instead of focusing on

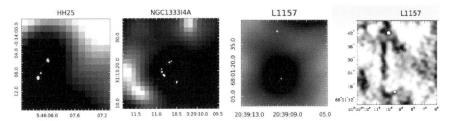

Figure 2. Three left panels: EVLA maps of methanol masers at 44 GHz (white dots) overlaid upon W1 (HH25 and NGC1333I4A) and W4 (L1157) WISE images. Right panel: methanol masers overlaid upon the map of the $5_0 - 4_0 A^+$ thermal emission in L1157.

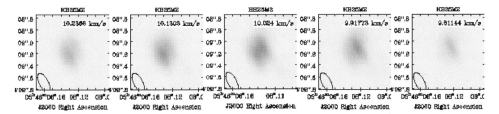

Figure 3. Example of a double spot: channel map of HH25M2 at 44 GHz.

much stronger MMIs in MSFRs? The answer is that the study of Class I methanol masers in LMSFRs might be more straightforward compared to the study of the "classical" MMIs in MSFRs, because, in contrast to MSFRs, LMSFRs are widespread and many of them are located only 200–300 pc from the Sun; they are less heavily obscured in optical and IR wave ranges, and there are many isolated low-mass Young Stellar Objects (YSOs). We continue to study MMIs in LMSFRs in order to better understand Class I methanol masers. Here we present the results of the observations of three maser sources performed at 44 GHz with the EVLA in the B configuration as well as CARMA observations of L1157 in the thermal lines of methanol $5_K - 4_K$ at 241 GHz.

3. New results

Spatial resolution of about $1.5''$, achievable at 44 GHz with the VLA in the D configuration, proved to be insufficient to resolve individual maser spots and measure their sizes and brightness temperatures. Therefore in 2013 we reobserved three LMMIs, HH25, NGC1333I4A, and L1157 at 44 GHz using EVLA in the B configuration, which provides an angular resolution of about $0.2''$ at 44 GHz. In addition, we observed L1157 in the $5_K - 4_K$ thermal lines of methanol at 241 GHz with the antenna array CARMA in the C configuration with an angular resolution of $\approx 1''$.

A collection of the overall maps of the three observed sources at 44 GHz is shown in Fig. 2. The maps show that each source consists of several spots. Hereafter M1 means the strongest spot in the region, M2 is the second strongest spot etc. Deconvolved spot sizes vary from $\sim 0.10''$–$0.15''$ for the stronger spots to $\sim 0.10''$–$0.3''$ for the weaker spots (30–45 AU and 30-90 AU, respectively).

The brightness temperatures of the strongest spots are as high as 5×10^5 K.

Maser spots in NGC1333I4A form an arc around a NIR object clearly seen in the W1 and W2 WISE maps.

The maps of individual spots show that many of them can be decomposed into two unresolved compact subspots. In these cases the brightest subspot is denoted subspot a, the second brightest subspot, subspot b. An example of such double spot is shown in Fig. 3, which exhibits the channel map of the second brightest spot in HH25 (HH25M2). Among the spots that demonstrate double structures are L1157M1 and M2, NGC1333I4AM2 etc.

An interesting result is the detection of unresolved spots demonstrating broad ($\gtrsim 3 - 5$ km s^{-1}) spectral lines. Their fluxes are about 0.1–0.2 Jy, which corresponds to brightness temperatures $\gtrsim 1000$ K. Thus, in spite of large linewidths, these objects are probably weak masers.

4. Discussion

According to the most popular maser model, compact maser spots arise in extended turbulent clumps because in a turbulent velocity field the coherence lengths l along some lines are increased. If masers are associated with turbulence, the map appearance depends on the maser regime. Saturated regime of maser amplification is characterized by a large number of spots of comparable intensity, while the unsaturated maser amplification results in a small number of bright spots (Strelnitski *et al.* 2017). The map of maser emission in L1157 with only two maser spots (Fig. 2) favors the unsaturated regime of maser amplification, which was studied by Sobolev *et al.* (1998).

One of the main parameters of the model by Sobolev *et al.* is τ_0, the absolute value of optical depth at the center of the inverted line when there is no turbulence in the cloud. From the intensities of thermal lines $5_K - 4_K A^+$ toward M1, observed with CARMA, we estimated $N_{\mathrm{CH_3OH}}$ ($\sim 10^{16}$ cm^{-2}) and τ_0 at 44 GHz (~ 12). From Table 1 of Sobolev *et al.* (1998) we estimated that the optical depth at 44 GHz toward M1 is $\tau^{44} \sim 7 - 8$ and T_{br} at this frequency $\sim 10^4$ K, much lower than the observed one. However, an increase of $N_{\mathrm{CH_3OH}}$ by a factor of less than 2 makes it possible to achieve the observed brightness temperature. Thus, the turbulence model is in "marginal" agreement with the observations.

SVK acknowledges the support of the Russian Foundation for Basic Research (project no. 15-02-07676).

References

Bae J.-H., Kim, K.-T., Youn S.-Y., Kim W.-J., Byun D.-Y., & Kang H., Oh C. S. 2011, *ApJS* 196, 21

Haschick A. D., Menten K. M., & Baan W. 1990, *ApJ* 354, 556

Kalenskii S. V., Promyslov V. G., Slysh V. I., Bergman P., & Winnberg A. 2006, *Astron. Rep.* 50, 289

Kalenskii S. V., Johansson L. E. B., Bergman P., Kurtz S., Hofner P., Walmsley C. M., & Slysh V. I. 2010a, *MNRAS* 405, 613

Kalenskii S. V., Kurtz S., Slysh V. I., Hofner P., Walmsley C. M., Johansson L. E. B., & Bergman P. 2010b, *Astron. Rep.* 54, 932

Kalenskii S. V., Kurtz S., Bergman P. 2013, *Astron. Rep.* 57, 120

Kang M., Lee J.-E., Choi M., Choi Y., Kim K.-T., Di Francesco J., Park Y.-S. 2013, *ApJS* 209, 25

Lyo, A.-R., Kim, J., Byun, D.-Y., & Lee, H.-G. 2014, *AJ* 148, 80

Menten K. M. 1991a, *ApJ* 380, L75

Menten, K. M. 1991b, *ASP Conference Series*, 16, 119

Sobolev, A. M., Vallin, B. K., & Watson, W. D. 1998, *ApJ* 498, 763

Strelnitski, V. S., Holder, B. P., Shishov, V. I., & Nezhdanova, N. I. 2017, *Astron. Astrophys Transactions*, in press

Astrophysical Masers:
Unlocking the Mysteries of the Universe
Proceedings IAU Symposium No. 336, 2017
A. Tarchi, M.J. Reid & P. Castangia, eds.

© International Astronomical Union 2018
doi:10.1017/S1743921317010511

Infrared variability, maser activity, and accretion of massive young stellar objects

Bringfried Stecklum[1], Alessio Caratti o Garatti[2], Klaus Hodapp[3], Hendrik Linz[4], Luca Moscadelli[5] and Alberto Sanna[6]

[1]Thüringer Landessternwarte Tautenburg,
Sternwarte 5, D-07778 Tautenburg, Germany
email: stecklum@tls-tautenburg.de

[2]Dublin Institute for Advanced Studies, Dublin, Ireland

[3]Institute for Astronomy, Hilo, USA

[4]Max-Planck Institut fr Astronomie, Heidelberg, Germany

[5]INAF, Firenze, Italy

[6]Max-Planck Institut fr Radioastronomie, Bonn, Germany

Abstract. Methanol and water masers indicate young stellar objects. They often exhibit flares, and a fraction shows periodic activity. Several mechanisms might explain this behavior but the lack of concurrent infrared (IR) data complicates the identification of its cause. Recently, 6.7 GHz methanol maser flares were observed, triggered by accretion bursts of high-mass YSOs which confirmed the IR-pumping of these masers. This suggests that regular IR changes might lead to maser periodicity. Hence, we scrutinized space-based IR imaging of YSOs associated with periodic methanol masers. We succeeded to extract the IR light curve from NEOWISE data for the intermediate mass YSO G107.298+5.639. Thus, for the first time a relationship between the maser and IR variability could be established. While the IR light curve shows the same period of ∼34.6 days as the masers, its shape is distinct from that of the maser flares. Possible reasons for the IR periodicity are discussed.

Keywords. masers, ISM: molecules, dust, stars: formation, individual ([TGJ91] S255 NIRS 3, [PFG2013] I22198-MM2-S)

---

1. Introduction

Class II methanol masers are signs of luminous young stellar objects (YSOs) (Breen *et al.* 2013). The strong infrared (IR) radiation from massive YSOs (MYSOs) heats up dust in their immediate environment which causes molecules, originally frozen onto the grains, to sublimate. Both, the high-column density of molecules in the gas phase as well as the strong mid-IR radiation due to the thermal dust emission are thought to be essential for the excitation of these masers (Sobolev *et al.* 1997). They often show flare activity, and a few dozens of them vary periodically within ∼30 to 800 days (Goedhart *et al.* 2014, Szymczak *et al.* 2015). In the absence of complementary data, in particular time-resolved IR photometry, this variability remained enigmatic. Thus, it is not surprising that various models were brought up to explain periodic masers, which rely on vastly differing mechanisms, and have no constraints yet other than to reproduce the maser light curves. These comprise masers in the atmosphere of evaporating icy planets orbiting OB stars (Slysh *et al.* 1998), modulation of the radio continuum seed radiation due to variable colliding binary winds (van der Walt 2011) or eclipsing massive binaries (Maswanganye *et al.* 2015), variation of the IR pumping radiation due to periodic accretion from a circumbinary disk (Araya *et al.* 2010), protostellar pulsations at high accretion rates (Inayoshi *et al.* 2013), or heating by accretion flow shocks in binary systems (Parvenov & Sobolev 2014). Recent conclusions from the statistics of both maser strength and variability point to a correlation between maser and MYSO luminosity while the latter appears to be anti-correlated with the maser variability. This suggests that IR flux variations may drive maser flares (Szymczak *et al.* 2017).

2. Accretion bursts and IR pumping of Class II methanol masers

Episodic accretion bursts are well known among low- and intermediate mass YSOs. They are caused when matter, piled-up at the inner circumstellar disk for various reasons, is being dumped onto the young star within rather a short time. For the first events found, the resulting luminosity increase was discovered in the optical as brightness rise of up to 5 mags. They were classified into two categories, FUors and EXors, according to the properties of their prototype objects (Herbig 1977, Herbig 1989). In the meanwhile, it turned out that these just represent the tip of the iceberg, seen in the latest stage of protostellar evolution. Recent IR surveys revealed that embedded YSOs are inherently variable, with the deeply embedded ones varying the most (Contreras Peña *et al.* 2017a). Spectroscopy of these variables confirmed their eruptive nature, showing a mixture of FUor/EXor-like features (Contreras Peña *et al.* 2017b). If disk-mediated accretion is a pathway to form OB stars, episodic accretion should occur for their precursors, too.

Thus, such events are likely accompanied by flares of Class II methanol masers as a consequence of the IR pumping theory. For this reason, we initiated NIR imaging of the massive star forming region S255IR two weeks after the flare of the associated 6.7 GHz methanol maser was reported (Fujisawa *et al.* 2015). It revealed the outburst of NIRS3, a $\sim$20 M$_\odot$ MYSO (SIMBAD designation ([TGJ91] S255 NIRS 3, Stecklum *et al.* 2016). Our extensive observing campaign yielded major properties of the MYSO burst (Caratti o Garatti *et al.* 2017). These results confirmed that OB stars may form via disk accretion and that their disks are prone to instabilities leading to accretion bursts, which trigger flares of Class II methanol masers. At about the same time as the burst of NIRS3 went off, a similar event was detected in the submm/mm which showed maser flares as well (Hunter *et al.* 2017, this volume). This provided independent support for the conclusions drawn above.

For what concerns the temporal behavior

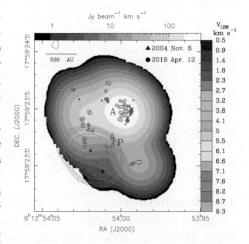

Figure 1. Map of the 6.7 GHz methanol masers toward NIRS3. Triangles and circles represent maser spots before and after the burst, respectively. The velocity-integrated emission of the 6.7 GHz masers (gray scale) and the JVLA 5 GHz radio continuum emission(red contours) are also shown. See online paper for color figure.

of the methanol masers during the burst of NIRS3, it was found that the 5.9 km/s component started to fade $\sim$15 d after reaching its peak level while a new feature at 6.5 km/s emerged which reached its peak flux $\sim$20 d later than the 5.9 km/s component (K. Fujisawa, priv. comm.). Our EVN and JVLA observations after the burst revealed that the main pre-burst maser cluster vanished and a new, extended region of 6.7 GHz maser emission surfaced further out (500 to 1000 au) (Moscadelli *et al.* 2017, Fig.1). From the delay between the rise of the old and

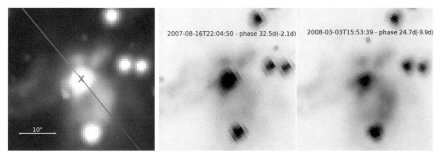

2007-08-16T22:04:50 - phase 32.5d(-2.1d)   2008-03-03T15:53:39 - phase 24.7d(-9.9d)

10"

Figure 2. Left - IRAC RGB image (channels 3, 2, 1) of G107 (cross). The line indicates the rotation axis of the circumstellar disk and the sense of the outflow (red-/blue-shifted). See online paper for color figure. **Center/right** - 4.5 μm images for the two epochs at phases of 32.5 and 24.7 d, respectively. The out of phase variability of the YSO and the nebulosity is obvious.

new maser components and their projected separation a lower limit to the propagation speed of the excitation could be derived for the first time which amounts to $\sim$0.15c. There are several possible reasons for the subluminal velocity. Since the energy transport by photons rests on absorption and re-emission as well as scattering, it will become slower at increasing optical depth. The heating of dust beyond the snow line will be delayed by the endothermic sublimation of volatiles frozen onto grains. Dust growth in the dense YSO environment will have a similar effect due to the increase in grain heat capacity.

Notably, the radio continuum stayed constant during the burst and eventually started to rise $\sim$300 d after the flare detection (Cesaroni *et al.*, A&A subm.).

3. The first IR light-curve for a periodic maser source

The lack of complementary data, in particular time-resolved IR photometry, hinders to disclose the driving mechanisms for periodic masers. The IR pumping of Class II methanol masers suggests that cyclic luminosity variations of YSOs will cause maser periodicity. In order to confirm this claim, we tried to establish IR light curves of periodic maser sources from IRAC and NEOWISE photometry. No successful ground-based attempts in this respect were reported so far. Unfortunately, useful data is very scarce since luminous YSOs are generally saturated in space-based IR imaging. This is not the case for G107.298+5.639 (G107 for short, SIMBAD designation [PFG2013] I22198-MM2-S), a less luminous, intermediate-mass YSO (Sánchez-Monge *et al.* 2010), situated at 750 ± 27 pc (Hirota *et al.* 2008), which nevertheless excites periodic methanol and water masers (Fujisawa *et al.* 2014, Szymczak *et al.* 2016).

Spitzer/IRAC observations at two epochs (PI G. Fazio, Fig. 2) clearly reveal variability of both the YSO and the nebulosity associated with the blue-shifted outflow lobe. The out of phase brightness change of the latter indicates a light echo.

From NEOWISE data, we established the light curve of G107 (Fig. 3) which has the same period as that of the masers (34.6 d, Fujisawa *et al.* 2014). However, its saw tooth shape is distinct from that of the maser flares, and resembles light curves of Cepheids. While the 3.6 μm and 4.5 μm IRAC fluxes of the two epochs preceding WISE are weaker, their ratios fit the normalized WISE light curve. This points to a quite stable period over $\sim$10 years as well as long-term brightening. Notably, G107 does not fit the period-luminosity relation for pulsating massive YSOs (Inayoshi *et al.* 2013). Moreover, the skewness of the light curve contradicts the eccentric binary accretion model (Artymowicz & Lubow 1996). So the question on the driving mechanism for the periodicity is still open.

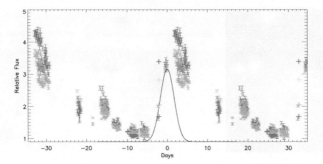

Figure 3. NEOWISE multi-color (W1,W2, W3, W4) light curve of G107, folded by the 34.6 d period, and normalized to minimum brightness. Zero represents the date of the maser flare peak. Large crosses mark IRAC flux ratios, normalized to the low state. The Gaussian fit to the methanol maser light curve (Fujisawa *et al.* 2014) is shown for comparison. The apparent jumps at phase of ∼4 d are due to long-term brightening. See online paper for color figure.

4. Conclusions

The change of the maser population of NIRS3 due to its burst will not be permanent. Since the burst ceased, they will redistribute according to the cooling of the circumstellar environment. This provides a unique opportunity to study the dynamics of maser excitation. The synergy between radio and IR observations is essential to disclose the reasons of maser variability. While the ongoing NEOWISE mission as well as the VVV(X) survey are extremely helpful in this respect, their cadence is not sufficient. Dedicated IR imaging capability is required to follow-up maser flares in a target-of-opportunity fashion.

References

Araya, E. D., Hofner, P., Goss, W. M., Kurtz, S., Richards, A. M. S *et al.* 2010, *ApJ*, 717, L133
Artymowicz, P. & Lubow, S. 1996, *ApJ*, 467, L77
Breen, S. L., Ellingsen, S. P., Contreras, Y., Green, J. A. *et al.* 2013, *MNRAS*, 435, 524
Caratti o Garatti, A., Stecklum, B., Garcia Lopez, R., Eislffel, J. *et al.* 2017, *NatPhys*, 13, 276
Contreras Peña, C., Lucas, P. W., Minniti, D., Kurtev, R. *et al.* 2017a, *MNRAS*, 465, 3011
Contreras Peña, C., Lucas, P. W., Kurtev, R., Minniti, D. *et al.* 2017b, *MNRAS*, 465, 3039
Fujisawa, K., Takase, G., Kimura, S., Aoki, N., Nagadomi, Y. *et al.* 2014, *PASJ*, 66, 78
Fujisawa,K., Yonekura, Y., Sugiyama, K., Horiuchi, H., Hayashi, T. *et al.* 2015, *ATel*, #8286
Goedhart, S., Maswanganye, J. P., Gaylard, M. J. *et al.* 2014, *MNRAS*, 437, 1808
Herbig, G. 1977, *ApJ*, 217, 693
Herbig, G. 1989, *ESOC*, 33, 233
Hirota, T., Bushimata, T., Choi, Y. K., Honma, M., Imai, H. *et al.* 2008, *PASJ*, 60, 961
Hunter, T. R., Brogan, C. L., MacLeod, G. Cyganowski, C. J. *et al.* 2017, *ApJ*, 873, L29
Inayoshi, K., Sugiyama, K. Hosokawa, T., Motogi, K., & Tanaka, K. 2013, *ApJ*, 769, L20
Inno, I., Matsunaga, N., Romaniello, M., Bono, G., Monson, A. *et al.* 2015, *A&A*, 576, A30
Maswanganye, J. P. Gaylard, M. J., Goedhart, S. *et al.* 2015, *MNRAS*, 446, 2730
Moscadelli, L., Sanna, A., Goddi, C., Walmsley, M. C., Cesaroni, R. *et al.* 2017, *A&A*, 600, L8
Parfenov, S. Y.u. & Sobolev, A. M. 2014, *MNRAS*, 444, 620
Sánchez-Monge, A. S., Palau, A., Estalella, R., Kurtz, S., Zhang, Q. *et al.* 2010, *ApJ*, 721, L107
Sobolev, A. M., Cragg, D. M., & Godfrey, P. D. 1997, *A&A*, 324, 211
Stecklum, B., & Caratti o Garatti, A. Cardenas, M. C. Greiner, J. *et al.* 2016, *ATel*, #8732
Slysh, V. I., Val'tts, I. E., Kalenskii, S. V., Larionov, G. M. *et al.* 1998, *ASP Conf.Ser.*, 144, 379
Szymczak, M., Wolak, P., & Bartkiewicz, A. 2015, *MNRAS*, 448, 2284
Szymczak, M., Olech, M., Wolak, P., Bartkiewicz, A., & Gawroski, M. 2016, *MNRAS*, 459, L56
Szymczak, M., & Olech, M. Sarniak, R. Wolak, P. Bartkiewicz, A *et al.* 2017, *arXiv*, 1710.04595
van der Walt, D. J. 2011, *AJ*, 141, 152

Astrophysical Masers:
Unlocking the Mysteries of the Universe
Proceedings IAU Symposium No. 336, 2017
A. Tarchi, M.J. Reid & P. Castangia, eds.

© International Astronomical Union 2018
doi:10.1017/S1743921317010523

On the origin of methanol maser variability: Clues from long-term monitoring

M. Szymczak, M. Olech,† R. Sarniak, P. Wolak and A. Bartkiewicz

Centre for Astronomy, Faculty of Physics, Astronomy and Informatics,
Nicolaus Copernicus University, Grudziadzka 5, 87-100 Torun, Poland
email: msz@astro.umk.pl

Abstract. High-mass young stellar objects (HMYSO) displaying methanol maser flux variability probably trace a variety of phenomena such as accretion events, magnetospheric activity, stellar flares and stellar wind interactions in binary systems. A long-term monitoring of the 6.7 GHz methanol line in a large sample of HMYSOs has been undertaken to characterize the variability patterns and examine their origins. The majority of the masers show significant variability on time-scales between a week and a few years. High amplitude short flares of individual features occurred in several HMYSOs. The maser features with low luminosity tend to be more variable than those with high luminosity. The variability of the maser features increases when the bolometric luminosity the powering star decreases. Statistical analysis of basic properties of exciting objects and the variability measures supports an idea that burst activity of methanol masers is driven mainly by changes in the infrared pumping rate.

Keywords. masers – stars: formation – ISM: clouds – radio lines: ISM

1. Introduction

Early information on the variability of 6.7 GHz methanol masers came from searching for new sources. Caswell *et al.* (1995) observed 245 objects on 4-5 occasions over 1.5 yr and found that 75% of features were not significantly variable while noticeable variability of amplitude <2 occurred in 48 sources. Their main suggestion was that the maser variations are related to changes in the maser path length or pump rate. Goedhart *et al.* (2004) monitored 54 sources over a period of 4.3 yr at $1-2$ week intervals or shorter. Diverse variability patterns on time-scales of a few days to several years were reported for a majority of the targets including periodic ($132-520$ day) variability. A flare activity on time-scales of a few months together with long-term variations on a time-scale of years was reported in G351.78−0.54 (MacLeod & Gaylard 1996). Changes in the maser pumping or disturbances with the maser regions were postulated as possible causes of that peculiar variability. Episodic and short (<6 day) flares of only one feature in the spectrum of G33.641−0.228 were detected (Fujisawa *et al.* 2014). These flares likely arise in a region of size much smaller than 70 au and can be induced by energy release in magnetic reconnection. Here, we present some of the results of long-term monitoring of the 6.7 GHz maser line for a large sample of HMYSOs. A full description of the sample and data presentation can be found in Szymczak *et al.* (2018).

2. Observations and results

A sample of 166 maser sources with peak 6.7 GHz flux density greater than 5 Jy was drawn from the Torun methanol source catalogue (Szymczak *et al.* 2012). Each target was

† He was the speaker. Indeed, the expected presenter of this talk, Prof. M. Szymczak, could not finally attend the Symposium.

M. Szymczak *et al.*

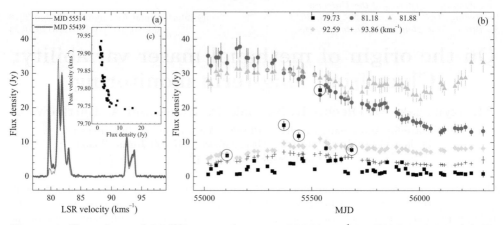

Figure 1. Short flares of 6.7 GHz maser feature at 79.73 kms^{-1} in G28.305−0.387. **(a)** The spectra taken before and during the flare are shown with the thin and thick lines, respectively. **(b)** The light curves of the maser features. The flaring events at 79.73 kms^{-1} are marked with open circles. Inset **(c)** shows the change in peak velocity of the flaring feature versus its flux density.

observed at least once a month between 2009 June and 2013 February using the Toruń 32 m radio telescope. Several circumpolar sources were observed at 2−3 day intervals. The data were dual polarisation taken with frequency switching mode. The spectral resolution was 0.09 kms^{-1} and typical rms noise level was 0.20−0.35 Jy. Absolute flux density calibration was accurate to within 10% (Szymczak *et al.* 2014).

A wide range of variability patterns from non-varying to very complex changes in the shape and intensity of the spectra was observed. The types of behaviour such as monotonic increases or decreases, aperiodic, quasi-periodic and periodic variations seen in the sources of our sample are similar to those reported in Goedhart *et al.* (2004).

2.1. *Short flares*

In five HMYSOs we detected rapid flares with relative amplitudes higher than two. In most cases these events appeared as outliers in the light curve of individual maser feature, while other features showed smooth and slight variations. Figure 1 shows the light curves of selected maser features of G28.305−0.387. Five bursts with a relative amplitude of 2−13 occurred at 79.73 kms^{-1}. For instance at MJD 55439 the flux density of this feature was 25.2 Jy while it was only 2.2 Jy 25 days earlier. No significant changes were seen in the other features. We cannot uncover a profile of those bursts due to sparse observations. The peak velocity of the flaring feature was measured by a Gaussian fitting. During the bursts the peak velocity decreased by 0.12−0.18 kms^{-1} relative to the velocity during a quiescent state. Blending of two maser features with slightly different velocities appears as the most natural explanation of this relation. One maser feature has high velocity and relative stable intensity while low velocity feature experiences bursting activity.

Probably the same type of flares was observed in G33.64−0.21 where typical rise and fall times were one and five days, respectively (Fujisawa *et al.* 2014). The flux density rose by a factor of 7 and fell exponentially. Similarly as in G28.305−0.387 the peak velocity of the flaring feature slightly changed during the bursts and returned to its pre-flare value. The VLBI data suggest that an active region where the flares arise is much less than 70 au. Fujisawa *et al.* (2014) proposed that the bursts are powered by energy released in impulsive events of magnetic field reconnection; the particles accelerated in solar-like flares heat the environment of reconnection region and can influence rapidly the

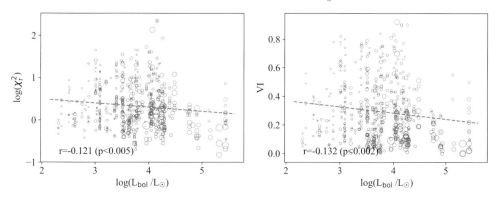

Figure 2. Reduced χ^2 (left) and variability index (right) of maser features vs. the bolometric luminosity of exciting star. The size of symbol is proportional to the square of 6.7 GHz maser luminosity of feature. The dashed lines represent the best-fitting results.

gas where the maser emission arises. Our survey indicates that short flares of the maser emission in restricted velocity range are not unique to G33.621−0.228. These bursts were observed for one feature of the spectrum or synchronously occurred for more features. We suggest that similar events may occur in more HMYSOs when monitored with high cadence.

2.2. *Variability measures vs maser and bolometric luminosities*

To quantify the variability we used the variability index as defined in Szymczak *et al.* (2014) and the reduced value of χ^2. We investigated the dependence of these variability measures on the luminosity of maser features and luminosity of exciting star. The distances were taken from trigonometric parallaxes (Reid *et al.* 2014) or else calculated from the Galactic rotation model (Reid *et al.* 2009). The kinematic distance ambiguity was resolved using the data published by Green & McClure-Griffiths (2011). For sources with known distances we have calculated the luminosity of each maser feature with well determined variability indices assuming the emission is isotropic. To determine the bolometric luminosity of powering star we used the SED fitting tool, SED models (Robitaille *et al.* 2006; 2007) and photometric data in the wavelength range from 3.4μm to 1.1 mm available from the public databases (see Sarniak *et al.* this volume).

Figure 2 displays the values of the reduced χ^2 and variability index (VI) versus the bolometric luminosity of exciting HMYSO for all the features not affected by the effects of confusion and offset observation. There is a weak anti-correlation between the star bolometric luminosity and both variability measures. It is clear that the variability of maser features increases in the sources powered by less luminous HMYSOs. We note that the significance of this correlation may be lowered by two factors: (i) the bolometric luminosity used is a measure of the total luminosity of the protocluster while the maser flux density likely depends on the intrinsic luminosity of a single HMYSO, (ii) the kinematic distance errors of the studied objects are much higher than those of a minority of maser sources with known trigonometric distances.

The size of symbols in Figure 2 is proportional to the square of the isotropic luminosity of maser features. There is a tendency that the maser features of higher luminosity are associated with the objects of higher stellar luminosity. It is possible that luminous exciting star provides higher pumping rate via infrared photons or longer path of maser amplification (Urquhart *et al.* 2013).

2.3. *Origin of variability*

In several periodic sources in the sample we observed a time delay (3−8 days) of the bursts between individual maser features. This indicates a radiative mechanism of pumping of the maser emission and a linear separation between group of maser clouds up to ∼1400 au which is comparable to the source size observed with VLBI. Thus changes in the pump rate appear as plausible cause of the maser variability. Recently observed giant flare in the 6.7 GHz maser emission toward S255 NIRS3 preceded by an increase in infrared luminosity (Caratti o Garatti *et al.* 2017; Moscadelli *et al.* 2017) possibly due to accretion event, provides a strong support for radiative pumping and indicates that changes in the infrared radiation drive the maser variability. High resolution observations suggest that the infrared radiation can enlarge the size of the excited region providing longer amplification path (Moscadelli *et al.* 2017).

Extraordinary outburst in G24.329+0.144 detected in our monitoring appears to be similar to that reported in S255 NIRS3 where the variability is driven by global changes in the pump rate. However, the burst peaks of two maser features showing the highest relative amplitude were delayed by ∼2.5 months and their shapes significantly changed (Wolak *et al.* this volume). Such large delays cannot be easily explained but suggest that local changes of physical conditions in the maser regions play a role. There are several sources in the sample which exhibited uncorrelated variability in a few spectral features. These may be caused by tiny fluctuations of the parameters which influence the maser optical depth or flux of seed photons. Correlations between the velocity fields and alignment of the structures along the line of sight in the environment of HMYSO may also cause changes in coherence velocity and path of maser amplification.

Acknowledgements

We would like to acknowledge support from the National Science Centre, Poland through grant 2016/21/B/ST9/01455.

References

Caratti o Garatti A., Stecklum, B., Garcia Lopez, R., *et al.* 2017, *NatPh*, 13, 276
Caswell, J. L., Vaile, R. A., & Ellingsen, S. P. 1995, *PASA*, 12, 37
Fujisawa, K., Sugiyama, K., Aoki, N., *et al.* 2014 *PASJ*, 66, 109
Goedhart, S., Gaylard, M. J., & van der Walt, D. J. 2004, *MNRAS*, 355, 553
Green, J. A. & McClure-Griffiths, N. M. 2011, *MNRAS*, 417, 2500
MacLeod, G. C. & Gaylard, M. J. 1996, *MNRAS*, 280, 868
Moscadelli, L., Sanna, A., Goddi, C., *et al.* 2017, *A&A*, 600, L8
Reid, M. J., Menten, K. M., Zheng, X. W., *et al.* 2009, *ApJ*, 700, 137
Reid, M. J., Menten, K. M., Brunthaler, A., *et al.* 2014 *ApJ*, 783, 130
Robitaille, T. P., Whitney, B. A., Indebetouw, R., Wood, K., & Denzmore, P. 2006 *ApJS*, 167, 256
Robitaille, T. P., Whitney, B. A., Indebetouw, R., Wood, K. 2007 *ApJS*, 169, 328
Szymczak, M., Wolak, P., Bartkiewicz, A., & Borkowski, K. M. 2012, *AN*, 333, 634
Szymczak, M., Wolak, P., & Bartkiewicz, A. 2014, *MNRAS*, 439, 407
Szymczak, M., Olech, M., Sarniak, R., Wolak, P., & Bartkiewicz, A. 2018, *MNRAS*, in press, arXiv:1710.04595
Urquhart, J. S., Moore, T. J. T., Schuller, F., *et al.* 2013, *MNRAS*, 431, 1752

Astrophysical Masers:
Unlocking the Mysteries of the Universe
Proceedings IAU Symposium No. 336, 2017
A. Tarchi, M.J. Reid & P. Castangia, eds.

© International Astronomical Union 2018
doi:10.1017/S1743921317011516

Long-term and highly frequent monitor of 6.7 GHz methanol masers to statistically research periodic flux variations around high-mass protostars using the Hitachi 32-m

Koichiro Sugiyama[1,2], Y. Yonekura[2], K. Motogi[3], Y. Saito[2], T. Yamaguchi[4], M. Momose[2,4], M. Honma[5], T. Hirota[1], M. Uchiyama[6], N. Matsumoto[1], K. Hachisuka[5], K. Inayoshi[7], K. E. I. Tanaka[8], T. Hosokawa[9] and K. Fujisawa[10]

[1] Mizusawa VLBI Observatory, National Astronomical Observatory of Japan (NAOJ), 2-21-1 Osawa, Mitaka, Tokyo 181-8588, Japan
email: koichiro.sugiyama@nao.ac.jp

[2] Center for Astronomy, Ibaraki University, 2-1-1 Bunkyo, Mito, Ibaraki 310-8512, Japan

[3] Graduate School of Sciences and Technology for Innovation, Yamaguchi University, 1677-1 Yoshida, Yamaguchi, Yamaguchi 753-8512, Japan

[4] College of Science, Ibaraki University, 2-1-1 Bunkyo, Mito, Ibaraki 310-8512, Japan

[5] Mizusawa VLBI Observatory, NAOJ, 2-12 Hoshigaoka-cho, Mizusawa-ku, Oshu, Iwate 023-0861, Japan

[6] Advanced Technology Center, NAOJ, 2-21-1 Osawa, Mitaka, Tokyo 181-8588, Japan

[7] Department of Astronomy, Columbia University, 550 W. 120th Street, New York, NY 10027, USA

[8] Department of Astronomy, University of Florida, Gainesville, FL 32611, USA

[9] Department of Physics, Graduate School of Science, Kyoto University, Sakyo-ku, Kyoto 606-8502, Japan

[10] The Research Institute for Time Studies, Yamaguchi University, 1677-1 Yoshida, Yamaguchi, Yamaguchi 753-8511, Japan

Abstract. We initiated a long-term and highly frequent monitoring project toward 442 methanol masers at 6.7 GHz (Dec > -30 deg) using the Hitachi 32-m radio telescope in December 2012. The observations have been carried out daily, monitoring a spectrum of each source with intervals of 9–10 days. In September 2015, the number of the target sources and intervals were redesigned into 143 and 4–5 days, respectively. This monitoring provides us complete information on how many sources show periodic flux variations in high-mass star-forming regions, which have been detected in 20 sources with periods of 29.5–668 days so far (e.g., Goedhart *et al.* 2004). We have already obtained new detections of periodic flux variations in 31 methanol sources with periods of 22–409 days. These periodic flux variations must be a unique tool to investigate high-mass protostars themselves and their circumstellar structure on a very tiny spatial scale of 0.1–1 au.

Keywords. Stars: massive — stars: formation — masers.

1. Introduction

Periodic flux variability of the methanol masers was first discovered in G 009.62+ 00.19 E, and the periodic flux variability of the methanol masers has been detected in 20 sources so far (including quasi-periodic ones), with their periods from 29.5 to 668 days (e.g., Goedhart *et al.* 2004). Patterns of the variability have been classified into two categories: continuous, and intermittent with a quiescent phase. Such periodic flux variability was also observed in other masers, e.g., water in IRAS 22198+6336 (Szymczak *et al.* 2016), and the variations were synchronized with those of methanol masers in the same sources. The periodic variability, therefore, must be a common phenomenon at around high-mass (proto-)stars, but appears in limited conditions. Because of their short timescale, the periodic variability is potentially important in studying high-mass protostars (HMPSs) and their circumstellar structure on spatial scales of 0.1–1 au, which are estimated under the condition of Keplerian rotation. These area must be the most important to understand the evolutionary track of HMPSs through the mass accretion rates toward onto the stellar surface (Hosokawa & Omukai 2009), however it is impossible to spatially resolve such an area at the distance of HM star-forming regions even by using a future instrument as the extended ALMA (Atacama Large Millimeter/submillimeter Array).

Four models have been proposed for interpretations of the periodic flux variability, possibly caused by global variation on a central engine: a colliding-wind binary (van der Walt 2011), a stellar pulsation (Inayoshi *et al.* 2013), a circumbinary accretion disk (Araya *et al.* 2010), and a rotation of spiral shocks within a gap region in a circumbinary disk (Parfenov & Sobolev 2014). The first model is based on changes in the flux of seed photons, while the remaining three ones are based on changes in the temperature of dust grains at the masing regions.

In previous observations of the methanol masers, flux monitoring for statistically understanding the number of periodic sources was conducted toward only ∼200 sources (e.g., Goedhart *et al.* 2004; Szymczak *et al.* 2015), although all the methanol maser sample consists of more than 1,000 sources (e.g., Breen *et al.* 2015, and references therein).

2. Observations

We initiated a long-term, highly frequent, and unbiased monitoring project using the Hitachi 32-m radio telescope (Yonekura *et al.* 2016) on 30 December, 2012 toward a large sample of the 6.7-GHz methanol masers (442 sources) with declination > −30 deg. In order to cover a wide range of periods from shorter than one month to longer than one year, we designed the observations as follows: until August 2015, the observations were carried out daily, monitoring a spectrum of each of the 442 sources with intervals of 9–10 days. Since September 2015, the number of the target sources were reduced to 143. Sources with modulation index (the standard deviation divided by the averaged value of flux densities), larger that 0.3 were selected. The observations had been continued daily, achieving the intervals in each source of 4–5 days to complete the research for periodic sources with periods shorter than one month. The latter data, combined with the former ones, are used to search for periodic sources with periods longer than one year. For details on the backend setup used for the observations, see Sugiyama *et al.* (2017).

3. Results

By February 2017, we had detected many sources in which periodic flux variations would have been expected, such as G 036.70+00.09 in right-panel of Fig. 1. The 6.7 GHz methanol masers in G 036.70+00.09 presented a periodic variation with continuous pattern in multiple spectral features. To quantitatively evaluate periodicity, we adopted the Lomb-Scargle periodogram (Lomb 1976; Scargle 1982). This is the most reliable method to search for periodicity in flux variations of the methanol masers (Goedhart *et al.* 2014).

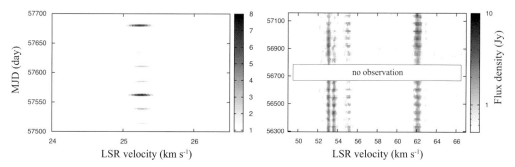

Figure 1. Dynamic spectrum of the 6.7 GHz methanol maser in periodic sources newly detected in our monitor. The horizontal and vertical axis is the local standard of rest velocity and the observation modified julian day, respectively. Gray color shows the flux density. (Left) G 014.23−00.50 with a period of 24 days. (Right) G 036.70+00.09 with a period of 53 days.

Here, we regard an oversampling factor of 4 and frequencies with false-alarm probability $\leqslant 10^{-4}$ as significant, as in Goedhart *et al.* (2014). Other two criteria to certify as a periodic source were set as follows: evaluated periods were accepted in the case of being detectable at least three periodic cycles during the entire observation period of ∼1550 days, meaning detectable periods were shorter than ∼520 days; the signal to noise ratio at the maximum timing was beyond 7.

As a result, we detected periodic flux variations in 42 sources, and for those 31 sources newly detected to host periodic variations, the periods of 22–409 days. Dynamic spectra as examples of the newly detected periodic sources, G 014.23−00.50 and G 036.70+00.09 with a period of 24 and 53 days (Sugiyama *et al.* 2017; Sugiyama *et al.* 2015), are shown in Fig. 1. These sources were classified into the pattern of periodic variability as intermittent and continuous, respectively.

These periodic sources are compiled with previous ones as histogram in terms of periods in Fig. 2. Left-panel in Fig. 2 shows the histogram classified into duration of monitor for the methanol masers: previous, our Hitachi but from September 2015, and Hitachi using all the data. On the other hand, right-panel in Fig. 2 is the same but classified into patterns of the periodic variability: continuous (sinusoidal), intermittent, and continuous but in a part of spectral features. This classification in terms of the pattern must be useful to distinguish which theoretical models are suitable to cause each periodic variation (van der Walt *et al.* 2016), such as the continuous pattern can be caused by the kappa mechanism in a stellar pulsation of HMPSs (Inayoshi *et al.* 2013).

On the basis of these data base for the periodic flux variations, we will verify a period-luminosity (P-L) relation, which is theoretically predicted in the stellar pulsation model. The P-L relation must be a unique tool to indirectly understand physical parameters, such as a mass, radius, and an accretion rate on the stellar surface, impossible to be reached by any observational instruments. In order to better constrain our results, we have initiated another related project of parallax measurements with VERA (VLBI Exploration of Radio Astrometry) for periodic sources, in which the source distance had not been measured by parallax. These measurements are necessary to precisely estimate the luminosity of sources causing periodic flux variations. We have also proceeded with the Hitachi 32-m monitor since June 2017 to complete periodic sources with periods up to two years, and the monitor will be finished by the beginning of 2019 yr.

The authors are grateful to Naoko Furukawa for substantial contributions to this monitoring project, and all the staff and students at Ibaraki University. We would also like to thank all the LOC and SOC in IAU Symposium 336 for their fruitful organizing.

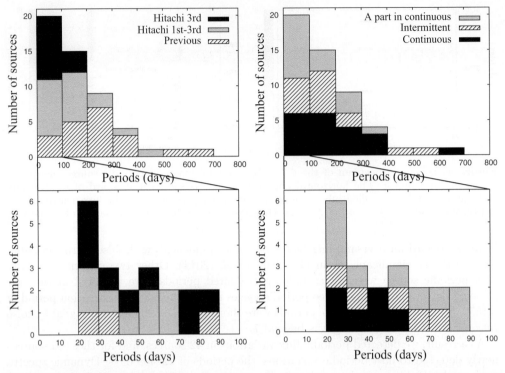

Figure 2. Histogram compiling all the periodic sources in terms of periods. (Left) Classified into duration of monitor for the methanol masers: previous, our Hitachi but from September 2015, and Hitachi using all the data shown by shaded, gray, and black boxes, respectively. The lower-panel is close-up to periods up to 100 days. (Right) Classified into patterns of the periodic variability: continuous (sinusoidal), intermittent, and continuous but in a part of spectral features shown by black, shaded, and gray boxes, respectively.

References

Araya, E. D., Hofner, P., Goss, W. M., *et al.* 2010, *ApJL*, 717, L133

Breen, S. L., Fuller, G. A., Caswell, J. L., *et al.* 2015, *MNRAS*, 450, 4109

Goedhart, S., Gaylard, M. J., & van der Walt, D. J. 2004, *MNRAS*, 355, 553

Goedhart, S., Maswanganye, J. P., Gaylard, M. J., & van der Walt, D. J. 2014, *MNRAS*, 437, 1808

Hosokawa, T. & Omukai, K. 2009, *ApJ*, 691, 823

Inayoshi, K., Sugiyama, K., Hosokawa, T., Motogi, K., & Tanaka, K. E. I. 2013, *ApJL*, 769, L20

Lomb, N. R. 1976, *Ap&SS*, 39, 447

Parfenov, S. Y. & Sobolev, A. M. 2014, *MNRAS*, 444, 620

Scargle, J. D. 1982, *ApJ*, 263, 835

Sugiyama, K., Yonekura, Y., Motogi, K., *et al.* 2015, *PKAS*, 30, 129

Sugiyama, K., Nagase, K., Yonekura, Y., *et al.* 2017, *PASJ*, 69, 59

Szymczak, M., Wolak, P., & Bartkiewicz, A. 2015, *MNRAS*, 448, 2284

Szymczak, M., Olech, M., Wolak, P., Bartkiewicz, A., & Gawroński, M. 2016, *MNRAS*, 459, L56

van der Walt, D. J. 2011, *AJ*, 141, 152

van der Walt, D. J., Maswanganye, J. P., Etoka, S., Goedhart, S., & van den Heever, S. P. 2016, *A&A*, 588, A47

Yonekura, Y., Saito, Y., Sugiyama, K., *et al.* 2016, *PASJ*, 68, 74

Astrophysical Masers:
Unlocking the Mysteries of the Universe
Proceedings IAU Symposium No. 336, 2017
A. Tarchi, M.J. Reid & P. Castangia, eds.

© International Astronomical Union 2018
doi:10.1017/S1743921317009978

Isotopic SiO Maser Emission from the BAaDE Survey

M. J Claussen[1], M. R. Morris[2], Y. M. Pihlström[3], L. O. Sjouwerman[4] and the BAaDE Team

[1] National Radio Astronomy Observatory
P.O Box O, Socorro, NM 87801 USA
email: `mclausse@nrao.edu`

[2] Div. of Astronomy & Astrophysics, UCLA, Los Angeles CA 90095 USA
email: `morris@astro.ucla.edu`

[3] Dept. of Physics & Astronomy, Univ. New Mexico, Albuquerque, NM 87131 USA
email: `ylva@unm.edu`

[4] National Radio Astronomy Observatory
P.O. Box O, Socorro, NM 87801 USA
email: `lsjouwer@nrao.edu`

Abstract. The Bulge Asymmetries and Dynamical Evolution (BAaDE) project aims to map the positions and velocities of up to ∼20,000 late-type stars with SiO maser emission along the full Galactic plane, with a large concentration in the Galactic Bulge and inner Galaxy. Both $J = 1 \rightarrow 0$ and $J = 2 \rightarrow 1$ transitions using the Very Large Array (VLA) and the Atacama Large Millimeter Array (ALMA) are being observed.

In the VLA observing setup, in addition to the ^{28}SiO, $v = 1$ and $v = 2 J = 1 \rightarrow 0$ maser transitions, the bandwidth was wide enough to include the $J = 1 \rightarrow 0$ transitions of the rare isotopologues of the SiO molecule in both the ground and vibrationally excited states: ^{29}SiO, $v = 0$, ^{30}SiO, $v = 0$, ^{29}SiO, $v = 1$, and ^{29}SiO, $v = 2$. Approximately 10% of the initial ∼3500 targets of the project show maser emission from at least one of these lines. Some of these stars (with isotopic maser emission) show high radial velocities which implies that they are indeed in the Galactic Bulge or inner Galaxy (i.e. not foreground objects).

We present line profiles, refined detection statistics, and the implications of the detection of the isotopic maser emission on pumping schemes that have been previously presented.

Keywords. masers, stars: AGB and post-AGB, survey

1. Introduction

The Bulge Asymmetries and Dynamical Evolution (BAaDE) project aims to survey up to ∼34,000 SiO maser stars along the Galactic Plane, concentrating in the Bulge and Inner Galaxy. The primary purpose is to study the dynamics and kinematics of the Galactic Bulge and Bar. The stars are color-selected from the MSX survey on their 8 - 15 μm color. Previous studies have shown that given a specific color selection, a detection rate of SiO masers greater than 50% is achieved. BAaDE was begun in 2013 on the VLA, using the **D** and **C** configurations, with a spectral setup shown in Fig. 1 and with 250 kHz resolution (∼1.7 km s^{-1}). The central velocities of detected masers can thus be estimated to ∼0.1 km s^{-1}. The VLA part of the BAaDE survey was completed earlier this year. The detection rate for SiO masers is ∼70%.

The VLA observing setup was arranged for observing not only the ^{28}SiO masers at 7mm (four vibrational states), but also the ^{29}SiO $v = 0$ (42.88 GHz), ^{29}SiO $v = 1$ (42.58 GHz) and the ^{30}SiO $v = 0$ (42.37 GHz) in the lowest rotational transition.

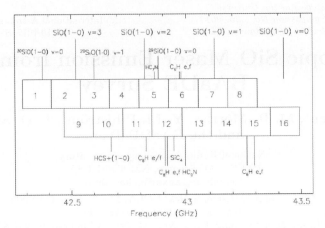

Figure 1. The spectral setup at the VLA for the BAaDE survey.

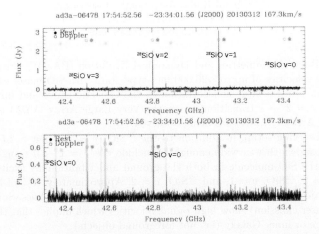

Figure 2. A spectrum of all the emission lines detected toward a target with LSR velocity +167.3 km s^{-1}.

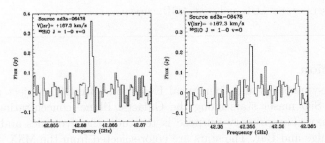

Figure 3. "Zoom-in" spectra of the isotopic SiO emission from the target with LSR velocity +167.3 km s^{-1}.

2. Background

Since the early 1990's it has been clear that there have been problems with the pumping schemes for isotopic maser species (e.g. Cernicharo, Bujarrabal, & Lucas 1991). Olofsson *et al.*, as far back as 1981, introduced the hypothesis of line-overlap pumping, and several papers in the 1990's investigated the importance of line overlaps (e.g. Gonzalez-Alfonso & Cernicharo 1997 developed a non-local radiative transfer code). Herpin & Baudry 2000

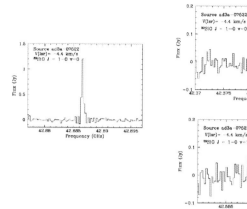

Figure 4. "Zoom-in" spectra of the isotopic SiO emission from a different target, with LSR velocity -4.4 km s^{-1}.

used LVG code and 40 rotational levels for each of $v = 0$ to 4 vibrational states and each of the isotopic species and found that they could explain a lot of the observed line intensities and made new predictions. For ^{29}SiO, they found that infrared line overlaps could explain the observations.

In the mid-to-late 2000s, VLBI observations of isotopic lines (Soria-Ruiz *et al.* 2005, 2007) found that the VLBI maps of the ^{29}SiO $v = 0, J = 1 \rightarrow 0$ lines in evolved stars were similar in nature to ^{28}SiO maps, i.e. they showed tangential amplification. These investigators suggested that a larger sample of VLBI maps of the isotopic masers will help to understand the line-overlap pumping of the SiO masers.

3. Example Spectra

Fig. 2 and Fig. 3 show some example spectra; Fig. 2 shows a spectrum with all lines detected for a target with an LSR velocity of $+167.3$ km s^{-1}. Fig. 3 shows the ^{29}SiO $J = 1 \rightarrow 0, v = 0$ and the ^{30}SiO $J = 1 \rightarrow 0, v = 0$ emission in "zoom-in" spectra (i.e. with the full resolution of the survey). Fig. 4 shows similar "zoom-in" spectra, with ^{29}SiO, ^{30}SiO, $v = 0$, and ^{29}SiO, $v = 1$ emission (for a different target, at -4.4 km s^{-1} LSR velocity).

4. Statistics and Summary

Of $\sim$3600 targets in the 2013 VLA observations, $\sim$360 or 10% show isotopic maser emission. This 10% reflects the ^{29}SiO $v = 0$ transition (i.e. detected in $\sim$360 objects. The ^{30}SiO $v = 0$ transition is detected in $\sim$4% of the targets ($\sim$ 150 objects), and the ^{29}SiO $v = 1$ transition is detected in $\sim$0.8% of the targets ($\sim$30 objects). If these statistics hold over the entire VLA-BAaDE sample ($\sim$20,000 targets) then we would have:

- $\sim$2000 objects detected in ^{29}SiO $v = 0$
- $\sim$800 objects detected in ^{30}SiO $v = 0$
- $\sim$150 objects detected in ^{29}SiO $v = 1$

Some fraction of these objects will be observable with VLBI techniques.

The BAaDE survey is already very successful. Lorant Sjouwerman has a detailed paper on the survey in this volume. Luis Henry Quiroga Nunez (JIVE) presented a poster (Poster 43, and thus a poster paper in this volume) on BAaDE and GAIA comparisons (poster 43) and Michael Stroh (UNM) presented a poster (Poster 45, and thus a poster paper in this volume) on a comparison of 43 GHz (BAaDE-VLA) and 86 GHz masers ($J = 2 \rightarrow 1$) (BAaDE-ALMA).

We have examined the silicon isotopic SiO masers from the first (2013) part of BAaDE at 7 mm wavelength − ^{29}SiO $v = 0$, ^{29}SiO $v = 1$, and ^{30}SiO $v = 0$ in the ground rotational state. By number, the detection rates for these isotopic masers is ∼10%, with the ^{29}SiO $v = 0$ maser being the most prevalent, and generally the strongest, of these three isotopic species. With this sample, either BAaDE data or follow-up VLA and VLBI observations, will provide a large number of evolved SiO maser stars to study the pumping of SiO masers, including the effects of line overlaps.

References

Cernicharo, J., Bujarrabal, V., & Lucas, R. 1991, *A&A*, 249, L27

Gonzalez-Alfonso, E.& Cernicharo, J. 1997, *A&A* 322, 938

Herpin, F. & Baudry, A. 2000, *A&A* 359, 1117

Olofsson, H., Hjalmarson, A., & Rydbeck, O. E. H. 1987, *A&A*, 100, L30

Soria-Ruiz, R., Alcolea, J., Colomer, F., Bujarrabal, V., & Desmurs, J.-F. 2007, *A&A*, 468, L1

Soria-Ruiz, R., Colomer, F., Alcolea, J., Bujarrabal, V., Desmurs, J.-F., & Marvel, K. B. 2005, *A&A*, 432, L39

Astrophysical Masers:
Unlocking the Mysteries of the Universe
Proceedings IAU Symposium No. 336, 2017
A. Tarchi, M.J. Reid & P. Castangia, eds.

© International Astronomical Union 2018
doi:10.1017/S1743921317011486

Constraining Theories of SiO Maser Polarization:
Analysis of a $\pi/2$ EVPA Change

T. L. Tobin[1,2], A. J. Kemball[1] and M. D. Gray[2]

[1]Department of Astronomy, University of Illinois at Urbana-Champaign
1002 W. Green Street, Champaign, IL 61801, USA
[2]Jodrell Bank Centre for Astrophysics, Alan Turing Building, University of Manchester
Manchester M13 9PL, UK

Abstract. The full theory of polarized SiO maser emission from the near-circumstellar environment of Asymptotic Giant Branch stars has been the subject of debate, with theories ranging from classical Zeeman origins to predominantly non-Zeeman anisotropic excitation or propagation effects. Features with an internal electric vector position angle (EVPA) rotation of $\sim \pi/2$ offer unique constraints on theoretical models. In this work, results are presented for one such feature that persisted across five epochs of SiO $\nu = 1, J = 1 - 0$ VLBA observations of TX Cam. We examine the fit to the predicted dependence of linear polarization and EVPA on angle (θ) between the line of sight and the magnetic field against theoretical models. We also present results on the dependence of m_c on θ and their theoretical implications. Finally, we discuss potential causes of the observed differences, and continuing work.

Keywords. masers, polarization, magnetic fields, stars: AGB and post-AGB

1. Introduction

Although theories have endeavored to explain the polarization of SiO masers originating from the near circumstellar environments (NCSE) of Asymptotic Giant Branch (AGB) stars, no theoretical consensus has yet been reached. Prominent theories as to the origin of SiO $\nu = 1, J = 1 - 0$ maser polarization ascribe it to the local magnetic field (Goldreich *et al.* 1973 , Elitzur 1996) or a change in the anisotropy of pumping radiation conditions or other non-Zeeman effects (Asensio Ramos *et al.* 2005, Watson 2009).

However, these theories differ in their ability to explain rotations of the EVPA by $\sim \pi/2$ within a single maser feature. In some theories, such as Goldreich *et al.* (1973) (hereafter GKK), the Electric Vector Position Angle (EVPA) is governed by the angle, θ, between the magnetic field and the line of sight. When θ is small, the linear polarization would be parallel to the projected magnetic field. However, when θ becomes larger than the Van Vleck angle ($\sim 55°$), the polarization would be perpendicular to the projected magnetic field. In this case, a rotation of the EVPA across a feature could be due to a slight change in the direction of the magnetic field with respect to the line of sight, spanning the Van Vleck angle.

This rotation could also be due to a change in the direction of the projected magnetic field across the spatial extent of the maser feature in the image plane. In this case, the EVPA would again be defined by the direction of the projected magnetic field in the sky, but the angle of the projected magnetic field would rotate within the masing material (Soker & Clayton 1999).

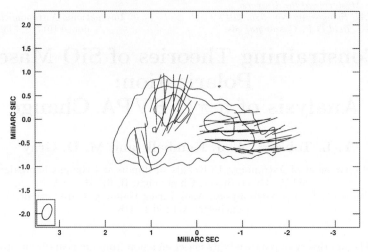

Figure 1. Target maser feature in epoch BD46AO. Contours denote frequency-averaged Stokes I with levels of $\{-10, -5, 5, 10, 20, 40, 80, 160, 320\} \times \sigma$, where $\sigma_{AO} = 1.6430$ mJy beam^{-1}. Vectors denote the frequency-averaged linear polarization, with 1 mas in vector length corresponding to 4 mJy beam^{-1}.

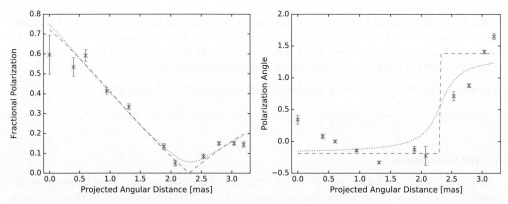

Figure 2. Fractional linear polarization (left) and relative EVPA (right) as a function of projected angular distance for epoch BD46AP. In both plots, 'X's with errors indicate the data. (Note: errors from absolute calibration of EVPA are not included, since the shape of the profile is the focus.) The best fit of the fractional linear polarization from GKK with $K = 0$ is the dashed line, while the fit with non-zero K is the dotted line.

Alternately, if the polarization is mainly governed by anisotropy of pumping radiation conditions, such a rotation could indicate a change in those conditions across the maser feature (Asensio Ramos *et al.* 2005). Here, we discuss our analysis of a maser feature with an internal EVPA rotation of $\sim \pi/2$ that persists across five epochs of observations, and our application of several tests of SiO maser polarization theories to the feature.

2. Observations

For this analysis, we used five epochs of the long-term, full-polarization SiO $\nu = 1 J = 1 - 0$ (43 GHz) VLBA campaign of the Mira variable, TX Cam. These observations have been previously analyzed for total intensity and kinematics (Diamond & Kemball 2003, Gonidakis *et al.* 2010), and linear polarization (Kemball *et al.* 2009). This work focuses on epoch codes BD46AN, BD46AO, BD46AP, BD46AQ, and BD46AR. For further information on the observations themselves, please see Diamond & Kemball (2003).

In addition, the linear polarization of our target maser feature was analyzed for one epoch (BD46AQ) in Kemball *et al.* (2011). As was done in that work, we reduce the data using the method described in Kemball & Richter (2011), to obtain accurate measurements of the low levels of circular polarization. This work in particular expands on previous work by increasing the number of epochs analyzed with accurate circular polarization and applying additional tests of maser polarization theory to the data.

3. Discussion

GKK and Linear Polarization. GKK cite an asymptotic solution for fractional Q and U polarization, Y and Z, respectively, in the regime $\Delta\omega \gg g\Omega \gg R \gg \Gamma$, as $Y = \frac{3\sin^2\theta - 2}{3\sin^2\theta}$, $Z = K$ for $\sin^2\theta \geqslant \frac{1}{3}$, and $Y = -1$, $Z = 0$ for $\sin^2\theta \leqslant \frac{1}{3}$, where Stokes V is assumed to be zero and K is some number such that $Y^2 + Z^2 \leqslant 1$. Typical applications of this theory assume $K = 0$. In the first plot in Figure 2, we fit the predicted linear polarization fraction, m_l, as a function of projected angular distance to the prediction by GKK. To do this, we assumed θ was a quadratic function of projected angular distance, d, and fit for the first- and second-order coefficients, and d_f, the value of d at which θ is the Van Vleck angle, following Kemball *et al.* (2011): $\theta = p_0(d^2 - d_f^2) + p_1(d - d_f) +$ arcsin $\sqrt{2/3}$. For completeness, we fit for both $K = 0$ and non-zero K. Notably, while this profile fit some epochs better than others, it provides a remarkably good fit.

The expected EVPA profiles are derived directly from the best fit functions to m_l, and then fit to the measured EVPA with a simple vertical offset, as we are not accounting for absolute EVPA. The results can be seen in the second plot in Figure 2.

In this case, $K = 0$ GKK predicts a that the EVPA profile will be a strict step function, whereas the data show a smooth rotation of the linear polarization. Generally, adding a non-zero K smooths out the rotation. However, it also causes the extremal angles to be approached asymptotically and can result in less of a net rotation. In contrast, most epochs of our data actually show a rotation of slightly more than $\pi/2$. Notably, such an investigation of EVPA rotation was also conducted by Vlemmings & Diamond (2006) for an H_2O maser of W43A, although the complex Zeeman structure of the water transition prevents the results from being directly analogous.

Zeeman Circular Polarization. Although GKK assumed Stokes $V = 0$, others have expanded on this theory by deriving the behavior of circular polarization due to Zeeman splitting. Elitzur (1996) predicted that the $m_c \propto 1/\cos\theta$. Gray (2012) predicted that m_c is roughly proportional to $\cos\theta$ but it may not be a purely linear relation. Finally, Watson & Wyld (2001) predicted a more complex, peaked function for $m_c(\cos\theta)$. The left plot in Figure 3 shows measured m_c as a function of $\cos\theta$ as determined from the $K = 0$ GKK fit to the linear polarization fraction profile. Although there is scatter at higher $\cos\theta$, our data appears most consistent with the prediction from Gray (2012).

Non-Zeeman Circular Polarization. Wiebe & Watson (1998) suggested that circular polarization may be a result of conversion from linear polarization due to non-Zeeman effects such as changing optical axes in the medium or a change in the magnetic field orientation along the line of sight. This type of non-Zeeman circular polarization would be limited by $m_c < m_l^2/4$ (Wiebe & Watson 1998). As shown in the second plot of Figure 3, the vast majority of our data are not consistent with this limit. Wiebe & Watson (1998) suggest that, individual points may fall outside this limit, but the average values should be consistent if the circular polarization is arising via this mechanism. Even averaging

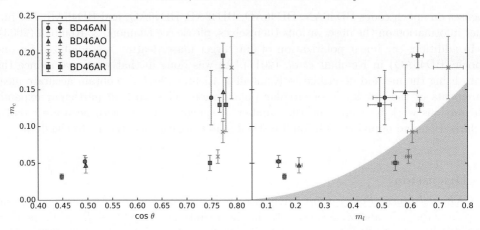

Figure 3. Fractional circular polarization, m_c, as a function of $\cos\theta$, as determined by the K=0 GKK fit (left) and fractional linear polarization, m_l (right). Points shown have m_c $S/N > 3$. Grey shading in right plot denotes region consistent with $m_c < m_l^2/4$.

our values over epoch, not a single epoch is consistent with this limit. This is consistent with the findings of Cotton *et al.* (2011).

Alternative Theories. Other explanations of this EVPA rotation include a curvature of the magnetic field itself within the masing material. Local changes in the direction of the magnetic field such as this have been predicted in Soker & Clayton (1999). In this case, the EVPA would be tracing the projected magnetic field as it rotates. However, if this was the case, we wouldn't expect to see the m_l profile that so closely resembles GKK.

Another possibility is that, instead of resulting from interaction with magnetic fields, the change in EVPA is a result of changing anisotropy conditions. Asensio Ramos *et al.* (2005) propose that a change the anisotropic pumping conditions could cause a rotation in the EVPA. However, a more extensive parameter space and lack of concurrent m_l predictions prevent application of a definitive test.

Acknowledgements

This material is based upon work supported by the National Science Foundation Graduate Research Fellowship Program under Grant No. DGE - 1144245.

References

Asensio Ramos, A., Landi Degl'Innocenti, E., & Trujillo Bueno, J. 2005, *ApJ*, 625, 985
Cotton, W. D., Ragland, S., & Danchi, W. C. 2011, *ApJ*, 736, 96
Diamond, P. J. & Kemball, A. J. 2003, *ApJ*, 599, 1372
Elitzur, M. 1996, *ApJ*, 457, 415
Goldreich, P., Keeley, D. A., & Kwan, J. Y. 1973, *ApJ*, 179, 111.
Gonidakis, I., Diamond, P. J., & Kemball, A. J. 2010, *MNRAS*, 406, 395
Gray, M. 2012, *Maser Sources in Astrophysics*, (Cambridge, UK: Cambridge University Press)
Kemball, A. J., Diamond, P. J., Gonidakis, I., *et al.* 2009, *ApJ*, 698, 1721
Kemball, A. J., Diamond, P. J., Richter, L., Gonidakis, I., & Xue, R. 2011, *ApJ*, 743, 69
Kemball, A. J. & Richter, L. 2011, *A&A*, 533, A26
Soker, N. & Clayton, G. C. 1999, *MNRAS*, 307, 993
Vlemmings, W. H. T. & Diamond, P. J. 2006, *ApJ*, 648, L59
Watson, W. D.2009, *RevMexAA (Serie de Conferencias)*, 36, 113
Watson, W. D. & Wyld, H. W. 2001, *ApJ*, 558, L55
Wiebe, D. S. & Watson, W. D. 1998, *ApJ*, 503, L71

Astrophysical Masers:
Unlocking the Mysteries of the Universe
Proceedings IAU Symposium No. 336, 2017
A. Tarchi, M.J. Reid & P. Castangia, eds.

© International Astronomical Union 2018
doi:10.1017/S1743921317010730

Pumping regimes of Class I methanol masers

A. M. Sobolev and S. Yu. Parfenov

Astronomical Observatory, Ural Federal University,
Lenin Ave. 51, Ekaterinburg 620083, Russia
email: Andrej.Sobolev@urfu.ru

Abstract. In the current paper we describe results of an extensive and refined analysis which shows that the beaming leads to considerable changes in the model line ratios and brightness estimates. For example, beaming shifts the locus of the brightest masers to the lower values of the gas densities. Recent theoretical paper by Leurini *et al.* (2016) presented extensive consideration of the Class I methanol maser (MMI) pumping. Their study allowed to distinguish only 3 of 4 MMI pumping regimes found in Sobolev *et al.* (2005) and Sobolev *et al.* (2007) on the basis of analysis of observational data combined with theoretical considerations. The regime when the line from the $J_{-2} - (J - 1)_{-1}$ E series is the brightest was missing in Leurini *et al.* (2016) results. This may be explained by considering the fact that the authors did not take into account considerable beaming effects.

Keywords. line: formation – masers – ISM: molecules – stars: formation

In our earlier papers Sobolev *et al.* (2005) and Sobolev *et al.* (2007) we have provided observational examples and results of preliminary theoretical analysis that Class I methanol maser (MMI) pumping has at least 4 regimes defined by the series of lines containing the brightest maser line. Since then extensive surveys of the southern maser sources confirmed this conclusion.

Survey of the 36 and 44 GHz lines (lowest frequency lines of the $J_{-1} - (J - 1)_0$ E and $J_0 - (J - 1)_1$ A^+ series) by Voronkov *et al.* (2014) have shown that these masers are bright and widespread and only 23% of detected maser features is detected in both transitions. These lines define the most widespread 1st and 2nd regimes of MMI pumping.

The lines $4_{-1} - 3_0$ E at 36.1 GHz and $5_{-1} - 4_0$ E at 84.5 GHz are likely to be weak masers under normal conditions of the massive star forming region (Berulis *et al.* 1991). Frequently, the maser nature of the lines in this regime is difficult to prove observationally. Anyhow, there are cases when the line profiles contain narrow spikes and the maser nature is proved interferometrically. The sources Sgr B2 and G1.6-0.025 (Salii *et al.* 2002) can be considered as representatives of this maser regime.

Lines of the $J_0 - (J - 1)_1$ A^+ series become prevalent in the other maser regime. Numerous sources in star forming regions manifest definitely maser lines arising in the $7_0 - 6_1$ A^+ and $8_0 - 7_1$ A^+ transitions at 44.1 and 95.2 GHz, respectively. Masers in the northern sources DR21 W, NGC 2264 and OMC-2 and numerous southern sources represent this regime. Theoretical analysis of the pumping shows that the lines of the $J_0 - (J - 1)_1$ A^+ series become brightest in the models with rather high beaming and moderate column densities.

Survey of the 25 GHz line from the $J_2 - J_1$ E series by Voronkov *et al.* (2007) have shown that these masers are widespread but mostly rather weak. However, bright example of OMC-1 clearly shows existence of this 3rd regime.

Survey of the 9 GHz line from the $J_{-2} - (J - 1)_{-1}$ E series by Voronkov *et al.* (2010) have shown that though these masers are rare there are clear examples of existence of the 4th MMI pumping regime (Voronkov *et al.* 2006; Voronkov *et al.* 2011).

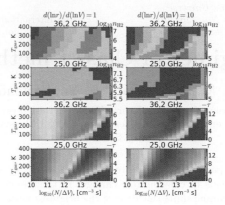

Figure 1. Effect of the beaming on the density, n_{H2}, column density and optical depth of the strongest masers at 36 and 25 GHz. Upper two panels show n_{H2} at which the maximum maser optical depth, τ_{max}, is reached. Lower two panels show τ_{max}.

In the current paper we describe results of extensive and refined analysis which tells that the beaming leads to considerable changes in the model line ratios and brightness estimates. For example, beaming shifts the locus of the brightest masers to the lower values of the gas densities (see Fig.1). The most refined present model of collisional excitation of methanol from Parfenov *et al.* (2016) was used.

The recent study of Leurini *et al.* (2016) of the MMI pumping allowed to distinguish only 3 of 4 MMI pumping regimes found in Sobolev *et al.* (2005) and Sobolev *et al.* (2007) on the basis of analysis of observational data combined with theoretical considerations. Leurini *et al.* (2016) results miss the regime when the line from the $J_{-2} - (J-1)_{-1}\ E$ series is the brightest. This can be explained by the fact that these authors did not take into account considerable beaming effects: clear example of the model with the beaming which matches observational data is published in Voronkov *et al.* (2006).

So, the beaming should be taken into account if one considers conditions of the MMI excitation. This is in accordance with suggestions from the paper "Modelling of Cosmic Molecular Masers: Introduction to Computation Cookbook" (Sobolev *et al.* 2012).

Acknowledgements

The study was supported by the program 211 of the Government of the Russian Federation, agreement №02.A03.21.0006.

References

I. I. Berulis, S. V. Kalenskij, A. M. Sobolev, & V. S. Strelnitskij 1991, *Astron. Astrophys. Trans.*, 1, 231
S. Leurini, K. M. Menten, & C. M. Walmsley 2016, *A&A*, 592, 31
S. Yu. Parfenov, D. A. Semenov, A. M. Sobolev, *et al.* 2016, *MNRAS*, 460, 2648
S. V. Salii, A. M. Sobolev, & N. D. Kalinina 2002, *Astron. Rep.*, 46, 955
A. M. Sobolev, A. B. Ostrovskii, M. S. Kirsanova, *et al.* 2005, *Proc. IAU*, 227, 174
A. M. Sobolev, D. M. Cragg, A. B. Ostrovskii, *et al.* 2007, *Proc. IAU*, 242, 81
A. M. Sobolev & M. D. Gray 2012, *Proc. IAU*, 287, 13
M. A. Voronkov, K. J. Brooks, A. M. Sobolev, *et al.* 2006, *MNRAS*, 373, 411
M. A. Voronkov, K. J. Brooks, A. M. Sobolev, *et al.* 2007, *Proc. IAU*, 242, 182
M. A. Voronkov, J. L. Caswell, S. P. Ellingsen, & A. M. Sobolev 2010, *MNRAS*, 405, 2741
M. A. Voronkov, A. J. Walsh, & J. L. Caswell 2011, *MNRAS*, 413, 2339
M. A. Voronkov, J. L. Caswell, S. P. Ellingsen, *et al.* 2014, *MNRAS*, 439, 2584

Astrophysical Masers:
Unlocking the Mysteries of the Universe
Proceedings IAU Symposium No. 336, 2017
A. Tarchi, M.J. Reid & P. Castangia, eds.

© International Astronomical Union 2018
doi:10.1017/S1743921317010481

Time-dependent numerical modelling of hydroxyl masers

J. P. Maswanganye[1], D. J. van der Walt[1] and S. Goedhart[2,1]

[1]CSR, North-West University, Potchefstroom Campus, Private Bag X6001, Potchefstroom 2520
email: `24827142@nwu.ac.za`

[2]SKA SA, 3rd Floor, The Park, Park Rd, Pinelands 7405, South Africa

Abstract. The statistical rate equations are used to model the OH masers to see if they will always have a one-to-one correspondence with the variation of dust temperature. It is concluded that one has to be careful to argue that the masers will always follow the dust temperature variation profile, and it is possible that different maser transitions from the same molecule respond differently to the same dust temperature variations.

Keywords. stars: formation, masers, methods: numerical, ISM: molecules

1. Motivation

The origin of the periodic variability in the two brightest class II methanol masers (MMIIs), at 6.7- and 12.2-GHz, is yet to be determined but there are currently five hypotheses which had been proposed. Three of these five hypotheses use the dust temperature variation as the origin of periodicity (Array *et al.*, 2010, Inayoshi *et al.*, 2013 and Parfenov & Sobolev, 2014). In these hypotheses, it is assumed that the dust temperature variability has a one-to-one correspondence with maser variability. It is, therefore, the aim of this work to investigate if there is always a one-to-one correspondence between the dust temperature variations and maser variability. The OH molecule is used because it is less complex compared to methanol.

2. Overview

The rate at which level population N_i of a molecule in a molecular cloud changes can be modelled by the standard statistical rate equations (Elitzur & Netzer, 1985). For OH, some transitions overlap, which may populate and de-populate the involved level populations. Therefore, the angle-averaged intensity should be modified to incorporate line overlap and Elitzur & Netzer (1985) derived it to be:

$$\bar{J}_a = S_a \left(1 - \beta_a\right)\left(1 - f_a\right) + \left(1 - \beta_{ab}\right) f_a S_{ab} + W_d \beta_{ab} I_d + W_{\mathrm{HII}} \beta_{ab} I_{\mathrm{HII}}, \tag{2.1}$$

where f_a, W_d (W_{HII}), I_d (I_{HII}) and S_{ab} (S_a) are the fraction of the line width profile which overlaps with the line profile of transition b, geometric dust (HII region) dilution factor, radiative intensity from the dust (HII region) and source function for overlapping (non-overlapping) line, respectively. See Bujarrabal *et al.* (1980) for the overlapping lines.

3. Results and discussion

In this investigation, the 24-level OH data, from the Leiden Atomic and Molecular Database (LAMDA) (Schöier *et al.*, 2005), are used in finding the solution of the statistical rate equations with Heun's method. At each second, the previous level populations

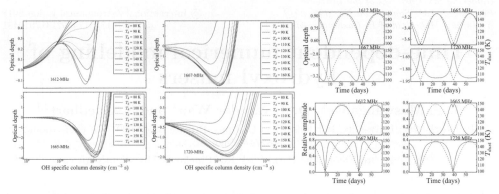

Figure 1. The *left* and *middle* panels are how the optical depths vary as a function of a specific column density at different dust temperatures (from 10 to 160 K). The dust's (HII region's) geometric dilution factor, molecular hydrogen density, fractional molecular abundance of OH, electron (kinetic) temperature, and emission measure were fixed at 0.1 (1.0×10^{-3}), 10^5 cm^{-3}, 10^{-5}, 9.0×10^3 (100) K and 3.0×10^8 cm^{-6} pc, respectively. *Right* panel shows how the optical depth (top) and relative amplitude (solid line) changes to an absolute cosine-like dust temperature light curve (dashed line).

$N_i(t_{j-1})$ were used to calculate the t_j level populations, $N_i(t_j)$. The convergence criterion was such that $\frac{|N_i(t_j) - N_i(t_{j-1})|}{N_i(t_j)} < 10^{-7}$ (Cragg, Sobolev & Godfrey, 2005).

After testing the implementation of the code, the set of physical parameters which could result in one or more of the four rotational ground state transitions of OH in masing were searched at the specific column density of 1.0×10^6 to 2×10^{20} cm^{-3} s. The negative optical depths in Fig. 1 (*left* and *middle* panels) imply that the transitions are masing. At the specific column density of 1.5×10^{11} cm^{-3} s, the 1612-MHz line is not masing but the rest are masing. At this specific column density, the dust temperature was changed with time and independently of whether the system is in equilibrium or not. The 1665- and 1667-MHz lines do not show a one-to-one correspondence to the dust temperature variations (*right* panel in Fig. 1). On the other hand, the 1720-MHz maser shows a one-to-one correspondence to the dust temperature variations. Such behaviour had been observed even with different dust temperature profiles and at different specific column densities.

Therefore, we conclude that it is possible that the dust temperature variations do not have a one-to-one correspondence with the maser variation and different masing lines can behave differently to the same dust temperature profile. It is then reasonable to argue that different maser species of different molecules can respond differently from the same changes of the dust temperature.

References

Araya E. D. *et al.* 2010, *ApJ*, 717, L133

Bujarrabal V., Guibert J., Rieu N. Q. & Omont A. 1980, *A&A*, 84, 311

Cragg D. M., Sobolev A. M. & Godfrey P. D. 2005, *MNRAS*, 360, 533

Elitzur M. & Netzer N. 1985, *ApJ*, 291, 464

Inayoshi K., Sugiyama K., Hosokawa T., Motogi K. & Tanaka K. E. I. 2013, *ApJ*, 769, L20

Parfenov S. Yu. & Sobolev A. M. 2014, *MNRAS*, 444, 620

Schöier F. L., van der Tak F. F. S., van Dishoeck E. F. & Black J. H. 2005, *A&A*, 432, 369

Astrophysical Masers:
Unlocking the Mysteries of the Universe
Proceedings IAU Symposium No. 336, 2017
A. Tarchi, M.J. Reid & P. Castangia, eds.

© International Astronomical Union 2018
doi:10.1017/S1743921318000819

Rapid burst of 6.7 GHz methanol maser in the high mass star region G33.641-0.228

Kārlis Bērziņš, Ivar Shmeld and Artis Aberfelds

Engineering Research Institute Ventspils International Radio Astronomy Centre (VIRAC),
Ventspils University College, Inzenieru 101, LV-3601, Ventspils, Latvia
email: karlis.berzins@venta.lv

Abstract. We report another 6.7 GHz methanol maser burst in the high mass star region G33.641-0.228. The flare is in its second component at $v_{\mathrm{LSR}} = 59.6$ km s^{-1} and was observed in August-September 2016 by VIRAC radio telescope RT-32 in Irbene, Latvia. Several bursts of the second spatial component of G33.641-0.228 have been reported previously by Fujisawa *et al.* The maximum peak flux density of the source was measured to reach 343 Jy that is 13 times increase from its ground level. Significant oscillations were discovered during the decay phase indicating a more complex burst mechanism that cannot be explained by a simple heating of the region.

Keywords. ISM: molecules – lines and bands – jets and outflows – magnetic fields

1. Introduction

It is generally thought that 6.7 GHz methanol masers are related to the interstellar medium around high mass stars. Some of them are variable. It is common to observe slow maser variations with typical timescales of months (e.g. Aberfelds *et al.* 2017). The CH$_3$OH maser source G33.641-0.228 (IRAS 18509+0027) is particularly interesting because it has been observed that only one of its spectral features is bursting, while other components remain constant. The burst of this object was first noticed and described by Fujisawa *et al.* 2012. The methanol maser source was first discovered by Szymczak *et al.* 2000. Its spatial structure has been studied with VLBI (Fujisawa *et al.* 2012, Bartkiewicz *et al.* 2016).

It has been reported that the second methanol maser component of this source has had bursts already several times (Fujisawa *et al.* 2014). Another burst of this component (at $v_{\mathrm{LSR}} = 59.6$ km s^{-1}) was observed in August-September of 2016 by the VIRAC 32 m radio telescope. The decay phase was monitored on daily basis during the burst activity.

2. Data and Analysis

After the modernization of the VIRAC radio telescopes in 2016, the G33.641-0.228 was chosen as a good observational candidate for studies of rapid maser variability. Starting from its very first observation in August 25th (Figure 1), an intense source burst (>340 Jy) was detected followed by oscillating decrease. The observations were carried out using the Irbene 32 meter fully steerable radio telescope RT-32 with a cryogenic receiver at 6.7 GHz. The *Rohde & Swartz FSW43* spectrometer was used at the back end to collect the spectral data. Data for only one polarization component were collected. All spectral data were calibrated to absolute intensity flux values within a relative error of 10%. During the decay phase, the burst was observed once per day limited by the source visibility at the RT-32 site. We measured that the intensity flux of the second component

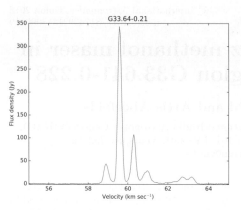

Figure 1. The intensity spectra of
G33.641-0.228 on August 25, 2016.

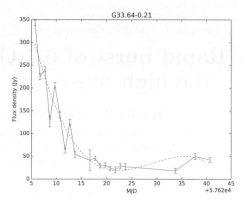

Figure 2. Oscillating afterburst lightcurve
(solid line) with 15σ errorbars comparing
to smooth decay data fit (dashed line).

of the 6.7 GHz methanol maser in G33.641-0.228 reached the level of at least 343 Jy and decreased afterwards. This is about 13 times above its usual level of 26 Jy.

3. Results and Discussion

The light curve of the burst of the 6.7 GHz maser in G33.641-0.228 (Figure 2) shows a clear oscillating behaviour putting an upper limit to its variability time-scale of the order of hours. Note that other spectral components do not experience such increase of flux intensity meaning that this phenomenon is related to only a limited spatial region. At this stage, we can rule out a simple model of thermal heating of the maser region gaining its thermal energy from a flare-up from the central star because this would be inconsistent with the rapid heating and cooling of a region on time-scales of several hours (less than a day) as detected by our observations. Instead, relatively narrow, well-directed outbursts from the central star causing rapid variations of the magnetic field are thought to be responsible for this kind of oscillating behaviour of the light curve of a single maser component. Regular observations of this source are needed for better understanding its nature and in order to catch further bursts. During the bursts, cooperative observations with better time sampling would be very useful.

Acknowledgements

This work is financed by ERDF project No. 1.1.1.1/16/A/213, being implemented in Ventspils University College.

References

Aberfelds, A., Shmeld, I., & Bērziņš, K. 2017, these Proceedings
&Bartkiewicz, A., Szymczak, M., van Langevelde, H. J. 2016, *A&A* 587, 104
Fujisawa, K., Sugiyama, K., Aoki, N., Hirota, T., Mochizuki, N., Doi, A., Honma, M., Kobayashi,
    H., Kawaguchi, N., Ogawa, H., Omodaka, T., & Yonekura, Y. 2012, *PASJ* 64, 17
Fujisawa, K., Aoki, N., Nagadomi, Y., Kimura, S., Shimomura, T., Takase, G., Sugiyama, K.,
    Motogi, K., Niinuma, K., Hirota, T., & Yonekura, Y. 2014, *PASJ* 66, 109
Szymczak, M., Hrynek, G., & Kus, A. J. 2000, *A&AS* 143, 269

Astrophysical Masers:
Unlocking the Mysteries of the Universe
Proceedings IAU Symposium No. 336, 2017
A. Tarchi, M.J. Reid & P. Castangia, eds.

© International Astronomical Union 2018
doi:10.1017/S1743921318000431

Analysis of bipolar outflow parameters, magnetic fields and maser activity relationship in EGO sources

O. S. Bayandina[1], I. E. Val'tts[1], P. Colom[2], S. E. Kurtz[3], G. M. Rudnitskij[4] and N. N. Shakhvorostova[1]

[1] Astro Space Center, P.N. Lebedev Physical Institute of RAS,
Profsoyuznaya st. 84/32, 117997, Moscow, Russia
email: bayandina@asc.rssi.ru

[2] LESIA, Observatoire de Paris-Meudon, CNRS, UPMC, Université Paris-Diderot,
5 place Jules Janssen, 92195, Meudon, France

[3] Centro de Radioastronomía y Astrofísica, Universidad Nacional Autónoma de México,
Apdo. Postal 3-72, 58089, Morelia, México

[4] Moscow State University, Sternberg Astronomical Institute,
Universitetskii Prospekt 13, Moscow, 119992, Russia

Abstract. The interferometric and single-dish observations of the Extended Green Objects sample have been carried out in order to check the possible common pumping mechanism of class I methanol maser (cIMM) and OH(1720 MHz) maser and their identification with a front of bipolar outflow as a source of interstellar shock stimulating collisional pumping of the molecules. High spatial and spectral resolution observations of OH masers allow us to investigate structure, kinematics, and magnetic field configuration of the inner region of the source, i.e., the outflow ejection region. Analysis of magnetic field strength in a disk area is crucial to understanding of the outflow origin.

Keywords. ISM: evolution, masers, magnetic fields

1. Introduction

Recently a promising new sample of about 300 massive young stellar objects (MYSOs) was identified on the base of excess of extended emission in the 4.5μm – green in the IRAC/*Spitzer* camera color scale (GLIMPSE-I survey, Churchwell *et al.* 2009). These objects were named EGO, i.e., Extended Green Objects (Cyganowski *et al.* 2008). The sample has been chosen for our study keeping in mind EGOs' association with active protostellar outflows, cIMMs (Cyganowski *et al.* 2009) and OH(1720 MHz) masers (Litovchenko *et al.* 2012).

2. Observations and Data Reduction

• The Karl Jansky Very Large Array (JVLA) (NRAO, USA): interferometric observations of 20 EGOs were carried out in 2013 using C-configuration with an angular resolution of about $12''$ and spectral resolution of 0.34 km/s in all four OH lines at frequencies of 1612, 1665, 1667, and 1720 MHz.

• The Nançay decimetric radio telescope (NRT) (Observatoire de Paris, France): high-spectral resolution polarimetric single-dish study were made in 2015 with a $3.5' \times 19'$ beam and spectral resolution of 0.07 km/s at frequencies of 1665, 1667, and 1720 MHz.

Position, V_{LSR}, integrated and peak flux density for each maser feature detected at JVLA in the RCP and LCP data-cubes were obtained for the EGOs. In order to identify

Zeeman patterns, we searched through detected maser features for groups with opposite circular polarization that coincide spatially to within OH spot sizes in the frames of the positional uncertainties (following the method discussed in, e.g., Fish *et al.* 2003). Magnetic field strength was calculated based on OH lines Zeeman splitting values estimated in Davies (1974). The polarization parameters such as the degree of circular and linear polarization, flux density in linear polarization, and polarization angles were obtained from NRT data.

3. Conclusions

- With the JVLA in the direction of 20 EGOs maser emission at 1665/1667 MHz was detected in 50% of the sample.
- Spatial association of cIMMs and OH(1720 MHz) masers was not detected – it may indicate the absence of the conditions necessary for excitation of OH(1720 MHz) masers in the interaction region of bipolar outflow and interstellar medium.
- OH masers at 1665 and 1667 MHz were detected within $\sim 0.1''$ from the continuum source – ejector of bipolar outflow found in e.g. Cyganowski *et al.* (2011), Towner *et al.* (2017).
- The magnetic field strength was obtained for spatially coincident possible Zeeman pairs, which were identified in 50% of 10 EGOs: the values range from -8.4 to $+13.2$ mG, that indicates the possible predominance of strong magnetic fields in the OH maser spots in EGOs.
- The velocity gradient indicating an association of the OH maser spots with the rotating discs, previously observed in thermal lines of CS and NH_3 molecules, was observed in $\sim$40% of the sources.
- Orientation of OH maser spot cluster in 70% of cases is perpendicular to the plane of propagation of the bipolar outflow.
- The direction of magnetic field and outflow is almost parallel: an analysis of linear polarization angle distribution in OH spot clusters shows it in the majority of the cases (assuming a magnetic field perpendicular to the polarization angle).

References

Churchwell, E., Babler, B. L., Meade, M. R., Whitney, B. A., Benjamin, R., Indebetouw, R., Cyganowski, C., Robitaille, T. P., Povich, M., Watson, C., & Bracker, S. 2009, *PASP*, 121, 213

Cyganowski, C. J., Whitney, B. A., Holden, E., Braden, E., Brogan, C. L., Churchwell, E., Indebetouw, R., Watson, D. F., Babler, B. L., Benjamin, R., Gomez, M., Meade, M. R., Povich, M. S., Robitaille, T. P., & Watson, C. 2008, *AJ*, 136, 2391

Cyganowski, C. J., Brogan, C., Hunter, T. R., & Churchwell, E. 2009, *ApJ*, 702, 1615

Cyganowski, C. J. & Brogan, C. L., Hunter T. R., Churchwell E. 2011, *ApJ*, 743, 56

Davies, R. D. 1974, in: F.J. Kerr & S. C. Simonson III (eds.), *IAU Symp. 60, Galactic Radio Astronomy* (Dordrecht: Reidel), 275

Fish, V. L., Reid, M. J., Argon, A. L., & Menten, K. M. 2003, *ApJ*, 596, 328

Litovchenko, I. D., Bayandina, O. S., Alakoz, A. V., Val'tts, I. E., Larionov, G. M., Mukha, D. V., Nabatov, A. S., Konovalenko, A. A., Zakharenko, V. V., Alekseev, E. V., Nikolaenko, V. S., Kulishenko, V. F., & Odintsov, S. A. 2012, *Astron. Rep.*, 56, 536

Reach, W. T., Rho, J., Tappe, A., Pannuti, T. G., Brogan, C. L., Churchwell, E. B., Meade, M. R., Babler, B., Indebetouw, R., & Whitney, B. A. 2006, *AJ*, 131, 1479

Towner, A. P. M., Brogan, C. L., Hunter T. R., Cyganowski, C. J., McGuire, B. A., Indebetouw, R., Friesen, R. K., & Chandler, C. J. 2017, *ApJSS*, 230, 22

Gabriele Surcis and Bringfried Stecklum (photo credit: S. Poppi)

Galaxies and Supermassive Black Holes

Astrophysical Masers:
Unlocking the Mysteries of the Universe
Proceedings IAU Symposium No. 336, 2017
A. Tarchi, M.J. Reid & P. Castangia, eds.

© International Astronomical Union 2018
doi:10.1017/S1743921318000753

Extragalactic maser surveys

C. Henkel[1,2], J.-E. Greene[3] and F. Kamali[1]

[1] Max-Planck-Institut für Radioastronomie, Auf dem Hügel 60, 53121 Bonn, Germany
[2] Astron. Dept., King Abdulaziz University, P.O. Box 80203, Jeddah 21589, Saudi Arabia
[3] Department of Astrophysical Sciences, Princeton University, Princeton, NJ 08544, USA
email: chenkel@mpifr-bonn.mpg.de

Abstract. Since the IAU (maser-)Symposium 287 in Stellenbosch/South Africa (Jan. 2012), great progress has been achieved in studying extragalactic maser sources. Sensitivity has reached a level allowing for dedicated maser surveys of extragalactic objects. These included, during the last years, water vapor (H_2O), methanol (CH_3OH), and formaldehyde (H_2CO), while surveys related to hydroxyl (OH), cyanoacetylene (HC_3N) and ammonia (NH_3) may soon become (again) relevant. Overall, with the upgraded Very Large Array (VLA), the Atacama Large Millimeter/submillimeter Array (ALMA), FAST (Five hundred meter Aperture Synthesis Telescope) and the low frequency arrays APERTIF (APERture Tile in Focus), ASKAP (Australian Square Kilometer Array Pathfinder) and MeerKAT (Meer Karoo Array Telescope), extragalactic maser studies are expected to flourish during the upcoming years. The following article provides a brief sketch of past achievements, ongoing projects and future perspectives.

Keywords. masers, galaxies: active, galaxies: ISM, galaxies: Seyfert, radio lines: galaxies

1. The past few decades

Maser lines allow us to detect and to localize tiny hotspots with exceptional activity across the Universe. These can be used to highlight regions with enhanced star formation, allow for determinations of proper motion out to distances of several Mpc, and help us to constrain the morphology and distance of galaxies. Before addressing recent or ongoing extragalactic maser surveys, here we briefly summarize achievements obtained till the time of the Stellenbosch meeting. We then proceed with the Local Group of galaxies (Sect. 2), H_2O megamaser projects (Sects. 3–6), the search for a correlation between H_2O and OH megamasers (Sect. 7), formaldehyde (Sect. 8), methanol, HC_3N, and HCN (Sect. 9), and future perspectives (Sect. 10).

Extragalactic 1.7 GHz (18 cm) OH masers were detected as early as in the mid seventees (e.g., Gardner & Whiteoak 1975), soon followed by the detection of a bright 22 GHz H_2O maser in M 33 by Churchwell *et al.* (1977). While line luminosities were high by Galactic standards, they were not surpassing their more local cousins by many orders of magnitude. This drastically changed, when the first "megamasers" were detected, this time in opposite order, starting with 22 GHz H_2O in NGC 4945 (Dos Santos & Lepine 1979) and followed by 1.7 GHz OH in Arp 220 (Baan *et al.* 1982). During the following years the interest mainly focused on OH, because these megamasers were rapidly associated with a so far not well known class of galaxies, the ULIRGs (UltraLuminous InfraRed Galaxies). This culminated in the large survey carried out by Darling & Giovanelli (2002), leading to a total number of ≈100 detected OH megamaser sources. With respect to 22 GHz H_2O megamasers, it took more than 15 years until it became clearer what they represented and in which class of galaxies they could be found. However, even today detection rates are typically below 10%. Crucial discoveries were the detection of satellite lines well off the systemic velocity in NGC 4258 (Nakai *et al.* 1993), the discovery of velocity drifts of the near systemic maser components of NGC 4258 (Haschick et al. 1994; Greenhill et al. 1995), the mapping of the H_2O maser features in NGC 4258 (Miyoshi et al. 1995), and the discovery that 22 GHz H_2O masers are most commonly found in Seyfert 2 and LINER (Low Ionization Nuclear Emission Line Region) galaxies (Braatz et al. 1996, 1997). All this indicated that some of the luminous water vapor masers, the so-called disk-masers, are forming Keplerian parsec or even sub-parsec scale accretion disks around their galaxy's supermassive nuclear engines, allowing for a determination of the disk's geometry, the mass of the nuclear engine, and the angular diameter distance to the parent galaxy.

Another molecular transition, the 4.8 GHz $J = 1$ K-doublet line of formaldehyde (H_2CO). was also proposed to be masing, greatly surpassing any Galactic counterpart in luminosity, Commonly seen in absorption, quasi-thermal 4.8 GHz emission lines form at densities of at least several 10^5 cm^{-3}, while maser emission is rarely seen in the Galaxy (e.g., Ginsburg et al. 2015). 4.8 GHz H_2CO emission was detected toward Arp 220, IC 680, and IR 15107+0724 (Baan et al. 1986; Baan et al. 1993; Araya et al. 2004) and, in the initial paper, the line from Arp 220 was interpreted as a maser likely amplifying the non-thermal redio background of the source.

2. The Local Group

The Large and Small Magellanic Clouds (LMC and SMC) allow for studies of masers commonly observed in the Galaxy, but under conditions of low metallicity and strong UV-radiation fields. During the past few years, Breen et al. (2013), Imai et al. (2013), and Johanson et al. (2014), mainly using the Australia Compact Array (ATCA), were successfull in finding new maser spots related to star formation. In the LMC, now we have 27 H_2O masers originating from 15 regions of star formation, and 3 such masers from late type stars. With the study of Breen et al. (2013), four new 22 GHz H_2O masers were detected in the SMC, thus tripling the number obtained three decades before by Scalise & Braz (1982). Most of these sources can greatly help, through proper motion measurements, to constrain the individual rotation of the two galaxies as well as to measure their orbital motion around the Milky Way.

Proper motion is also the main motivation to systematically measure the less conspicuous members of the Local Group at 6.7 and 22 GHz, searching for methanol and water vapor masers using the Sardinia Radio Telescope (SRT). Results from this ongoing project are summarized by the contribution of A. Tarchi.

In the Andromeda galaxy (M 31) Sjouwerman et al. (2010) and Darling (2011) detected first 6.7 GHz CH_3OH and 22 GHz H_2O masers, respectively. After this encouraging start, however, no new maser sources could be identified during the following years (see Darling et al. 2016, who surveyed ≈500 additional sources with compact 24μm emission in M 31). The unknown proper motion of M 31 is the main obstacle on the way to a basic understanding of Local Group dynamics. Therefore, detecting many such maser lines with suitable flux density for interferometric studies is of high importance.

3. H_2O megamaser detection surveys

Already before the Stellenbosch meeting, dedicated 22 GHz H_2O maser surveys have been carried out. These include unsuccessful searches for Fanaroff-Riley I (FR I) galaxies (Henkel et al. 1998) and relatively nearby low luminosity type I and type II QSOs (Bennert et al. 2009; König et al. 2012) as well as the detection of maser emission in an FR II galaxy (Tarchi et al. 2003). Additional surveys lead to the detection of a type 2 QSO at

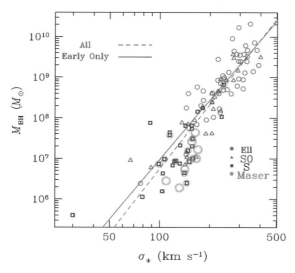

Figure 1. Relationship between stellar velocity dispersion and central mass (M_{BH}, taken from Greene *et al.* 2016). Dashed line: Fit to the entire sample. Solid line: Early type galaxies only. Small open circles: ellipticals; small open triangles: lenticulars; small open squares: spirals; single circles surrounding an open square or double circles: disk-maser galaxies.

redshift $z = 0.66$ (Barvainis & Antonucci 2005) and to the detection of a gravitationally lensed type I QSO at $z = 2.64$ (Impellizzeri *et al.* 2008; Castangia *et al.* 2011; McKean *et al.* 2011).

Searching for 22 GHz H_2O masers mainly toward Narrow-Line Seyfert 1 galaxies (NLS1s), which appear to contain relatively low mass nuclear emgines but luminosities compatible with their broad-line counterparts, Tarchi *et al.* (2011) reported two new detections, thus leading to a total of five known masers in this type of galaxies. The more recent survey by Hagiwara *et al.* (2013a) did not yield new positive results.

Another survey lacking new detections, but also with high relevance for our understanding of the H_2O megamaser phenomenon and the calibration of nuclear mass determinations, has been carried out by van den Bosch et al. (2016). They observed galaxies, where the gravitational sphere of influence of the central engine is extended enough to be resolvable by present day optical or near infrared (NIR) facilities. The detection of H_2O disk-masers in such galaxies would have a great impact. The radio data would provide a reliable black hole mass, which could then be compared with those derived by other potentially less accurate methods using optical or NIR data. The non-detections imply that NGC 4258 was and still remains the only such calibrator, where nuclear masses derived from H_2O and other methods, applicable to a larger number of galaxies, could be compared. Furthermore, most galaxies observed by van den Bosch *et al.* (2016)) were of early type, with estimated nuclear masses $M_{BH} \gtrsim 10^8 \, M_\odot$. Apparently, there is a rather narrow window for the occurrence of H_2O megamasers, which appear to be confined to nuclear regions with $10^6 \, M_\odot < M_{BH} < 10^8 \, M_\odot$.

There are also recent 22 GHz H_2O surveys with detections. Beside the RadioAstron measurements of known sources with ultrahigh angular resolution, discussed by W. A. Baan in this volume, a single-dish study to be mentioned in this context is that of Wagner (2013). He targeted Seyfert 2 galaxies with high X-ray luminosity and high hydrogen column density, also including a few OH-absorbers. His detection rate, 4 out of 37 galaxies, exceeds 10% and therefore provides a good example on how to select promising sources. Yamauchi *et al.* (2017) chose galaxies with an absorbed 2 keV continuum, a strong Fe

6.4 keV line and significant infrared emission. They detected three galaxies in a sample of 10 targets, an exceptionally high detection rate, with the caveat that their sample is comparatively small. Finally, Zhang *et al.* (2012) and Liu *et al.* (2017)) showed that almost exclusively Seyfert 2 galaxies with high radio continuum luminosities are exhibiting H_2O maser emission. This led to a successful pilot study involving 18 sources (see Zhang *et al.* 2017) and the contribution by J.S. Zhang in this volume.

4. The MCP

At the core of the present extragalactic 22 GHz H_2O maser research stands the Megamaser Cosmology Project (MCP) with the goals (1) to study accretion disk morphology, (2) to determine nuclear black hole masses, and (3) to constrain the Hubble constant to a precision of a few percent, avoiding any standard candles but using instead a direct geometric approach (see Herrstein *et al.* 1999 for the first application of this technique). Introduced at the Stellenbosch meeting by Henkel *et al.* (2012), three MCP articles had been published by that time, two on UGC 3789 and one on black hole masses (Reid *et al.* 2009; Braatz *et al.* 2010; Kuo *et al.* 2011). In addition, Greene et al. (2010) analyzed obtained black hole masses as a function of associated bulge masses. More recently, the number of MCP publications has tripled (Reid *et al.* 2013, Kuo *et al.* 2013, 2015, Pesce *et al.* 2015, Gao *et al.* 2016, 2017). These articles present detailed high resolution maps of a total of nine galaxies, most of them exhibiting the common disk-maser fingerprint. Detailed results are presented by J. A. Braatz, in this volume. In addition, maser disk physics and upper magnetic field limits were discussed. Following Pesce *et al.* (2015), the spiral shock model proposed by Maoz & McKee (1998) is not consistent with the measured properties of the maser disks, since the so-called high velocity features, hundreds of km s^{-1} off the parent galaxy's systemic velocity, do not show any of the predicted systematic velocity drifts. Limits obtained searching for Zeeman splitting typically reach 200-300 mG (1σ), while the corresonding limit for NGC 1194 is 73 mG.

Beside disk morphology, M_{BH}, and geometric distance, there are three major lines of research making use of the data supplied by the MCP: (1) A comparison of the resulting supermassive black hole masses with those derived by gas or stellar dynamics, bulge mass, total galactic mass, or mass within the central kiloparsec; (2) a comparison of position angle and inclination of the respective galaxies as a function of radius, starting from the outer large scale disk and proceeding to the smallest scales, provided by the H_2O disk-masers, representing a nuclear disk viewed edge-on; (3) a test of the AGN paradigm, involving a supermassive nuclear engine, a jet, and, perpendicular to it, an accretion disk.

(1) Comparisons of the mass of the nuclear engine with those of other galactic parameters reveal poor correlations, thus leading to results which drastically differ from the tight correlation obtained for massive elliptical galaxies. Furthermore it appears (see Fig. 1) that the masses of the supermassive nuclear engines of megamaser galaxies are below the expected correlation, while similarly sized spiral galaxies studied at optical or NIR wavelengths do not show this effect. Greene *et al.* (2016) and Läsker *et al.* (2016)) offer two most likely explanations for this effect: Either the non-maser selected galaxies miss the low mass end of the BH distribution due to an inability to resolve their spheres of influence or the disk-maser galaxies preferentially occur in lower BH-mass environments.

(2) The disk-maser galaxies with their known nuclear morphology allow for a unique comparison of position angles between largest and smallest scales. This reveals that the megamaser disks are neither aligned with the large scale disks of their parent galaxies nor with the morphology encountered at a galactocentric radius close to 100 pc (Greene *et al.* 2010, 2013; Pjanka *et al.* 2016). Fig. 2 shows correlations between the large and $r \approx$

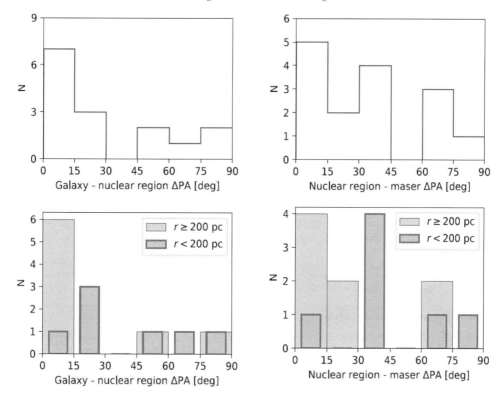

Figure 2. Position angle (PA) differences between the angular momenta of the large scale disk versus the central region at radius $r \approx 200$ pc (left) and the central region versus the pc-scale megamaser disk (right). The two lower panels also differentiate between nuclear regions $r < 200$ pc and >200 pc, still indicating a possible alignment between the large scale disks and the $r > 200$ pc regions, but not with the $r < 200$ pc environment. Only the smaller one of two possible position angles is shown (the other one is $180°$ minus the given angle).

200 pc scale (left panels) and between the $r \approx 200$ pc and ≈ 1 pc disk maser orientations. A differentiation between $r > 200$ pc and < 200 pc is also presented (lower panels).

(3) To become detectable, the 22 GHz H_2O maser line requires gas fulfilling certain boundary conditions, like kinetic temperatures $T_{kin} \gtrsim 300$ K and very high densities, $n(H_2) \gtrsim 10^7$ cm^{-3} (e.g., Kylafis & Norman 1991). Thus, the H_2O line is confined to a specific physical environment and is certainly not telling the entire story. To gain deeper insights, radio continuum data are an excellent additional probe because they can be obtained with similar angular resolution and are also not affected by dust extinction. With the geometry being known, i.e. with nuclear disks viewed edge-on, radio continuum data are ideal to check the AGN paradigm: Are there jets and are these really two-sided, as it is expected for jets ejected parallel to the plane of the sky? Can we follow individual blobs, thus directly determining their speed? Can we detect emission from inside the maser disks? And are there correlations with the nuclear mass and/or the size of the maser disks? With this motivation dedicated continuum measurements at 33 GHz (Very Large Array, VLA) and 5 GHz (Very Long Baseline Array, VLBA) have been carried out. Some preliminary VLBA results related to this ongoing project are presented by F. Kamali in this volume. Fig. 3 shows the correlations between the 33 GHz VLA continuum flux densities and the inner and outer maser disk radii.

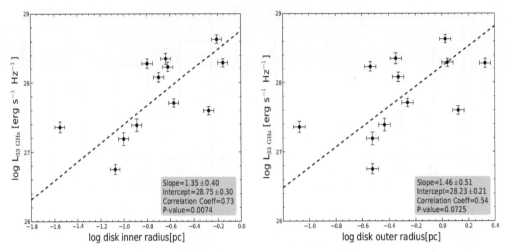

Figure 3. Maser disk inner (left) and outer (right) radius versus VLA 33 GHz continuum luminosity. Slope, intercept, correlation coefficient and p-values (likelihood that there is no correlation) are also given. From Kamali *et al.* (2017).

5. A caveat when analyzing 22 GHz H$_2$O maser emission

Arp 220 with its extreme infrared luminosity ($L_{IR} \gtrsim 10^{12}\,L_\odot$) is often taken as the prototypical ULIRG. With the characteristic ratio of 10^{-9} between (isotropic) 22 GHz H$_2$O and infrared luminosity (e.g., fig. 9 in Henkel *et al.* 2005), we could thus expect strong maser emission with $L_{H_2O} \approx 1000\,L_\odot$, yielding a flux density of $\approx$20 mJy, if evenly distributed over a velocity range of 400 km s^{-1}. Even in case of a deviation from this rule by a full factor of ten, 22 GHz H$_2$O maser lines are characterized by narrow spikes (e.g., Fig. 4), which would likely compensate for this effect. Furthermore, Arp 220 is characterized by a star formation rate amounting to a few 100 M$_\odot$ yr^{-1} (e.g., Kennicutt 1998). Therefore there should be many star formation related H$_2$O masers, so many, that deviations from the 10^{-9}-rule might be minimized. In this respect it is noteworthy that 22 GHz maser emission from this merging galaxy pair has only been detected recently (Zschaechner *et al.* 2016). The reason is absorption by the NH$_3$ $(J, K) = (3,1)$ transition. The NH$_3$ (3,1) line at 22.234506 GHz with levels 165 K above the ground state and the H$_2$O line at 22.235080 GHz, 645 K above the ground state, are separated by only $\Delta V \approx$ 8 km s^{-1}. Zschaechner *et al.* (2016) find approximately –60 Jy km s^{-1} for the western and +60 Jy km s^{-1} for the eastern nucleus of Arp 220 (the absolute values are the same within $\approx$5%), thus canceling but a negligible fraction of the total signal if viewed by a single-dish telescope. Toward the western nucleus NH$_3$ absorption dominates, while toward the eastern one emission, likely due to H$_2$O, is mainly seen. Apparently, when observing ULIRGs, high resolution data are mandatory, because in such galaxies gas densities and kinetic temperatures tend to be high, thus providing suitable conditions for the presence of non-metastable ammonia transitions interfereing with the 22 GHz maser emission from H$_2$O if there is a continuum which can be absorbed. For an evaluation of the intensity of the NH$_3$ (3,1) line, then the detection of other non-metastable cm-wave NH$_3$ inversion transitions will be mandatory.

6. Other H$_2$O maser lines

With the 22 GHz H$_2$O line detected in 178 galaxies (see the contribution by J. Braatz), there are also other promising lines to be observed. Humphreys *et al.* (2005) (NGC 3079)

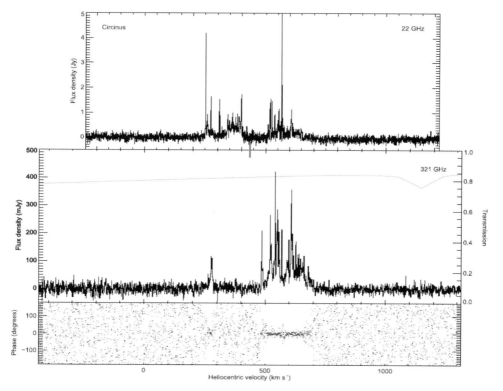

Figure 4. 22 GHz (upper panel) and 321 GHz (lower panel) H_2O maser spectra from the Circinus galaxy. The bottom panel provides a phase plot for the calibrated 321 GHz spectrum. The short vertical line connecting upper and lower panel denotes the systemic velocity. From Pesce *et al.* (2016)

and Cernicharo *et al.* 2006 (Arp 220) opened this field by observing the 183 GHz $3_{13} \rightarrow 2_{20}$ transition which is unlike the $6_{16} \rightarrow 5_{23}$ transition at 22 GHz not connecting states 645 K, but levels only 200 K above the ground state. Thus emission is expected to be more widespread than at 22 GHz, but is also affected by atmospheric obscuration, so that a non-zero redshift helps to enhance sensitivity. In recent years several additional studies have been carried out (Hagiwara *et al.* 2013b, 2016; Galametz *et al.* 2016, Humphreys *et al.* 2016, Pesce *et al.* 2016), focussing on this 183 GHz line but also on the 321 GHz $10_{29} \rightarrow 9_{36}$ transition, $\approx$1850 K above the ground state and thus only tracing extremely highly excited gas. Studied sources (see also the contribution by D. Pesce in this volume) are the Circinus galaxy, where the 321 GHz line is covering a similar velocity range as the 22 GHz line (perhaps even a slightly wider one, see Fig. 4), and NGC 4945, where the 183 GHz line is covering the entire velocity range, while the higher excitation gas sampled by the 321 GHz transitions is only seen at the upper end of the galaxy's velocity range, near $V \approx 700 \, \mathrm{km \, s^{-1}}$. What is still missing here are high resolution observations of these lines allowing for a detailed comparison with the 22 GHz data.

7. 1.7 GHz OH versus 22 GHz H_2O

OH and H_2O megamasers appear to be mutually exclusive, likely because they trace very different physical environments. Wiggins *et al.* (2016) searched with the Green Bank Telescope (GBT) for luminous H_2O maser emission in OH megamaser galaxies and

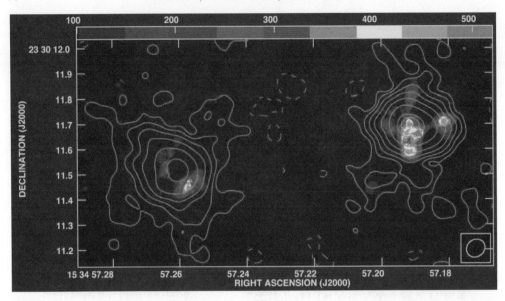

Figure 5. A composite map of the continuum (contours) and the 4.8 GHz formadehyde emission of Arp 220 (from Baan *et al.* 2017). Contour levels are 0.3 × (−1, 1, 2, 4, 8, 16, 32, 64, and 80) mJy beam^{-1} with peak flux densities of ∼13 and 30 mJy beam^{-1} for the eastern and western nucleus, respectively. A temperature scale in units of mJy km s^{-1} beam^{-1} for the line emission is given at the upper edge of the image. The synthesized beam is indicated in the lower right.

confirmed, after IC 694, with II Zw 96 a second such object. Measuring instead OH in galaxies with strong H_2O masers using the Effelsberg telescope and the GBT, two of the sample galaxies were detected in OH, but in absorption. Details of this latter survey can be found in the contribution by E. Ladu, in this volume.

8. Formaldehyde (H_2CO)

Mangum *et al.* (2008, 2013) observed, with the Green Bank 100-m telescope, 56 star forming galaxies in the 4.8 and 14.5 GHz K-doublet transitions of formaldehyde (H_2CO). Both transitions were detected in 13 targets, mostly in absorption but sometimes also in emission. Applying Occam's Razor and assuming quasi-thermal and not maser radiation in case of the emission lines, all data could be nicely fitted with a Large Velocity Gradient model, yielding densities in the range $10^{4.5...5.5}$ cm^{-3}. Nevertheless, high resolution data from three galaxies with emission lines indicate that brightness temperatures can greatly exceed in some regions those expected in case of thermal emission, reaching values of $10^{4...5}$ K (Baan *et al.* 2017). The three galaxies are IC 860, IR 15107+0724 and Arp 220 (see Fig. 5 for an image). As a consequence, it would be interesting to evaluate in how far this affects the analysis by Mangum *et al.* (2080, 2013), which is mainly but not entirely based on H_2CO 4.8 and 14.5 GHz absorption lines.

9. Methanol, HC$_3$N, and HCN

In the Galaxy, the so-called Class II 6.7 GHz transition exhibits the strongest methanol maser lines, being directly associated with sites of massive star formation (e.g., Menten 1991). Searches for this line in extragalactic space were, however, not very successful, only providing maser detections in the Magellanic Clouds and the Andromeda galaxy (see Sect. 2). After unsuccessfully searching for Class II lines much stronger than those encountered in the Galaxy, Class I masers, being clearly less conspicuous in the Galaxy and less directly associated with sites of massive star formation, have become part of a recent survey. And here, surprisingly, emission much more luminous than the corresponding Galactic masers could be found (Ellingsen *et al.* 2014, 2017a; Chen *et al.* 2015, 2016; McCarthy et al. 2017). What makes these detections peculiar is that the masers are not associated with the very center of their parent galaxies, but are instead detected in the outskirts of their nuclear environments. A detailed account of these new findings, including the possible detection of an HC_3N $J = 4{\rightarrow}3$ maser (Ellingsen *et al.* 2017b), is given by the contributions of S. Ellingsen, X. Chen, and T. McCarthy in this volume. The presence of a weakly masing HCN $J = 1{\rightarrow}0$ line right at the center of the whirlpool galaxy M 51, in a region with relatively weak CO emission, has been proposed by Matsushita *et al.* (2015).

10. The bright future

Interestingly, and possibly for the first time in this series of meetings, extragalactic OH was not a topic of much discussion. However, this will likely change in the forthcoming years. With Apertif (Aperture Tile in Focus), MeerKat (Meer Karoo Array Telescope), ASKAP (Australia Square Kilometer Array Pathfinder) and FAST (Five hundred meter Aperture Spherical Telescope), new OH surveys sometimes piggybacking on Hɪ measurements, will cover significant parts of the sky. This will greatly help in obtaining new detections of luminous OH masers and to reach eventually an independent estimate of the number of merger galaxies as a function of redshift.

With the Very Large Array (VLA) and possibly also with the Atacama Large Millimeter/submillimeter Array (ALMA), new extragalactic Class I methanol maser surveys can be carried out. Here we are still near the start. While the standard lines near 36 and 44 GHz have been measured in a small number of galaxies, higher frequency Class I masers detectable with ALMA have, to our knowledge, not even been touched.

With respect to 22 GHz H_2O masers, we note that the MCP is close to completion. We can expect a final Hubble constant deduced from this survey with an uncertainty of only a few percent during the next one or two years. However, this will be by no means the end of H_2O megamaser research. Not only other H_2O lines are and will remain attractive. The 22 GHz maser line itself may remain at the center of interest. The reason is that detection rates of future 22 GHz maser surveys have the potential to go up dramatically. So far, average detection rates are well below 10%. Detection rates of disk-masers are even close to or less than 1%. All this may change with the introduction of new criteria, either led by X-ray spectroscopy, by the radio continuum luminosity, or by a combination of the two methods. At the same time, the James Webb Space Telescope (JWST), ALMA and NOEMA (NOrthern Extended Millimeter Array) may provide new constraints to the masses of nuclear engines, to be compared with those deduced from the maser-disk data.

References

Araya, E., Baan, W. A., & Hofner, P. 2004, *ApJS*, 154, 541
Baan, W. A., Wood, P. A. D., & Haschick, A. D. 1982 *ApJ*, 260, L49
Baan, W. A., Güsten, R., & Haschick, A. D. 1986, *ApJ*, 305, 830
Baan, W. A., Haschick, A. D., & Uglesich, R. 1993, *ApJ*, 415, 140
Baan, W. A., An, T., Klöckner, H.-R., & Thomasson, P. 2017, *MNRAS*, 469, 916
Barvainis, R. & Antonucci, R. 2005, *ApJ*, 628, L89
Bennert, N., Barvainis, R., Henkel, C., & Antonucci, R. 2009, *ApJ*, 695, 276

Braatz, J. A., Wilson, A. S., & Henkel, C. 1996, *ApJS*, 106, 51

Braatz, J. A., Wilson, A. S., & Henkel, C. 1997, *ApJS*, 110, 321

Braatz, J. A., Reid, M. J., & Hunphreys, E. M. L., *et al.* 2010, *ApJ*, 718, 657

Breen, S. L., Lovell, J. E. J., Ellingsen, S. P., *et al.* 2013, *MNRAS*, 432, 1382

Castangia, P., Impellizzeri, C. M. V., McKean, J. P., *et al.* 2011, *A&A*, 529, A150

Cernicharo, J., Pardo, J. R., & Weiß, A. 2006, *ApJ*, 646, L49

Chen, X., Ellingsen, S. P., Baan, W. A., *et al.* 2015, *ApJ*, 800, L2

Chen, X., Ellingsen, S. P., Zhang, J. S., *et al.* 2016, *MNRAS*, 459, 357

Churchwell, E., Witzel, A., Huchtmeier, W., *et al.* 1977, *A&A*, 54, 969

Darling, J. 2011, *ApJ*, 732, L2

Darling, J. & Giovanelli, R. 2002, *AJ*, 124, 100

Darling, J., Gerard, B., Amiri, N., & Lawrence, K. 2016, *ApJ*, 826, 24

Dos Santos, P. M. & Lepine, J. R. D. 1979, *Nature*, 278, 34

Ellingsen, S. P., Chen, X., & Qiao, H.-H., *et al.* 2014, *ApJ*, 790, L28

Ellingsen, S. P., Chen, X., Breen, S. L., & Qiao, H.-H. 2017a, *MNRAS*, 472, 604

Ellingsen, S. P., Chen, X., Breen, S. L., & Qiao, H.-H. 2017b, *ApJ*, 841, L14

Galametz, M., Zhang, Z.-Y., Immer, K., *et al.* 2016, *MNRAS*, 462, L36

Gao, F., Braatz, J. A., Reid, M. J., *et al.* 2016, *ApJ*, 817, 128

Gao, F., Braatz, J. A., Reid, M. J., *et al.* 2017, *ApJ*, 834, 52

Gardner, F. F. & Whiteoak, J. B. 1975, *MNRAS*, 173, 77p

Ginsburg, A., Walsh, A., Henkel, C., *et al.* 2015, *A&A* 584, L7

Greene, J. E., Peng, C. Y., Kim, M. *et al.* 2010, *ApJ*, 721, 26

Greene, J. E., Seth, A., & den Brok, M., *et al.* 2013, *ApJ*, 771, 121

Greene, J. E., Seth, A., Kim, M., *et al.* 2016, *ApJ*, 826, L32

Greenhill, L. J., Henkel, C., Becker, R., Wilson, T. L., & Wouterloot, J. G. A. 1995, *A&A*, 304, 21

Hagiwara, Y., Doi, A., & Hachisuka, K. 2013a, *ASP Conf. Ser.*, 476, 295

Hagiwara, Y., Miyoshi, M., Doi, A., & Horiuchi, S. 2013b, *ApJ*, 768, L38

Hagiwara, Y., Horiuchi, S., Doi, A., Miyoshi, M., & Edwards, P. G. 2016, *ApJ*, 827, 69

Haschick, A. D., Baan, W. A., & Peng, E. W. 1994, *ApJ*, 437, L35

Henkel, C., Wang, Y. P., Falcke, H., Wilson, A. S., & Braatz, J. A. 1998, *A&A*, 335, 463

Henkel, C., Peck, A. B., Tarchi, A., *et al.* 2005, *A&A*, 436, 75

Henkel, C., Braatz, J. A., Reid, M. J., *et al.* 2012, *IAUS*, 336, 301

Herrnstein, J. R., Moran, J. M., Greenhill, L. J., *et al.* 1999, *Nature*, 400, 539

Humphreys, E. M. L., Greenhill, L. J., & Reid, M. J. 2005, *ApJ*, 634, L133

Humphreys, E. M. L., Vlemmings, W. H. T., & Impellizzeri, C. M. V., *et al.* 2016, *A&A*, 592, L13

Imai, H., Katayama, Y., Ellingsen, S. P., & Hagiwara, Y. 2013, *PASJ*, 65, 28

Impellizzeri, C. M. V., McKean, J. P., Castangia, P., *et al.* 2008, *Nature*, 456, 927

Johanson, A. K., Migenes, V., & Breen, S. L. 2014, *ApJ*, 781, 78

Kamali, F., Henkel, C., Brunthaler, A., *et al.* 2017, *A&A*, 605, A84

Kennicutt, R. C. 1998, *ARAA*, 36, 189

König, S., Eckart, A., Henkel, C., & García-Marín, M. 2012, *MNRAS*, 420, 2263

Kuo, C. Y., Braatz, J. A., Condon, J. J., *et al.* 2011, *ApJ*, 727, 20

Kuo, C. Y., Braatz, J. A., Reid, M. J., *et al.* 2013, *ApJ*, 767, 154

Kuo, C. Y., Braatz, J. A., Lo, K. Y., *et al.* 2015, *ApJ*, 800, 26

Kylafis, N. D. & Norman, C. A. 1991, *ApJ*, 373, 525

Läsker, R., Greene, J. E., Seth, A., *et al.* 2016, *ApJ*, 825, 3

Liu, Z. W., Zhang, J. S., Henkel, C., *et al.* 2017, *MNRAS*, 466, 1608

Mangum, J. G., Darling, J., Menten, K. M., & Henkel, C. 2008, *ApJ*, 673, 832

Mangum, J. G., Darling, J., Henkel, C., & Menten, K. M. 2013, *ApJ*, 766, 108

Maoz, E. & McKee, C. F. 1998, *ApJ*, 494, 218

Matsushita, S., Trung, D.-V., Boone, F., *et al.* 2015, *ApJ*, 799, 26

McCarthy, T. P., Ellingsen, S. P., Chen, X., *et al.* 2017, *ApJ*, 846, 156

McKean, J. P., Impellizzeri, C. M. V., Roy, A. L., *et al.* 2011, *MNRAS*, 410, 2506

Menten, K. M. 1991, *ApJ*, 380, L75

Miyoshi, M., Moran, J., Herrnstein, J., *et al.* 1995, *Nature*, 373, 127

Nakai, N., Inoue, M., & Miyoshi, M. 1993, *Nature*, 361, 45

Pesce, D. W., Braatz, J. A., Condon, J. J., *et al.* 2015, *ApJ*, 810, 65

Pesce, D. W., Braatz, J. A., & Impellizzeri, C. M. V. 2016, *ApJ*, 827, 68

Pjanka, P., Greene, J. E., Seth, A. C., *et al.* 2017, *ApJ*, 844, 165

Reid, M. J., Braatz, J. A., Condon, J. J., *et al.* 2009 *ApJ*, 695, 287

Reid, M. J., Braatz, J. A., Condon, J. J., *et al.* 2013, *ApJ*, 767, 154

Scalise, E. & Braz, M. A. 1982, *A&A*, 87, 528

Sjouwerman, L. O., Murray, C. E., Pihlström, Y. M., Fish, V. L, & Araya, E. D. 2010, *ApJ*, 724, 158

Tarchi, A., Henkel, C., Chiaberge, M., & Menten, K. M. 2003, *A&A*, 407, L33

Tarchi, A., Castangia, P., Columbano, A., Panessa, F., & Braatz, J. A. 2011, *A&A*, 532, A125

van den Bosch, R. C. E., Greene, J. E., Braatz, J. A., Constantin, A., & Kuo, C.-Y. 2016, *MNRAS*, 455, 158

Wagner, J. 2013, *A&A*, 560, A12

Wiggins, B. K., Migenes, V., & Smidt, J. M. 2016, *ApJ*, 816, 55

Yamauchi, A., Miyamoto, Y., Nakai, N., *et al.* 2017, *PASJ*, 69, L6

Zhang, J. S., Henkel, C., Guo, Q., & Wang, J. 2012, *A&*, 538, A152

Zhang, J. S., Liu, Z. W., Henkel, C., Wang, J. Z., & Coldwell, G. V. 2017, *ApJ*, 836, L20

Zschaechner, L. K., Ott, J., Walter, F., *et al.* 2016, *ApJ*, 833, 41

Astrophysical Masers:
Unlocking the Mysteries of the Universe
Proceedings IAU Symposium No. 336, 2017
A. Tarchi, M.J. Reid & P. Castangia, eds.

© International Astronomical Union 2018
doi:10.1017/S1743921318000133

Progress toward an accurate Hubble Constant

Sherry H. Suyu[1,2,3]

[1] Max-Planck-Institut für Astrophysik, Karl-Schwarzschild-Str. 1, 85748 Garching, Germany
email: `suyu@mpa-garching.mpg.de`

[2] Institute of Astronomy and Astrophysics, Academia Sinica, P.O. Box 23-141, Taipei 10617, Taiwan

[3] Physik-Department, Technische Universität München, James-Franck-Straße 1, 85748 Garching, Germany

Abstract. The Hubble constant is a key cosmological parameter that sets the present-day expansion rate as well as the age, size, and critical density of the Universe. Intriguingly, there is currently a tension in the measurements of its value in the standard flat ΛCDM model – observations of the Cosmic Microwave Background with the Planck satellite lead to a value of the Hubble constant that is lower than the measurements from the local Cepheids-supernovae distance ladder and strong gravitational lensing. Precise and accurate Hubble constant measurements from independent probes, including water masers, are necessary to assess the significance of this tension and the possible need of new physics beyond the current standard cosmological model. We present the progress toward an accurate Hubble constant determination.

Keywords. gravitational lensing, masers, (cosmology:) cosmic microwave background, (cosmology:) cosmological parameters, (cosmology:) distance scale

1. Introduction

The standard cosmological model "flat ΛCDM", consisting of dark energy (with density characterized by a cosmological constant Λ) and cold dark matter (CDM) in a spatially flat Universe, has emerged in the past decade. This simple model has yielded excellent fit to various cosmological observations, including the temperature anisotropies in the cosmic microwave background (CMB) and galaxy density correlations in baryon acoustic oscillations (BAO). Recent CMB experiments, particularly the Wilkinson Microwave Anisotropy Probe (WMAP; Komatsu *et al.* 2011, Hinshaw *et al.* 2013) and the Planck satellite (Planck Collaboration 2014, 2016), and BAO surveys (e.g., Anderson *et al.* 2014, Ross *et al.* 2015, Kazin *et al.* 2014) have provided stringent constraints on cosmological parameters in the flat ΛCDM model.

The Hubble constant (H_0) is a key cosmological parameter that sets the present-day expansion rate as well as the age, size, and critical density of the Universe. Intriguingly, Planck's value of $H_0 = 67.8 \pm 0.9 \, \mathrm{km \, s^{-1} \, Mpc^{-1}}$ (Planck Collaboration 2016), from Planck temperature data and Planck lensing under the flat ΛCDM model, is lower than recent direct measurements based on the distance ladder, of $73.24 \pm 1.74 \, \mathrm{km \, s^{-1} \, Mpc^{-1}}$ (Riess *et al.* 2016) and of $74.3 \pm 2.1 \, \mathrm{km \, s^{-1} \, Mpc^{-1}}$ (Freedman *et al.* 2012). Planck does not directly measure H_0, but rather enables its indirect inference through measurements of combinations of cosmological parameters given assumptions of the background cosmological model. On the other hand, Planck's H_0 value is similar to the results of some of the megamaser measurements (e.g., $H_0 = 68.9 \pm 7.1 \, \mathrm{km \, s^{-1} \, Mpc^{-1}}$ from Reid *et al.* 2013, and $H_0 = 66.0 \pm 6.0 \, \mathrm{km \, s^{-1} \, Mpc^{-1}}$ from Gao *et al.* 2016).

A 1% *direct* measurement of the Hubble constant would help address the possible tension with the CMB value which, if significant, would point towards deviations from the standard flat ΛCDM and new physics. For example, when one relaxes the flatness or Λ assumption in the CMB analysis, strong parameter degeneracies between H_0 and other cosmological parameters appear, and the degenerate H_0 values from the CMB become compatible with the local H_0 measurements from the distance ladder (Planck Collaboration 2016, Riess *et al.* 2016). Thus, a 1% measurement of H_0 is important for understanding the nature of dark energy, neutrino physics, the spatial curvature of the Universe and the validity of General Relativity (e.g., Hu 2015, Suyu *et al.* 2012, Weinberg *et al.* 2013). For example, the dark energy figure of merit of any survey that does not directly measure H_0 improves by $\sim 40\%$ if H_0 is known to 1%.

In the following, we describe a few of the various ways of measuring the Hubble constant, and refer the readers to recent reviews on the Hubble constant for a more comprehensive overview.

2. Cosmic Microwave Background

The CMB radiation has a thermal black body spectrum, with small temperature fluctuations in different directions of the sky. The power spectrum of the CMB temperature anisotropies displays acoustic peaks, and the shape of the power spectrum provides information on our cosmological model.

In particular, the ratios of the peak heights, such as from the first three peaks, place constraints on $\Omega_m h^2$ and $\Omega_b h^2$ (e.g., Hu *et al.* 1997), where Ω_m is the total matter density, Ω_b is the baryon density, and h is the dimensionless parameter of the Hubble constant $H_0 = 100h \, \mathrm{km \, s^{-1} \, Mpc^{-1}}$. Without further model assumptions, one cannot extract h precisely as it is not directly constrained by the CMB observations and is thus degenerate with other cosmological parameters.

By assuming that our Universe is the standard spatially-flat 6-parameter ΛCDM cosmology, then the (angular) locations of the acoustic peaks allow the determination of the acoustic scale, θ_*, which is the ratio of the comoving size of the sound horizon at the time of last-scattering (r_s) and the angular diameter distance at which we are observing the fluctuations (D_A), i.e., $\theta_* = r_s/D_A$. Since D_A and r_s depend on cosmological model parameters, the measurement of θ_* places tight constraint on the following combination of parameters:

$$\Omega_m h^{3.2} (\Omega_b h^2)^{-0.54} \tag{2.1}$$

(e.g., Percival *et al.* 2002, Planck Collaboration 2014).

By combining equation (2.1) with the $\Omega_m h^2$ and $\Omega_b h^2$ constraints, the Planck Collaboration obtained a precise measurement of the Hubble constant of $h = 0.678 \pm 0.009$ (Planck Collaboration 2016). We stress that this assumes the flat ΛCDM model (for which equation (2.1) is derived), and the value of H_0 from the CMB changes markedly when the assumption of the flat ΛCDM model is relaxed.

3. Megamasers

Water masers in orbit around an active galactic nucleus (AGN) provide a geometric approach to measuring the Hubble constant. By observing the Doppler shifts in the maser lines, one could measure the velocity v_r of the masers in orbit around the AGN, and also their angular positions θ_r from the central AGNs. Furthermore, by observing the change in velocities of the "systemic masers" (which are the masers located in front of the AGN for nearly edge-on maser disks), the acceleration a_r could be determined. This provides a measurement of the physical size of the disk r, since $a_r = v_r^2/r$. This physical size could then be compared to the angular size, to derive the angular diameter distance to the maser: $D = r/\theta_r = v_r^2/(a_r\theta_r)$. Through the distance-redshift relation, the angular diameter distance then provides a measurement of H_0.

The Megamaser Cosmology Project (MCP) aims to determine H_0 precisely via measurements of geometric distances to galaxies in the Hubble flow. The Hubble constant based on the analysis of four megamaser galaxies in the MCP is $H_0 = 69.3 \pm 4.2\,\mathrm{km\,s^{-1}\,Mpc^{-1}}$, from UGC 3789 with $H_0 = 76\pm8\,\mathrm{km\,s^{-1}\,Mpc^{-1}}$ (Reid et al. 2013 with updates), NGC 6264 with $H_0 = 68 \pm 9\,\mathrm{km\,s^{-1}\,Mpc^{-1}}$ (Kuo et al. 2013), NGC 6323 with $H_0 = 73\pm26\,\mathrm{km\,s^{-1}\,Mpc^{-1}}$ (Kuo et al. 2015) and NGC 5765b with $H_0 = 66\pm6\,\mathrm{km\,s^{-1}\,Mpc^{-1}}$ (Gao et al. 2016). There are several more megamaser galaxies in the MCP sample that are being analysed. We refer to the contributions by J. Braatz for more details on the MCP.

4. Extragalactic Distance Ladder

The distance ladder has provided precise measurements of the Hubble constant. In fact, it was the method used in the *Hubble Space Telescope* (*HST*) Key Project that yielded the Hubble constant with 10% uncertainty (Freedman et al. 2001), resolving the "factor-of-two" controversy in the Hubble constant that lasted decades.

By measuring distances (d) to faraway objects in the Hubble flow (where peculiar velocities are negligible) and also their recessional velocities (v) via redshifts, the Hubble constant can be inferred through Hubble's law $v = H_0 d$. However, distance measurements to such faraway objects are difficult to obtain directly. Thus, a practical approach is to measure absolute distances to nearby objects (e.g., through parallax), and then use methods to measure relative distances (such as supernovae) to further away objects. This builds a "ladder" to obtain distances to faraway object in the Hubble flow.

In the Carnegie Hubble Program, Freedman et al. (2012) calibrated the Cepheid distance scale using mid-infrared observations. Combining this with data from the *HST* Key Project, Freedman et al. (2012) obtained $H_0 = 74.3\,\mathrm{km\,s^{-1}\,Mpc^{-1}}$ with a systematic uncertainty of $\pm2.1\,\mathrm{km\,s^{-1}\,Mpc^{-1}}$.

In the SH0ES program, Riess et al. (2016) more than doubled the sample of reliable Type Ia supernovae (SNe Ia) having a Cepheid-calibrated distance, allowing a reduction in the systematic uncertainties in the H_0 measurement. Using the distance measurement to the maser galaxy NGC 4258, Cepheids in the Large Magellanic Cloud, the Milky Way and M31, and ~ 300 SNe Ia at $z < 0.15$, Riess et al. (2016) obtained $H_0 = 73.24 \pm 1.74\,\mathrm{km\,s^{-1}\,Mpc^{-1}}$.

As an alternative to the traditional Cepheids distance ladder, Beaton et al. (2016) are carrying out the Carnegie-Chicago Hubble Program with the goal of reaching a 3% measurement of H_0. This distance ladder uses RR Lyrae variables, the tip of the red giant branch and SNe Ia, providing an independent cross-check of the traditional Cepheids distance ladder.

5. Strong lensing time delays

Strong gravitational lensing occurs when there is a chance alignment of a massive object along the line of sight to a background source. The light from the background source gets deflected by the gravitational field of the foreground object such that multiple distorted images of the background source appear around the foreground lens. When the source is one that varies in its luminosity, such as an activie galactic nucleus or supernova, the variability of the source manifests in each of the multiple images but delayed in time due to the different light paths (e.g., Vanderriest *et al.* 1989, Kochanek *et al.* 2006, Courbin *et al.* 2011).

Strong gravitational lenses with measured time delays between the multiple images provide a competitive approach to measuring the Hubble constant, completely independent of the distance ladder (Refsdal 1964). The time delay (Δt) depends on the "time-delay distance" ($D_{\Delta t}$) and the lens mass distribution. This time-delay distance is a combination of the three angular diameter distances in lensing (observer-source distance $D_{\rm s}$, observer-lens distance $D_{\rm d}$, and lens-source distance $D_{\rm ds}$). As a result, $D_{\Delta t}$ is inversely proportional to H_0 and depends weakly on other cosmological constants. In addition to $D_{\Delta t}$, the angular diameter distance to the lens $D_{\rm d}$, which is more sensitive to dark energy parameters, can be extracted from the lens system if the velocity dispersion of the foreground lens galaxy is measured and combined with the time delays (Jee *et al.* 2015).

The H0LiCOW collaboration (Suyu *et al.* 2017) obtained exquisite imaging and spectroscopic observations on 5 lensed quasars with time delays from the COSMOGRAIL collaboration (e.g., Courbin *et al.* 2011, Tewes *et al.* 2013) and radio monitoring (Fassnacht *et al.* 1999, 2002). Using the time-delay monitoring/measurements (Bonvin *et al.* 2017), lens environment studies with wide-field imaging and spectroscopy (Rusu *et al.* 2017, Sluse *et al.* 2017), and lens mass modeling with *HST* observations, Wong *et al.* (2017) measured the time-delay distance to the lens system HE0435−1223 with a precision of 7.6% via a blind analysis. Combining this with two other time-delay distance measurements (to B1608+656 and RXJ1131−1231; Suyu *et al.* 2010, 2013, 2014) and considering flat ΛCDM cosmology, Bonvin *et al.* (2017) measured $H_0 = 71.9^{+2.4}_{-3.0}\,\mathrm{km\,s^{-1}\,Mpc^{-1}}$. In addition, the exquisite data sets obtained on the H0LiCOW lenses allow studies of the lens galaxies and quasars, including the connection between black holes and their host galaxies (Ding *et al.* 2017a, 2017b).

There are two more lenses in the H0LiCOW sample that are currently being analyzed. Four more with time delays from COSMOGRAIL are getting follow-up observations. Wide-field imaging surveys including the Dark Energy Survey (Dark Energy Survey Collaboration 2005, 2016) and the Hyper-Suprime Cam Survey (Aihara *et al.* 2017) are yielding new lensed quasar systems (e.g., Agnello *et al.* 2015, Lin *et al.* 2017, Ostrovski *et al.* 2017, Sonnenfeld *et al.* 2017), with the first one being monitored and its delays measured (Courbin *et al.* 2017). With hundreds of new lens systems expected from current and future surveys, a 1% measurement of H_0 from lensing time delays will be achievable (Jee *et al.* 2016, Shajib *et al.* 2017).

6. Summary

Stakes are high to assess the significance of the current tension in the H_0 values from some of the cosmological probes. Independent methods to measure H_0 with 1% precision and accuracy are necessary to overcome systematic effects for verifying or falsifying the standard cosmological paradigm.

Acknowledgements

SHS thanks E. Komatsu for helpful discussions on the Hubble constant from observations of the Cosmic Microwave Background, A. Riess for information on the SH0ES program, and the organizers for the wonderful symposium. SHS gratefully acknowledges the support from the Max Planck Society through the Max Planck Research Group.

References

Agnello, A., Treu, T., Ostrovski, F., *et al.* 2015, *MNRAS*, 454, 1260

Aihara, H., Arimoto, N., Armstrong, R., *et al.* 2017, ArXiv e-prints (1704.05858)

Anderson, L., Aubourg, É., Bailey, S., *et al.* 2014, *MNRAS*, 441, 24

Beaton, R. L., Freedman, W. L., Madore, B. F., *et al.* 2016, *ApJ*, 832, 210

Bonvin, V., Courbin, F., Suyu, S. H., *et al.* 2017, *MNRAS*, 465, 4914

Courbin, F., Chantry, V., Revaz, Y., *et al.* 2011, *A&A*, 536, A53

Courbin, F., Bonvin, V., Buckley-Geer, E., *et al.* 2017, ArXiv e-prints (1706.09424)

Dark Energy Survey Collaboration *et al.* 2016, *MNRAS*, 460, 1270

Ding, X., Liao, K., Treu, T., *et al.* 2017a, *MNRAS*, 465, 4634

Ding, X., Treu, T., Suyu, S. H., *et al.* 2017b, *MNRAS*, 472, 90

Fassnacht, C. D., Pearson, T. J., Readhead, A. C. S., Browne, I. W. A., Koopmans, L. V. E., Myers, S. T., & Wilkinson, P. N. 1999, *ApJ*, 527, 498

Fassnacht, C. D., Xanthopoulos, E., Koopmans, L. V. E., & Rusin, D. 2002, *ApJ*, 581, 823

Freedman, W. L., Madore, B. F., Scowcroft, V., Burns, C., Monson, A., Persson, S. E., Seibert, M., & Rigby, J. 2012, *ApJ*, 758, 24

Freedman, W. L., Madore, B. F., Gibson, B. K., *et al.* 2001, *ApJ*, 553, 47

Gao, F., Braatz, J. A., Reid, M. J., *et al.* 2016, *ApJ*, 817, 128

Hinshaw, G., Larson, D., Komatsu, E., *et al.* 2013, *ApJS*, 208, 19

Hu, W. 2005, in Astronomical Society of the Pacific Conference Series, Vol. 339, Observing Dark Energy, ed. S. C. Wolff & T. R. Lauer, 215

Hu, W., Sugiyama, N., & Silk, J. 1997, *Nature*, 386, 37

Jee, I., Komatsu, E., & Suyu, S. H. 2015, *JCAP*, 11, 033

Jee, I., Komatsu, E., Suyu, S. H., & Huterer, D. 2016, *JCAP*, 4, 031

Kazin, E. A., Koda, J., Blake, C., *et al.* 2014, *MNRAS*, 441, 3524

Kochanek, C. S., Morgan, N. D., Falco, E. E., McLeod, B. A., Winn, J. N., Dembicky, J., & Ketzeback, B. 2006, *ApJ*, 640, 47

Komatsu, E., Smith, K. M., Dunkley, J., *et al.* 2011, *ApJS*, 192, 18

Kuo, C. Y., Braatz, J. A., Reid, M. J., Lo, K. Y., Condon, J. J., Impellizzeri, C. M. V., & Henkel, C. 2013, *ApJ*, 767, 155

Kuo, C. Y., Braatz, J. A., Lo, K. Y., *et al.* 2015, *ApJ*, 800, 26

Lin, H., Buckley-Geer, E., Agnello, A., *et al.* 2017, *ApJL*, 838, L15

Ostrovski, F., McMahon, R. G., and Connolly, A. J., *et al.* 2017, *MNRAS*, 465, 4325

Percival, W. J., Sutherland, W., Peacock, J. A., *et al.* 2002, *MNRAS*, 337, 1068

Planck Collaboration *et al.* 2014, *A&A*, 571, A16

—. 2016, *A&A*, 594, A13

Refsdal, S. 1964, *MNRAS*, 128, 307

Reid, M. J., Braatz, J. A., Condon, J. J., Lo, K. Y., Kuo, C. Y., Impellizzeri, C. M. V., & Henkel, C. 2013, *ApJ*, 767, 154

Riess, A. G., Macri, L. M., Hoffmann, S. L., *et al.* 2016, *ApJ*, 826, 56

Ross, A. J., Samushia, L., Howlett, C., Percival, W. J., Burden, A., & Manera, M. 2015, *MNRAS*, 449, 835

Rusu, C. E., Fassnacht, C. D., Sluse, D., *et al.* 2017, *MNRAS*, 467, 4220

Shajib, A. J., Treu, T., & Agnello, A. 2017, ArXiv e-prints (1709.01517)

Sluse, D., Sonnenfeld, A., Rumbaugh, N., *et al.* 2017, *MNRAS*, 470, 4838

Sonnenfeld, A., Chan, J. H. H., Shu, Y., *et al.* 2017, ArXiv e-prints (1704.01585)

Suyu, S. H., Marshall, P. J., Auger, M. W., Hilbert, S., Blandford, R. D., Koopmans, L. V. E., Fassnacht, C. D., & Treu, T. 2010, *ApJ*, 711, 201

Suyu, S. H., Treu, T., Blandford, R. D., *et al.* 2012, ArXiv e-prints (1202.4459)

Suyu, S. H., Auger, M. W., Hilbert, S., *et al.* 2013, *ApJ*, 766, 70

Suyu, S. H., Treu, T., Hilbert, S., *et al.* 2014, *ApJL*, 788, L35

Suyu, S. H., Bonvin, V., Courbin, F., *et al.* 2017, *MNRAS*, 468, 2590

Tewes, M., Courbin, F., Meylan, G., *et al.* 2013, *A&A*, 556, A22

The Dark Energy Survey Collaboration. 2005, ArXiv Astrophysics e-prints (astro-ph/0510346)

Vanderriest, C., Schneider, J., Herpe, G., Chevreton, M., Moles, M., & Wlerick, G. 1989, *A&A*, 215, 1

Weinberg, D. H., Mortonson, M. J., Eisenstein, D. J., Hirata, C., Riess, A. G., & Rozo, E. 2013, *Phys. Rep.*, 530, 87

Wong, K. C., Suyu, S. H., Auger, M. W., *et al.* 2017, *MNRAS*, 465, 4895

Astrophysical Masers:
Unlocking the Mysteries of the Universe
Proceedings IAU Symposium No. 336, 2017
A. Tarchi, M.J. Reid & P. Castangia, eds.

© International Astronomical Union 2018
doi:10.1017/S1743921317010249

A Measurement of the Hubble Constant by the Megamaser Cosmology Project

James Braatz[1], James Condon[1], Christian Henkel[2,3], Jenny Greene[4], Fred Lo[1], Mark Reid[5], Dominic Pesce[1,6], Feng Gao[7], Violette Impellizzeri[8], Cheng-Yu Kuo[9], Wei Zhao[7], Anca Constantin[10], Lei Hao[7] and Eugenia Litzinger[11,12]

[1] National Radio Astronomy Observatory, 520 Edgemont Road, Charlottesville, VA 22903, USA

[2] Max-Planck-Institut für Radioastronomie, Auf dem Hügel 69, D-53121 Bonn, Germany

[3] Astronomy Department, Faculty of Science, King Abdulaziz University, P.O. Box 80203, Jeddah 21589, Saudi Arabia

[4] Department of Astrophysics, Princeton University, Princeton, NJ, 08544, USA

[5] Harvard-Smithsonian Center for Astrophysics, 60 Garden Street, Cambridge, MA 02138, USA

[6] Department of Astronomy, University of Virginia, 530 McCormick Road, Charlottesville, VA 22904, USA

[7] Shanghai Astronomical Observatory, Chinese Academy of Sciences, 200030 Shanghai, China

[8] Joint ALMA Office, Alonso de Cordova 3107, Vitacura, Santiago, Chile

[9] Department of Physics, National Sun Yat-Sen University, No. 70, Lianhai Road, Gushan Dist., 804 Kaohsiung City, Taiwan, R.O.C.

[10] Department of Physics and Astronomy, James Madison University, Harrisonburg, VA 22807, USA

[11] Institut für Theoretische Physik und Astrophysik, Universität Würzburg, Emil-Fischer-Str. 31, D-97074 Würzburg, Germany

[12] Dr. Remeis Sternwarte & ECAP, Universität Erlangen-Nürnberg, Sternwartstrasse 7, D-96049 Bamberg, Germany

Abstract. The Megamaser Cosmology Project (MCP) measures the Hubble Constant by determining geometric distances to circumnuclear 22 GHz H_2O megamasers in galaxies at low redshift (z < 0.05) but well into the Hubble flow. In combination with the recent, exquisite observations of the Cosmic Microwave Background by WMAP and Planck, these measurements provide a direct test of the standard cosmological model and constrain the equation of state of dark energy. The MCP is a multi-year project that has recently completed observations and is currently working on final analysis. Based on distance measurements to the first four published megamasers in the sample, the MCP currently determines $H_0 = 69.3 \pm 4.2$ km s^{-1} Mpc^{-1}. The project is finalizing analysis for five additional galaxies. When complete, we expect to achieve a ~4% measurement. Given the tension between the Planck prediction of H_0 in the context of the standard cosmological model and astrophysical measurements based on standard candles, the MCP provides a critical and independent geometric measurement that does not rely on external calibrations or a distance ladder.

Keywords. masers, distance scale

1. Introduction

Exquisite observations of the Cosmic Microwave Background (CMB) at $z \simeq 1100$ by WMAP and Planck establish a framework for precision cosmology, but observations of the CMB on their own do not uniquely determine all fundamental cosmological parameters. Complementary observations in the local universe at $z \simeq 0$ can constrain critical parameters such as the equation of state for dark energy and the number of families of relativistic particles. Nevertheless, it is possible to use CMB results to *predict* certain observable cosmological parameters, including the Hubble Constant, but only in the context of a specific cosmological model. In this way, the base ΛCDM model of cosmology combined with Planck CMB observations makes the precise prediction: $H_0 = 67.8 \pm 0.9$ km s^{-1} Mpc^{-1} (Planck Collaboration 2016). Astrophysical measurements of H_0 then make a powerful test of the model. Recent observations based on standard candles determine $H_0 = 73.24 \pm 1.74$ km s^{-1} Mpc^{-1} (Riess *et al.* 2016) and $H_0 = 74.3 \pm 2.6$ km s^{-1} Mpc^{-1} (Freedman *et al.* 2012), in tension with the Planck result. Meanwhile, measurements from BAO + SN Ia determine $H_0 = 67.3 \pm 1.0$ km s^{-1} Mpc^{-1} (Alam *et al.* 2017), in line with the Planck prediction, and observations of gravitational lensing determine $H_0 = 71.9 ^{+2.4}_{-3.0}$ km s^{-1} Mpc^{-1} (Bonvin *et al.* 2017). The tension between predicted values of H_0 from the CMB and observed values based on standard candles motivates a concerted effort in observational cosmology to measure H_0 using multiple, independent methods to minimize systematic uncertainties, with a current goal to achieve agreement at the $\sim 3\%$ level.

The Megamaser Cosmology Project (MCP) is an international, multi-year project to measure the Hubble Constant using observations of 22 GHz water vapor megamasers in the circumnuclear accretion disks of Active Galactic Nuclei (AGN). The MCP determines geometric distances to galaxies in the range ~ 50-200 Mpc, directly in the Hubble Flow. The measurement of H_0 by the MCP is a one-step process that requires no external calibrations or distance ladders.

In addition to the determination of H_0, the MCP also measures gold-standard masses of supermassive black holes in AGNs, maps the geometry and orientation of the thin molecular accretion disk, and probes physical conditions of the disks on scales of tenths of a pc. Other contributions in this volume, including those by Henkel *et al.* and Pesce *et al.*, discuss aspects of these studies. In this paper, we focus on the progress of the MCP measurement of H_0.

2. A Survey for H_2O Disk Megamasers

Magnificent observations of the prototypical H_2O megamaser disk in NGC 4258 provide a precise distance to the galaxy (Humphreys *et al.* 2013) and make it an important anchor for the extragalactic distance scale, but at only 7.54 Mpc (Riess *et al.* 2016) it is too close for a direct measurement of H_0. However, the MCP has been conducting surveys to identify similar maser disks in the Hubble Flow. The MCP surveys target obscured AGNs, mainly Seyfert 2 galaxies, with $z < 0.05$ selected from SDSS, 6dF, and 2MRS optical spectroscopy. Additional smaller samples of galaxies have been included as survey targets based on X-ray or IR properties indicative of obscured AGN. The MCP and its pilot programs have surveyed about 2800 galaxies over nearly 10 years, and detected about 3%. The project then pursues followup studies for detections showing evidence of disk rotation.

Figure 1 shows the total number of extragalactic 22 GHz water masers detected each year. Altogether, there are 178 galaxies currently known to host 22 GHz H_2O masers. Of them, about 150 are in AGN and the others are associated with star formation in the host galaxies. About 37 of the megamasers in AGN show spectral profiles characteristic of edge-on disk masers, and 10 of those have been targeted for distance measurement observations.

MCP surveys use the sensitive Green Bank Telescope (GBT) to detect new masers. The survey strategy is to cover each candidate galaxy with a short 10-minute observation

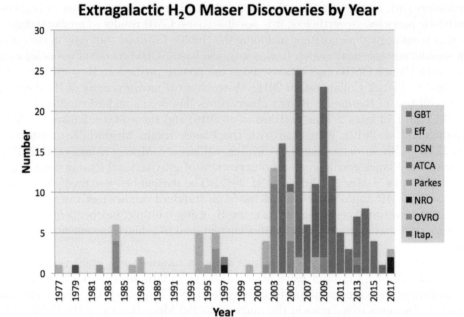

Figure 1. Extragalactic 22 GHz H_2O Maser detections, by year. The MCP used the Green Bank Telescope to conduct the large survey for new megamasers. About half of all known extragalactic 22 GHz water masers were discovered as part of the MCP or its pilot programs.

and, for detections and marginal cases, obtain a longer integration to better characterize the extent of the maser emission. The GBT has detected $\sim 65\%$ of the known 22 GHz maser galaxies, most of those through the MCP surveys. The drop off in GBT detections since 2016, evident in Figure 1, reflects the end of the MCP surveys and refocusing of resources on observations needed to measure distances to the selected disk maser galaxies.

Masers originating in an edge-on accretion disk have a characteristic shape to the spectral profile, with 3 distinct sets of maser features grouped in velocity. Maser components in the central velocity group are termed systemic masers and form along the line of sight to the central black hole, while red- and blue-shifted "high-velocity" masers originate near the disk midline, rotating away from and toward the observer, respectively. Ideal candidates for distance measurements have at least one maser component bright enough ($\gtrsim 80$ mJy) to enable VLBI self-calibration, clean Keplerian rotation profiles evident from the VLBI mapping, a large number of maser components within each group (red, blue, systemic) spread over a wide velocity range, bright and distinct systemic features that enable reliable tracking of accelerations, and accelerations in the systemic features sufficiently large ($\gtrsim 1$ km s^{-1} yr^{-1}) to track over ~ 2 years of monitoring. The MCP web page† includes an online catalog of all known extragalactic 22 GHz water maser galaxies.

3. Spectral Monitoring and VLBI observations

Once a suitable maser disk is identified, two types of additional observations are required to determine the distance: spectral monitoring and VLBI imaging. Spectral monitoring is used to measure line-of-sight accelerations of individual maser components. For each galaxy targeted for a distance measurement, we require at least two years of

† https://safe.nrao.edu/wiki/bin/view/Main/MegamaserCosmologyProject

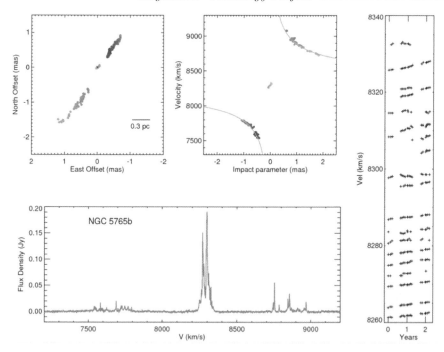

Figure 2. The H_2O Megamaser in NGC 5765b (also see Gao *et al.* 2016). The top left panel shows the VLBI maser map and the top center panel shows a P-V diagram. The solid lines on the P-V diagram represent a Keplerian fit to the rotation curve. The bottom panel shows a representative GBT spectrum, the three groups of maser features identifying the characteristic spectral profile of an edge-on maser disk. The right panel shows results of GBT spectral monitoring. Each symbol on the right-panel plot marks the velocity of a maser peak in the systemic part of the spectrum. The positive slopes evident in the evolution of the maser velocities represent the centripetal acceleration as maser clouds orbit the central supermassive black hole.

spectra, observed monthly. We use ∼3 hour integrations with the GBT to obtain each spectrum. The weather in Green Bank prohibits sensitive K-band data in the summer months, so most of our targets have a summer gap in the monitoring sequence. In one case, CGCG 074-064, we supplemented the GBT observations with VLA data taken during the summer. We determine accelerations of each individual maser feature using a global least-squares fit (e.g. see Reid *et al.* 2013 and Gao *et al.* 2016). Where possible we fit accelerations for the full time sequence of spectra, but variability and line blending force fitting to shorter time sequences in most cases.

In addition to the spectral monitoring, we obtain sensitive VLBI maps to measure the distribution of maser spots in the disk. We use the VLBA augmented by the GBT, phased VLA, and for high-declination targets also the Effelsberg telescope. We observe multiple tracks to reach the required sensitivity, and combine the data prior to imaging. When possible, we self-calibrate to the brightest maser component in the spectrum, thereby improving the efficiency of the observation.

4. An Example: the MCP Galaxy NGC 5765b

Figure 2 presents the set of observations for one of our MCP galaxies, NGC 5765b. See Gao *et al.* (2016) for a full presentation of the data and the determination of the distance to this megamaser system. The bottom left panel shows a representative GBT spectrum. The panel on the right highlights results from the ∼2 year spectral monitoring campaign. Each symbol on the plot marks the velocity of an individual maser peak in the systemic part of the spectrum. The positive slopes of the individual tracks mark the redward drift of each maser component, and the slope is equal to the acceleration of that maser cloud as it orbits the central supermassive black hole. In NGC 5765b, the bright systemic features enable precise tracking of individual maser drifts. While the velocity drifts in NGC 5765b are easily identifiable throughout the entire velocity range of the systemic group, in some other galaxies the fainter and blended systemic maser lines make acceleration measurements a challenge. For most galaxies in our sample, the measured accelerations of the systemic features is the limiting factor in the determination of the galaxy distance. We also measure the line-of-sight accelerations of the red- and blue-shifted high-velocity components (not shown). Those have measured values near zero since the acceleration is directed in the plane of the sky for maser clouds at tangential locations in the orbit, near the disk midline.

The top left panel in Figure 2 shows the VLBI map of maser positions, clearly presenting a thin, edge-on, rotating disk. The location of each spot in the map is determined with a 2D Gaussian fit to the unresolved maser detection in the corresponding VLBI spectral channel. The uncertainty of each maser position measurement is therefore approximately equal to the beam size divided by twice the signal-to-noise in that spectral channel. The top central panel shows the position-velocity diagram derived from the VLBI map. The solid lines represent a Keplerian fit to the rotation of high-velocity maser clouds, and demonstrate that the disk is in Keplerian rotation.

5. Model Fitting and Determination of H_0

For each megamaser system in our sample, we fit a warped disk model to the observed maser data to determine the distance, and H_0. We determine the probability density function for H_0 using a Markov Chain Monte Carlo (MCMC) method. In addition to H_0, the model fits for the mass of the black hole, the (x_0, y_0) position of the dynamical center, recession velocity of the dynamical center, the position angle and inclination of the disk, and up to 4 parameters to describe the warping in the position angle and inclination angle directions. The model fitting approach is described in Reid *et al.* (2013). We thus obtain an H_0 measurement individually for each maser galaxy in the sample. Systematic errors in disk modeling are not likely to be correlated among different galaxies, so the combined measurement of H_0 in this paper is determined by a weighted mean of the individual measurements.

The MCP has so far published distances to four galaxies. Here we update the published value to the first of these, UGC 3789 (Reid *et al.* 2013) by extending the MCMC run to 3 billion trials to reach convergence. Our updated measurement based on UGC 3789 is $H_0 = 76 \pm 8$ km s^{-1} Mpc^{-1}. The other measurements come directly from the published values: $H_0 = 68 \pm 9$ km s^{-1} Mpc^{-1} (Kuo *et al.* 2013), $H_0 = 73 \pm 26$ km s^{-1} Mpc^{-1} (Kuo *et al.* 2015), and $H_0 = 66 \pm 6$ km s^{-1} Mpc^{-1} (Gao *et al.* 2016). The weighted average of these values is our current best estimate: $H_0 = 69.3 \pm 4.2$ km s^{-1} Mpc^{-1}.

The MCP has completed all observations and is currently finalizing the processing and analysis on five additional megamaser galaxies that will contribute to the final result. When complete, we expect the final measurement to achieve a total uncertainty of ∼4%.

References

Alam, S., Ata, M., Bailey, S. *et al.* 2017, *MNRAS*, 470, 2617
Bonvin, V., Coubin, F., Suyu, S. *et al.* 2017, *MNRAS*, 465, 4914

Freedman, W. L., Madore, B. F., Gobson, B. K., *et al.* 2012, *ApJ*, 758, 24

Gao, F., Braatz, J. A., Reid, M. J., *et al.* 2016, *ApJ*, 817, 128

Humphreys, E. M. L., Reid, M. J., Moran, J. M., *et al.* 2013, *ApJ*, 775, 13

Kuo, C. Y., Braatz, J. A., Reid, M. J., *et al.* 2013, *ApJ*, 767, 155

Kuo, C. Y., Braatz, J. A., Lo, K. Y., *et al.* 2015, *ApJ*, 800, 26

Planck Collaboration, Ade, P. A. R., Aghanim, N., *et al.* 2016, *A&A*, 594, 13

Reid, M. J., Braatz, J. A., Condon, J. J., *et al.* 2013, *ApJ*, 767, 154

Riess, A. G., Macri, L. M., Hoffmann, S. L., *et al.* 2016, *ApJ*, 826, 56

Astrophysical Masers:
Unlocking the Mysteries of the Universe
Proceedings IAU Symposium No. 336, 2017
A. Tarchi, M.J. Reid & P. Castangia, eds.

© International Astronomical Union 2018
doi:10.1017/S1743921317009814

A systematic observational study of radio properties of H_2O megamaser Seyfert 2s: A Guide for H_2O megamaser surveys

J. S. Zhang[1], Z. W. Liu[1] and C. Henkel[2,3]

[1] Center for Astrophysics, Guangzhou University,
Guangzhou 510006, China
email: jszhang@gzhu.edu.cn

[2] Max-Planck-Institut für Radioastronomie, Auf dem Hügel 69,
D-53121 Bonn, Germany

[3] Astron. Dept., King Abdulaziz University, P.O. Box 80203,
Jeddah 21589, Saudi Arabia

Abstract. Analyzing archival data from different telescopes, H_2O megamaser Seyfert 2s appeared to exhibit higher nuclear radio luminosities than non-masing Seyfert 2s (Zhang *et al.* 2012). This has been confirmed by our follow-up study on multi-band (11, 6, 3.6, 2, 1.3 cm) radio properties of maser host Seyfert 2s, through systematic Effelsberg observations (Liu *et al.* 2017). The nuclear radio luminosity was supposed to be a suitable indicator to guide future AGN maser searches. Thus we performed a pilot survey with the Effelsberg telescope on H_2O maser emission toward a small sample of radio-bright Seyfert 2 galaxies with relatively higher redshift (>0.04). Our pilot survey led to one new megamaser source and one additional possible detection, which reflects our success in selecting H_2O megamaser candidates compared to previous observations (higher detection rate, larger distance). Our successful selection technique choosing Seyfert 2s with radio-bright nuclei may provide good guiding for future H_2O megamaser surveys. Therefore we are conducting a large systematic survey toward a big Seyfert 2 sample with such radio-bright nuclei. Detections of luminous H_2O masers at large distance (z>0.04) may hold the great potential to increase our knowledge on the central highly obscured but still very enigmatic regions of active Seyfert galaxies (Zhang *et al.* 2017).

Keywords. radio, properties, megamasers, Seyfert 2s

1. Introduction

H_2O megamasers are found to be mostly located in heavily obscured nuclear region of Seyfert 2s or LINER galaxies (Braatz *et al.* 1997, Zhang *et al.* 2006, Greenhill *et al.* 2008, Zhang *et al.* 2010). Due to their extremely high luminosities ($L_{H_2O} > 10\,L_\odot$, assuming isotropic radiation), the ultimate energy source of H_2O megamasers is believed to be an AGN, which is supported by all interferometrical studies of megamasers so far carried out (Lo 2005). It was proposed that the nuclear radio continuum may provide "seed" photons, which can be amplified by foreground maser clouds to produce strong megamaser emission (E.g., Braatz *et al.* 1997; Henkel *et al.* 1998; Herrnstein *et al.* 1998). In addition, the isotropic luminosity of the nuclear radio continuum is believed to be an indicator of AGN power (e.g., Diamond-Stanic *et al.* 2009). Therefore, maser emission is expected to be stronger in radio-bright nuclei. The first statistical study on this proposition from Braatz *et al.* (1997) supported this trend, but with limited statistical significance due to the small sample (among the 16 known megamaser sources at that time, only nine had

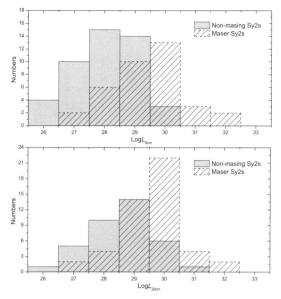

Figure 1. Radio luminosity distributions for maser and non-masing Sy2s (in $\mathrm{erg\,s^{-1}\,Hz^{-1}}$).

the complementary radio data). Now, the situation has greatly improved, so that we can perform a systematic study on the radio properties of H_2O megamaser host galaxies.

2. Analysis and results

As our first step, the radio continuum data of all published H_2O maser galaxies at that time (85 sources) and a complementary Seyfert sample devoid of detected maser emission were investigated and analyzed. Our analysis indicated that maser host Seyfert 2 galaxies have higher nuclear radio continuum luminosities (at 6 cm and 20 cm), exceeding those of the comparison Seyfert 2 sample by factors of order 5 (Figure 1). This supports the previous proposition that the nuclear H_2O megamaser emission is correlated with the nuclear radio emission and the nuclear radio luminosity may be a suitable indicator to guide future AGN maser searches (Zhang *et al.* 2012).

However, the uncertainties of this initial analysis were still quite large. Measured data (at 20 cm and 6 cm) were mostly taken from different telescopes for both maser sources and non-masing sources. Even if data were taken from the same telescope, measurements were normally performed at different epochs. And there are only a few data at other radio bands for both samples (e.g., 3.6 cm, 2.0 cm). A more complete radio dataset was urgently needed. While an interferometric study would be a better choice, systematic studies of the lower resolution radio continuum with single-dish telescopes are still worthwhile, because they are not affected by missing flux.

As the second step, we therefore performed systematic Effelsberg multi-band radio continuum observations (11 cm, 6.0 cm, 3.6 cm, 2.0 cm and 1.3 cm) within a tiny time span in January 2014 toward the H_2O megamaser Seyfert 2s and the control Seyfert 2 sample without detected maser emission. Our analysis shows that a difference in radio luminosity (at all bands) is statistically significant, i.e. that the maser Seyfert 2 galaxies tend to have higher radio luminosities by a factor of 2 to 3 than the non-masing ones, commonly reaching values above a critical threshold of $10^{29}\,\mathrm{erg\,s^{-1}\,Hz^{-1}}$ (Figure 2). The difference between maser and non-masing Seyfert 2s is supported by observations in each

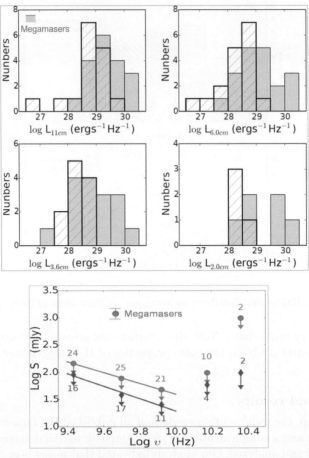

Figure 2. Upper panels: Comparisons of radio luminosities for megamaser and non-masing Sy2s. Bottom: The SED for maser Sy 2s and non-masing Sy2s. Trend lines at low frequencies (i.e. 11 cm, 6.0 cm and 3.6 cm) are also presented for both samples, from Liu *et al.* (2017).

wavelength band (bottom panel in Figure 2), i.e., the mean flux density of maser Seyfert 2s is larger than those of the non-masing ones at each band, roughly by a factor of 2. In addition, the black hole mass and the accretion rate were estimated for our maser Seyfert 2s and non-masing Seyfert 2s, assuming the radio luminosity as an isotropic tracer of AGN power. It shows that the accretion rates of maser Seyfert 2s are nearly one order magnitude larger than those of non-masing Seyfert 2s (Liu *et al.* 2017). This may provide a possible connection between H_2O megamaser formation and AGN activity, as well as suitable constraints on future megamaser surveys.

From previous analysis of archival radio data (Zhang *et al.* 2012) and systematic observations with the Effelsberg telescope (Liu *et al.* 2017), we found strong evidence for a scenario where H_2O megamasers locate in radio-bright Seyfert 2s. Nuclear radio luminosity is therefore supposedly a suitable indicator to guide future AGN maser searches. To test this, we conducted, as a third step, a pilot survey toward a small sample of Seyfert 2s with radio-bright nuclei, using the Effelsberg-100 m telescope. Our 18 targets were selected from a large SDSS-DR7 Seyfert 2 sample containing 4035 sources ($0.04 < z < 0.1$, Coldwell *et al.* 2013), with a 20 cm luminosity threshold of 10^{29} erg s^{-1} Hz^{-1} (Liu *et al.* 2017). Even though our survey was, due to the large distance of the targets, extremely

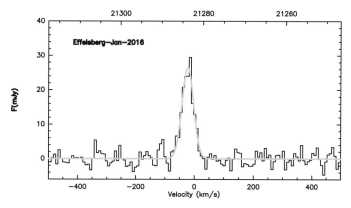

Figure 3. The smoothed average spectrum with the fitting line of the new megamaser source SDSS 102802.9+104630.4, with a redshift of 0.044776, taken from Zhang *et al.* (2017).

shallow, it led to one new strong megamaser source ($L_{H2O} > 1000 \, L_\odot$, Figure 3) and one additional tentative detection with an (isotropic) luminosity of several $100 \, L_\odot$, which reflects our success in selecting H_2O megamaser candidates compared to previous observations (Zhang *et al.* 2017). While our detection rate of 5.5% or 11% (the latter including a tentative detection) appears to be higher than in most other surveys (the uncertainty lies in the limited number of studied sources), the distance to our targets is also larger, thus making the success of this pilot survey particularly noteworthy. Our successful selection technique may provide good guiding for future H_2O megamaser surveys, i.e., Seyfert 2s with radio-bright nuclei.

Thus we will continue our H_2O megamaser survey toward that large Seyfert 2 sample, with our radio luminosity criterium of $\log L > 29 \, \mathrm{erg \, s^{-1} \, Hz^{-1}}$. More H_2O megamaser detections at a large distance (z>0.04) can be expected from such a large systematic survey of radio-bright Seyfert 2 galaxies. While presumably leading to a better determination of the upper (isotropic) luminosity limit of such masers, these will shed new light onto the central highly obscured but very enigmatic nuclear regions of active Seyfert galaxies.

Acknowledgements

This work is supported by the Natural Science Foundation of China (No. 11473007, 11590782). Thank the Effelsberg staff much for their kind help during our observations.

References

Braatz, J. A., Wilson, A. S., & Henkel, C. 1997, *ApJS*, 110, 321
Coldwell, G. V., Gurovich, S., & Díaz, T. J. 2013, *MNRAS*, 437, 1199
Greenhill, L. J., Tilak, A., & Madejski, G. 2008, *ApJ*, 686, L13
Henkel, C., Wang, Y. P., Falcke, H., Wilson, A. S., & Braatz, J. A. 1998, *A&A*, 335, 463
Herrnstein, J. R., Greenhill, L. J., Moran, J. M., *et al.* 1998, *ApJ*, 497, L69
Liu, Z. W., Zhang, J. S., Henkel, C. *et al.* 2017, *MNRAS*, 466, 1608
Lo, K. Y. 2005, *ARA&A*, 43, 625
Diamond-Stanic, A. M., Rieke, G. H., & Rigby, J. R. 2009, *ApJ*, 698, 623
Zhang, J. S., Henkel, C. Kadler, M. *et al.* 2006, *A&A*, 450, 933
Zhang, J. S., Henkel, C., Guo, Q., Wang, H. G., & Fan, J. H. 2010, *ApJ*, 708, 1528
Zhang, J. S., Henkel, C., Wang, J., & Guo, Q. 2012, *A&A*, 538, 152
Zhang, J. S., Liu, Z. W., Henkel, C., Wang, J. *et al.* 2017, *ApJL*, 836, L20

Astrophysical Masers:
Unlocking the Mysteries of the Universe
Proceedings IAU Symposium No. 336, 2017
A. Tarchi, M.J. Reid & P. Castangia, eds.

© International Astronomical Union 2018
doi:10.1017/S1743921317010614

Water maser emission in hard X-ray selected AGN

Francesca Panessa[1], Paola Castangia[2], Andrea Tarchi[2], Loredana Bassani[3], Angela Malizia[3], Angela Bazzano[1] and Pietro Ubertini[1]

[1]INAF - Istituto di Astrofisica e Planetologia Spaziali di Roma (IAPS), Via del Fosso del Cavaliere 100, 00133 Roma, Italy
[2]INAF-Osservatorio Astronomico di Cagliari (OAC-INAF), Via della Scienza 5, 09047 Selargius (CA), Italy
[3]Istituto di Astrofisica Spaziale e Fisica Cosmica (IASF-INAF), Via P. Gobetti 101, 40129 Bologna, Italy

Abstract. Water megamaser emission is powerful in tracing the inner region of active nuclei, mapping accretion disks and providing important clues on their absorption properties. From the X-ray spectra of AGN it is possible to estimate the intrinsic power of the central engine and the obscuring column density. The synergy between X-ray and water maser studies allows us to tackle the AGN inner physics from different perspectives. For a complete sample of AGN selected in the 20-40 keV energy range, we have investigated the presence of water maser emission and its connection to the X-ray emission, absorption and accretion rate. The hard X-ray selection of the sample results in a water maser detection rate much higher than those obtained from optically-selected samples.

Keywords. galaxies: active, radio lines: galaxies, X-rays: galaxies

1. Introduction

Our knowledge of the geometry of the inner structure of Active Galactic Nuclei (AGN) is still very limited. We still need to identify the boundaries between the accretion disk, the Broad Line Region and the molecular torus. The clumpy nature of the obscuring torus has been established by means of X-ray and IR arguments, however recent interferometric infrared data have found evidence of a poloidal dusty structure, further complicating the already fuzzy picture of the obscuring material around super massive black holes (see Ramos Almeida & Ricci 2017 for a review).

The combination of hard X-ray and water maser studies of AGN offers a unique opportunity to improve our knowledge of the inner structure of AGN, in particular of the correct distribution of the column density, a key ingredient of the accretion history of the Universe. On one hand, hard X-rays (above 10 keV) are very effective in finding nearby AGN (both unabsorbed and absorbed) since they are transparent to obscured regions/objects, i.e. those that could be missed at other frequencies such as optical, UV, and even X-rays. On the other hand, water maser emission is able to trace the molecular gas content and dynamics in nearby AGN (Greenhill *et al.* 2003). However, so far only a small fraction of AGN have a detected 22 GHz water maser emission line, this fraction ranges between $\sim 7\%$ from a galaxy survey (Braatz *et al.* 2007) up to $\sim 26\%$ from an optically selected sample of nearby AGN (Panessa & Giroletti 2013).

Recently, a possible relation between the water maser and the X-ray emission have been reported. Indeed, statistical studies on a sample of 42 H2O maser galaxies have shown that 95% are heavily obscured with column density $N_H > 10^{23}$ cm^{-2} and 60% of them are

Compton thick ($N_H > 10^{24}$ cm^{-2}). The fraction of Compton thick sources increases when considering only masers in accretion disks (76% have been found to be Compton thick, Greenhill *et al.* 2008, see also Castangia *et al.* 2013). In addition, a tentative correlation has been found between maser isotropic luminosity and unabsorbed X-ray luminosity (Kondratko *et al.* 2006).

In this work, we present preliminary results on the fraction of water maser emission in a hard X-ray selected sample of AGN and on the correlation between water maser and unabsorbed hard X-ray luminosity.

2. The Sample

The soft gamma-ray sky has being surveyed by INTEGRAL/IBIS and by Swift/BAT in the last 15 years at energies greater than 10 keV. The various all sky catalogues released so far contain large fractions of active galaxies, i.e. $\sim$ 30% among INTEGRAL/IBIS and up to 70% among Swift/BAT sources (see for example Bird *et al.* 2010, Bird *et al.* 2016, Baumgartner *et al.* 2013 and references therein). Together these samples provide the most extensive list of soft gamma-ray selected active galaxies known to date.

In this work, we focus on the large sample of AGN extracted from INTEGRAL/IBIS, considering the sample of 272 AGN discussed by Malizia *et al.* (2012) added with 108 sources that have been discovered or identified with active galaxies afterwards (Malizia *et al.* 2016). This set of 380 soft gamma-ray selected AGN is fully characterized in terms of optical class/redshift and X-ray properties including information on the X and soft gamma-ray fluxes and high energy absorption. This sample is not complete nor uniform and so to overcome this limitation we have considered a subset of AGN (all included in the sample of 380 objects) which represents instead a complete sample. This sample is thoroughly discussed in Malizia *et al.* (2009) and is made of 87 galaxies detected in the 20-40 keV band.

In order to look for possible water maser counterparts, we have consulted the catalogues maintained on the Web site of the Megamaser Cosmology project (MPC, https://safe.nrao.edu/wiki/bin/view/Main/MegamaserCosmologyProject) which is the largest and most comprehensive catalog of all galaxies surveyed for water maser emission at 22 GHz (Reid *et al.* 2009 and Braatz *et al.* 2010). The literature has also been searched for report of water maser observations/detection, to integrate and complete our information. We have found water maser information for 40% (total sample) and 75% (complete sample) of the sources.

3. Results

We have found that the fraction of detected water maser emission in the total sample of 380 AGN is around 15$\pm$3 %, increasing up to 19$\pm$5 % in the complete sample. If we consider only optically selected type 2 sources, this fraction raises up to 22$\pm$5 % for the total sample and 31$\pm$10 % for the complete sample. These results suggest that the hard X-ray selection is very efficient in selecting water maser sources and that the low fractions observed so far are likely due to observational biases (Panessa *et al.* in preparation). In Figure 1, we show the water maser versus unabsorbed 20-100 keV luminosity correlation. This correlation is, however, only marginally significant (correlation coefficient R=0.43) and more data are needed to confirm it.

F. Panessa *et al.*

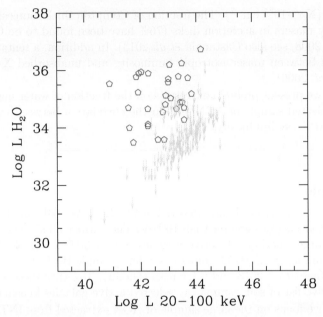

Figure 1. The water maser emission luminosity versus the unabsorbed 20-100 keV luminosity of the total sample. Blue polygons are the water maser detections and cyan arrows are upper limits (Panessa *et al.* in preparation).

References

Baumgartner, W. H., Tueller, J., Markwardt, C. B., *et al.* 2013, *ApJS*, 207, 19

Bird, A. J., Bazzano, A., Malizia, A., *et al.* 2016, *ApJS*, 223, 15

Bird, A. J., Bazzano, A., Bassani, L., *et al.* 2010, *ApJS*, 186, 1

Braatz, J. A., Reid, M. J., Humphreys, E. M. L., *et al.* 2010, *ApJ*, 718, 657

Braatz, J., Kondratko, P., Greenhill, L., *et al.* 2007, Astrophysical Masers and their Environments, 242, 402

Castangia, P., Panessa, F., Henkel, C., Kadler, M., & Tarchi, A. 2013, *MNRAS*, 436, 3388

Greenhill, L. J., Tilak, A., & Madejski, G. 2008, *ApJL* 686, L13

Greenhill, L. J., Booth, R. S., Ellingsen, S. P., *et al.* 2003, *ApJ*, 590, 162

Kondratko, P. T., Greenhill, L. J., & Moran, J. M. 2006,*ApJ*, 652, 136

Malizia, A., Landi, R., Molina, M., *et al.* 2016, *MNRAS*, 460, 19

Malizia, A., Bassani, L., Bazzano, A., *et al.* 2012, *MNRAS*, 426, 1750

Malizia, A., Stephen, J. B., Bassani, L., *et al.* 2009, *MNRAS*, 399, 944

Panessa, F. & Giroletti, M. 2013, *MNRAS*, 432, 1138

Ramos Almeida, C. & Ricci, C. 2017, Nature Astronomy, 1, 679

Reid, M. J., Braatz, J. A., Condon, J. J., *et al.* 2009, *ApJ*, 695, 287

Astrophysical Masers:
Unlocking the Mysteries of the Universe
Proceedings IAU Symposium No. 336, 2017
A. Tarchi, M.J. Reid & P. Castangia, eds.

© International Astronomical Union 2018
doi:10.1017/S1743921317010031

Extragalactic class I methanol maser: A new probe for starbursts and feedback of galaxies

Xi Chen[1,2] and Simon P. Ellingsen[3]

[1] Center for Astrophysics, GuangZhou University, Guangzhou, 510006, China
email: `chenxi@gzhu.edu.cn`

[2] Shanghai Astronomical Observatory, Chinese Academy of Sciences, Shanghai 200030, China
email: `chenxi@shao.ac.cn`

[3] School of Physical Sciences, University of Tasmania, Hobart, Tasmania, Australia
email: `simon.ellingsen@utas.edu.au`

Abstract. We report progress on research relating to 36.2 GHz extragalactic class I methanol masers, including a review of published work and new observations at high angular resolution. These observations reveal that extragalactic class I masers are excited in shocked gas and maybe associated with starbursts, galactic-scale outflows from active galactic nuclei (AGNs) feedback, or the inner-end region of the galactic bar. The current observational results suggests that extragalactic class I methanol masers provide a new probe for starbursts and feedback in active galaxies.

Keywords. masers – stars: formation – ISM: molecules – galaxies

1. Introduction

There are more than 30 methanol transitions known to exhibit maser emission in the radio frequency range (Ellingsen *et al.* 2012). These masers are widely excited in star forming regions (SFRs) within our Galaxy and are recognised to be one of the most effective tracers for investigating SFRs (e.g., Ellingsen *et al.* 2006). The methanol maser transitions are empirically classified into two types, which have been labelled as class I or class II transitions (e.g., Batrla & Menten 1988; Menten 1991). Class I methanol masers usually arise from multiple locations within a SFR, spread over areas of 0.1–1.0 parsec in extent (Voronkov *et al.* 2006). While the class II methanol masers are located close to (within ∼1″) the high-mass young stellar object (YSO ; Caswell *et al.* 2010). More than 1000 methanol maser sources from class I and class II transitions have been detected in our Galaxy (e.g., Green *et al.* 2009; Chen *et al.* 2014; Yang *et al.* 2017).

Compared to the rich and active methanol maser phenomena in our Galaxy, discoveries of extragalactic methanol masers are limited to the class II transitions at 6.7 GHz or 12 GHz, which have only been detected in the Large Magellanic Cloud (LMC; Green 2008; Ellingsen et al. 2010) and Andromeda (M31; Sjouwerman *et al.* 2010). A number of extragalactic surveys for class II methanol masers are described in the literature (Ellingsen *et al.* 1994, Phillips *et al.* 1998 and Darling *et al.* 2003). These searches observed over one hundred sources showing OH and H_2O megamaser galaxies, and/or (Ultra-) Luminous Infrared emission, but no detections were made in any of these surveys.

More than 20 years ago Sobolev (1993) suggested that the $4_{-1} \rightarrow 3_0$ E 36.2 GHz class I methanol transition may be more likely to be observed as an extragalactic maser than any of the class II transitions. Recently, widespread methanol maser emission (from over 300 positions) in the 36.2 GHz class I transition have been detected toward the central Molecular Zone (CMZ) of the Milky Way (Yusef-Zadeh *et al.* 2013), providing evidence in support of Sobolev's conjecture. In contrast, in the same region there are only three positions where 6.7 GHz class II methanol masers were detected by the Parkes methanol multibeam survey (Caswell *et al.* 2010). This is consistent with the hypothesis that class I methanol transitions can be excited on large scales in the central regions of luminous galaxies (such as starburst galaxies and merging galaxies), where widespread shocks and enhanced cosmic rays may significantly boost methanol abundances on large scales. To test this hypothesis, we have performed a number of surveys for the extragalactic class I methanol masers at 36.2 GHz transition using interferometers, Australia Telescope Compact Array (ATCA) and Jansky Very Large Array (JVLA), and single dishes Green Bank Telescope (GBT). In this paper, we mainly focus on reporting the observing results from these surveys.

2. Overview of extragalactic class I methanol maser surveys

2.1. *First survey with the ATCA*

The first systemic survey for extragalactic class I methanol masers was undertaken with the ATCA in the H168 array configuration (angular resolution of $\sim 7''$). The survey targets were a sample of about ten OH maser emission selected galaxies. We detected emission in the 36.2 GHz methanol transition towards the central regions of two starburst galaxies, NGC 253 and Arp 220 from this survey (see Ellingsen *et al.* 2014 and Chen *et al.* 2015 for details). Interestingly, emission from $7_{-2} \to 8_{-1}$ E 37.7 GHz class II methanol transition was also detected in Arp 220. These observations represent the first detections of methanol emission from the 36.2 and 37.7 GHz from sources beyond the Milky Way.

Comparing the line width (a few 10s of km s^{-1}) of the detected methanol emission with other thermal molecular lines detected in the two galaxies, they are at least a factor of $2 - 3$ times narrower. In addition, the integrated emission we observed from the 36.2 GHz transition is at least 30 times stronger than that predicted for thermal methanol emission (based on previous observations of thermal methanol transitions in these sources). Figure 1 shows that the detected methanol emission from the 36.2 GHz (and for Arp 220 also the 37.7 GHz) transition is significantly offset from the centre of the two galaxies where the strongest thermal molecular emission is typically observed. In combination, these findings strongly suggest that the detected 36.2 GHz methanol emission in the two galaxies is produced by masing. The luminosity of the 36.2 GHz methanol emission in Arp 220 is well in excess of a million times stronger than that of typical star formation masers in this transition in the Milky Way, and so represents the first detection of a methanol megamaser.

As shown in Figure 1 (right panel), the 36.2 GHz methanol masers towards Arp 220 show that there is a good spatial alignment of the methanol and diffuse, soft X-ray emission in this system. Two hypotheses have been proposed which can explain this phenomenon. One is that the enhancement of the methanol abundance is due to the effect of the high-energy cosmic rays in these regions. This has been proposed to explain the production of widespread 36.2 GHz methanol masers in the CMZ of the Milky Way (Yusef-Zadeh *et al.* 2013) and NGC 253 (Ellingsen *et al.* 2014). Another hypothesis is that the X-ray plume in Arp 220 traces a starburst-generated superwind region wherein rapidly outflowing and inflowing shock-heated gases are produced. In this scenario a critical assumption is that the widespread shocks within the plume region play an important role in producing widespread 36.2 GHz class I methanol emission.

2.2. *Single dish survey with the Green Bank Telescope*

In order to clarify the pumping conditions required for extragalactic class I methanol masers (either outflow-driven shocks, cosmic rays or both), a systematic survey for the

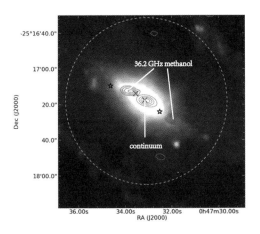 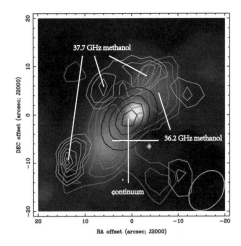

Figure 1. The 36 GHz continuum emission and 36.2 GHz methanol emission from towards NGC 253 (*left:* Ellingsen *et al.* 2014) and Arp 220 (*right:* Chen *et al.* 2015) detected with ATCA observations in H168 array. **Left panel**: for NGC 253 with 36 GHz continuum emission (contours at 20%, 40%, 60%, and 80% of 95 mJy beam^{-1}) and 36.2 GHz methanol emission (contours at 20%, 40%, 60%, and 80% of 310 mJy km s^{-1} beam^{-1}). The background image is from Spitzer IRAC data at the 3.6, 4.5 and 8.0 μm bands. **Right panel**: for Arp 220 with 36 GHz continuum emission (contours at 20, 40, 60, and 80 σ; 1σ=0.35 mJy beam^{-1}), and integrated methanol emission from both the 36.2 GHz and the 37.7 GHz transitions. Both methanol emission contours are starting at 2.5σ with increments of 1σ (1σ=0.4 Jy km s^{-1} beam^{-1}). The background image is the X-ray emission image extracted from McDowell *et al.* (2003).

36.2 GHz methanol masers was performed with the GBT towards 16 galaxies with both extended X-ray emission and megamaser emission (from water or OH). Extended X-ray emission may trace the physical environment of outflow-driven shocks or high-energy cosmic rays, allowing us to investigate these phenomena. In this survey, large baseline ripples in the GBT spectra limited the results to tentative detections towards 11 of the target galaxies (Chen *et al.* 2016). These tentative detections show that the peak flux densities of the methanol emission is weak (in a range of $2 - 8$ mJy). The majority of them show a single component, with typical linewidth of ~ 200 km s^{-1}. Most of them have a peak velocity located within or near the edge of the velocity range of the HI emission and OH or H$_2$O megamasers, suggesting that they are associated with the host galaxies. However, significant offsets of the methanol spectral features with respect to the systemic velocity are seen in Mrk 231. Figure 2 shows a comparison of the spectra of the methanol emission with those of the dense gas tracer HCO^{+} 1–0 (89.2 GHz) for three galaxies. It can be clearly seen that in Mrk 231 two methanol spectral features are observed and these are blue- and red-shifted by approximately 1000 km s^{-1} with respect to the systemic velocity. The velocity offsets are consistent with the high-velocity outflow components revealed by millimetre CO observations (Cicone et al. 2012), supporting the hypothesis that class I methanol megamasers might be produced in galactic-scale outflows.

Investigation of possible relationships between the tentative methanol emission and the emission in other wavelength ranges of host galaxies indicates no correlation between the methanol and X-ray luminosities of the host galaxies (as shown in left panel of Figure 3), suggesting that the methanol megamasers are not related to galactic-scale high-energy cosmic rays. Whereas there are good correlations between the methanol luminosity, and that of the IR and radio luminosities of the host galaxies, i.e., L$_{methanol}$ $\propto$ L$_{IR}$ $\propto$ L$_{36GHz}$ (as shown in two right panels of Figure 3), suggesting that the methanol

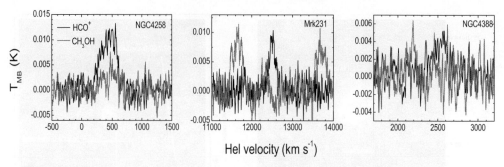

Figure 2. Comparison of spectra of the 36.2 GHz class I methanol detected in the GBT survey with HCO$^+$ 1–0 spectra for three galaxies.

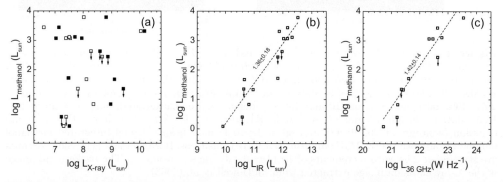

Figure 3. Relationships between the luminosity of the 36.2 GHz methanol megamasers detected from the GBT survey and that observed in other wavelength ranges for the host galaxies: (a) methanol vs. soft-x ray emission (0.5–2 keV; marked with open squares) and hard-x ray emission (2–10 keV; marked with filled squares) relations; (b) methanol vs. infrared; (c) methanol vs. radio (36 GHz) relation. Upper limits for the methanol luminosity are indicated by downward arrows for the four non-detections. Dotted lines are fits to the data labeled by their slopes.

emission may be related to the starburst activity of the host galaxy. Combining these investigations, we argue that class I methanol masers appear to be excited in galaxies with significant starburst and strong molecular outflow, and provide a potential probe of starburst activity and feedback process via wind-driven outflows of these active galaxies.

2.3. *Higher resolution observation with the JVLA towards NGC253*

Recently, follow-up higher angular resolution and sensitive observations of 36.2 GHz methanol towards NGC253 were undertaken with the ATCA in the EW367 and 1.5A arrays (Ellingsen *et al.* 2017; see also Ellingsen's paper in this proceeding). The combination of the previous ATCA H168 array and new EW367 and 1.5A array data reveal that the methanol shows compact emission from seven positions at an angular resolution of ∼ 1″ (labeled MM1–MM7; see Ellingsen et al. 2017). With the angular resolution of the combined ATCA array data the derived brightness temperature is a few K (or 220 K) using the peak flux density (or the integrated intensity) for the brightest components (Ellingsen *et al.* 2017). Hence, considering the brightness temperature alone, these observations do not require that the methanol emission is from a maser (although there are other lines of evidence that leave little doubt that it is).

To investigate the compact components of the 36.2 GHz methanol transition, we used the JVLA in the A-array configuration to make observations of this line towards NGC253 (Chen *et al.* in prep.). The JVLA A-array observations have an angular resolution of

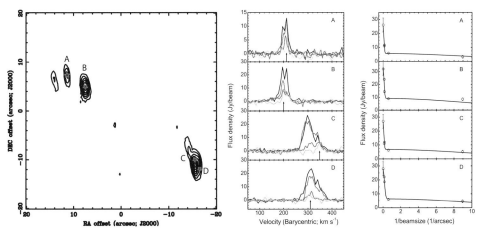

Figure 4. 36.2 GHz methanol emission detected with the ATCA and JVLA. **Left panel**: Overlaid the positions of the compact methanol emission (stars) detected with JVLA in the A-array configuration on the integrated methanol emission image (contours) obtained using the ATCA in the EW367 configuration. **Middle panel**: The peak flux densities of the four compact components detected at different angular resolutions with the ATCA in the H168, EW367 and 1.5A array configurations, and JVLA in the A-Array configuration. **Right panel**: The change in peak flux density of the four compact components versus the inverse of the beamsize. The solid lines represent the combined Gaussian profiles of both the extended and compact components with angular sizes of 6 arcseconds and 0.06 arcsenconds, in the distribution. Note, the zero-spacing data is obtained from the Shanghai Tianma 65-m telescope.

$\sim 0.1''$, one order of magnitude greater than the previous ATCA data. Compared to the previous ATCA observations, the JVLA data show four compact methanol components located within the regions of more extended emission detected with the ATCA (see left panel of Figure 4). The peak flux densities of the compact components detected with the JVLA are in the range of 3 – 9 mJy/beam, corresponding to typical brightness temperatures of >3000 K, which is far higher than the typical temperature ($\sim$ 100 K) of thermal methanol emission regions. These data provide unambiguous confirmation of methanol maser emission in extragalactic sources.

Figure 4 (middle and right panels) shows the peak flux densities of the methanol emission detected at different angular resolutions by the ATCA and JVLA. We can see that the flux density rapidly decreases as the angular scale decreases from a few arcseconds to 2 arcseconds, but the flux density does not change significantly when the angular scale decreases less than 2 arcseconds. This shows that the methanol emission consists of both extended and compact structures, with typical scales of 6 arcsecs and 60 mas, respectively.

3. Other new detections

Using the ATCA and JVLA, we have also made detections of 36.2 GHz methanol masers towards some other galaxies. We list some of these other detections here:

• Methanol maser emission with a narrow linewith (< 10 km s^{-1}) has been detected towards NGC4945 (see McCarthy *et al.* 2017 and McCarthy's paper in this proceeding). The maser emission may be associated with molecular inflow, which is observed in HF absorption.

• Both extended and compact methanol maser emission are detected towards the Seyfert galaxy NGC 1068 with the ATCA and JVLA (Chen *et al.* in prep.). The detected maser emission is located at the edge of the galactic bar, suggesting that the shock produced in the region between the bar edge and inflow or outflow gas of the galaxy may excite the maser emission.

• JVLA observations in the D-array configuration have detected the methanol maser emission from four positions along the galactic bar of nearby galaxy Maffei2 (Chen *et al.* in prep.). The velocity of the methanol emission is consistent with the hot dense gas tracer NH_3, suggesting that the methanol maser might be associated with shocks in star forming clouds along the galactic bar.

4. Summary

We have performed searches for 36.2 GHz extragalactic class I methanol masers toward a number of galaxies with the ATCA, GBT and JVLA. Significant detections have been made towards some galaxies (including NGC253, Arp220, NGC 4945, NGC1068 and Maffei2). Follow-up higher resolution observations with the JVLA have revealed that the methanol emission consists of both the extended and compact components. However their pumping mechanism is still unclear. Current observations suggest that the masers might be excited in shocked gas triggered in the starburst region, bar edge, or AGN wind-driven outflow and inflow. We argue that class I methanol megamasers may provide a new tool for investigating the starburst and feedback processes of active galaxies.

5. Acknowledgements

This work was supported by the National Natural Science Foundation of China (11590781).

References

Batrla, W. & Menten, K. M. 1988, *Ap. Lett.*, 329, 117
Caswell, J. L., Fuller, G. A., Green, J. A., *et al.* 2010, *MNRAS*, 404, 1029
Chen, X., Ellingsen, S. P., Baan, W. A., *et al.* 2015, *Ap. Lett.*, 800, 2
Chen, X., Ellingsen, S. P., Gan, C. G., *et al.* 2014, *ChSBu*, 59, 1066
Chen, X., Ellingsen, S. P., Zhang, J. S., *et al.* 2016, *MNRAS*, 459, 357
Cicone C., Feruglio C., Maiolino R., *et al.* 2012, *A&A*, 543, A99
Darling, J., Goldsmith, P., Li, D., & Giovanelli, R. 2003, *AJ*, 125, 1177
Green, J. A., Caswell, J. L., Fuller, G. A., *et al.* 2009, *MNRAS*, 392, 783
Green, J. A., Caswell, J. L., Fuller, G. A., *et al.* 2008, *MNRAS*, 385, 948
Ellingsen, S. P. 2006, *ApJ*, 638, 241
Ellingsen, S. P., Breen, S. L., Caswell, J. L., Quinn, L. J., & Fuller, G. A. 2010, *MNRAS*, 404, 779
Ellingsen, S. P., Chen, X., Breen, S. L., & Qiao, H.-H. 2017, *MNRAS*, 472, 604
Ellingsen, S. P., Chen, X., Qiao, H. Q., *et al.* 2014, *Ap. Lett.*, 790, 28
Ellingsen, S. P., Norris, R. P., Whiteoak, J. B., *et al.* 1994, *MNRAS*, 267, 510
Ellingsen, S. P., Sobolev, A. M., Cragg, D. M., *et al.* 2012, *Ap. Lett.*, 759, 5
McCarthy, T. P., Ellingsen, S. P., Chen, X. *et al.* 2017, *Ap. Lett.*, 846, 156
McDowell, J. C., Clements, D. L., Lamb, S. A., *et al.* 2003, *ApJ*, 591, 154
Menten, K. M. 1991, *Ap. Lett.*, 380, 75
Phillips, C. J., Norris, R. P., Ellingsen, S. P., & Rayner, D. P. 1998, *MNRAS*, 294, 265
Sjouwerman, L. O., Murray, C. E., Pihlströ, Y. M., Fish, V. L., & Araya, E. D. 2010, *Ap. Lett.*, 724, 158
Sobolev, A. M. 1993, *Astron. Lett.*, 19, 293
Voronkov, M. A., Brooks, K. J., Sobolev, A. M., *et al.* 2006, *MNRAS*, 373, 411
Yang, W. J., Xu, Y., Chen, X., *et al.* 2017, *ApJS*, 231, 20
Yusef-Zadeh, F., Cotton, W., Viti, S., Wardle, M., & Royster, M. 2013, *Ap. Lett.*, 764, L19

Astrophysical Masers:
Unlocking the Mysteries of the Universe
Proceedings IAU Symposium No. 336, 2017
A. Tarchi, M.J. Reid & P. Castangia, eds.

© International Astronomical Union 2018
doi:10.1017/S1743921317010687

Class I Methanol Maser Emission in NGC 4945

Tiege P. McCarthy[1,2], **Simon P. Ellingsen**[1], **Xi Chen**[3,4],
Shari L. Breen[5], **Maxim A. Voronkov**[2] **and Hai-hua Qiao**[6,4]

[1]School of Physical Sciences, University of Tasmania, Hobart, Tasmania 7001, Australia
email: tiegem@utas.edu.au
email: simon.ellingsen@utas.edu.au

[2]Australia Telescope National Facility, CSIRO, PO Box 76, Epping, NSW 1710, Australia

[3]Center for Astrophysics, GuangZhou University, Guangzhou 510006, China

[4]Shanghai Astronomical Observatory, Chinese Academy of Sciences, Shanghai 200030, China
email: chenxi@shao.ac.cn

[5]Sydney Institute for Astronomy (SIfA), School of Physics, University of Sydney, Sydney, NSW 2006, Australia
email: shari.breen@sydney.edu.au

[6]National Time Service Center, Chinese Academy of Sciences, Xi'An, Shaanxi 710600, China
email: qiaohh@shao.ac.cn

Abstract. We have detected maser emission from the 36.2 GHz ($4_{-1} \rightarrow 3_0 E$) methanol transition towards NGC 4945. This emission has been observed in two separate epochs and is approximately five orders of magnitude more luminous than typical emission from this transition within our Galaxy. NGC 4945 is only the fourth extragalactic source observed hosting class I methanol maser emission. Extragalactic class I methanol masers do not appear to be simply highly-luminous variants of their galactic counterparts and instead appear to trace large-scale regions where low-velocity shocks are present in molecular gas.

Keywords. masers, galaxies: starburst, stars: formation

1. Introduction

Methanol maser emission has been observed in over 1200 sources throughout our Galaxy (e.g., Ellingsen *et al.* 2005; Caswell *et al.* 2010, 2011; Green *et al.* 2010, 2012; Voronkov *et al.* 2014; Breen *et al.* 2015). The rich masing spectrum of this species makes it a powerful tool for investigating high-mass star-formation. Two different pumping mechanisms, collisional and radiative, divide the various transitions of methanol into class I and class II masers respectively. Class I methanol masers within our Galaxy are typically associated with the interface of molecular clouds, low-velocity shocks from cloud-cloud interaction, extended green objects (EGO) or expanding H II regions (Kurtz *et al.* 2004; Cyganowski *et al.* 2009, 2012; Voronkov *et al.* 2010, 2014). Class I methanol masers are observed towards low- and high-mass star-formation regions, however, the class II transitions are exclusively associated with high-mass star formation (Breen *et al.* 2013). Galactic class I methanol masers are generally observed in numerous individual spots across each star-formation regions, in comparison class II methanol masers are closely associated with the location of the young-stellar objects (Ellingsen 2006; Breen *et al.* 2010; Caswell *et al.* 2010).

Extragalactic methanol masers are much less commonly observed than their Galactic counterparts. Confirmed detections of extragalactic class I masers have previously only been reported in three sources (NGC 253, NGC 1068 and Arp 220; Ellingsen *et al.* 2014;

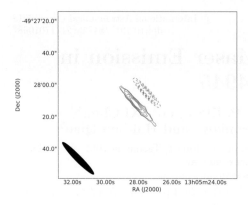 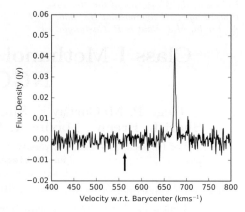

Figure 1. Left: 36.2-GHz emission methanol emission (solid contours 45%, 50%, 60%, 70%, and 80% of the 256 mJy km s^{-1} beam^{-1}) and the 7mm continuum emission (dotted contours 15%, 30%, 45%, 60%, and 80% of the 385 mJy km s^{-1} beam^{-1}). The black ellipse describes the synthesised beam size for our observations. Right: 36.2-GHz spectrum from the region of peak emission within our spectral line cube (imaged at 1 km s^{-1}). Vertical arrow indicates the systemic velocity of NGC 4945.

Wang *et al.* 2014; Chen *et al.* 2015), and extragalactic class II masers in two local group galaxies (Sjouwerman *et al.* 2010, for M31; Green *et al.* 2008 and Ellingsen *et al.* 2010, for the LMC). The extragalactic class II masers appear to be highly luminous versions of their Galactic equivalents, in contrast, the extragalactic class I methanol masers do not appear to result from the same phenomenon as those observed towards star-formation regions in our Galaxy.

Widespread emission from the 36.2-GHz $4_{-1} \rightarrow 3_0$ E methanol maser transition is observed towards the centre of our galaxy (Yusef-Zadeh *et al.* 2013). The integrated luminosity of this emission is > 5600 Jy km s^{-1} spread over more than 350 unique sites within a 160×43 pc region. Yusef-Zadeh *et al.* (2013) suggested that the mechanism producing an abundance of methanol in the central region of the Milky Way is photo-desorption of methanol from cold dust grains. It is unlikely that such a phenomenon is exclusive to our Galaxy and, therefore, should be present in the central regions of other galaxies.

NGC4945 is a nearby (3.7 ± 0.3 Mpc; Tully *et al.* 2013) barred-spiral galaxy, with a hybrid starburst and AGN nucleus. The starburst process in NGC 4945 is the primary source of energy for exciting photo-ionized gas, with infrared observations revealing that the AGN is heavily obscured (Spoon *et al.* 2000, 2003). We report observations towards this source at both the 36.2-GHz class I and 37.7-GHz class II methanol maser transitions.

2. Observations

The observations utilised the Australia Telescope Compact Array (ATCA) on 2015 August 25 and 26. The ATCA was configured in the EW352 array, with a minimum baseline of 31 metres and maximum baseline of 352 metres. Limited hour angle coverage, combined with this east-west array, caused an elongation of the synthesised beam (26×4 arcseconds at 36.2 GHz). The hybrid mode CFB 1M/64M was used for the Compact Array Broadband Backend (Wilson *et al.* 2011). We centred both 2048 MHz bands on 36.85-GHz and configured two zoom bands in the 64 MHz IF that covered the rest frequencies of 36.169265 and 37.703700 GHz, for the $4_{-1} \rightarrow 3 - 0$E and $7_{-2} \rightarrow 8_{-1}$E masing transitions respectively. The velocity range of our observations is -350 to 1200 km s^{-1}(barycentric) with a spectral resolution of 0.26 km s^{-1} for the 36.2-GHz transition. The FWHM of the primary beam of the ATCA antennas at 36.2-GHz is approximately 70 arcseconds which corresponds to a linear scale of 1200 pc a the assumed distance of 3.7 Mpc for NGC4945. As our observations consisted of only a single pointing, centred on the galactic nucleus, we are only sensitive to maser emission within 600 pc of the centre of the galaxy.

All data reduction was completed using MIRIAD, following standard techniques and procedures for 7-mm ATCA spectral line observations. PKS B1934-648 and PKS B1253-055 were used as the amplitude and bandpass calibrators respectively. Two minute observations of J1326-5256 were interleaved with the 10 minute source scans in order to perform phase calibration. Phase and Amplitude self-calibration were implemented using the continuum emission from the central region of NGC 4945. We used continuum subtraction to isolate the continuum emission component from the spectral line emission. The velocity range we imaged over was 200 to 1000 km s^{-1} barycentric (molecular line emission in NGC 4945 is observed between 300 and 800 km s^{-1}; Ott *et al.* 2001) with a spectral resolution of 1 km s^{-1} and average RMS noise of ~2.2 mJy beam^{-1} in each channel.

3. Results

Emission from the 36.2-GHz methanol transition was detected towards NGC 4945, along with 7-mm continuum emission. No emission was detected from the methanol 37.7-GHz line. This result was expected, and the transition was observed 'for free' as it falls in the same IF as the 36.2-GHz transition. The 36.2-GHz emission (coordinate of 3:05:28.093 and -49:28:12.306 in right ascension and declination respectively) is observed offset by 10 arcsec from the galactic nucleus, perpendicular from the position angle of the galactic disk and the 7-mm continuum emission is observed at the nucleus of NGC 4945 (see Figure 1). The angular offset of the 36.2-GHz emission corresponds to 174 ± 14 pc at the assumed distance of 3.7 Mpc. The 36.2-GHz emission has a peak flux density of 43.8 mJy at a position of 674 km s^{-1}, with the emission spread over a velocity range of 50 km s^{-1}, from 660 to 710 km s^{-1} with a total integrated flux density of 256 mJy km s^{-1}.

Despite the elongation of our beam, we can be confident in the observed offset from the galactic nucleus as it is perpendicular to the major-axis of elongation. Initial analysis of a second epoch of observations from 2017 June 30 have confirmed that both the position and offset measured from the original observations are accurate.

4. Discussion and Implications

It is important to first convincingly attribute the observed emission to a maser process. There are numerous pieces of evidence supporting this, the strongest of which is the narrow line width of the primary component of the emission (~ 8 km s^{-1}). Narrow line widths are a characteristic property of maser emission within Galactic sources and have also observed in NGC 253 (Ellingsen *et al.* 2014, 2017b). In addition to the emission being narrower than typical thermal emission, it is also offset from any previously reported thermal emission in the source. From our observed integrated intensity of 0.258 Jy km s^{-1} we determine an isotropic luminosity of 4.44×10^7 Jy km s^{-1} kpc^2 for the methanol emission in NGC 4945. This is approximately a factor of two less than the isotropic luminosity of NGC 253 (9.0×10^7 Jy km s^{-1} kpc^2; Ellingsen *et al.* 2014). The combination of these factors strongly indicates that the 36.2-GHz emission we are observing is a maser.

The maser emission in NGC 4945 is the fourth confirmed detection of class I methanol maser emission in an extragalactic source. This small sample size has made it difficult to determine what morphological features or phenomena these extragalactic methanol masers are associated with. High-resolution observations of NGC 253 have determined that the two regions of methanol maser emission in this source appear to be associated with inner interface of the galactic bar either side of the galactic nucleus (Ellingsen *et al.* 2017b). Presently the dynamics of the galactic bar in NGC 4945 are poorly understood. However, a position angle of 33 degrees, and azimuth angle of 40 degrees (with respect to the galactic plane) is reported by Ott *et al.* (2001). The position of our maser emission does not appear to be associated with a bar of these properties, however, the accuracy of the bar dynamics is still under discussion.

Monje *et al.* (2014) reports the possible presence of a molecular inflow in NC 4945, detected in hydrogen fluoride (HF) absorption. The properties of this possible inflow, upper limit on radius of 200 pc and velocity range of $560 - 720$ km s^{-1}, match with those of the methanol maser emission. This detection is only tentative, and the observed properties could be explained by other phenomena. Ellingsen *et al.* (2017b) speculate that extragalactic class I methanol maser emission is associated with large-scale molecular infall, therefore, association between the maser emission observed towards NGC 4945 and this possible molecular inflow would be consistent with this explanation.

References

Breen, S. L., Caswell, J. L., Ellingsen, S. P., & Phillips, C. J. 2010, *MNRAS*, 406, 1487
Breen, S. L., Ellingsen, S. P., Contreras, Y., *et al.* 2013, *MNRAS*, 435, 524
Breen, S. L., Fuller, G. A., Caswell, J. L., *et al.* 2015, *MNRAS*, 450, 4109
Caswell, J. L., Fuller, G. A., Green, J. A., *et al.* 2010, *MNRAS*, 404, 1029
Caswell, J. L., Fuller, G. A., Green, J. A., *et al.* 2011, *MNRAS*, 417, 1964
Chen, X., Ellingsen, S. P., Baan, W. A., *et al.* 2015, *ApJL*, 800, L2
Cyganowski, C. J., Brogan, C. L., Hunter, T. R., & Churchwell, E. 2009, *ApJ*, 702, 1615
Cyganowski, C. J., Brogan, C. L., Hunter, T. R., *et al.* 2012, *ApJL*, 760, L20
Ellingsen, S. P. 2006, *ApJ*, 638, 241
Ellingsen, S. P., Breen, S. L., Caswell, J. L., Quinn, L. J., & Fuller, G. A. 2010, *MNRAS*, 404, 779
Ellingsen, S. P., Chen, X., Breen, S. L., & Qiao, H.-H. 2017b, *MNRAS*, 472, 604
Ellingsen, S. P., Chen, X., Qiao, H.-H., *et al.* 2014, *ApJL*, 790, L28
Ellingsen, S. P., Shabala, S. S., & Kurtz, S. E. 2005, *MNRAS*, 357, 1003
Green, J. A., Caswell, J. L., Fuller, G. A., *et al.* 2008, *MNRAS*, 385, 948
Green, J. A., Caswell, J. L., Fuller, G. A., *et al.* 2010, *MNRAS*, 409, 913
Green, J. A., Caswell, J. L., Fuller, G. A., *et al.* 2012, *MNRAS*, 420, 3108
Kurtz, S., Hofner, P., & Álvarez, C. V. 2004, *ApJS*, 155, 149
Monje, R. R., Lord, S., Falgarone, E., *et al.* 2014, *ApJ*, 785, 22
Ott, M., Whiteoak, J. B., Henkel, C., & Wielebinski, R. 2001, *A&A*, 372, 463
Sjouwerman, L. O., Murray, C. E., Pihlström, Y. M., Fish, V. L., & Araya, E. D. 2010, *ApJL*, 724, L158
Spoon, H. W. W., Koornneef, J., Moorwood, A. F. M., Lutz, D., & Tielens, A. G. G. M. 2000, *A&A*, 357, 898
Spoon, H. W. W., Moorwood, A. F. M., Pontoppidan, K. M., *et al.* 2003, *A&A*, 402, 499
Tully, R. B., Courtois, H. M., Dolphin, A. E., *et al.* 2013, *AJ*, 146, 86
Voronkov, M. A., Caswell, J. L., Ellingsen, S. P., Green, J. A., & Breen, S. L. 2014, *MNRAS*, 439, 2584
Voronkov, M. A., Caswell, J. L., Ellingsen, S. P., & Sobolev, A. M. 2010, *MNRAS*, 405, 2471
Wang, J., Zhang, J., Gao, Y., *et al.* 2014, Nature Communications, 5, 5449
Wilson, W. E., Ferris, R. H., Axtens, P., *et al.* 2011, *MNRAS*, 416, 832
Yusef-Zadeh, F., Cotton, W., Viti, S., Wardle, M., & Royster, M. 2013, *ApJL*, 764, L19

Astrophysical Masers:
Unlocking the Mysteries of the Universe
Proceedings IAU Symposium No. 336, 2017
A. Tarchi, M.J. Reid & P. Castangia, eds.

© International Astronomical Union 2018
doi:10.1017/S1743921318000704

Sardinia Radio Telescope (SRT) observations of Local Group dwarf galaxies

A. Tarchi[1], P. Castangia[1], G. Surcis[1], A. Brunthaler[2], K. M. Menten[2], M. S. Pawlowski[3], A. Melis[1], S. Casu[1], M. Murgia[1], A. Trois[1], R. Concu[1], C. Henkel[2,4] and J. Darling[5]

[1] INAF-Osservatorio Astronomico di Cagliari, Via della Scienza 5, I-09047 Selargius
email: atarchi@oa-cagliari.inaf.it

[2] Max-Planck-Institut für Radioastronomie, Auf dem Hügel 71,53121 Bonn, Germany
[3] Department of Physics and Astronomy, University of California, Irvine, CA 92697, USA
[4] Astronomy Departement, King Abdulaziz University, P.O. Box 80203, Jeddah 21589, Saudi Arabia
[5] Center for Astrophysics and Space Astronomy, Department of Astrophysical and Planetary Sciences, University of Colorado, 389 UCB, Boulder, CO 80309-0389, USA

Abstract. The dwarf galaxies in the Local Group (LG) reveal a surprising amount of spatial structuring. In particular, almost all non-satellite dwarfs belong to one of two planes that show a very pronounced symmetry. In order to determine if these structures in the LG are dynamically stable or, alternatively, if they only represent transient alignments, proper motion measurements of these galaxies are required. A viable method to derive proper motions is offered by VLBI studies of 22-GHz water (and 6.7-GHz methanol) maser lines in star-forming regions.

In 2016, in the framework of the Early Science Program of the Sardinia Radio Telescope (SRT), we have conducted an extensive observational campaign to map the entire optical body of all the LG dwarf galaxies that belong to the two planes, at C and K band, in a search for methanol and water maser emission.

Here, we outline the project and present its first results on 3 targets, NGC 6822, IC 1613, and WLM. While no luminous maser emission has been detected in these galaxies, a number of interesting weaker detections has been obtained, associated with particularly active star forming regions. In addition, we have produced deep radio continuum maps for these galaxies, aimed at investigating their star forming activity and providing an improved assessment of star formation rates in these galaxies.

Keywords. galaxies: dwarf, Local Group, telescopes, techniques: spectroscopic, radio lines: galaxies, masers, radio continuum: galaxies

1. The SRT in a (pea)nutshell

The Sardinia Radio Telescope (SRT, Bolli *et al.* 2015, Prandoni *et al.* 2017) is a 64-m antenna located at San Basilio (Sardinia, Italy). The shaped surface of the primary mirror minimizes standing waves at the secondary focus, improving the quality of the baselines in spectroscopic measurements.

The SRT is equipped with three receivers: a K-band (18-26.5 GHz) 7-beam array in the Gregorian focus, provided with a mechanical derotator; a C-band (5.7-7.7 GHz) single feed in the Beam Wave Guide (BWG) focus; a coaxial dual-feed L and P band (1.3-1.8 GHz and 0.305-0.41 GHz, respectively) receiver in the primary focus. A number of additional packages, with different characteristics, are under construction and/or development aimed at covering also the S, Q, and W bands.

The SRT backend set offers: a broad-band ROACH 2-based digital backend, SARDARA, with bandwidths of up to 2 GHz, 16k channels, full Stokes information, for

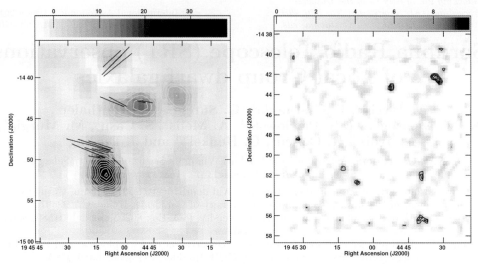

Figure 1. *Left:* SRT C-band total intensity $21' \times 21'$ image of NGC 6822 resulting from the spectral average of the bandwidth between 6000 and 7200 MHz. The FWHM beam is 2.9 arcmin. The noise level is 0.65 mJy/beam. Contour levels are 0.65 mJy/beam $\times$ (3, 4, 6, 8, 10, 15, ... ,80). Electric field polarization vectors are overimposed. The length of the vectors is proportional to the polarization percentage (with 10% being a bar of $\sim$ 2.1 arcmin), while their orientation represents the polarization angle. The error on the polarization angle is less than $10°$, and the fractional polarization is above $3\sigma_{FPOL}$. *Right:* SRT K-band total intensity $18' \times 18'$image of NGC 6822 resulting from the spectral average of the bandwidth between 21570 and 22770 MHz. The FWHM beam is 0.9 arcmin. The noise level is 2.0 mJy/beam. Contour levels are 2.0 mJy/beam $\times$ (3, 3.5, 4, 4.5, 5, 6, 7).

which multi-feed capabilities and zoom modes are going to be available soon; a narrow-band digital spectrometer, XARCOS, with bandwidths between 62.5 and 0.5 MHz, 2048 channels (yielding a maximum frequency resolution of 0.25 kHz), and full Stokes and 7-feed information provided. Additional backends are also available, including a Digital Filter Banks (DFB) and a Digital Base Band Converter (DBBC), mainly used for pulsar and VLBI measurements, respectively, and a Total Power analogue backend.

2. The SRT project ESP0003

After the completion of the Engineering Commissioning (Bolli *et al.* 2015) and Scientific Commissioning (Prandoni *et al.* 2017) in 2013 and 2016, respectively, and the participation of the SRT at the regular EVN sessions (see, e.g., Sanna *et al.* 2017; P. Castangia, in this book), regular single-dish operations were officially launched by an Early Science Program (ESP). The ESP lasted from February to August 2016 and comprised 13 large scientific projects, out of which four involved spectroscopic measurements, including the ESP project described in this contribution, labeled ESP0003 and titled 'Proper motions and star formation activity of Local Group dwarf galaxies'.

2.1. *Project motivation*

The project takes motivation from the discovery that almost all non-satellites dwarf galaxies in the Local Group belong to one of two planar, extremely symmetric, structures, termed LG plane 1 and LG plane 2 (Pawlowski, Kroupa, & Jerjen 2013; Ibata *et al.* 2013). If all these structures in the LG are dynamically stable then the galaxies will move within the associated plane, which allows to predict the expected proper motion of individual satellites in such structures (Pawlowski & Kroupa 2013; Pawlowski, McGaugh & Jerjen 2015); otherwise the structures would only represent transient alignments which would disperse as the dwarf galaxies continue their independent motions. In order to probe these two different scenarios, proper motions of these galaxies would be fundamental.

For nearby galaxies, a viable method to derive proper motions is offered by VLBI studies of 22-GHz water maser lines in star-forming regions. Indeed, VLBI observations in phase-referencing mode of water maser spots allow us to measure the 3D motions of the host galaxies with respect to distant quasars. In addition, distance measurements are possible by applying the rotational parallax method to the detected maser spots with relevant implications on the cosmological parameters and the total mass of matter (luminous and dark) of the LG. So far, such studies have been successfully performed on a limited number of maser spots detected in the galaxies M 33 and IC 10 (e.g. Greenhill *et al.* 1993; Brunthaler *et al.* 2005; 2007). Alternatively, also methanol maser lines can, in principle, be used for proper motion studies.

Within the aforementioned framework, it is particularly important to detect as many water/methanol maser sources as possible in any galaxy belonging to the LG and, in particular, in the LG dwarf galaxies belonging to the LG planar structures 1 and 2.

2.2. Sample and data reduction

For project ESP0003, we have thus mapped the full spatial extension of all non-satellite LG dwarf galaxies belonging to the two LG planes 1 and 2 in a search for new maser sources.

The sample is comprised of 14 Local Group dwarf galaxies 'associated' with Planes 1 and 2 (Tarchi *et al.* in prep).

About 200 hours in total (C and K bands) were allocated to the project. The SRT was used in conjunction with the SARDARA backend (see previous section) with a bandwidth of 1.5 GHz and 16384 channels, yielding a channel spacing of 91 kHz ($\equiv$ 4 km/s @ 6.7-GHz and 1.2 km/s @ 22-GHz). Full polarization information were recorded. Maps were performed using On-the-Fly technique along RA and DEC. The data were reduced and analyzed using the proprietary Single-dish Spectral-polarimetry Software (SCUBE; Murgia *et al.* 2016).

2.3. Preliminary results and discussion

So far, we produced the epoch-averaged cubes at C and K band for three galaxies, NGC 6822, IC 1613, and WLM. Spectral cubes were also averaged in frequency to produce radio continuum maps for the three galaxies. Fig. 1 shows the total intensity radio continuum maps of NGC 6822 at C and K band. The maps reveal a number of compact sources at both frequencies, and extended emission at C-band. Polarized emission was also imaged at C-band (see Fig. 1, left panel). Spectral index studies are ongoing to assess the nature of these sources associated to either star formation activity or background objects (Tarchi *et al.* in prep). In the spectral line cubes, we searched for the 6.7-GHz methanol and 22-GHz water maser emission, using the 'maser finder' method developed by S. Curiel and described in Surcis *et al.* (2011).

No maser emission above 5 sigma has been detected in the three targets analyzed.

For the water masers, this is consistent with the expectations. Using the equation by Brunthaler *et al.* (2006; their Eq. 2), that includes the water maser luminosity function, the number of expected (water) masers in a galaxy can be computed using, as inputs, the star formation rate (SFR) of the galaxy and the maser luminosity threshold of the image cube. The noise in the K-band SRT cube was of order 90 mJy/beam. Thus, the number of expected water masers for our targets was smaller than 0.1. On the other side, for IC 10 (also a LG dwarf galaxy) an expected number of 0.3 is derived using the same relation. However, the number of masers found in this galaxy so far is actually two, indicating that possibly the SFR values are not always precise for individual galaxies and/or maser flares may alter the computed expected maser numbers. Indeed, promisingly, a number of tentative maser features (with SNR between 3 and 4) have been detected in all three galaxies of our sample and await confirmation.

2.4. *Future steps*

The spectropolarimetric observations at C and K bands, presented in this study, of the full optical body of three targets of our sample of non-satellite Local Group dwarfs clearly shows the capabilities offered by the SRT to perform, along with maser surveys and monitoring programs, sensitive searches for masers in extended extragalactic (and Galactic) sources.

The data reduction for the other eleven dwarf galaxies in the sample observed with the SRT has yet to be completed in order to derive a more statistically-relevant view of the presence/absence of maser sources in these category of galaxies with relatively low star formation rates. In addition, from the radio continuum maps, spectral index and radio-emission based SFRs will be derived for all galaxies providing clues on the nature of the compact (and extended) emission.

Single-dish follow-ups are also planned of the tentative maser sources detected by us, for the first time, in the three galaxies studied so far, whose confirmation may provide promising targets for proper motion VLBI studies.

Acknowledgments

The SRT ESP activities were made possible thanks to the invaluable support of the entire SRT Operations Team. The development of the SARDARA backend has been funded by the Autonomous Region of Sardinia (RAS) using resources from the Regional Law 7/2007 with the research project CRP-49231. The SRT is funded by MIUR, ASI, and RAS and is operated as National Facility by INAF.

References

Bolli, P., Orlati, A., Stringhetti, L., Orfei, A., Righini, S., *et al.* 2015, *J. Asron. Instr.* Vol. 4, Nos. 3 & 4, 1550008

Brunthaler, A., Reid, M. J., Falcke, H., Greenhill, L. J., & Henkel, C. 2005, *Science* 307, 1440

Brunthaler, A., Henkel, C., de Blok, W. J. G., *et al.* 2006, *A&A* 457, 109

Brunthaler, A., Reid, M. J., Falcke, H., Greenhill, L. J., Henkel, C., & Menten, K. M. 2007, *A&A* 462, 101

Greenhill, L. J., Moran, J. M., Reid, M. J., Menten, K. M., & Hirabayashi H. 1993, *ApJ* 406, 482

Ibata, Rodrigo A., Lewis, Geraint F., Conn, Anthony R., Irwin, Michael J., McConnachie, Alan W., *et al.* 2013, *Nature* 493, 62

Murgia, M., Govoni, F., Carretti, E., Melis, A., Concu, R., *et al.* 2016, *A&A* 461, 3516

Pawlowski, M. S. & Kroupa, P. 2013, *MNRAS* 435, 2116

Pawlowski, M. S. & Kroupa, P. 2013, *MNRAS* 435, 2116

Pawlowski, M. S., McGaugh, S. S., & Jerjen, H. 2015, *MNRAS* 453, 1047;

Prandoni, I., Murgia, M., Tarchi, A., Burgay, M., Castangia, P., *et al.* 2017, *A&A* 608, 40

Sanna, A., Moscadelli, L., Surcis, G., van Langevelde, H. J., Torstensson, K. J. E.,& Sobolev, A. M. 2017, *A&A* 603, 94

Surcis, G., Vlemmings, W. H. T., Curiel, S., Hutawarakorn Kramer, B., Torrelles, J. M., & Sarma, A. P. 2011, *A&A* 527, 48

Astrophysical Masers:
Unlocking the Mysteries of the Universe
Proceedings IAU Symposium No. 336, 2017
A. Tarchi, M.J. Reid & P. Castangia, eds.

© International Astronomical Union 2018
doi:10.1017/S1743921317010201

Methanol Masers in the Andromeda Galaxy

Ylva M. Pihlström[1,2] and Loránt O. Sjouwerman[2]

[1]Department of Physics and Astronomy, University of New Mexico,
MSC07 4220, Albuquerque, NM 87111, USA
email: ylva@unm.edu

[2]National Radio Astronomy Observatory,
P.O. Box O, 1003 Lopezville Road Socorro, NM 87801, USA
email: lsjouwer@nrao.edu

Abstract. Is M31 going to collide with the Milky Way, or spiral around it? Determining the gravitational potential in the Local Group has been a challenge since it requires 3D space velocities and orbits of the members, and most objects have only had line-of-sight velocities measured. Compared to the less massive group members, the transverse velocity of M31 is of great interest, as after the Milky Way, M31 is the most dominant constituent and dynamic force in the Local Group. Proper motion studies of M31 are preferentially done using masers, as continuum sources are much weaker, and are enabled through the high angular resolution provided by VLBI in the radio regime. The challenges of achieving high astrometric accuracy at high VLBI frequencies (> 20 GHz) makes observations at lower frequencies attractive, as long as sufficient angular resolution is obtained. In particular, we have discovered 6.7 GHz methanol masers in M31 using the VLA, and here we will address their feasibility as VLBI proper motion targets using a set of global VLBI observations.

Keywords. masers, techniques: high angular resolution, galaxies: individual (M31), (galaxies:) Local Group, radio lines: galaxies

1. Introduction

One main observational task in cosmology is to determine the distribution of dark matter in galaxies and galaxy groups. Locally, this is measured by estimating the masses of the individual components of the Local Group. Determining the gravitational potential in the Local Group has been a challenge since it requires full three-dimensional velocities of the objects. Except for close Local Group components with optically measured proper motions (the LMC, the SMC, the Canis Major dwarf galaxy and other components closer than ∼150 kpc; e.g., Kallivayalil *et al.* 2009; Piatek *et al.* 2008; Vieira *et al.* 2010 and references therein), only line-of-sight velocities have been available. This situation is improving after, for example, Brunthaler *et al.* (2005; 2007) successfully measured proper motions of both M33 and IC10 using VLBI observations of H₂O masers. Other methods to determine the gravitational potential have applied statistical approaches (e.g., van der Marel & Guhathakurta 2008), but trigonometric measurements would put more stringent limits on the potential.

A major uncertainty in the mass distribution of the Local Group still lies in the undetermined transverse velocity of M31. With the Milky Way, M31 is the most dominant constituent and dynamic force in the Local Group (together with M31's satellite M33). Obtaining the transverse velocity of M31 would resolve the largest unknown in the modeling (e.g., Peebles *et al.* 2001; Loeb et al. 2005; Cox & Loeb 2008). With the high angular resolution provided by VLBI in the radio regime, proper motion studies of masers with M31 should be feasible, similar to what has been done for M33 (Brunthaler *et al.* 2007).

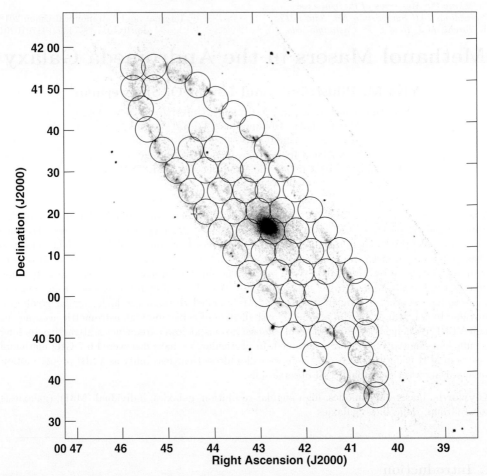

Figure 1. VLA 6.7 GHz pointings overlaid onto an MSX 8μm infrared image of M31. The size of the circles represents the primary beam of the VLA pointings. The cross marker placed in the eastern row of circles denotes the position of the methanol maser seen in Fig. 2.

High frequency masers like the 22 GHz H$_2$O transition are optimally suited for proper motion studies due to the high angular resolution that can be achieved. However, with the small primary beam it has been difficult to find masers in M31 since the galaxy subtends a large angle on the sky (40$'$ × 150$'$ in the infrared). Targeted M31 maser searches were therefore aimed to find H$_2$O masers near Hɪɪ regions with Hα, radio continuum emission, or CO emission, without any detections until very recently (Darling 2011; Huchtmeier, Eckart & Zensus 1988; Imai *et al.* 2001). Although 22 GHz masers are preferred for proper motion studies, other maser transitions may be an alternative when no H$_2$O masers are available. We have discovered Class II 6.7 GHz methanol masers in M31 (Sjouwerman *et al.* 2010), and here we are using global VLBI data to investigate their feasibility as VLBI proper motion targets in M31 and in other galaxies that may lack sufficiently bright H$_2$O maser detections.

2. Detection of Methanol Masers in M31

We have performed a systematic survey for 6.7 GHz methanol masers in M31. Selected regions of M31 were observed with the VLA during 2009 until the VLA shut down in

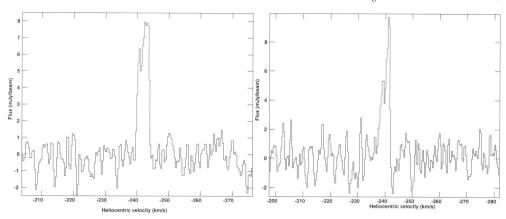

Figure 2. Spectra of one of the Class II 6.7 GHz methanol masers observed in M31, displaying a more or less constant flux density over a period of three years. The spectrum on the left was observed in 2010, and the one on the right in 2013, both using the upgraded VLA.

mid-January 2010. Several tentative detections at the 5-7σ level have been made. One of these candidates was selected for a follow-up study using the upgraded VLA in July 2010, confirming that the feature was real (see Sjouwerman *et al.* 2010). Later on, additional pointings covering most of the M31 angular extent was performed with the upgraded VLA system (the complete set of pointing positions is shown in Fig. 1). In September 2013, we did another observation with the VLA confirming the maser is still present, and this showed the maser is stable, with the same flux density as in 2010 (Fig. 2). The maser peak flux density of about 8 mJy/beam compares to masers on the high-end tail of the Class II 6.7 GHz methanol maser distribution in the Milky Way (Goldsmith, Pandian & Deshpande 2008). The brightest Galactic methanol maser with a very accurate distance measurement is W3(OH): ~3700 Jy at 2.0 kpc (Pestalozzi, Minier & Booth 2005; Hachisuka *et al.* 2006). At this distance, the M31 detection would measure ~1300 Jy, while the brightest maser in the LMC would be ~3000 Jy (e.g., Beasley *et al.* 1996). Thus, it is plausible that M31 may host even brighter methanol masers than the initial detection (Sjouwerman et al. 2010). However, given the lower star formation rate of M31 compared to the Milky Way, it is equally plausible that all of the brightest methanol masers have been detected in the VLA observations.

3. Follow up VLBI Observations

To test the feasibility of the detected Class II methanol masers as Very Long Baseline Interferometry (VLBI) targets, observations using a global array consisting of the European VLBI Network (EVN), the Very Long Baseline Array (VLBA), and the phased Very Long Array (VLA) was performed. The data were taken over four consecutive days, with in total 24 hours of observing time. To obtain an accurate source position, phase referencing using a nearby calibrator within $< 1.5°$ was applied. The cycle time was three minutes.

A preliminary analysis of the data does not confidently detect the maser feature. Given the previously measured VLA flux density, we are confident that an unresolved maser should be detectable with the achieved VLBI sensitivity. During the VLBI observations, the phased VLA also collects a pseudo-continuum data set for the individual VLA baselines. Using this VLA-only data set, despite a much coarser spectral resolution, it could be confirmed that the maser was still present, although it may have reduced somewhat in its brightness.

The difficulty of confidently detecting the maser may be due to the emission distribution, implying that it may be resolved on the scales of the VLBI baselines. However, if the M31 6.7 GHz methanol maser emission would be contained within a similar spatial distribution as the methanol masers in Galactic star forming regions, the emission should be unresolved. Galactic star forming regions are often accompanied by other signs of star formation, including H_2O maser emission, $H\alpha$ emission, and HII region signatures. Toward the location of the methanol maser there are no other typical signs of star formation confirmed, which could indicate a different maser formation environment in M31. However, the velocity width, luminosity, and more or less invariable peak flux density is consistent with other Class II methanol masers associated with Galactic star forming regions.

Additional work on the data to confirm whether the 6.7 GHz maser feature is indeed non-detected will be performed, including more rigorous radio frequency interference removal and combining all the four days of observations.

Acknowledgements

The European VLBI Network is a joint facility of independent European, African, Asian, and North American radio astronomy institutes. The National Radio Astronomy Observatory is a facility of the National Science Foundation operated under cooperative agreement by Associated Universities, Inc.

References

Beasley, A. J., Ellingsen, S. P., Claussen, M. J., & Wilcots, E. 1996, ApJ 459, 600
Brunthaler, A., Reid, M. J., Falcke, H., Henkel, C., & Menten, K. M. 2007, A&A 462, 101
Brunthaler, A., Reid, M. J., Falcke, H., Greenhill, L. J., & Henkel, C. 2005, Science 307, 1440
Cox, T. J. & Loeb, A. 2008, MNRAS 386, 461
Darling, J. 2011, ApJLetters 732, L2
Goldsmith, P. F., Pandian, J. D., & Deshpande, A. A. 2008, ApJ 680, 1132
Hachisuka, K., Brunthaler, A., Menten, K. M., Reid, M. J., Imai, H., Hagiwara, Y., Miyoshi, M., Horiuchi, S., & Sasao, T. 2006, ApJ 645, 337
Huchtmeier, W. K., Eckart, A., & Zenzus, A. J. 1988, A&A 200, 26
Imai, H., Ishihara, Y., Kameya, O., & Nakai, N. 2001, PASJ 53, 489
Kallivayalil, N., Besla, G., Sanderson, R., & Alcock, C. 2009, ApJ 700, 924
Loeb, A., Reid, M. J., Brunthaler, A., & Falcke, H. 2005, ApJ 633, 894
van der Marel, R. P. & Guhathakurta, P. 2008, ApJ 678, 187
Peebles, P. J. E., Phelps, S. D., Shaya, E. J., & Tully, R. B. 2001, ApJ 557, 495
Pestalozzi, M. R., Minier, V., & Booth, R. S. 2005, A&A 432, 737
Piatek, S., Pryor C. & Olszewski, E. W. 2008, AJ 135, 1024
Sjouwerman, L. O., Murray, C. E., Pihlström, Y. M., Fish, V. L., & Araya, E. 2010, ApJLetters 724, L158
Vieira, K., Girard, T. M., van Altena, W. F., Zacharias, N., Casetti-Dinescu, D. I., Korchagin, V. I., Platais, I., Monet, D. G., Lopez, C. E., Herrera, D., & Castillo, D. J. 2010, AJ 140, 1934

Astrophysical Masers:
Unlocking the Mysteries of the Universe
Proceedings IAU Symposium No. 336, 2017
A. Tarchi, M.J. Reid & P. Castangia, eds.

© International Astronomical Union 2018
doi:10.1017/S1743921317009280

The Maser-Starburst connection in NGC 253

Simon P. Ellingsen

School of Physical Sciences, University of Tasmania, Hobart 7001, TAS, Australia

Abstract. NGC 253 is one of the closest starburst galaxies to the Milky Way and as such it has been studied in detail across the electromagnetic spectrum. Recent observations have detected the first extragalactic class I methanol masers at 36 and 44 GHz and the first extragalactic HC$_3$N (cyanoacetylene) masers in this source. Here we discuss the location of the masers with respect to key morphological features within NGC 253 and the association between the masers and the ongoing starburst.

Keywords. masers, galaxies: starburst, stars: formation

1. Introduction

Masers have proven to be a powerful tool for studying the kinematics of both galactic and extragalactic environments (e.g. Goddi *et al.* 2011; Miyoshi *et al.* 1995). Megamasers are extragalactic maser sources which have an isotropic luminosity a milliion times or greater than that of typical galactic maser emission from the same transitions. Megamaser emission from the 22 GHz transition of water and the 1667 MHz transition of OH have been known for more than 30 years (Dos Santos & Lepine 1979; Baan *et al.* 1982) and there are now more than 100 sources known (see the review talk by Henkel 2018, in these proceedings). The discovery of strong and common galactic methanol maser transitions in the 1980s and 1990s lead to searches for methanol megamasers, particularly from the 6.7 GHz transition, however, these were all unsuccessful (Ellingsen *et al.* 1994; Phillips *et al.* 1998; Darling *et al.* 2003).

The strongest Galactic class I methanol maser transitions are the $7_0 \rightarrow 6_1 A^+$ and $4_{-1} \rightarrow 3_0 E$ transitions (rest frequencies of 44 and 36 GHz respectively) and in the first international conference dedicated to astrophysical masers, held in 1992 in Arlington, Andrej Sobolev suggested that the 36 GHz transition may provide the best prospects for being detected in extragalactic sources. Furthermore, his subsequent paper (Sobolev 1993) made the prediction that emission with a strength of around 50 mJy would be present in the nearby starburst galaxy NGC 253. At the time that Sobolev's paper was written there were relatively few radio telescopes capable of making observations at 36 GHz and so the prediction remained untested until 2014, when observations by Ellingsen *et al.* (2014) detected 36 GHz methanol emission with a peak intensity of around 20 mJy towards NGC 253 (Fig. 1). The 36 GHz methanol emission in NGC 253 is 4 to 5 orders of magnitude stronger than that observed in typical Galactic star formation regions and offset from centre of the galaxy by 200–300 pc, along the direction of the galactic bar (Ellingsen *et al.* 2014).

2. NGC 253

NGC 253 is the largest galaxy in the Sculptor group, the closest group of galaxies beyond the local group. It is a barred spiral which we observe at a high inclination angle to our line-of-sight and has a star formation rate several times higher than that of the Milky

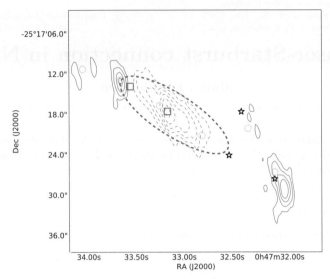

Figure 1. Integrated 36 GHz methanol maser emission (solid contours) and 7mm radio continuum (dashed contours) towards NGC 253. The thick-dashed ellipse marks the half-maximum intensity of the CMZ (Sakamoto *et al.* 2011). The squares mark the location of 22 GHz water masers (Henkel *et al.* 2004), the stars are supershells identified in the CO observations of Sakamoto *et al.* (2006), the circles are NH₃ cores from Lebrón *et al.* (2011).

Way (e.g. Ott *et al.* 2005), more than half of which is located in the central molecular zone (CMZ). The proximity of NGC 253 makes it possible to study the nuclear starburst at high resolution and sensitivity and it provides an important testbed for understanding the relationship between star formation in normal spirals, nuclear starbursts and merger-driven starburst systems (Leroy *et al.* 2015). Observations of the OH and water maser transitions towards NGC 253 have detected maser emission from each (Turner 1985; Henkel *et al.* 2004), however, the class I methanol emission is offset from the location of water masers (Fig. 1) and the relationship with the OH emission and absorption from the central region of NGC 253 is unclear due to the complexity and limited angular resolution available in the historical VLA observations of Turner.

3. Maser-starburst connection

The initial detection of the 36 GHz methanol emission in NGC 253 has recently been confirmed in follow-up observations made with both the very large array (VLA) (Gorski *et al.* 2017) and the Australia Telescope Compact Array (ATCA) (Ellingsen *et al.* 2017b). To date extragalactic class I methanol maser emission has been observed at high angular resolution towards three sources NGC 253, NGC 4945 (McCarthy *et al.* 2017) and Arp 220 (Chen *et al.* 2015); with tentative detections with single dish telescopes made towards a further 10 or so sources (Wang *et al.* 2014; Chen *et al.* 2016). For the three sources where there is high resolution information on the distribution of the methanol masers, they are observed significantly offset from the nucleus of the galaxy (100s-1000s of pc), so the natural question to ask is - what do the class I extragalactic masers trace?

It is widely thought that the bar-potential plays a critical role in driving the nuclear starburst in NGC 253 (e.g. García-Burillo *et al.* 2000) and interestingly, the south-western 36 GHz methanol maser emission lies at the point where the inner edge of the bar meets the central molecular zone (see Fig. 1). Sensitive, high-resolution observations of the molecular gas in NGC 253 show that this is a region of wide spread low-velocity shocks

(Meier *et al.* 2015). The 36 GHz class I transition is observed to be widespread in the CMZ of the Milky Way (Yusef-Zadeh *et al.* 2013) and Galactic class I methanol maser emission is known to be associated with molecular outflows (e.g. Voronkov *et al.* 2006; Cyganowski *et al.* 2009), expanding HII regions (e.g. Voronkov *et al.* 2010), cloud-cloud collisions (e.g. Sobolev 1992) and molecular cloud-SNR interactions (e.g. Pihlström *et al.* 2014). The two common factors which are present in all of these environments is the presence of cool molecular gas and low-velocity shocks.

Cool molecular gas is the basic fuel for star formation, while low-velocity shocks are the trigger for the star formation process. Hence, early- and intermediate-stage starburst galaxies must have both of these present on scales which are much larger than that of typical Galactic star formation regions or the CMZ of the Milky Way. However, the class I methanol maser emission in NGC 253 does not appear to be simply a scaled-up version of that typically observed in Galactic star formation regions. The ATCA observations show that rather than being a blend of a large number of compact emission regions, the majority of the 36 GHz methanol emission in NGC 253 is relatively diffuse and resolved out on spatial scales smaller than ~50 pc (Ellingsen *et al.* 2017b). This is similar to what is observed for OH and formaldehyde megamaser emission (Baan *et al.* 2017) and in contrast to the very compact emission seen in water megamasers (see the paper by Baan in these proceedings). Furthermore, the 44 GHz class I methanol maser emission in NGC 253 is two orders of magnitude weaker than the 36 GHz masers (Ellingsen *et al.* 2017b), which is very different from what is observed in typical Galactic star formation regions (Voronkov *et al.* 2014). When comparing the thermal molecular gas in NGC 253 with the 36 GHz methanol maser emission, Ellingsen *et al.* (2017a) observed the HC_3N (J=4-3) emission to be very similar in distribution, with some of the emission detected on $0.1''$ scales. These observations show that some of the HC_3N emission is masing. This is the first detection of maser emission in this transition in any source, Galactic or extragalactic, and further evidence that the class I methanol maser emission in NGC 253 is a different phenomenon from that observed in Galactic sources.

4. Conclusions

The 36 GHz class I methanol masers appear to be an exciting new tool for investigations of molecular gas in extragalactic sources. In NGC 253 the emission appears to trace regions where cool molecular gas is undergoing low-velocity shocks, such as at the interface between the galactic bar and the central molecular zone. Further investigation of the masing and molecular gas in other sources is required to determine their similarities and differences and hence obtain a more comprehensive understanding of the critical aspects which determine where strong extragalactic class I methanol masers are likely to be found. If the luminosity of the class I methanol maser emission scales with the star formation rate (as has been suggested by Chen *et al.* 2016), then it may be possible to detect much more distant sources in the 36 GHz transition and use this to investigate starburst galaxies.

Acknowledgements

I would like to thank the following collaborators who have made contributions to the work presented here, Xi Chen (Guangzhou University & Shanghai Astronomical Observatory), Shari Breen (University of Sydney), Hai-Hua Qiao (Shanghai Astronomical Observatory), Willem Baan (ASTRON), Tiege McCarthy (University of Tasmania) and Maxim Voronokov (CSIRO Astronomy and Space Science).

References

Baan W. A., An T., Klöckner H.-R., & Thomasson P., T, *MNRAS*, 469, 916

Baan W. A., Wood P. A. D., & Haschick A. D., 1982, *ApJ*, 260, L49

Chen X., Ellingsen S. P., Baan W. A., Qiao H.-H., Li J., An T., Breen S. L., 2015, *ApJ*, 800, L2

Chen X., Ellingsen S. P., Zhang J.-S., Wang J.-Z., Shen Z.-Q., Wu Q.-W., Wu Z.-Z., 2016, *MNRAS*, 459, 357

Cyganowski C. J., Brogan C. L., Hunter T. R., & Churchwell E., C, *ApJ*, 702, 1615

Darling J., Goldsmith P., Li D., & Giovanelli R., G, *AJ*, 125, 1177

Dos Santos P. M., Lepine J. R. D., 1979, *Nature*, 278, 34

Ellingsen S., Chen X., Breen S., Qiao H.-H., 2017a, *ApJ*, 841, L14

Ellingsen S. P., Chen X., Breen S. L., Qiao H.-H., 2017b, *MNRAS*, 472, 604

Ellingsen S. P., Chen X., Qiao H.-H., Baan W., An T., Li J., Breen S. L., 2014, *ApJ*, 790, L28

Ellingsen S. P., Norris R. P., Whiteoak J. B., Vaile R. A., McCullch P. M., Price M. G., 1994, *MNRAS*, 267, 510

García-Burillo S., Martín-Pintado J., Fuente A., Neri R., 2000, *A&A*, 355, 499

Goddi C., Moscadelli L., & Sanna A., S, *A&A*, 535, L8

Gorski M., Ott J., Rand R., Meier D. S., Momjian E., Schinnerer E., 2017, *ApJ*, 842, 124

Henkel C., 2018, in IAU Symposium, Vol. 336, Astrophysical Masers: Unlocking the Mysteries of the Universe, Tarchi A., Reid M. J., Castangia P., eds.

Henkel C., Tarchi A., Menten K. M., Peck A. B., 2004, *A&A*, 414, 117

Lebrón M., Mangum J. G., Mauersberger R., Henkel C., Peck A. B., Menten K. M., Tarchi A., Weiß A., 2011, *A&A*, 534, A56

Leroy A. K. *et al.*, 2015, *ApJ*, 801, 25

McCarthy T. P., Ellingsen S. P., Chen X., Breen S. L., Voronkov M. A., Qiao H.-h., 2017, *ApJ*, 846, 156

Meier D. S. *et al.*, 2015, *ApJ*, 801, 63

Miyoshi M., Moran J., Herrnstein J., Greenhill L., Nakai N., Diamond P., & Inoue M., I, *Nature*, 373, 127

Ott J., Weiss A., Henkel C., & Walter F., W, *ApJ*, 629, 767

Phillips C. J., Norris R. P., Ellingsen S. P., Rayner D. P., 1998, *MNRAS*, 294, 265

Pihlström Y. M., Sjouwerman L. O., Frail D. A., Claussen M. J., Mesler R. A., McEwen B. C., 2014, *AJ*, 147, 73

Sakamoto K. *et al.*, 2006, *ApJ*, 636, 685

Sakamoto K., Mao R.-Q., Matsushita S., Peck A. B., Sawada T., Wiedner M. C., 2011, *ApJ*, 735, 19

Sobolev A. M., 1992, *AZh*, 69, 1148

Sobolev A. M., 1993, *Astron. Lett.*, 19, 293

Turner B. E., 1985, *ApJ*, 299, 312

Voronkov M. A., Brooks K. J., Sobolev A. M., Ellingsen S. P., Ostrovskii A. B., Caswell J. L., 2006, *MNRAS*, 373, 411

Voronkov M. A., Caswell J. L., Ellingsen S. P., Green J. A., Breen S. L., 2014, *MNRAS*, 439, 2584

Voronkov M. A., Caswell J. L., Ellingsen S. P., Sobolev A. M., 2010, *MNRAS*, 405, 2471

Wang J., Zhang J., Gao Y., Zhang Z.-Y., Li D., Fang M., Shi Y., 2014, *Nature Comm.*, 5, 5449

Yusef-Zadeh F., Cotton W., Viti S., Wardle M., & Royster M., R, *ApJ*, 764, L19

Astrophysical Masers:
Unlocking the Mysteries of the Universe
Proceedings IAU Symposium No. 336, 2017
A. Tarchi, M.J. Reid & P. Castangia, eds.

© International Astronomical Union 2018
doi:10.1017/S1743921317010559

Spatially resolving the OH masers in M82

Megan Argo

Jeremiah Horrocks Institute, University of Central Lancashire, UK
email: `margo@uclan.ac.uk`

Abstract. With luminosities between those of typical Galactic OH masers and more distant OH megamasers, the masers in the nearby galaxy M82 are an interesting population which can be used to probe the physical conditions in the central starburst region of this irregular galaxy. Following on from previous low spatial resolution studies, here we present the initial results of two high-resolution observations separated by eight years. We find that some of the maser spots are resolved into multiple spatial components when observed with the EVN, as predicted by our previous studies, but that significantly less flux is recovered that that seen with the previous VLA observations. We conclude that some of this flux difference is likely due to variability but that, in common with the results seen in Arp220, there may also be a significant diffuse component.

Keywords. masers, galaxies: individual (M82)

1. Introduction

M82 is a nearby starburst galaxy, located 3.6 Mpc away (Freedman *et al.* 1994) in the constellation of Ursa Major. As it is inclined almost edge-on to our line of sight, little in the way of spiral structure can easily be seen in this galaxy, and the thick dust lanes caused by the starburst activity obscure much of the optical centre of the galaxy. Numerous radio studies of this galaxy have shown large amounts of gas, as well as over fifty compact radio sources which can be separated into H II regions and supernova remnants based on their radio spectra and/or their VLBI morphology (e.g. McDonald *et al.* 2002, Fenech *et al.* 2010, Muxlow *et al.* 2010, Brunthaler *et al.* 2010).

Bright OH main line emission was first detected using the Effelsberg telescope in the 1970s (Rieu *et al.* 1976), while Weliachew *et al.* (1984) looked at the HI and OH within the disk of M82 and made the suggestion that the masers, while bright, were not necessarily brighter than Galactic maser spots if each region contained many individual spots. Maser emission in the OH satellite lines was also detected (Seaquist *et al.* 1997).

Wills *et al.* (2000) used radio observations of the 21cm line of H I to investigate the motions of the atomic gas in this galaxy, finding evidence for motions that they modelled as being due to an inner bar. In 2002, in an effort to compare the atomic and molecular gas at the same spatial resolution, the OH lines at 1665 and 1667MHz were

Year	Array	Spatial resolution	Velocity resolution	Reference
2002	VLA	1.4"	17 km/s	Argo *et al.* (2007)
2004	EVN	0.03"	1.4 km/s	in prep
2006	VLA	1.4"	1.4 km/s	Argo *et al.* (2010)
2012	EVN	0.03"	1.4 km/s	in prep
2017	e-MERLIN	0.15"	1.4 km/s	in prep

Table 1. Observations of the masers in M82.

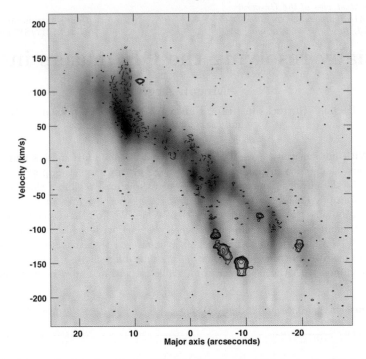

Figure 1. Position-velocity plot of the OH distribution in M82 from the VLA 2006 observations (contours) overlaid on the Hɪ distribution from Wills *et al.* (greyscale) across the central ∼1-kpc starburst of the galaxy. The two masers with significant velocity structure here at around −150 km/s also have significant spatial structure in the EVN 2012 observations in Fig. 2.

observed with the VLA in the same configuration as the Hɪ observations. These observations were designed to probe the absorption, rather than maser emission, but several previously-undetected OH main line masers were serendipitously discovered, even in the broad frequency channels of this observation (Argo *et al.* 2007). All of the maser spots were significantly brighter than typical Galactic masers, an effect likely due to the superposition of several masing clouds along a particular line of sight. This conclusion was supported by later VLA observations at higher velocity resolution, showing significant velocity structure in several of the maser spots (Fig 1; Argo *et al.* 2010).

In order to investigate the maser population further we obtained EVN observations at high spatial and velocity resolution, which showed that several of the maser spots are indeed resolved into multiple spatial components on scales of a few milliarcseconds, with one spot splitting into a spectacular 3.5-parsec ring at these resolutions (Argo *et al.* 2012).

The story was still incomplete, however. The total flux recovered in the EVN observations is much less than that observed in the VLA observations at the same velocity resolution. To test whether this flux reduction is due to variability, or to a diffuse component to which the EVN is not sensitive, we obtained a second epoch of EVN data, as well as intermediate spatial scale e-MERLIN observations at matched velocity resolution. Table 1 shows a summary of all observations of the OH masers carried out as part of this investigation.

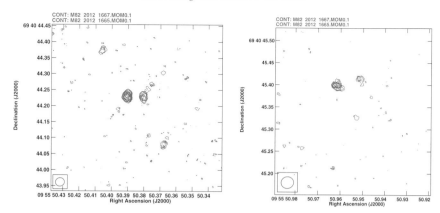

Figure 2. Comparison of 1665 (red) and 1667 MHz (blue) emission for the masers 50.37+44.3 (left panel) and 50.95+45.4 (right panel) in the 2012 EVN observations, illustrating the difference in line ratios for the different spatial components within these maser regions.

2. New observations

Since the EVN observations were carried out eight years apart, it is inevitable that there are significant changes in the hardware available between the two epochs. The 2004 observation used eight EVN stations, the 2012 observation included an additional three Russian and two Chinese stations. Both observations lasted for 18 hours, using the same phase calibrators (J0958+6533 and 0955+697A) with the same switching cycle, and the same velocity resolution. Despite the smaller beam in the 2012 observations, due to the longer baselines, this dataset was imaged with the same beam as the 2004 dataset so that the observations would be directly comparable.

The newer e-MERLIN observations were carried out in summer 2017 using the same phase calibrators as the EVN observations. They included all available antennas of the array, with 8 IFs covering a total bandwidth of 512 MHz, as well as additional spectral windows covering the HI line and each of the OH lines with the same spectral resolution as the EVN and VLA2006 observations. This dataset is currently being analysed and will be discussed further in an upcoming paper.

3. Results so far

In the earlier VLA studies, many of the masers have narrow spectral widths, however the maser regions located at 50.37+44.3 and 50.95+45.4 were both clearly resolved into multiple velocity components in the high-velocity resolution VLA study (see Fig. 1). The feature at 50.37+44.3 split into several velocity components, but was unresolved spatially at a resolution of 1.4", whereas 50.95+45.4 was resolved in both parameters with the spatial extent stretching over several arcseconds (Argo *et al.* 2010). While many of the masers remain compact at VLBI resolution, both of these features are clearly resolved into multiple spatial components, matching the prediction from the high velocity VLA observations. Interestingly, in the 2012 EVN data, the feature at 50.37+44.3 is extended over a larger physical region than 50.95+45.4, although it is important to note here that any extended diffuse maser emission would be resolved out in these observations. A careful analysis of the velocities of each of the VLBI components is underway, with comparisons being made between the two EVN epochs, and with the spectral peaks seen in 2006 with the VLA.

A comparison of the two VLBI datasets shows that there is considerable variation in the line ratios of the maser spots, with some being visible in only the 1665-MHz line, others only appearing in the 1667-MHz line, and some individual spots showing clear differences in line ratios between the two EVN epochs. The lack of continuum emission at the maser locations in the EVN images is expected from previous high-resolution VLBI observations of the galaxy (e.g. Fenech *et al.* 2010) - the long baselines resolve out all but the most compact emission within the galaxy. This results in limits to the amplification factors of between 40 and more than 200, again comparable with the results from Arp220 (Lonsdale *et al.* 1998).

4. To be continued

The results presented here are from the preliminary study of the high resolution data. The question of whether the apparent associations of any of the maser spots with continuum sources is being investigated by a careful comparison of the EVN images with continuum datasets at similar resolution; indications are that compact OH maser emission is offset from the continuum, demonstrating that high-gain maser action is responsible for at least part of the maser emission in M82. Observations at the same velocity resolution but intermediate angular resolution have recently been carried out with e-MERLIN. These observations should allow us to search for a more diffuse maser component such as that observed in Arp220.

One question it has so far been impossible to answer is that of variability: are changes in spot brightness between observations due to variability, or because each observation has used dramatically different parameters in either spectral or angular resolution, making both a comparison of brightnesses between epochs and a search for morphological changes impossible. The two EVN epochs, together with the new e-MERLIN data, should allow us to answer these questions.

References

Argo, M. K., Pedlar, A., Beswick, R. J., & Muxlow, T. W. B. 2007 *MNRAS*, 380, 596
Argo, M. K., Pedlar, A., Beswick, R. J., Muxlow, T. W. B., & Fenech, D. 2010 *MNRAS*, 402, 2703
Argo, M. K., Beswick, R. J., Muxlow, T. W.B., Fenech, D. M., van Langevelde, H. J., Gendre, M., & Pedlar, A. 2012 *Proceedings of the 11th European VLBI Network Symposium, Bordeaux, France*, arXiv:1301.4820
Brunthaler, A., Martí-Vidal, I., Menten, K. M., Reid, M. J., Henkel, C., Bower, G. C., Falcke, H., Feng, H., Kaaret, P., Butler, N. R., Morgan, A. N., & Weiß, A. 2010 *Astron. Astrophys,* 516, A27
Fenech, D., Beswick, R., Muxlow, T. W. B., Pedlar, A., & Argo, M. K. 2010 *MNRAS*, 408, 607
Freedman, W. L., Hughes, S. M., Madore, B. F., Mould, J. R., Lee, M. G., Stetson, P., Kennicutt, R. C., Turner, A., Ferrarese, L., Ford, H., Graham, J. A., Hill, R., Hoessel, J. G., Huchra, J., & Illingworth, G. D. 1994 *ApJ*, 427, 628
Lonsdale, C. J., Lonsdale, C. J., Diamond, P. J., & Smith, H. E. 1998 *Astrophys. J. Letters*, 493, L13
McDonald, A. R., Muxlow, T. W.B., Wills, K. A., Pedlar, A., & Beswick, R. J. 2002 *MNRAS*, 334, 912
T. W. B. Muxlow, R. J. Beswick, S. T. Garrington, A. Pedlar, D. M. Fenech, M. K. Argo, J. van Eymeren, M. Ward, A. Zezas, & A. Brunthaler *MNRAS*, 404, L109
Nguyen-Q-Rieu; Mebold, U., Winnberg, A., Guibert, J., & Booth, R. 1976 *Astron. Astrophys.*, 52, 467
Seaquist, E. R., Frayer, D. T., & Frail, D. A. 1997 *ApJL*, 487, L131
Weliachew, L., Fomalont, E. B., & Greisen, E. W. 1984 *Astron. Astrophys.*, 137, 335
Wills, K. A., Das, M., Pedlar, A., Muxlow, T. W. B., Robinson, T. G. 2000 *MNRAS*, 316, 33

Astrophysical Masers:
Unlocking the Mysteries of the Universe
Proceedings IAU Symposium No. 336, 2017
A. Tarchi, M.J. Reid & P. Castangia, eds.
© International Astronomical Union 2018
doi:10.1017/S1743921317009966

AGN accretion disk physics using H_2O megamasers

Dominic Pesce[1,2], James Braatz[2], James Condon[2], Feng Gao[2,3], Christian Henkel[4,5], Violette Impellizzeri[2,6], Eugenia Litzinger[7,8], K. Y. Lo[2] and Mark Reid[9]

[1]Department of Astronomy, University of Virginia, 530 McCormick Road, Charlottesville, VA 22904, USA

[2]National Radio Astronomy Observatory, 520 Edgemont Road, Charlottesville, VA 22903, USA

[3]Shanghai Astronomical Observatory, Chinese Academy of Sciences, 200030 Shanghai, China

[4]Max-Planck-Institut für Radioastronomie, Auf dem Hügel 69, D-53121 Bonn, Germany

[5]Astronomy Department, Faculty of Science, King Abdulaziz University, P.O. Box 80203, Jeddah 21589, Saudi Arabia

[6]Joint Alma Office, Alsonso de Cordova 3107, Vitacura, Santiago, Chile

[7]Institut für Theoretische Physik und Astrophysik, Universität Würzburg, Emil-Fischer-Str. 31, D-97074 Würzburg, Germany

[8]Dr. Remeis Sternwarte & ECAP, Universität Erlangen-Nürnberg, Sternwartstrasse 7, D-96049 Bamberg, Germany

[9]Harvard-Smithsonian Center for Astrophysics, 60 Garden Street, Cambridge, MA 02138, USA

Abstract. Many accretion disks surrounding supermassive black holes in nearby AGN are observed to host 22 GHz water maser activity. We have analyzed single-dish 22 GHz spectra taken with the GBT to identify 32 such "Keplerian disk systems," which we used to investigate maser excitation and explore the possibility of disk reverberation. Our results do not support a spiral shock model for population inversion in these disks, and we find that any reverberating signal propagating radially outwards from the AGN must constitute <10% of the total observed maser variability. Additionally, we have used ALMA to begin exploring the variety of sub-mm water megamasers that are also predicted, and in the case of the 321 GHz transition found, to be present in these accretion disks. By observing multiple masing transitions within a single system, we can better constrain the physical conditions (e.g., gas temperature and density) in the accretion disk.

Keywords. accretion disks, masers, galaxies: active

1. Introduction

The Megamaser Cosmology Project (MCP) is an effort to make geometric distance measurements to H_2O megamaser-hosting galaxies in the Hubble flow to constrain H_0 to a precision of several percent (Braatz *et al.* 2013). Using the "megamaser technique," first employed by Herrnstein *et al.* (1999) on the galaxy NGC 4258, the MCP has published distances to the galaxies UGC 3789 (Reid *et al.* 2013), NGC 6264 (Kuo *et al.* 2013), NGC 6323 (Kuo *et al.* 2015), and NGC 5765b (Gao *et al.* 2016), and additional distance measurements are under way.

The MCP has conducted a large single-dish survey of Seyfert 2 galaxies using the Robert C. Byrd Green Bank Telescope (GBT)†, with the goal of discovering suitable

† The National Radio Astronomy Observatory is a facility of the National Science Foundation operated under cooperative agreement by Associated Universities, Inc.

megamaser candidates for distance measurements. In Pesce *et al.* (2015; hereafter P15) we identified a list of 32 "clean" megamaser disk systems from the MCP survey and the literature. Dynamically "clean" disk megamaser systems are those for which the gravitational potential acting on the maser clouds is dominated by the central SMBH (i.e., there are no outflow or jet components), and all such masers have characteristic single-dish spectral profiles marked by three distinct groups of maser features centered around the systemic velocity of the host galaxy ("triple-peaked" profiles). The central group ("systemic masers") coincides roughly with the recession velocity of the galaxy, and the masing arises along a line of sight through the disk to the central AGN. The two sets of "high-velocity" features, redshifted and blueshifted from the galaxy recession velocity, arise from the midline of the accretion disk along lines of sight that are tangent to the orbital motion.

2. Testing a spiral shock model

Maoz & McKee (1998; hereafter MM98) developed a model for maser excitation that sought primarily to explain two seemingly unusual properties of the high-velocity maser features in NGC 4258: (1) the features show periodic radial placement within the disk (see, e.g., Argon *et al.* 2007), and (2) the redshifted features are stronger by roughly an order of magnitude than the blueshifted features (see, e.g., Humphreys *et al.* 2008). In the MM98 model, the only regions of the disk that exhibit masing are the trailing edges of spiral shock waves (i.e., the post-shock regions behind the spiral shocks). We then observe high-velocity features wherever the line of sight is tangent to a spiral shock, because this geometry maximizes the gain path for maser amplification.

The MM98 model makes two testable predictions for the behavior of the high-velocity maser features in disk systems: (1) the redshifted features should appear systematically stronger than the blueshifted features, and (2) all high-velocity features should exhibit velocity drifts towards lower orbital velocities. Both of these predictions stem from the geometrical configuration of spiral shocks in a disk, which cause the redshifted features to originate from a region of the disk that lies in front of the midline while the reverse is true for the blueshifted features (see Figure 1 in MM98). Because the majority of the disk contains noninverted gas, the blueshifted photons pass through a region of velocity-coherent material that presents an absorption opportunity not seen by the redshifted photons. This preferential absorption of blueshifted photons then explains the observed red-blue flux asymmetry. As the (trailing) spiral shock rotates through the disk, the segment of the spiral that lies tangent to the line of sight (and thus the region of masing gas) moves radially outward with time. In a Keplerian disk, the orbital velocity is a monotonically decreasing function of radius, so the rotation of the spiral shock causes the observed maser feature velocities to drift with time. Specifically, the redshifted features are predicted to exhibit a negative velocity drift while the blueshifted features show a positive velocity drift.

Both Bragg *et al.* (2000) and Humphreys *et al.* (2008) found that the accelerations of high-velocity maser features in NGC 4258 are inconsistent with the MM98 model. We have further tested whether the model holds for disk maser systems in general by measuring the flux asymmetry and velocity drifts of high-velocity features in the MCP sample. Figure 1 shows a histogram of the redshifted-to-blueshifted flux ratios for all disk maser systems in our sample; a detailed description of the analysis is presented in P15. We can see that the mean of the flux ratio distribution shows no statistically significant deviation from unity, and that NGC 4258 is a clear outlier. In Table 2 of P15 we list the

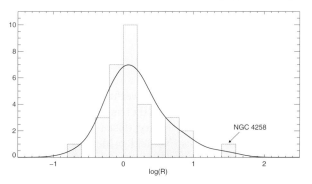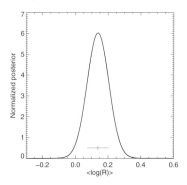

Figure 1. *Left*: a histogram of $\log(R)$ for the disk maser sample, where $R \equiv F_{\mathrm{red}}/F_{\mathrm{blue}}$ is the ratio of the redshifted to blueshifted flux. The bin containing NGC 4258 is marked, and we can see that it lies off to one side of the bulk distribution. The overplotted black curve shows the histogram smoothed using a Gaussian kernel; see P15 for details. *Right*: The normalized posterior distribution for the mean of $\log(R)$, computed as described in P15. The best-fit value of 0.136 ± 0.07 (1σ uncertainty), which is consistent with the parent distribution having a mean flux ratio of unity, is marked.

velocity drifts measured from the 11 most well-monitored maser systems in our sample; we found no evidence for the systematic velocity drifts predicted by the MM98 model.

3. A search for disk reverberation

The unique geometry and simple dynamics of Keplerian disk systems enable us to convert spectral information (i.e., velocity of a maser feature) directly into spatial information (i.e., radial location within the disk), without requiring a high-resolution VLBI map to spatially resolve the maser distribution. If a signal of some sort that affects the maser intensity is propagating radially through the disk at a fixed velocity – such as might be expected if, e.g., a flare from the AGN causes a change in the maser pumping efficiency – we can thus track the progress of that signal using spectral monitoring observations alone. Furthermore, by measuring the progression of that signal through the maser spectrum we could potentially make an independent measurement of the SMBH mass. For a signal propagating with a speed v_{s} through the disk, corresponding to a drift $\dot{v}$ through the maser spectrum, the SMBH mass can be obtained using

$$M_{\mathrm{BH}} = -\frac{v_{\mathrm{s}}\,(v - v_0)^3}{2G\dot{v}}. \tag{3.1}$$

Here, v_0 is the recession velocity of the dynamic center and v is the observed velocity of the signal (i.e., the velocity that one would read off of the spectrum). If we know the value of v_{s} – such as would be the case for a signal propagating at c, for instance – then we can measure the quantities v_0, v, and $\dot{v}$ from the spectral monitoring and thereby derive the mass of the SMBH. Such a mass measurement can then further be combined with VLBI data to determine an independent distance to the system.

We have searched for a radially-propagating signal using the time-series spectra for our 6 most well-monitored sources. The details of our signal-extraction procedure are given in P15, and we put upper limits on the strength of any such propagating signal at a typical 3σ level of $\sim$1-2 mJy. Because the high-velocity features in these maser spectra typically display variability at the $\sim$tens of mJy level, we can therefore say that any contribution to this variability from radially-propagating signals must occur at the $\lesssim$10% level. The variability that we see is therefore most likely dominated by local processes, rather than some global influence from the central engine.

4. (Sub)millimeter H$_2$O megamasers

Water megamasers emitting at 22 GHz have proven to be powerful tools for studying AGN, cosmology, and SMBHs. Population inversion of the 22 GHz transition occurs under a range of physical conditions (i.e., gas temperature, gas density, dust temperature) that are also predicted to invert a number of other transitions (Neufeld & Melnick 1991; Yates *et al.* 1997). Many of these transitions emit at (sub)millimeter wavelengths, and several are expected to have line luminosities comparable to or greater than what is observed at 22 GHz (see, e.g., Gray *et al.* 2016). By observing multiple masing transitions in a single Keplerian disk system, we can constrain the physical conditions in the accretion disk as a function of orbital radius.

In Pesce *et al.* (2016), we presented an ALMA detection of 321 GHz water megamaser emission towards NGC 4945 and an updated calibration of the previously-discovered (see Hagiwara *et al.* 2013) 321 GHz water megamaser in Circinus. The 321 GHz masers in Circinus are about an order of magnitude weaker in flux density than the 22 GHz masers, though the isotropic luminosity is actually ∼4 times stronger at 321 GHz. In NGC 4945 the 321 GHz masers are down in flux density by about two orders of magnitude from their 22 GHz counterparts, with an isotropic luminosity that is smaller by roughly one order of magnitude. Though neither of these systems contains a "clean" Keplerian disk, they serve to illustrate the feasibility of extending megamaser science to the (sub)millimeter regime. Future ALMA observations will seek to map out these systems and to discover many more.

References

Argon, A. L., Greenhill, L. J., Reid, M. J., Moran, J. M., & Humphreys, E. M. L. 2007, *ApJ*, 659, 1040

Braatz, J., Reid, M., Kuo, C.-Y., *et al.* 2013, Advancing the Physics of Cosmic Distances, 289, 255

Bragg, A. E., Greenhill, L. J., Moran, J. M., & Henkel, C. 2000, *ApJ*, 535, 73

Gao, F., Braatz, J. A., Reid, M. J., *et al.* 2016, *ApJ*, 817, 128

Gray, M. D., Baudry, A., Richards, A. M. S., *et al.* 2016, *MNRAS*, 456, 374

Hagiwara, Y., Miyoshi, M., Doi, A., & Horiuchi, S. 2013, *ApJL*, 768, L38

Herrnstein, J. R., Moran, J. M., Greenhill, L. J., *et al.* 1999, *Nature*, 400, 539

Humphreys, E. M. L., Reid, M. J., Greenhill, L. J., Moran, J. M., & Argon, A. L. 2008, *ApJ*, 672, 800-816

Kuo, C. Y., Braatz, J. A., Reid, M. J., *et al.* 2013, *ApJ*, 767, 155

Kuo, C. Y., Braatz, J. A., Lo, K. Y., *et al.* 2015, *ApJ*, 800, 26

Maoz, E. & McKee, C. F. 1998, *ApJ*, 494, 218

Neufeld, D. A. & Melnick, G. J. 1991, *ApJ*, 368, 215

Pesce, D. W., Braatz, J. A., Condon, J. J., *et al.* 2015, *ApJ*, 810, 65

Pesce, D. W., Braatz, J. A., & Impellizzeri, C. M. V. 2016, *ApJ*, 827, 68

Reid, M. J., Braatz, J. A., Condon, J. J., *et al.* 2013, *ApJ*, 767, 154

Yates, J. A., Field, D., & Gray, M. D. 1997, *MNRAS*, 285, 303

Astrophysical Masers:
Unlocking the Mysteries of the Universe
Proceedings IAU Symposium No. 336, 2017
A. Tarchi, M.J. Reid & P. Castangia, eds.

© International Astronomical Union 2018
doi:10.1017/S1743921317010596

A new jet/outflow maser in the nucleus of the Compton-thick AGN IRAS 15480-0344

Paola Castangia[1], Andrea Tarchi[1], Alessandro Caccianiga[2], Paola Severgnini[2], Gabriele Surcis[1] and Roberto Della Ceca[2]

[1] INAF-Osservatorio Astronomico di Cagliari, Via della Scienza 5
09047, Selargius (CA), Italy
email: pcastangia@oa-cagliari.inaf.it

[2] INAF-Osservatorio Astronomico di Brera, Via Brera 28
20121, Milano, Italy

Abstract. Investigations of H_2O maser galaxies at X-ray energies reveal that most harbor highly absorbed AGN. Possible correlations between the intrinsic X-ray luminosity and the properties of water maser emission have been suggested. With the aim of looking into these correlations on a more solid statistical basis, we have search for maser emission in a well-defined sample of Compton-thick AGN. Here we report the results of the survey, which yielded a surprisingly high maser detection rate, with a particular focus on the newly discovered luminous water maser in the lenticular (field) S0 galaxy IRAS 15480-0344. Recently, VLBI observations have been obtained to image the line and continuum emission in the nucleus of this galaxy. The radio continuum emission at VLBI scales is resolved into two compact components that are interpreted as jet knots. Based on the single-dish profile, the variability of the maser emission, and the position of the maser spots with respect to these continuum sources, we favor of a jet/outflow origin for the maser emission, consistent with similar cases found in other radio-quiet AGN. This scenario is consistent with the hypothesis of the presence of strong nuclear winds recently invoked to explain the main characteristics of field S0 galaxies.

Keywords. Masers, galaxies: active, galaxies: nuclei.

1. Introduction

Active galactic nuclei (AGN) associated with water masers are usually characterized by high levels of absorption both in the optical (i.e. they are spectroscopically classified as Sy2 or LINERS) and in the X-rays (e.g. Braatz *et al.* 1997; Madejski *et al.* 2006; Zhang *et al.* 2006). In the X-rays, in particular, the measured columns densities are usually above 10^{23} cm^{-2}, with a large fraction of sources in the Compton-thick (CT) regime ($N_H > 10^{24}$ cm^{-2}), especially those hosting H_2O masers associated with a nuclear accretion disc (Greenhill *et al.* 2008; Castangia *et al.* 2013).

The maser phenomena in AGN are thought to be intimately connected to X-ray emission, since photons at these energies are the most promising candidates to provide the necessary excitation mechanism for the maser emission (Neufeld *et al.* 1994). In this picture, a relationship between X-ray luminosity and the intensity of the maser emission is expected. Hints of such a correlation have been claimed (Kondratko, Greenhill & Moran 2006), although the level of significance is low, possibly due to the small number of sources used in the analysis. Larger and well-defined samples of AGN, with good X-ray data, are required to better study the fundamental relationship between X-ray properties and maser emission in this class of sources.

Far-infrared (FIR; 20–$100\,\mu m$) emission is also thought to play an important role in water maser emission. FIR emission has been proved to be an efficient indicator of the

129

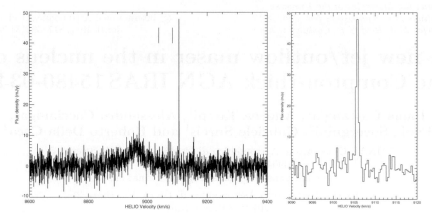

Figure 1. *Left panel:* GBT spectrum of the H_2O maser emission in IRAS15480. The vertical lines mark the recessional velocity of the galaxy and the CO centroid velocity, 9084 and 9039 km s^{-1}, respectively (Castangia *et al.* 2016). *Right panel:* Zoom of the narrow emission feature.

presence of extragalactic water masers, both those associated with star formation and those associated to AGN (Henkel *et al.* 2005; Castangia *et al.* 2008; Darling 2011). In particular, a sample of galaxies with IRAS point source flux density >50 Jy, yielded a maser detection rate of 23%, among the highest ever obtained in extragalactic maser searches (Henkel *et al.* 2005; Surcis *et al.* 2009). Hence, "ad hoc" samples of AGN, selected on the basis of their X-ray and FIR emission, have the potential, not only to shed light on the properties of the innermost region of AGN, but also constitute a promising group of targets to search for water maser emission.

With the purpose of studying the X-rays/maser connection, we searched for 22 GHz water maser emission in a well defined sample of 36 CT AGN, selected in the local Universe through a combination of mid-IR (*IRAS*) and X-ray (*XMM-Newton*) data (for details, see Severgnini *et al.* 2012). All the galaxies in the sample were already observed at 22 GHz in previous surveys, and water maser emission was detected in 17/36 of them. We re-observed some of the non-detected sources in order to improve their upper limit on the maser luminosity. These new observations lead to the discovery of a new luminous water maser in the lenticular (S0) galaxy IRAS 15480-0344 (hereafter IRAS15480) confirming the exceptionality of this sample (Castangia *et al.* 2016). Indeed, with the detection of IRAS15480, the maser detection rate of the entire sample becomes 50%, making the CT AGN sample used here one of the most prolific samples for finding extragalactic water masers (Castangia *et al.* in prep.). Here we present the H_2O maser in IRAS15480 and discuss its origin in light of new VLBI images of the radio continuum emission in the nuclear region of the galaxy.

2. The maser in IRAS15480

IRAS15480 is an isolated lenticular galaxy that harbors a Seyfert 2 nucleus (Young *et al.* 1996). On April 7, 2012, we detected 22 GHz water maser emission in IRAS15480 using the Green Bank Telescope (GBT). The spectrum consists of two main features: a broad blueshifted component, with a full width at half maximum (FWHM) linewidth of $\sim$90 km s^{-1} and a narrow (FWHM<1 km s^{-1}) one close to the systemic velocity of the galaxy (Fig. 1; for details, see Castangia *et al.* 2016). The total isotropic luminosity is $\sim$200 L$_\odot$. Single-dish spectra at two subsequent epochs revealed the stability of the narrow component, whose properties (velocity, linewidth and peak flux density)

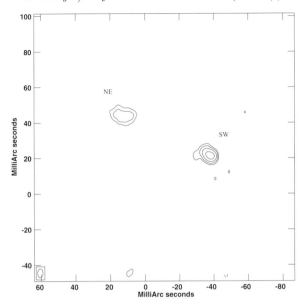

Figure 2. EVN map of the radio continuum emission in IRAS15480 at 5 GHz (Castangia *et al.* in prep.). Contour levels are (-1, 1, 2, 4, 8...)×0.2 mJy/beam (5σ).

remained constant within the errors in the three epochs. On the contrary, the peak and integrated flux density of the broad emission feature decreased to half of their initial values in approximately one month. Following the discovery, we performed interferometric observations with the Very Long Baseline Array (VLBA) and confidently detected the narrow component, which is located in the nuclear region of the galaxy. A weaker maser feature has also been detected in the velocity range of the broad component. The spatial separation between the maser spots is ∼15 pc. The different single-dish profiles and variability, together with the spatial separation between the two features suggest a composite origin for the maser emission in IRAS15480. On the basis of its large linewidth and strong spectral variability, we believe that the broad component might originate from the interaction of a jet with the interstellar medium of the host galaxy. The small linewidth and the absence of high velocity features, instead, indicate a possible association with a nuclear outflow or wind for the narrow component. Although we favour a jet/outflow origin for the maser in IRAS15480, the possibility that all of the maser emission is produced in a slowly rotating ($V_{\rm rot} \sim 110\,{\rm km\,s^{-1}}$) accretion disc, with a rather large radius (15 pc), however, cannot be completely ruled out.

3. The radio continuum emission from the nucleus of IRAS15480

The Very Large Array (VLA) detected radio emission towards the nucleus of IRAS15480 at 1.7 and 8.4 GHz with measured flux densities of 42 mJy (NVSS; Condon *et al.* 1998) and 11 mJy (Schmitt *et al.* 2001), respectively. The 8.4 GHz VLA map is, to date, the highest resolution image of the nuclear region of IRAS15480 and shows an unresolved radio source with less than 60 pc diameter (Schmitt *et al.* 2001). In order to study the nuclear region of IRAS15480 with an angular resolution comparable with that of our VLBA spectral line maps, we observed the nucleus of IRAS15480 with the European VLBI network (EVN) at 1.7 and 5 GHz, in the period between February and March 2015. The EVN images display two bright sources (Fig. 2): a slightly resolved source located in the southwest (SW) and a second more extended one at distance of $\sim$30 pc along P.A.$\sim$ 60° (NE). Interestingly, one of the two maser spots (the one corresponding to the broad blueshifted line) is coincident with source NE, while the narrow line emission at the systemic velocity seems to originate in a region between the two continuum components (Castangia *et al.* in prep.). The flat spectral index and high brightness temperature of the southwestern source suggest that its radio emission might be interpreted as synchrotron self-absorbed emission from the base of a jet. The presence of the northeastern source with a steeper spectral index reinforces this interpretation and points towards a "compact-jet" scenario to explain the radio emission from the nucleus of IRAS15480, as have been proposed for other Seyfert galaxies (e. g. Caccianiga *et al.* 2001; Giroletti & Panessa 2009; Bontempi *et al.* 2012). In this context, the association of the blueshifted maser spot with the northeastern source would be in agreement with our initial interpretation that part of the maser emission originates from a jet-cloud interaction (e. g. NGC1068, Gallimore *et al.* 2001; Mrk348, Peck *et al.* 2003), indicating that we may have added a new source to the few confirmed jet/outflow masers reported so far (Tarchi *et al.* 2012, and references therein).

References

Braatz, J. A. & Wilson, A. S., Henkel 1997, *ApJS*, 110, 321

Bontempi, P., Giroletti, M., Panessa, F., Orienti, M., & Doi, A 2012, *MNRAS*, 426, 588

Caccianiga, A., Marchã, M. J. M., Thean, A., Dennett-Thorpe, J. 2001 *MNRAS*, 328, 867

Castangia, P., Tarchi, A., Henkel, C., & Menten, K. M. 2008 *A&A*, 479, 111

Castangia, P., Panessa, F., Henkel, C., Kadler, M., & Tarchi, A. 2013 *MNRAS*, 436, 3388

Castangia, P., Tarchi, A., Caccianiga, A., Severgnini, P., & Della Ceca, R. 2016 *A&A*, 586, 89

Condon, J. J., Cotton, W. D., Greisen, E. W., Yin, Q. F., Perley, R. A., Taylor, G. B., & Broderick, J. J. 1998 *AJ*, 115, 1693

Darling, J. 2011, *ApJ*, 732, 2

Gallimore, J. F., Henkel, C., Baum, S. A., Glass, I. S., Claussen, M. J., Prieto, M. A., & Von Kap-herr, A. 2001, *ApJ*, 556, 694

Giroletti, M. & Panessa, F 2009, *ApJ*, 706, 260

Greenhill, L. J., Tilak, A., & Madejski, G. 2008, *ApJ*, 686, 13

Henkel, C., Peck, A. B., Tarchi, A., Nagar, N. M., Braatz, J. A., Castangia, P., & Moscadelli, L. 2005, *A&A*, 436, 75

Kondratko, P. T., Greenhill, L. J., & Moran, J. M. 2006, *ApJ*, 652, 136

Madejski, G., Done, C., Życki, P. T., & Greenhill, L. 2006, *ApJ*, 636, 75

Neufeld, D. A., Maloney, P. R., & Conger, S. 1994, *ApJ*, 436, 127

Peck, A. B., Henkel, C., Ulvestad, J. S., Brunthaler, A., Falcke, H., Elitzur, M., Menten, K. M., & Gallimore, J. F. 2001, *ApJ*, 590, 149

Schmitt, H. R., Ulvestad, J. S., Antonucci, R. R. J., & Kinney, A. L 2001, *ApJS*, 132, 199

Severgnini, P., Caccianiga, A., & Della Ceca, R. 2012, *A&A*, 542, 46

Surcis, G., Tarchi, A., Henkel, C., Ott, J., Lovell, J., & Castangia, P. 2009, *A&A*, 502, 529

Young, S., Hough, J. H., Efstathiou, A., Wills, B. J., Bailey, J. A., Ward, M. J., & Axon, D. J. 1996, *MNRAS*, 281, 1206

Tarchi, A. 2012, in: R. S. Booth, E. M. L. Humphreys, & W. H. T. Vlemmings (eds.), *Cosmic Masers - from OH to H0*, Proc. IAU Symposium No. 287 (Stellembosh: CUP), p. 323

Zhang, J. S., Henkel, C., Kadler, M., Greenhill, L. J., Nagar, N., Wilson, A. S., & Braatz, J. A. 2006, *A&A*, 450, 933

Astrophysical Masers:
Unlocking the Mysteries of the Universe
Proceedings IAU Symposium No. 336, 2017
A. Tarchi, M.J. Reid & P. Castangia, eds.

© International Astronomical Union 2018
doi:10.1017/S1743921317009383

On the low detection efficiency of disk water megamasers in Seyfert 2 AGN

Alberto Masini[1,2] and Andrea Comastri[1]

[1]INAF-Osservatorio Astronomico di Bologna, via Gobetti 93/3, 40129 Bologna, Italy

[2]Dipartimento di Fisica e Astronomia, Università di Bologna, via Gobetti 93/2, 40129 Bologna, Italy

email: `alberto.masini4@unibo.it`

Abstract. Disk megamasers are a unique tool to study active galactic nuclei (AGN) sub-pc environment, and precisely measure some of their fundamental parameters. While the majority of disk megamasers are hosted in heavily obscured (i.e., Seyfert 2, Sy2) AGN, the converse is not true, and disk megamasers are very rarely found even in obscured AGN. The very low detection rate of such systems in Sy2 AGN could be due to the geometry of the maser beaming, which requires a strict edge-on condition. We explore some other fundamental factors which could play a role in a volume-limited survey of disk megamasers in Sy2 galaxies, most importantly the radio luminosity.

Keywords. galaxies: active, masers, galaxies: general

---

1. Introduction

Many works found a striking correspondence between the presence of a megamaser disk and Compton-thick absorption in the X-ray band (e.g., Greenhill *et al.* 2008, Castangia *et al.* 2013, Masini *et al.* 2016), implying a tight link between megamaser disks and Sy2 AGN, with the masing disk tracing the toroidal obscuring medium invoked by AGN unification models. However, since the maser beaming angle is quite narrow (15°-20°) and maser amplification occurs preferentially along the disk equatorial plane, disks are detected when they are almost edge-on. This makes them hard to detect even in Sy2 AGN, with an observed detection efficiency $\lesssim 1\%$ (Braatz *et al.* in prep). Although not corrected for the survey sensitivity, this value is worth to be explored.

2. Covering factors and Radio luminosity

Assuming that a maser disk is detected if the line of sight angle ranges between $(90 \pm 10)°$ with respect to the polar axis, its half-opening angle, defined as the angle between the symmetry axis of the system and the edge of the disk, is $\theta_{\rm disk} = 80°$. A Sy2 galaxy, on the other hand, is detected when the torus intercepts the line of sight, and we can analogously define its opening angle $\theta_{\rm tor}$. The average Sy2 covering factor (CF) can range between that of the maser disk itself ($\theta_{\rm tor} = \theta_{\rm disk}$) and a CF = 1 (i.e. a sphere, $\theta_{\rm tor} = 0$), where the covering factor of a toroidal structure with an half-opening angle θ is defined as CF $= \sin(\pi/2 - \theta)$. The probability of detecting a maser disk in a Sy2 AGN, P, is then given by the ratio of the maser disk covering factor with respect to the torus one:

$$P = \frac{\rm CF_{disk}}{\rm CF_{tor}} = \frac{\sin(\pi/2 - \theta_{\rm disk})}{\sin(\pi/2 - \theta_{\rm tor})}. \qquad (2.1)$$

This probability gives the intrinsic detection efficiency (i.e., when correcting for survey completeness) of disk masers in Sy2 AGN. In the left panel of Figure 1, the maser disk

133

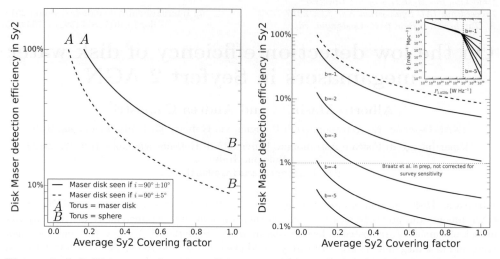

Figure 1. *Left.* Disk maser detection efficiency as a function of average Sy2 covering factor, assuming two different thickness for the maser disk. *Right.* Effect of different slopes b of the 1.4 GHz LF (inset) on the intrinsic detection efficiency.

detection efficiency is shown as a function of the average Sy2 covering factor (making $\theta_{\rm tor}$ vary, assuming $\theta_{\rm disk} = 80°$ and $\theta_{\rm disk} = 85°$; black solid and dashed lines, respectively). With this simple geometric argument, we get that the *intrinsic* fraction of maser disks in Sy2 AGN is always higher than $\sim 10\%$. This means that, in a volume-limited survey, the intrinsic fraction of Sy2 AGN hosting a disk maser should be $> 10\% - 20\%$, depending on the effective beaming angle of the maser disk, in contrast with observations.

An hard X-ray luminosity threshold $L_{\rm X}$ might be a crucial parameter in order to explain the observed low detection efficiency, and a proper hard X-ray selection may boost the observed detection efficiency in radio surveys to $20\% - 30\%$ (see F. Panessa, this volume).

The radio power may be another crucial factor to be considered, since there is evidence that megamaser galaxies are a factor of $2-3$ radio brighter than non-megamaser galaxies (Liu *et al.*, 2017; Zhang *et al.*, 2017). In particular, adopting the radio luminosity function (LF) at 1.4 GHz of Pracy *et al.* (2016), one can compute the probability for a Sy2 galaxy to have a radio power higher than, e.g., $2P_*$. Depending on the unconstrained high-end of the LF, different probabilities are computed, which lower the intrinsic fraction of maser disks in Sy2 or, equivalently, boost the detection efficiency of radio surveys (right panel of Figure 1). Measuring the exact value of the intrinsic detection efficiency of disk megamasers and the shape of the high frequency radio LF will shed new light on the average Sy2 covering factor.

References

Castangia, P., Panessa, F., Henkel, C., Kadler, M., & Tarchi, A. 2013, *MNRAS*, 436, 3388
Greenhill, L. J., Tilak, A., & Madejski, G. 2008, *ApJL*, 686, L13
Liu, Z. W., Zhang, J. S., Henkel, C., et al. 2017, *MNRAS*, 466, 1608
Masini, A., Comastri, A., Baloković, M., et al. 2016, *A&A*, 589, A59
Pracy, M. B., Ching, J. H. Y., Sadler, E. M., et al. 2016, *MNRAS*, 460, 2
Zhang, J. S., Liu, Z. W., Henkel, C., Wang, J. Z., & Coldwell, G. V. 2017, *ApJL*, 836, L20

Astrophysical Masers:
Unlocking the Mysteries of the Universe
Proceedings IAU Symposium No. 336, 2017
A. Tarchi, M.J. Reid & P. Castangia, eds.

© International Astronomical Union 2018
doi:10.1017/S1743921317010043

Searching for warped disk AGN candidates

E. Fedorova[1], B. I. Hnatyk[1], V. I. Zhdanov[1] and A. Vasylenko[2]

[1]Astronomical Observatory of Taras Shevchenko National University of Kyiv,
Observatorna str. 3, 04053 Kiev, Ukraine, email: efedorova@ukr.net

[2]Main Astronomical Observatory of the NASU, Zabolotnogo 27, 03680, Kiev, Ukraine

Abstract. Mapping the maser emission of subnuclear regions of active galactic nuclei (AGN) enable us to determine some interesting details of the geometry of the accretion disks (AD) under the condition that they have "maser skin". Additional information about disk warp in the innermost zone near the central black hole (BH) can be disclosed by means of modeling the shape of the relativistically broadened iron emission lines in the energy range 6-7 keV. Here we analyze the influence of the AD geometry (warp) on the shape of the set of relativistically broadened emission lines, as well as consider some examples of AGNs identified by maser mapping techinque as warped and having the complex shape of iron lines near 6.4 keV.

Keywords. AGN, accretion: warped accretion disk

It was shown by Bardeen & Peterson (1975) that Lense-Thirring effect in accretion systems with misaligned angular momentum of AD and BH spin leads to the deformation of the disk geometry, so-called disk warp/tilt. Some physical factors which can cause such misalignment were considered recently by Chakraborty & Bhattacharya (2017), including instability due to comparably young age of the system, inhomogeneity or clumpiness of the accretion disk, as well as subnuclear star formation processes.

Following Ingram *et al.* (2016), observational appearances of such warped disks include: 1. if disk is precessing, it is possible to see specific timing features in X-ray lightcurves; 2. if hot surface spots are present, sub-pc scale warp can be traced from VLBI observations (Greenhill 2003, Kartje *et al.* 1999); 3. the shape of emission lines and continuum spectra formed in the vicinity of the central BH depends significantly on the disk geometry (Reynolds *et al.* 2009), that is why the warp can be traced out from X-ray spectral data.

Leaving the explanations of the possible physical reasons of such misalignment, we consider here some observational appearances of AGN with warped accretion disks (WAD) and the methods enabling us to reconstruct the AD geometry in such systems. We analyze the effect of the disk warp on the shape of Fe-K emission lines, using the simplified model of WAD consisting of two flat disks with angle $\Delta\theta$ between them (Fig.1, the left panel). The resulting simulated line profiles are shown on the right panel of the Fig.1, in comparison with the regular one.

We applied this model to fit the observational data on Fe-K lines in several AGNs, where their shape is not interpretable within the regular AD model. These objects are: NGC 1194 (XMM-Newton/EPIC and Suzaku/XIS data were used), NGC 1068, IC 2560, NGC 5506, and the Circinus (XMM-Newton/EPIC data). The results of our fitting, and the information about the megamaser disks and BH masses are shown in the Table.1.

Analysis of the X-ray data on Fe-K lines in AGNs give us the possibility to disclose the peculiarities of the line profile which can be interpreted as a sign of WAD in that AGN. However, one should note that there can be alternative explanations (for instance, NGC 1194, Fedorova *et al.* (2015)) of such peculiarity. Combining the data of Fe-K line observations with the results of maser emission observations enable us to be more

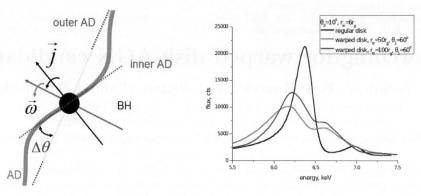

Figure 1. Left panel: WAD Model: two disks with angular momenta. Right panel: Simulated Fe-K line profiles for WADs with different transition radii R_{tr} vs. regular one. θ_0 and θ_1 are inner and outer disk inclination angles, and R_{in} is the inner disk radius.

Table 1. Warped disk candidates and their observational parameters.

Object	megam.disk, pc	M_{BH}, $10^7 M_\odot$	E_α, keV	E_β, keV	R_{in}^*	R_{tr}^*	θ_0	θ_1
NGC 1194	regular, 0.54-1.33[1]	6.6^2	6.4±0.03	7.01±0.11	7^{+10}_{-3}	500±150	44±10	7^{+10}_{-1}
NGC 1068	warped, 0.57-0.92[1,3]	1.5^4	6.4±0.02	6.95±0.07	7^{+6}_{-1}	86±5	24±10	14±4
IC 2560	warped, 0.09-0.34[5]	0.44^6	6.37±0.04	7.02±0.10	8^{+12}_{-2}	140±90	37±10	< 20
NGC 5506	not identified[7]	10^8	6.40±0.04	–	7^{+22}_{-1}	160±90	40±15	< 10
Circinus	warped, 0.11-0.4[9]	1.7^{10}	6.44±0.03	7.07±0.05	9^{+5}_{-3}	101±10	35±1	4±3

Notes: *in units of R_g; [1]Kartje *et al.* (1999); [2]Kuo (2011); [3]Caproni *et al.* (2005) (note that the tilt angle $\Delta\theta = \theta_0 - \theta_1 = 10^o \pm 14^o$ is in a good concordance with the value of $\Delta\theta \approx 11^o$ mentioned there); [4]Greenhill & Gwinn (1997); [5]Yamauchi *et al.* (2012); [6]Graham (2008); [7]Tarchi *et al.* (2011); [8]Uttley & McHardy (2005); [9]Greenhill (2003); [10]Gnerucci *et al.* (2013).

confident considering the WAD hypothesis as the more appropriate one. In the same time, it gives us possibility to reconstruct more detailed map of the AD, where the outer zones are reconstructed from maser emission mapping down to scales of 10^{-2} pc and the inners ones (usually below 10^{-3} pc) are traced out by fitting the Fe-K line profiles.

References

Bardeen J. H. & Peterson J. A. 1975, *ApJ*, 195, L15
Caproni A., Abraham Z., Cuesta H. J. M. 2005, *BJP*, 35(4B), 1167
Chakraborty D. & Bhattacharya S. 2017, *MNRAS*, 469, 3062
Fedorova E., Vasylenko A., Hnatyk B. I., & Zhdanov V. I. 2015, *Astron.Nachr.*, 364, 25
Gnerucci A., Marconi A., Capetti A., Axon D. J., & Robinson A. 2013, *A&A*, 549, A139
Graham A. W. 2008, *PASA*, 25, 167.
Greenhill L. 2003, *ApJ*, 48, 171
Greenhill L. & Gwinn C. R. 1997, *A&SS*, 248(1-2), 261
Ingram A., van der Klis M., Middleton M. *et al.* 2016, *MNRAS*, 461(2), 1967
Kartje J. F., Koenigl A., & Elitzur M. 1999, *ApJ*, 513, 180
Kuo C. Y. 2011, *ApJ*, 727(1), 20
Reynolds C. S., Nowak M. A., Markoff S. *et al.* 2009, *ApJ*, 691, 1159
Tarchi A., Castangia P., Columbano A. *et al.* 2011, *POS Science (NLS1)*, 031
Uttley P. & McHardy I. M. 2005, *MNRAS*, 363(2), 586
Yamauchi A., Nakai N., Ishihara Y., Diamond P., & Sato N, 2012, *PASJ*, 64(5), 103

Astrophysical Masers:
Unlocking the Mysteries of the Universe
Proceedings IAU Symposium No. 336, 2017
A. Tarchi, M.J. Reid & P. Castangia, eds.

© International Astronomical Union 2018
doi:10.1017/S1743921317009759

A survey for OH masers in H_2O maser galaxies with the Effelsberg and Green Bank radio telescopes

Elisabetta Ladu[1], Andrea Tarchi[2], Paola Castangia[2], Gabriele Surcis[2] and Christian Henkel[3,4]

[1]Dipartimento di Fisica, Università degli Studi di Cagliari,
S.P.Monserrato- Sestu km 0,700, I-09042 Monserrato (CA), Italy
email: `ladueli91@gmail.com`

[2]INAF-Osservatorio Astronomico di Cagliari, via della Scienza 5, 09047, Selargius (CA), Italy

[3]Max-Planck-Institut für Radioastronomie, Auf dem Hügel 71,53121 Bonn, Germany

[4]Astronomy Departement, King Abdulaziz University, P.O. Box 80203, Jeddah 21589, Saudi Arabia

Abstract. We present a search for OH maser emission in galaxies hosting H_2O masers with the 100-m Effelsberg radio telescope and the Green Bank Telescope (GBT). This survey is aimed at investigating the apparent rarity and/or possible mutual exclusion of megamaser emission from OH and H_2O in the same galaxy. Our study establishes new and better upper limits on the OH maser luminosity. Our work duplicates the number of H_2O masers searched for OH emission. No new maser detections have been found. OH absorption, both in the 1667 and 1665 MHz transitions, is instead detected in two galaxies of the sample, IC342 and NGG5793.

Keywords. masers, techniques:spectroscopic, surveys, ISM:molecules, galaxies:active

1. Introduction

Luminous extragalactic masers are traditionally referred to as megamasers and, so far, are mostly seen in radio lines of either hydroxyl or water vapour. These species have been very rarely observed in the same galaxy. This peculiarity gives rise to different hypotheses mainly related to the extreme and different physical conditions for OH and H_2O maser emission and/or different (nuclear) evolutionary phases of the host galaxies, e.g., Tarchi *et al.* (2011). In this work, we present a search for OH maser emission in galaxies hosting H_2O masers observed with the Effelsberg radio telescope and the GBT. This survey is aimed at providing new clues to the debated rarity of the simultaneous presence of megamaser emission from OH and H_2O in the same galaxy, see, e.g., Wagner (2013) and Wiggins *et al.* (2016).

2. Observation and Data Reduction

The sample, comprising all water maser sources visible by Effelsberg (Dec. > -30°) that were never searched for OH maser emission, was observed in a search for 1.6-GHz OH maser emission. A subset of the aforementioned sample, mainly constituted by those sources affected by RFI in the Effelsberg data, was then observed also with the GBT. The bandwidth is 20 MHz for Effelsberg and 12.5 MHz for the GBT respectively, centered at the recessional velocity of the galaxy. The analysis of the Effelsberg data was performed using the GILDAS software (IRAM), while for the GBT data, we used the software GBTIDL (NRAO).

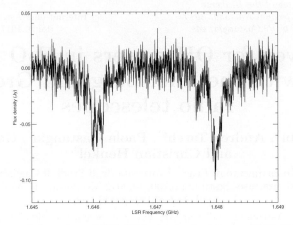

Figure 1. Spectrum of the galaxy NGC 5793 obtained with the GBT. The absorption lines are produced by the OH molecule at 1665 and 1667 MHz.

3. Results and Summary

In this work, we observed a total of 45 galaxies. For 10 of them, spectra were corrupted by RFI at the Effelsberg and GBT sites. Hence, our work establishes new upper limits on the OH maser luminosity for 35 galaxies, out of which 27 had no OH measurements reported, while for 8 galaxies we improved the upper limits. This work duplicates the number of extragalactic H_2O masers searched for OH emission. OH absorption features, both in the 1667 and 1665 MHz transitions, are detected for two galaxies in the sample: IC 342, for the first time, and NGC 5793 (see Fig. 1) confirming that reported, for the 1667-MHz transition only, by Hagiwara *et al.* (1997, and references therein). These detections provide promising targets for follow-up interferometric measurements useful to investigate the molecular gas in the (likely) nuclear regions of the hosts. No new OH maser detections have been found confirming the paucity of objects hosting (mega)maser emission from both molecular species and hinting at the necessity of extending the samples where maser emission from both molecular species are searched for and of detailed observations of the few objects known, so far, to host simultaneous hydroxyl and water megamaser emission. In this framework, an important role may be that played by the 64-m Sardinia Radio Telescope (SRT), since at L band the different RFI environment would allow for observations of those targets affected by RFI at the other sites. In addition, the sensitivity of the SRT at L and K bands makes it a very suitable instrument to pursue the aforementioned studies.

4. Acknowledgements

We are indebted to the operators at the 100-m telescope Effelsberg and the GBT for their assistance with the observations. AT and PC would also like to thank J. Braatz, J. Darling, and K. Willett for useful suggestions on the GBT observational setup.

References

Hagiwara, Y., Kohno, K., Kawabe, R., & Nakai, N. 1997, *PASJ*, 49, 171
Tarchi, A., Castangia, P., Henkel, C., Surcis, G., & Menten, K. M. 2011, *A&A*, 525A, 91
Wagner, J. 2013, *A&A*, 560A, 12
Wiggins, Brandon K. & Migenes, Victor, Smidt, & Joseph M. 2016, *ApJ*, 816, 55

Astrophysical Masers:
Unlocking the Mysteries of the Universe
Proceedings IAU Symposium No. 336, 2017
A. Tarchi, M.J. Reid & P. Castangia, eds.

© International Astronomical Union 2018
doi:10.1017/S174392131701095X

Radio continuum of galaxies with H_2O megamaser disks

F. Kamali[1], C. Henkel[1,2], A. Brunthaler[1], C. M. V. Impellizzeri[3,4], K. M. Menten[1], J. A. Braatz[3], J. E. Greene[5], M. J. Reid[6], J. J. Condon[3], K. Y. Lo[3], C. Y. Kuo[7], E. Litzinger[8,9] and M. Kadler[9]

[1] Max-Planck-Institut für Radioastronomie, Auf dem Hügel 69, 53121 Bonn, Germany
email: `fkamali@mpifr-bonn.mpg.de`
[2] Astron. Dept., King Abdulaziz University, P.O. Box 80203, Jeddah 21589, Saudi Arabia
[3] National Radio Astronomy Observatory, 520 Edgemont Road, Charlottesville, VA 22903, USA
[4] Joint ALMA Office, Alonso de Córdova 3107, Vitacura, Santiago, Chile
[5] Department of Astrophysical Sciences, Princeton University, Princeton, NJ 08544, USA
[6] Harvard-Smithsonian Center for Astrophysics, 60 Garden Street, Cambridge, MA 02138, USA
[7] Department of Physics, National Sun Yat-Sen University, No.70, Lianhai Road, Gushan Dist., Kaohsiung City 804, Taiwan (R.O.C.)
[8] Institut für Theoretische Physik und Astrophysik, Universität Würzburg, Campus Hubland Nord, Emil-Fischer-Str. 31, 97074 Würzburg, Germany
[9] Dr. Remeis-Observatory, Erlangen Centre for Astroparticle, Physics, University of Erlangen-Nüremberg, Sternwartstr. 7, 96049 Bamberg, Germany

Abstract. In our attempt to investigate the basic active galactic nucleus (AGN) paradigm requiring a centrally located supermassive black hole (SMBH), a close to Keplerian accretion disk and a jet perpendicular to its plane, we have searched for radio continuum in galaxies with H_2O megamasers in their disks. We observed 18 such galaxies with the Very Large Baseline Array in C band (5 GHz, ~2 mas resolution) and we detected 5 galaxies at $8\,\sigma$ or higher levels. For those sources for which the maser data is available, the positions of masers and those of the 5 GHz radio continuum sources coincide within the uncertainties, and the radio continuum is perpendicular to the maser disk's orientation within the position angle uncertainties.

Keywords. Galaxies: active – Galaxies: jets – Galaxies: nuclei - Galaxies: Seyfert – Radio continuum: galaxies

1. Introduction

Galaxies, where the 22 GHz H_2O maser line (from the H_2O vapor 6_{16} - 5_{23} rotational transition) traces their central accretion disk, provide a unique view into the pc to sub-pc region surrounding the SMBHs (a prototype is NGC 4258, see Herrnstein *et al.*1998). Observing the radio continuum (which traces outflows or jets launched by the central engine) with the same linear resolution as the maser disk, provides an opportunity to study the spatial relationship between the radio jets and megamaser disks in detail.

2. Sample and Observations

We initially observed a sample of 24 active galaxies with the Very Large Array (VLA) mostly in B configuration and in Ka-band (central frequency of 33 GHz, see Kamali *et al.* 2017). From the 21 detected galaxies, we observed 14 sources plus 4 other sources with the Very Large Baseline Array (VLBA) in C band and with 2 mas resolution, to map the radio continuum on the same linear resolution as that of the maser disk. The contour maps of the 5 detected sources are shown in Fig. 1.

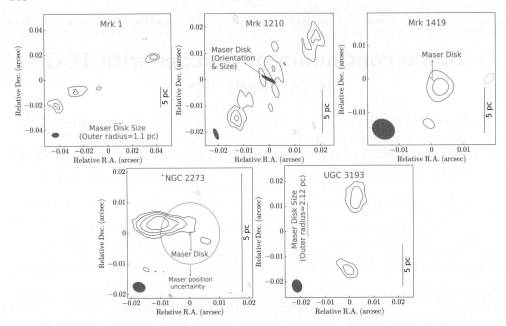

Figure 1. The 5 GHz contour maps for the five H_2O disk maser galaxies detected in the VLBA observations. The contour levels are ±3, ±6, ±12, ±24 times the $1\,\sigma$ rms ($\sim 35\,\mu$Jy/beam). The synthesized beam is shown in the lower left corner of each plot. The H_2O maser disk is also shown (for Mrk 1 and UGC 3193 only the size of the maser disk is available, for Mrk 1210 both size and orientation of the maser disk are known and for Mrk 1419 and NGC 2273 the position of the maser disk is also known). Disk maser data taken from Kuo *et al.* (2011), Braatz *et al.* (2015) and Zhao *et al.* in prep.

3. Results and outline

As seen in Fig. 1, when the maser position is available, it coincides with the 5 GHz radio continuum source within the maser position uncertainties. In addition, for those galaxies for which the maser disk orientation is measured, the maser disk is perpendicular to the extended radio emission within the position angle uncertainties. The low detection rate of radio continuum emission on pc to sub-pc scales in our study ($\sim$27%) could indicate that most of the radio emission observed on the kpc scale (in low luminosity AGNs (LLAGNs)) is due to star formation activity. Re-observations of the detected sources with a higher sensitivity, as well as observations with intermediate resolutions ($\sim$40 mas), are our next steps towards a better understanding of the radio continuum emission from the LLAGNs.

References

Herrnstein, J. R., Greenhill, L. J., Moran, J. M., Diamond, P. J., Inoue, M., Nakai, N., & Miyoshi, M. 1998, *ApJ*, 497L, 69H

Kamali, F., Henkel, C., Brunthaler, A., Impellizzeri, C. M. V., Menten, K. M., Braatz, J. A., Greene, J. E., Reid, M. J., Condon, J. J., Lo, K. Y., Kuo, C. Y., Litzinger, E., & Kadler, M. 2017, *A&A*, 605A, 84K

Kuo, C. Y., Braatz, J. A., Condon, J. J., Impellizzeri, C. M. V., Lo, K. Y., Zaw, I., Schenker, M., Henkel, C., Reid, M. J., & Greene, J. E. 2011, *ApJ*, 727, 20K

Braatz, J., Condon, J., Constantin, A., Gao, F., Greene, J., Hao, L., Henkel, C., Impellizzeri, V., Kuo, C.-Y., Litzinger, E., Lo, K. Y., Pesce, D., Reid, M., Wagner, J., & Zhao, W. 2015b, *IAUGA*, 2255730B

Zhao, W, Braatz, J. A, Condon, J. J., Lo, K. Y, Ried, M. J., Henkel, C., Pesce, D. W, Greene, J. E, Gao, F., Kuo, C. Y, Impellizzeri, C. M. V in prep.

Astrophysical Masers:
Unlocking the Mysteries of the Universe
Proceedings IAU Symposium No. 336, 2017
A. Tarchi, M.J. Reid & P. Castangia, eds.

© International Astronomical Union 2018
doi:10.1017/S1743921317011656

X-Ray Characteristics of Water Megamaser Galaxies

K. Leiter[1,2], M. Kadler[1], J. Wilms[2], J. Braatz[3], C. Grossberger[2,4], F. Krauß[5], A. Kreikenbohm[1,2], M. Langejahn[1,2], E. Litzinger[1,2], A. Markowitz[6,7] and C. Müller[8]

[1]Lehrstuhl für Astronomie, Universität Würzburg, Würzburg, Germany
[2]Dr. Karl Remeis Sternwarte & ECAP, Universität Erlangen-Nürnberg, Bamberg, Germany
[3]NRAO, Charlottesville, USA; [4] MPE, Garching, Germany
[5]GRAPPA & API, University of Amsterdam, Netherlands
[6]University of California (UCSD/CASS), San Diego, USA
[7]NCAC, Warsaw, Poland; [8] IMAPP, Radboud University, Netherlands

Abstract. We have compiled the X-ray characteristic properties for a unique and homogeneous sample of Type 2 AGN with water megamaser activity observed by *XMM-Newton* and for a control sample of non-maser galaxies, both analyzed in a uniform way. A comparison of the luminosity distributions confirms previous results (from smaller and/or less systematic studies) that water maser galaxies appear more luminous than non-maser sources. In addition, the maser phenomenon is associated with more complex X-ray spectra, higher column densities and higher equivalent widths of the Fe Kα line. Both a sufficiently luminous X-ray source and a high absorbing column density in the line of sight favor the appearance of the water megamaser phenomenon in AGN.

Keywords. masers, galaxies: active, galaxies: Seyfert, X-rays: galaxies; cosmological parameters

1. Introduction

Water maser galaxies are a rare subclass of Active Galactic Nuclei (AGN). They play a key role in modern cosmology, providing a unique way to measure geometrical distances to galaxies within the Hubble flow. Modern megamaser observational programs have the goal to measure the Hubble parameter with an accuracy of 3% and to provide a constraint on the equation of state of dark energy (Braatz *et al.* 2008; also Braatz, this volume). An increasing number of independent measurements of suitable water masers is providing the statistics necessary to decrease the uncertainties of such measurements. Studies at X-ray energies have the potential to yield important constraints on target-selection criteria for future maser surveys, promising increased detection rates of new megamaser galaxies.

2. *XMM-Newton* archival survey of megamaser galaxies

We have performed an extensive archival survey of *XMM-Newton* observations of water megamaser galaxies. Based on the sample of Zhang *et al.* (2012) we compiled a sample of maser (30 Type 2 AGN) and a control sample of non-maser (38 Type 2 AGN) galaxies that have been observed with *XMM-Newton*. To avoid biases potentially introduced by having significantly different redshift distributions between the two samples, we cut the maser sample at z = 0.02.

We aim to characterize the spectral features of maser and non-maser galaxies in a systematic study on a larger scale than previous small-sample analyses based on non-uniform data from different X-ray satellites. A comparison shows that megamaser galaxies have notably more complex X-ray spectra than non-maser sources. In Fig.1, we present three

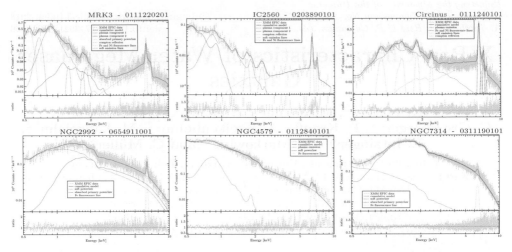

Figure 1. Three examples of *XMM-Newton* X-ray spectra of megamasers (*top panels*) and non-maser AGN (*bottom panels*).

characteristic X-ray spectra from each of the maser and non-maser samples, representing the typical distribution of the spectral components used in each sample. An overview of all maser and non-maser spectra of our study will be provided in Leiter *et al.* in prep.; see also Leiter *et al.* (2014). Typical models of maser X-ray spectra include a highly absorbed power law, multiple ionized plasma emission components, strong Fe lines and X-ray reflection. Furthermore, maser galaxies exhibit a higher fraction of X-ray reflection than non-maser galaxies. Purely reflection-dominated spectra are only observed among masers. On the other hand, non-maser spectra can typically be described by a dominating weakly or partially absorbed power law, ionized plasma emission and emission from neutral iron. Hence, the physical processes in maser sources are generally more complex than in non-maser sources. By estimating the parameters of the best-fit spectral models we can see that the maser phenomenon is also associated with higher column densities and Compton-reflection, which is correlated with a larger equivalent width of the Fe Kα line, and higher intrinsic luminosities. Generally, the maser fraction increases towards high intrinsic luminosities. These findings are in agreement with previous results that found higher absorbing column densities in maser galaxies (e.g. Castangia *et al.* 2013, Greenhill *et al.* 2008, Zhang *et al.* 2006).

Our results suggest that it is possible to define selection criteria for future maser searches based on the complexity of the X-ray spectral model, the absorbing column density, the Fe-line equivalent width, and the intrinsic X-ray luminosity of sources found in upcoming large area sky surveys at X-ray energies like e.g. the eROSITA survey.

References

Braatz, J. A., Ried, M. J., Greenhill, L. J., Condon, J. J., Lo, K. Y., Henkel, C., Gugliucci, N. E., & Hao, L. 2008, *ASPC*, 395, 103B

Castangia, P., Panessa, F., Henkel, C., Kadler, M., & Tarchi, A. 2013, *MNRAS*, 436, 3388-3398

Greenhill, L. J., Tilak, A., & Madejski, G. 2008, *ApJ*, 686: L13-L16

Leiter, K., Kreikenbohm, A., Litzinger, E., Krauss, F., Markowitz, A., Grossberger, C., Langejahn, M., Müller, C., Kadler, M., & Wilms, J. 2014, *xru*, confE, 272L

Zhang, J. S., Henkel, C., Guo, Q., & Wang, J. 2012, *A&A*, 538, A152

Zhang, J. S., Henkel, C., Kadler, M., Greenhill, L. J., Nagar, N., Wilson, A. S., & Braatz, J. A. 2006, *A&A*, 450, 933Z

Sherry Suyu (photo credits: S. Poppi)

The Structure of the Milky Way

Astrophysical Masers:
Unlocking the Mysteries of the Universe
Proceedings IAU Symposium No. 336, 2017
A. Tarchi, M.J. Reid & P. Castangia, eds.

Perspectives on Galactic Structure

Ortwin Gerhard

Max-Planck-Institute for Ex. Physics, Giessenbachstr. 1, D-85748 Garching, Germany
email: gerhard@mpe.mpg.de

The Milky Way is currently the subject of great observational effort. This includes both ESA's unique Gaia mission, as well as a multitude of ground-based surveys. Several of these are already returning data of unprecedented depth and quality for large numbers of Milky Way stars. These new data are likely to lead to a quantum step in our understanding of Milky Way structure and evolution. Because the new data will allow us to study our Galaxy at much greater resolution than possible in other galaxies, we also expect to greatly improve our understanding of disk galaxy formation in general.

In this talk I gave an overview of the stellar components and the dark matter distribution in our Galaxy, concentrating on its inner regions. The Milky Way is a barred galaxy whose central bulge has a box/peanut shape and consists of multiple stellar populations. With dynamical equilibrium models for the bulge, bar, and inner disk based on recent survey data stellar masses have been determined for the different Galactic components. According to these models, about two thirds of the Milky Way's stellar mass is located inside $R \sim 5.3$ kpc, and the bulge has $\sim 25\%$ of the total stellar mass.

The best estimate of the pattern speed puts corotation at $R \sim 6.1 \pm 0.5$ kpc, only ~ 2 kpc inside the solar orbit. Combining the mass of dark matter determined in the bulge with the Galactic rotation curve near the Sun leads to the inference that the dark matter mass distribution in the Galaxy must have an inner ~ 2 kpc core or shallow cusp. Incoming proper motion data from the VVV survey and from Gaia will tighten these results significantly.

Chemo-dynamical equilibrium models for the bulge and bar show a strongly barred distribution of the metal-rich stars, and a radially varying dynamics of the metal-poorer stars: these correspond to a thick disk-bar outside ~ 1 kpc, but change to an inner centrally concentrated component whose origin is presently not clear. APOGEE data have similarly shown an inside-out gradient in the surface density distributions of disk stars with metallicity. From on-going ground-based surveys and the Gaia data we expect large progress in charting the Galactic disk around the Sun, the Milky Way's spiral arms, and the chemo-kinematics of disk stars, as well as in understanding the dynamics and evolution of the Galactic disk.

Similarly, mapping the multiple substructures in the Galactic stellar halo from these new data, as well as mapping the Galactic gravitational potential and, thus, the gravitational mass distribution of the dark matter halo on large scales, are amongst the primary goals of the Gaia mission and of Galactic research. This work has already begun and significant progress is expected in the coming years. The ability of carrying this work out in depth based on star-by-star data will make the Milky Way a unique case for studying how similar galaxies form and evolve.

References

Bland-Hawthorn, J. & Gerhard, O. 2016, *AR&A*, 54, 529
Gerhard, O. 2017, *Rediscovering our Galaxy, IAU Symposium 334*, Cambridge University Press

Astrophysical Masers:
Unlocking the Mysteries of the Universe
Proceedings IAU Symposium No. 336, 2017
A. Tarchi, M.J. Reid & P. Castangia, eds.

© International Astronomical Union 2018
doi:10.1017/S1743921317009140

Structure and Kinematics of the Milky Way

Mark J. Reid

Harvard-Smithsonian Center for Astrophysics, 60 Garden Street, Cambridge, MA 02138, USA
email: reid@cfa.harvard.edu

Abstract. Maser astrometry is now providing parallaxes with accuracies of ±10 micro-arcseconds, which corresponds to 10% accuracy at a distance of 10 kpc! The VLBA BeSSeL Survey and the Japanese VERA project have measured ≈ 200 parallaxes for masers associated with young, high-mass stars. Since these stars are found in spiral arms, we now are directly mapping the spiral structure of the Milky Way. Combining parallaxes, proper motions, and Doppler veloci-ties, we have complete 6-dimensional phase-space information. Modeling these data yields the distance to the Galactic Center, the rotation speed of the Galaxy at the Sun, and the nature of the rotation curve.

Keywords. astrometry, Galaxy: structure, Galaxy: fundamental parameters, Galaxy: kinemat-ics and dynamics

1. Introduction

While we have imaged galaxies throughout the observable universe, we know very little of the structure of our Milky Way. The reasons for this shortcoming are that we are inside the Milky Way, dust absorbs optical light in its plane, and distances are very great and hard to measured. The Milky Way may have some of the characteristics of UGC 12158 shown in Fig. 1. But, we really don't know.

Imagine sending a spacecraft on a million-year journey out of the Milky Way to take a picture and send it back to us. While this is clearly impractical, it turns out that we can generate a map of the Milky Way's spiral structure in only a few years. The Bar and Spiral Structure Legacy (BeSSeL) Survey and the VLBI Exploration of Radio Astrometry (VERA) project are measuring trigonometric parallaxes of molecular maser sources in regions of high-mass star formation across vast regions of the Milky Way. These parallaxes are now tracing the spiral structure of the Milky Way in great detail (eg, Reid *et al.* 2014).

2. Spiral Structure

Combining published BeSSeL Survey and VERA parallaxes with preliminary results from the last two years of BeSSeL Survey, we now have located upwards of 200 maser sources associated with high-mass star forming regions across large portions of the Milky Way (see Fig. 2). At least six spiral arm *segments* are clearly visible. Note that some of these segments may connect far past the Galactic center, such as suggested for the Scutum and Outer arms (Dame & Thaddeus 2011, Anderson *et al.* 2015). However, we caution against assuming a log-periodic spiral pattern with a constant pitch angle to trace a spiral arm around the Milky Way (Honig & Reid 2015).

Recent contributions of parallax measurements have been to point out that the Perseus arm, previously thought to be one of two dominant Milky Way arms, has a significant "gap" in high-mass star formation between Galactic longitudes of about 50° to 90°

Figure 1. UGC 12158, a galaxy sharing characteristics with the Milky Way (HST image).

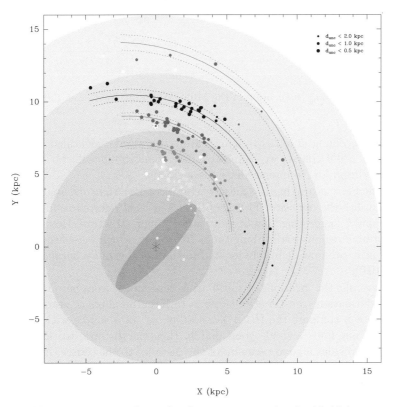

Figure 2. Parallax measurements for molecular masers associated with high-mass star formation across the Milky Way. These include published results from the VERA project and both published and unpublished results from the BeSSeL Survey. Starting from the outside, the Outer, Perseus, Local, Sagittarius, Scutum, and Norma spiral arms are indicated with lines.

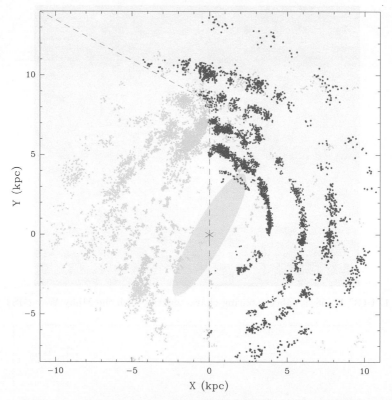

Figure 3. Visualization of the Milky Way from Reid *et al.* (2016).

(Choi *et al.*2014, Zhang *et al.*2013). This feature can also be identified in CO longitude-velocity emission. Of course, we cannot exclude some significant star formation in the gap, especially from low to intermediate mass clouds.

Whereas the Perseus arm has less massive star formation than one might have expected, the Local arm has the opposite. This arm has often been referred to in the literature as the "Orion spur," suggesting that it is a rather minor structure. However, many high-mass star forming regions toward longitudes 70° to 80°, which were suspected of being in the Perseus arm, are now definitely located in the Local arm by parallax measurements (Xu *et al.* 2016).

One can leverage the locations of spiral arms determined from parallax measurements to provide information to estimate distances to large numbers of star-forming regions. Basically, one can compare the (l, b, V) "coordinates" of a source with those traced by spiral arms. If one finds a single good match, then the distance to that section of the matching arm provides a good estimate of the distance to the source. In some cases, the (l, b, V) coordinates of the source may be consistent with those of two or more arms, and this can be used to distinguish among other distance indicators, such as from kinematic information. We have developed a Bayesian approach to combine all distance informa-tion when estimating the distance to individual sources (Reid *et al.* 2016). Using the Bayesian distance estimator for thousands of star-forming region sources in catalogs of molecular clumps, HII regions, infrared sources, and molecular masers, we can generate a visualization of the portion of the Milky Way where arm locations have been measured by parallax. Fig. 3 shows such a visualization.

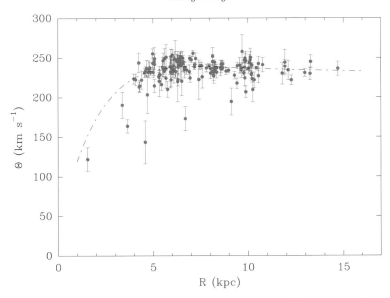

Figure 4. Rotation curve data from parallax measurements which supply full 6-dimensional phase-space information. The dashed line is a "universal" rotation curve (Persic, Salucci & Stel 1996) appropriate for spiral galaxies and fitted to the data.

In Fig. 3, dark blue points indicate star forming regions for which an arm associa-tion was likely. Cyan points indicate regions where an arm could not be assigned; these comprise a small fraction of sources visible from the northern hemisphere and could be inter-arm regions of star formation. The cyan points to the left of the dashed red lines are best observed from the southern hemisphere, and the locations of spiral arms there have yet to be determined. The locations of these sources come primarily from kinematic distances, which are not accurate enough to clearly trace spiral arms.

3. Kinematics

Given source coordinates, distances (from parallaxes), proper motions, and line-of-sight velocities, we have full 6-dimensional phase-space information for each source. These data are in a Heliocentric reference frame. One can easily transform to a Galactocentric frame, provided one knows the location of the Sun and its velocity vector as it orbits the Galaxy. All of these parameters can be estimated by fitting a simple model of rotation to the parallax data, yielding R_0=8.34 kpc and Θ_0=240 km s^{-1} (Reid *et al.* 2014). Adopting these parameter values, we can directly determine the rotation curve of the Milky Way as shown in Fig. 4. It is important to remember that this rotation curve is unique in that it uses 3-dimensional velocities (not only the line-of-sight component) and parallax distances (not more uncertain kinematic distances).

Fig. 4 shows that most high-mass star forming regions follow a rotation curve which is nearly flat between about 5 and 15 kpc from the Galactic center. There is an indication of a turn-down inside of 5 kpc, but more data are needed to properly characterize this portion of the rotation curve. Note that some outliers are expected as, for example, a result of super-bubbles giving rise to gas clouds with significant peculiar motions, which later form massive stars with molecular masers.

The rotation curve shown in Fig. 4 actually has a slight curvature for $R > 5$ kpc, which can explain a long-standing puzzle. The IAU recommended value for Θ_0 of 220 km s^{-1}

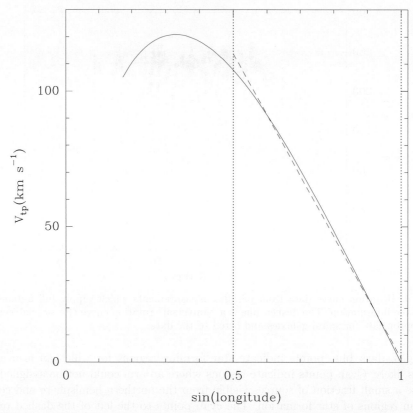

Figure 5. Maximum velocity for gas following a Persic "universal" rotation curve that best fits the parallax data as a function of sine of Galactic longitude (solid line) from Reid & Dame (2016). The dashed line is a straight line fit over the range $0.5 < \sin \ell < 1$, similar to that done for HI data by Gunn, Knapp & Tremaine (1979). The slope of the line of 220 km s^{-1} gives the rotation speed of the Galaxy at the Sun, Θ_0, if one assumes a flat rotation curve. The curvature in the rotation curve shown invalidates this assumption and explains why this method gives a value of Θ_0 lower than by direct fitting of the parallax data.

falls significantly below the 240 km s^{-1} value given by the rotation curve in Fig. 4. The IAU value was adopted largely based on fitting the maximum velocity seen in HI spectra as a function of $\sin \ell$, over the Galactic longitude range $0.5 < \sin \ell < 1$. The slope of a straight-line fit gives Θ_0, provided the rotation curve is flat over the fitting range (Gunn, Knapp & Tremaine 1979). Interestingly, if one instead uses the rotation curve shown in Fig. 4 to predict maximum velocities versus $\sin \ell$ and then fits those with a straight line, one recovers the IAU value of 220 km s^{-1}. This occurs even though the rotation curve never falls below 228 km s^{-1} over that longitude range and has an average value of 238 km s^{-1}. Thus, the discrepancy between the IAU and parallax-based values for Θ_0 can be attributed to assuming a flat rotation curve in the presence of some curvature (Reid & Dame 2016).

References

Anderson, L. D., Armentrout, W. P., Johnstone, B. M. *et al.* 2015, *ApJS*, 221, 26
Choi, Y. K., Hachisuka, K., Reid, M. J. *et al.* 2014, *ApJ*, 790, 99
Dame, T. M. & Thaddeus, P. 2011, *ApJL*, 734, L24
Gunn, J. E., Knapp, G. R., & Tremaine, S. D. 1979, *ApJ*, 84, 1181

Honig, Z. N. & Reid, M. J. 2015, *ApJ*, 800, 53

Persic, M., Salucci, P., & Stel, F. 1996, *MNRAS*, 281, 27

Reid, M. J., Menten, K. M., Brunthaler, A. *et al.* 2014, *ApJ*, 783, 130

Reid, M. J., Dame, T. M., Menten, K. M., *et al.* 2016, *ApJ*, 823, 77

Reid, M. J. & Dame 2016, *ApJ*, 832, 159

Xu, Y., Reid, M. J., Dame, T. M., *et al.* 2016, *Sci.Adv.*, V.2, No.9, 1600878

Zhang, B., Reid, M. J., Menten, K. M., *et al.* 2013, *ApJ*, 775, 79

Astrophysical Masers:
Unlocking the Mysteries of the Universe
Proceedings IAU Symposium No. 336, 2017
A. Tarchi, M.J. Reid & P. Castangia, eds.

© International Astronomical Union 2018
doi:10.1017/S1743921318000832

Structure of the Milky Way:
View from the Southern Hemisphere

Lucas J. Hyland[1], Simon P. Ellingsen[1] and Mark J. Reid[2]

[1]School of Physical Sciences, University of Tasmania
7005 Tasmania, Australia
email: Lucas.Hyland@utas.edu.au, Simon.Ellingsen@utas.edu.au

[2]Harvard–Smithsonian Centre for Astrophysics, Cambridge
MA 02138, USA
email: mreid@cfa.harvard.edu

Abstract. The exclusive association of Class II methanol masers with high mass star formation regions and in turn spiral arms, makes them ideal tracers of spiral structure. The bright and compact nature of masers also makes them good sources for Very Long Baseline Interferometry, with their fluxes visible on some of the longest terrestrial baselines. The success of the BeSSeL (Bar and Spiral Structure Legacy) project has demonstrated the use of masers in large scale high–precision trigonometric parallax surveys. This survey was then able to precisely map the spiral arms visible from the Northern Hemisphere and recalculate the fundamental Milky Way parameters R_0 and θ_0. The majority of the Milky Way is visible from the Southern Hemisphere and at the present time the Australian LBA (Long Baseline Array) is the only Southern Hemisphere array capable of taking high–precision trigonometric parallax data. We present the progress–to–date of the Southern Hemisphere experiment. We will also unveil a new broadband Southern Hemisphere array, capable of much faster parallax turnaround and atmospheric calibration.

Keywords. galaxies: structure, astrometry, masers, telescopes, techniques: interferometric

1. Introduction

Despite living inside the Milky Way Galaxy, until very recently we had very little information about the exact shape, size and number of spiral arms. This is due to our unique position inside the disk with little to no perspective on the distance to stars and objects in the sky above us. Distance remains the elusive quantity it has always been in all aspects of astrophysics with the only tried and true method to calculate distances being trigonometric parallax.

In 2004, the BeSSeL (Bar and Spiral Structure Legacy) project began with the aim to calculate the distance to masers associated with High Mass Star Formation, therefore tracing the spiral structure of the Milky Way. In 2008, Japan's VERA (VLBI Exploration of Radio Astrometry) project also joined in that endeavour, and together they have catalogued well over 100 parallaxes to water and Class II methanol masers in the Northern Hemisphere.

This collection of data allowed recalcualtion of the fundamental parameters of the Milky Way: $R_0 = 8.34 \pm 0.16 \, \text{kpc}$ and $\Theta_0 = 240 \pm 8 \, \text{km s}^{-1}$ and implied the Milky Way was ∼50% more massive than previously thought (Reid *et al.* 2009, 2009b, 2014). Reid *et al.* (2016) use a baysian approach to assign sources with non–parallax distances to spiral arms, leveraged by the high–accuracy parallax distances. However due to the lack of parallaxes in the Southern Hemisphere, the structure of the 3rd and 4th quadrants of the Milky Way remain uncertain.

2. Progress to date: V255

The Australian Long Baseline Array (LBA) is the only array in the Southern Hemisphere with the capabilities to perform high–precision VLBI. Due to this, the V255 LBA large project began March 2008 with the aim to observe and calculate parallaxes to as many 6.7 GHz class II methanol masers as possible. In the time since then, has collected complete data (four or more epochs) for 27 masers and partial data for a further 4 (Table 1). Largely successful, so far the V255 project has produced 3 parallaxes towards the sources G339.884-1.259, G305.200+0.019 and G305.202+0.208 (Krishnan *et al.* 2015, 2017).

Despite the ongoing V255 project, time allocation and technical short–fallings limit the potential and rate of parallax measurements. The LBA is only available for a few weeks per year, with the V255 project allocated ~3 days per year. The time required to gather enough data for 100 parallaxes is on the order of another 15 years. In addition, the times that are allocated are often non–optimal for sources, which can lead to a factor of 2 loss in parallax sensitivity.

Table 1. Sources observed to–date by the V255 project, epoch of observation and experiment code of epoch.

Source	Epochs	Experiments
G9.62+0.20	2008 Mar, Aug ; 2009 Feb, Sep, Dec; 2010 Oct	V255a,b,e,f,h,m
G8.68–0.73	2008 Mar, Aug ; 2009 Feb, Sep, Dec; 2010 Oct	V255a,b,e,f,h,m
NGC6334F	2008 Nov ; 2009 Feb, Sep ; 2010 Mar, Jul	V255c,d,g,i,j,z
NGC6334I(N)	2008 Nov ; 2009 Feb, Sep ; 2010 Mar, Jul	V255c,d,g,i,j,z
G329.066–0.308	2010 Jul, Oct ; 2011 Mar, Jul, Nov	V255k,l,n,o,p
G329.029–0.205	2010 Jul, Oct ; 2011 Mar, Jul, Nov	V255k,l,n,o,p
G329.031–0.198	2010 Jul, Oct ; 2011 Mar, Jul, Nov	V255k,l,n,o,p
MonR2	2011 Mar, Jul, Nov ; 2012 Mar	V255n,o,p,q
G188.95+0.89	2011 Mar, Jul, Nov ; 2012 Mar	V255n,o,p,q
G339.884–1.259	2012 Mar ; 2013 Mar, Jun, Aug, Nov ; 2014 Mar, Nov	V255q,r,s,t,u,v,w
G339.681–1.208	2012 Mar ; 2013 Aug, Nov ; 2014 Mar, Nov	V255q,t,u,v,w
G339.682–1.207	2012 Mar ; 2013 Aug, Nov ; 2014 Mar, Nov	V255q,t,u,v,w
G305.200+0.019	2013 Mar, Jun, Aug, Nov ; 2015 Mar	V255r,s,t,u,x
G305.202+0.208	2013 Mar, Jun, Aug, Nov ; 2015 Mar	V255r,s,t,u,x
G305.208+0.206	2013 Mar, Jun, Aug, Nov ; 2015 Mar	V255r,s,t,u,x
G263.250+0.514	2013 Mar, Jun, Aug, Nov ; 2015 Mar, Jul; 2017 Aug	V255r,s,t,u,x,y,ad
G287.371+0.644	2014 Mar, Nov ; 2015 Mar, Jul, Sep; 2017 Aug	V255v,w,x,y,z,ad
G291.274–0.709	2014 Mar, Nov ; 2015 Mar, Jul, Sep	V255v,w,x,y,z
G316.640–0.087	2014 Mar, Nov ; 2015 Jul, Sep	V255v,w,y,z
G316.811–0.057	2014 Mar, Nov ; 2015 Jul, Sep	V255v,w,y,z
G345.003–0.224	2015 Jul, Sep ; 2016 Mar, Jun	V255y,z,aa,ab,ac
G345.003–0.223	2015 Jul, Sep ; 2016 Mar, Jun	V255y,z,aa,ab,ac
G345.010+1.797	2015 Jul, Sep ; 2016 Mar, Jun	V255y,z,aa,ab,ac
G345.010+1.792	2015 Jul, Sep ; 2016 Mar, Jun	V255y,z,aa,ab,ac
G294.990–1.719	2016 Mar, Jun ; 2017 Mar, Aug	V255aa,ab,ac,ad
G298.262+0.739	2016 Mar, Jun ; 2017 Mar, Aug	V255aa,ab,ac,ad
G326.475+0.703	2016 Jun ; 2017 Mar, Aug	V255ab,ac,ad
G328.808+0.633	2016 Jun ; 2017 Mar, Aug	V255ab,ac,ad
G328.809+0.633	2016 Jun ; 2017 Mar, Aug	V255ab,ac,ad
G339.617–0.117	2017 Aug	V255ad
G340.050–0.250	2017 Aug	V255ad

The heterogeneous nature of the LBA leads to a very small mutual–frequency window around the observation frequency. As a spanned bandwidth is used to calculate delays due

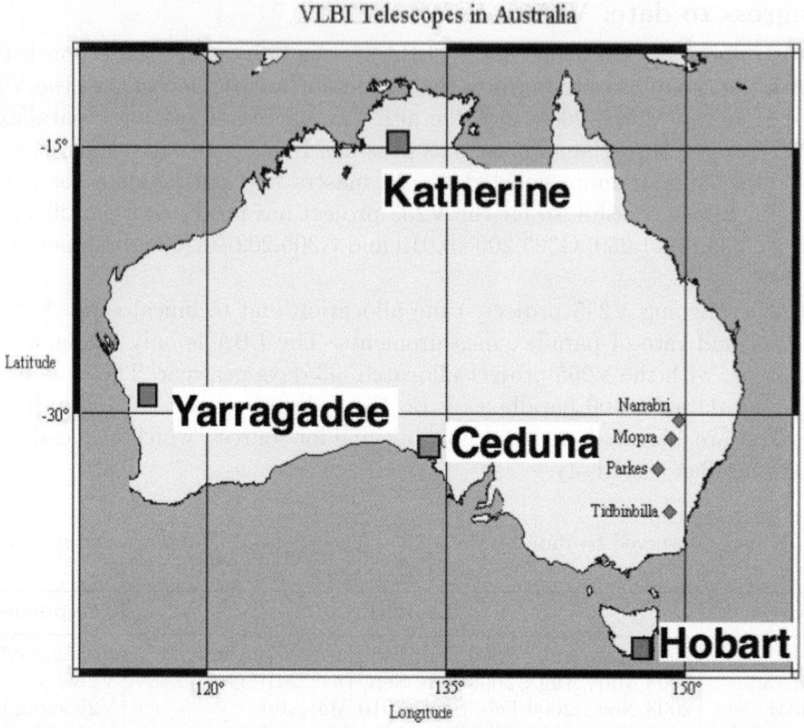

VLBI Telescopes in Australia

Figure 1. Positions of the radio telescopes in Australia. Empathised are the sites for the Hobart 12m, Katherine 12m, Yarragadee 12m and Ceduna 30m. Together these four telescopes form the ASCI Array.

to the troposphere in the current data–reduction pipeline, the small bandwidth available leads to large uncertainties.

The final and most severe issue is the lack of reliable ionospheric correction. The LBA uses Total Electron Content (TEC) maps derived from GPS observations in the same manner as the BeSSeL uses on the VLBA. However, due to the relative lack of GPS stations in the Southern Hemisphere, Krishnan *et al.* (2015) found that the static ionospheric delays due to the uncertainty in the maps contributed the most to final position uncertainty. The VLBA utilises additional linear–combination of delays to further reduce the ionospheric delays, however this requires a larger spanned bandwidth than the LBA possesses.

3. Future work: ASCI

The University of Tasmania (UTas) own and operate five radio telescopes. Ceduna 30m and Hobart 26m participate in VLBI as part of the LBA, whilst the remaining three 12m telescopes are dedicated to geodetic VLBI. Katherine, Yarragadee and Hobart 12m form the AuScope array (Lovell *et al.* 2013), equipped with S/X geodetic receivers operating ~200 days per year.

A collaboration involving UTas, Max Planck Institut für Radioastronomie, CSIRO Astronomy and Space Science, Geoscience Australia, Auckland University of Technology and Harvard–Smithsonian Centre for Astrophysics secured funds for the design and construction of three wideband receivers for the AuScope array. The receivers are part of the VLBI Global Observing System (VGOS; Sun *et al.* 2014) upgrade for the geodetic antennas.

This upgrade presents an opportunity to use the three 12m geodetic antennas for parallax observations – they are free ∼100 days per year and operation and correlation is undergone at UTas. Designed and built by *Callisto*, the receivers cover the frequency range $2.3 - 14$ GHz which includes the 6.7 GHz methanol line of interest. Also, this large frequency range is comparable with that used by BeSSeL on the VLBA ($3 - 7$ GHz) to correct for the troposphere and ionosphere. The cryogenics only cool the LNAs down to 40 K, which is optimised for geodetic VLBI. Due to this, Ceduna 30m will be added to the array to boost sensitivity and provide intermediate baselines (Figure 1). Together these four telescopes will form the AuScope–Ceduna Interferometer, or ASCI.

Rioja *et al.* (1997) show that there exists a plane of ionospheric delay present above each telescope. The authors write that to correct for this, a phase referencing technique called *MultiView* (previously called *cluster–cluster*) can be implemented. This involves cycling through multiple calibrators around the target within a duty cycle, thus solving for a phase plane solution that can be interpolated to the target position. This has been shown to significantly reduce systematic error due to ionospheric effects at low frequencies, where the effect is much greater (Rioja *et al.* 2017). Tests to utilise this method with the ASCI array are currently under–way with the current receivers, and will continue once the broadband receivers are installed.

References

Krishnan, V., Ellingsen, S. P., Reid, M. J., *et al.* 2017, *MRNAS*, 465, 1095

Krishnan, V., Ellingsen, S. P., Reid, M. J., *et al.* 2015, *ApJ*, 805, 129

Lovell, J. E. J., McCallum, J. N., Reid, P. B., *et al.* 2013, Journal of Geodesy, 87, 527

Reid, M. J., Menten, K. M., Brunthaler, A., & Moellenbrock, G. A. 2009, in Astronomy, Vol. 2010, astro2010: The Astronomy and Astrophysics Decadal Survey

Reid, M. J., Menten, K. M., Zheng, X. W., *et al.* 2009b, *ApJ*, 700, 137

Reid, M. J., Menten, K. M., Brunthaler, A., *et al.* 2014, *ApJ*, 783, 130

Reid, M. J., Dame, T. M., Menten, K. M., & Brunthaler, A. 2016, *ApJ*, 823, 77

Rioja, M. J., Stevens, E., Gurvits, L., *et al.* 1997, Vistas in Astronomy, 41, 213

Rioja, M. J., Dodson, R., Orosz, G., Imai, H., & Frey, S. 2017, *AJ*, 153, 105

Sun, J., Böhm, J., Nilsson, T., *et al.* 2014, Journal of Geodesy, 88, 449

Astrophysical Masers:
Unlocking the Mysteries of the Universe
Proceedings IAU Symposium No. 336, 2017
A. Tarchi, M.J. Reid & P. Castangia, eds.

© International Astronomical Union 2018
doi:10.1017/S1743921317008481

Interferometry of class I methanol masers, statistics and the distance scale

Maxim A. Voronkov[1,3], Shari L. Breen[2,3], Simon P. Ellingsen[3] and Christopher H. Jordan[4]

[1]CSIRO Astronomy and Space Science, PO Box 76, Epping, NSW 1710, Australia
email: maxim.voronkov@csiro.au

[2]Sydney Institute for Astronomy (SIfA), School of Physics, University of Sydney, Sydney, NSW 2006, Australia

[3]School of Mathematics and Physics, University of Tasmania, GPO Box 252-37, Hobart, Tas 7000, Australia

[4]International Centre for Radio Astronomy Research, Curtin University, Bentley, WA 6845, Australia

Abstract. The Australia Telescope Compact Array (ATCA) participated in a number of survey programs to search for and image common class I methanol masers (at 36 and 44 GHz) with high angular resolution. In this paper, we discuss spatial and velocity distributions revealed by these surveys. In particular, the number of maser regions is found to fall off exponentially with the linear distance from the associated young stellar object traced by the 6.7-GHz maser, and the scale of this distribution is 263±15 milliparsec. Although this relationship still needs to be understood in the context of the broader field, it can be utilised to estimate the distance using methanol masers only. This new technique has been analysed to understand its limitations and future potential. It turned out, it can be very successful to resolve the ambiguity in kinematic distances, but, in the current form, is much less accurate (than the kinematic method) if used on its own.

Keywords. masers – ISM: molecules – stars: formation

1. Introduction

Early studies of methanol masers empirically divided them into two classes (Batrla *et al.* 1987): class I are typically scattered around the presumed location of the young stellar object (YSO) over an area often compared with the primary beam of the 20-m class radio telescope, while class II are compact at arcsecond resolution and pinpoint the location of the YSO. These differences were traced to collisional and radiative pumping, respectively (see Voronkov *et al.* (2014) and references therein for further information). The Australia Telescope Compact Array (ATCA) participated in a number of survey programs, both targeted and blind, to search for and image common class I methanol masers (at 36 and 44 GHz) with high angular resolution (Voronkov *et al.* 2014; Jordan *et al.* 2015, 2017; and unpublished data 36-GHz data from the Methanol Multibeam (MMB) follow-up project). This resulted in a large sample of class I methanol masers studied at sub-arcsecond resolution, and provided us the basis for statistical analysis.

2. Velocity distribution

Due to complexity of spatial and kinematic structure of class I methanol masers, we decomposed emission into groups of Gaussian components coincident within 3σ in position and velocity and used those groups for statistical analysis instead of individual components (see Voronkov *et al.* (2014) for further details). Each component group was assigned an association with a 6.7-GHz maser (from the MMB project, see Green *et al.* (2017) and references therein), if found within an arcminute (if there is more than one 6.7-GHz maser in the vicinity we used the nearest). The middle of the velocity range spanned by the 6.7-GHz maser emission is often used as a reference velocity for the kinematic distance estimates. The comparison between this mid-range velocity and velocity of the associated class I emission revealed that the difference is largely (ignoring a few high-velocity components) a Gaussian variate with the standard deviation of 3.65±0.05 and 3.32±0.07 km s^{-1} for the 36 and 44-GHz masers, respectively (Voronkov *et al.* 2014). The difference between the two standard deviations, if real, may be a consequence of preferential orientation with respect to the line of sight direction for the 36 and 44-GHz masers which belong to two different transition series (Sobolev *et al.* 2007; see also the paper in this volume). The distribution of velocity offsets has also a small but significant mean of −0.57 km s^{-1} (class I masers are blue-shifted; uncertainties are 0.06 and 0.07 km s^{-1} for the 36 and 44-GHz masers, respectively). However, the velocities of class I masers show a better agreement with that of the thermal gas: the standard deviation of the velocity offsets with respect to CS is 1.5±0.1 km s^{-1} with insignificant mean offset (see Jordan *et al.* 2015, 2017). Therefore, the velocity of class I methanol masers seems to be a better tracer of the quiescent gas velocity than the velocity derived from the 6.7-GHz spectra. This conclusion is in agreement with the earlier results of Green & McClure-Griffiths (2011) on the velocity distribution of 6.7-GHz masers.

3. Spatial distribution

Despite general scatter around the presumed YSO location, the smaller spatial separations of class I maser emission are somewhat preferred with the region of influence of a YSO being less than 1 pc (see Fig. 29 of Kurtz *et al.* 2004). The large statistics and availability of homogenous 6.7-GHz data through the MMB survey allowed us to probe this distribution in detail. It turned out that the linear separations between class I maser components and associated 6.7-GHz masers have an exponential distribution with good accuracy (the scale is 263±15 milliparsec; see Voronkov *et al.* 2014). This relationship still needs to be understood in the context of the broader field.

4. New method to estimate distances

The empirical distribution of linear separations discussed in the previous section can, in principle, be turned into a statistical distance estimator from the measured angular separations. This method does not use velocities and, therefore, is applicable in the Galactic Centre direction where the kinematic method fails. Here we analyse the limitations of this new approach. First, the number of component groups for each individual source is rarely large. Therefore, it may be more practical to assume that the measured angular separation is the exponential variate for each individual source. Then, one can use the fact that, for the exponential distribution, the mean is an unbiased maximum likelyhood estimator of the scale parameter. And equating the measured mean angular separation ($\langle d_{arcsec} \rangle$) to the expected linear scale of 263±15 milliparsecs (Voronkov *et al.* 2014), one gets the following recipe for the distance $D_{kpc} \approx 54/\langle d_{arcsec} \rangle$. The numbers may be too low for a proper statistical test that the measured angular separations for a given source are indeed consistent with the exponential distribution. However, one could compute the standard deviation and compare it with the mean as a cross-check: they are identical for the exponential distribution. Note, the uncertainty of the expected linear

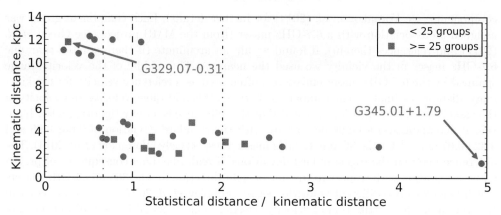

Figure 1. Comparison of the distance estimate based on the empirical spatial distribution of class I masers and the kinematic distance estimate based on the 6.7-GHz velocities. Each point represents a single 6.7-GHz maser from the MMB catalogue associated with at least 15 class I methanol maser component groups (sources with at least 25 such groups are shown by (red) squares). The vertical dash-dotted lines enclose the area where both methods agree (dashed vertical line represents the exact agreement) with the 95% confidence, under assumption that the new method is the only source of uncertainty and the class I methanol maser has 15 component groups.

scale quoted above has rather small contribution (about 5%) to the total error budget which is dominated by the uncertainty of $\langle d_{arcsec} \rangle$ in the case of small number statistics and, quite likely, by the systematics discussed below.

Second, the distribution of linear separations, which this method is based upon, has been obtained for an ensamble of sources. Applying it to individual sources introduces additional systematics due to possible hidden parameters and idiosyncrasies of the sources. In paricular, it is discussed by Voronkov *et al.* (2014) that more evolved sources show some tendency to be more spread out both spatially and in velocity. This evolutionary trend has been completely ignored in the analysis above. In addition, the ATCA follow-up of the 44-GHz masers found in the MALT45 blind survey (Jordan *et al.* 2017) and the new (unpublished) 36-GHz ATCA survey targeting MMB 6.7-GHz masers, both revealed many simple sources. This is an additional evidence suggesting that the source sample studied by Voronkov *et al.* (2014) may not be representative of the whole population of class I methanol masers (or, even, the subset of such masers located in high-mass star forming regions). Another important systematic factor is the association of class I methanol masers with a range of phenomena producing shocked gas, not just outflows (see discussion in Voronkov *et al.* 2010; 2014). The spatial distribution of class I masers conceivably depends on the exact mechanism, but details, or even the relative occurrence of different scenarios, are not clear at present.

To examine the accuracy of such statistical distances in practical terms, we compared them with kinematic distances based on the middle of the velocity range of the associated 6.7-GHz maser for the whole sample of Voronkov *et al.* (2014) where an association between class I and class II masers can be made and where the class I maser emission was decomposed into at least 15 component groups. Each point in Fig. 1 represents a single source. Statistical distances are expected to be more accurate for sources with large number of component groups which are shown as (red) squares. But, in general, points are expected to be confined between two dash-dotted lines on the diagram (agreement with the kinematic method with 95% confidence), or roughly a factor of 2 uncertainty in distances. However, it is evident from Fig. 1 that there are many outliers with grossly overestimated distances for nearby sources and underestimated for sources beyond 10 kpc. Inspection of individual sources in the former category reveals two main reasons for overestimated distances: the sensitivity cutoff due to primary beam (e.g. the most extreme outlier point corresponding to G345.01+1.79 which was found by Voronkov *et al.* (2014) to have emission beyond the half-power point of the primary beam) and questionable associations of the class I maser emission with the 6.7-GHz masers in complex sources. In both cases, the apparent spread of class I maser emission is underestimated causing the method to overestimate the distance. However, it is worth noting that there are many relatively compact sources which have less than 15 component groups. There seems to be a sysematic tendency to overestimate the distances to such sources, although some of them may indeed have been incorrectly assigned to the near kinematic distance. For distant sources, extreme outliers (which this method tries to bring closer) are likely to be largely genuine cases of sources wrongly assigned to the far kinematic distance (see, for example, the discussion on G329.07−0.31 in Voronkov *et al.* 2014). Therefore, despite having a relatively low accuracy on its own, this method can be used to assist resolving the kinematic distance ambiguity. It can also provide an extra constraint for the Bayesian approach suggested by Reid *et al.* (2016; see also the paper in this volume).

It is worth noting that this method can be extended to sources without a 6.7-GHz maser or other way to pinpoint the YSO location. The averaged position of all class I maser emission is close to the location of the 6.7-GHz maser for the majority of the sources in our sample (although there are several sources with pronounced asymmetry for which it is not the case) and can be used as a proxy. This is an additional source of systematic uncertainty, but may be acceptable as the statistical distances are the rough estimates anyway.

References

Batrla, W., Matthews, H. E., Menten, K. M., & Walmsley, C. M. 1987, *Nature*, 326, 49

Green, J. A., Breen, S. L., Fuller, G. A., McClure-Griffiths, N. M., Ellingsen, S. P., Voronkov, M. A., Avison, A., Brooks, K., Burton, M. G., Chrysostomou, A., Cox, J., Diamond, P. J., Gray, M. D., Hoare, M. G., Masheder, M. R. W., Pestalozzi, M., Phillips, C., Quinn, L. J., Richards, A. M. S., Thompson, M. A., Walsh, A. J., Ward-Thompson, D., Wong-McSweeney, D., & Yates, J. A. 2017, *MNRAS*, 469, 1383

Green, J. A. & McClure-Griffiths, N. M. 2011, *MNRAS*, 368, 1843

Jordan, C. H., Walsh, A. J., Breen, S. L., Ellingsen, S. P., Voronkov, M. A., & Hyland, L. J. 2017, *MNRAS*, 471, 3915

Jordan, C. H., Walsh, A. J., Lowe, V., Voronkov, M. A., Ellingsen, S. P., Breen, S. L., Purcell, C. R., Barnes, P., Burton, M. G., Cunningham, M. R., Hill, T., Jackson, J. M., Longmore, S. N., Peretto, N., & Urquhart, J. S. 2015, *MNRAS*, 448, 2344

Kurtz, S., Hofner, P., & Álvarez, C. V. 2004, *ApJS*, 155, 149

Reid, M. J., Dame, T. M., Menten, K. M., & Brunthaler, A. 2016, *ApJ*, 823, 77

Sobolev, A. M., Cragg, D. M., Ellingsen, S. P., Gaylard, M. J., Goedhart, S., Henkel, C., Kirsanova, M. S., Ostrovskii, A. B., Pankratova, N. V., Shelemei, O. V., van der Walt, D. J., Vasyunina, T. S., & Voronkov, M. A. 2007, in: J. M. Chapman & W. A. Baan (eds.), *Proc. IAU Symp. 242, Astrophysical Masers and their Environments* (Cambridge Univ. Press), p. 81

Voronkov, M. A., Caswell, J. L., Ellingsen, S. P., Green, J. A., & Breen, S. L. 2014, *MNRAS*, 439, 2584

Voronkov, M. A., Caswell, J. L., Ellingsen, S. P., & Sobolev, A. M. 2010, *MNRAS*, 405, 2471

Astrophysical Masers:
Unlocking the Mysteries of the Universe
Proceedings IAU Symposium No. 336, 2017
A. Tarchi, M.J. Reid & P. Castangia, eds.

© International Astronomical Union 2018
doi:10.1017/S1743921317011061

Maser Astrometry and Galactic Structure Study with VLBI

Mareki Honma[1,2,3], Takumi Nagayama[1], Tomoya Hirota[1,3],
Nobuyuki Sakai[1], Tomoaki Oyama[1], Aya Yamauchi[1],
Toshiaki Ishikawa[1], Toshihiro Handa[4], Ken Hirano[1],
Hiroshi Imai[4], Takaaki Jike[1], Osamu Kameya[1,3],
Yusuke Kono[1], Hideyuki Kobayashi[1], Akiharu Nakagawa[3],
Katsunori M. Shibata[1,3], Daisuke Sakai[1,5], Kazuyoshi Sunada[1],
Koichiro Sugiyama[1], Katsuhisa Sato[1], Toshihiro Omodaka[4],
Yoshiaki Tamura[1] and Yuji Ueno[1]

[1] National Astronomical Observatory of Japan, 023-0861, Iwate, Japan
[2] email: mareki.honma@nao.ac.jp

[3] Dept. of Astronomical Science, SOKENDAI, 023-0861, Iwate, Japan
[4] Dept. of Physics, Kagoshima University, 890-8580, Kagoshima, Japan
[5] Dept. of Astronomy, the University of Tokyo, 113-8654, Tokyo, Japan

Abstract. In this proceeding paper, we introduce the recent results of Galactic maser astrometry by mainly focusing on those obtained with Japanese VLBI array VERA. So far we have obtained parallaxes for 86 sources including preliminary results, and combination with the data obtained with VLBA/BeSSeL provides astrometric results for 159 sources. With these most updated results we conduct preliminary determinations of Galactic fundamental parameters, obtaining $R_0 = 8.16 \pm 0.26$ kpc and $\Theta_0 = 237 \pm 8$ km/s. We also derive the rotation curve of the Milky Way Galaxy and confirm the previous results that the rotation curve is fairly flat between 5 kpc and 16 kpc, while a remarkable deviation is seen toward the Galactic center region. In addition to the results on the Galactic structure, we also present brief overviews on other science topics related to masers conducted with VERA, and also discuss the future prospect of the project.

Keywords. maser, VLBI astrometry, VERA, the Milky Way Galaxy

1. Introduction

Phase-referencing VLBI observations of maser sources have been regularly producing accurate astrometric measurements relative to background QSOs at 10 micro-arcsecond level (e.g., see Reid & Honma 2014 and references therein). Currently two major groups have been conducting massive survey of maser parallaxes in the northern hemisphere, namely VLBA/BeSSeL (Reid *et al.* 2014 and references therein), and VERA (VLBI Exploration of Radio Astrometry, Kobayashi *et al.* 2003). The main targets for such astrometric observations are H_2O and CH_3OH masers in star-forming regions, with some cases H_2O and SiO masers in late-type stars. So far accurate parallaxes and proper motions have been measured for more than 100 maser sources, which provide critical information for determinations of the Galaxy's fundamental parameters, the rotation curve, structure of spiral arms, and so on.

Maser astrometry is unique and powerful for tracing the structure of the Milky Way Galaxy even in the GAIA era. Studies done so far already showed that maser astrometry with VLBI can achieve a parallax accuracy at 10 μas level or even better (e.g., see Reid & Honma 2014), which will be the only match-up of GAIA's target accuracy. Radio mission

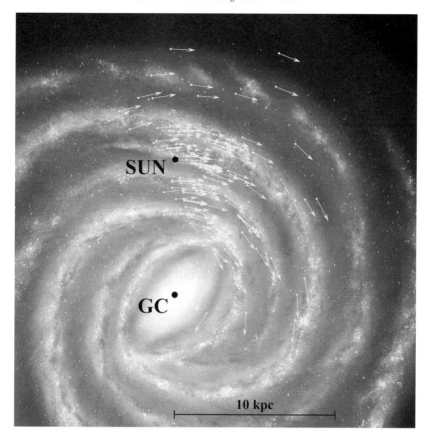

Figure 1. Maser source distribution overlaid on the face-on artistic view of the Milky Way. Points with an arrow show the sources for which accurate parallax and motion were measured with VLBI.

of maser sources is suitable for observing through the Galactic plane, where optical and near-infrared emissions suffer from severe dust obscuration. Furthermore, most common targets for maser astrometry are those associated with star-forming regions in the Milky Way, while the optical astrometry such as GAIA mainly observes normal stars. This makes the maser astrometry most suitable for tracing the structure and kinematics of the Galaxy's spiral arms, which can be best traced with the youngest populations. Also, in terms of galactic dynamics, these two populations (gases and stars) respond differently to the Galactic potential, as the stars are collisionless while star-forming regions are gaseous components and thus collisional. Therefore, in the GAIA era, it is essential to combine the optical astrometry and maser astrometry to obtain a comprehensive picture of the Milky Way Galaxy, and hence maser astrometry will be more important than ever.

For these reasons, we have been conducting maser astrometry with VERA, which is an array dedicated to maser astrometry, and in this presentation we introduce its most updated results.

2. Galaxy structure

We obtained the first parallax results with VERA in 2007 (Honma *et al.* 2007; Hirota *et al.* 2007), and since then we have been conducting maser astrometry program on regular basis with an average machine time of ~2000 hr per year. So far we have completed

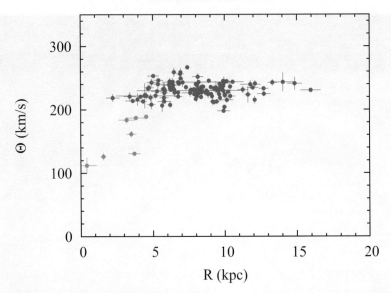

Figure 2. Rotation curve of the Milky Way determined from the maser sources with accurate parallaxes and proper motions. While the sources within 4 kpc show relatively large deviation (most probably due to the Galactic bar), the rotation curve is fairly flat between 5 kpc and 16 kpc.

observations of about180 sources, and now have been conducting the data analyses of them. As of October 2017, we have obtained parallaxes for 86 sources (of which roughly half of the results are already published and the rest to come quite soon). By combining the BeSSeL results (e.g., Reid *et al.* 2014), we compiled a list of 159 maser sources with accurate parallaxes and proper motions. Among them 144 sources are associated with star-forming regions, which we use for preliminary analysis of the Galactic structure in the present paper. For comparisons, in the previous studies, Honma *et al.*(2012) had 52 sources, and Reid *et al.*(2014) had 103 sources available at the time of their publications.

Figure 1 shows the distribution of the maser sources in the plan view of the Milky Way. The sources are mostly located at galactic latitudes between $l = 0°$ and $240°$, which is caused by the geographic locations of the arrays, i.e., both VLBA and VERA are located in the northern hemisphere. The source distances from the Sun reach well beyond 10 kpc, and hence VLBI astrometry is readily tracing the Galaxy-scale structure at a 10-kpc scale. One can see association of the maser sources with spiral arms such as the local arm and Perseus Arm, showing that the maser sources are a good tracer of spiral arms. In figure 1, the proper motions are also shown with an arrow attached to each source. Overall distribution of the motion vectors clearly demonstrates that the Milky Way Galaxy is in a circular rotation as a first-order representation. Masers associated with spiral arms provide a unique opportunity to investigate the kinematics/dynamics of the spiral arms of the Galaxy. In fact, recently Sakai *et al.*(2015) argued that the non-circular motion associated with the Perseus arm can be modeled with a density-wave-type dynamical model for a spiral arm. Since the number of the spiral-arm masers with accurate distances is still limited, this should be investigated further by increasing the number of maser sources.

From the proper motion distribution, one can determine the dynamical center of the Milky Way, which provides an accurate dynamical measurement of the Galactic center distance R_0 as well as the mean angular velocity at the Sun $\Omega_\odot$. The latter is linked to the fundamental constant Θ_0 as $\Theta_0 = R_0 \Omega_\odot - V_\odot$, where $V_\odot$ is the peculiar motion of the Sun in the direction of the Galactic rotation. Based on an MCMC analysis, we obtained preliminary results for these parameters as summarized in table 1. We obtained $R_0 = 8.16 \pm 0.26$ kpc, reaching at an accuracy of 3%. The value is slightly smaller than the 1985 IAU standard of 8.5 kpc, but being consistent with recent studies, such as Reid *et al.*(2014) from maser astrometry, and Boehle *et al.*(2016) from stellar motions around Sgr A*.

The Galactic rotation velocity of $\Theta_0 = 237 \pm 8$ km $^{-1}$ is larger than the IAU value of 220 km s^{-1} by 10%. The significance is still at 2-σ level, but the uncertainty mainly comes from the error in R_0 rather than $\Omega_\odot$. As seen in table 1, the angular velocity $\Omega_\odot$ itself is determined at a 1%-level. As far as the angular velocity of the LSR (Ω_0) is concerned, the IAU standards give $\Omega_0 = 220/8.5 = 25.9$ km s^{-1} kpc^{-1} while our results provide $\Omega_0 = \Omega_\odot - V_\odot/R_0 = 29.02 \pm 0.39$ km s^{-1} kpc^{-1}. Therefore, in terms of the angular velocity Ω_0, the difference is significant at 8-σ, suggesting that the Galactic rotation speed should be revised in near future.

The rotation curve determined from maser astrometry is also shown in figure 2. The figure shows that the rotation curve of the Milky Way is fairly flat in a galacto-centric distance between 5 and 16 kpc. Also notable is that the in the central part within 5 kpc from the Galaxy's center shows a large deviation from a flat rotation curve. Several sources with $R < 5$ kpc lie below the flat rotation curve, which is presumably due to the effect of the bar. Therefore, measuring more sources in this region will potentially trace the gas dynamics associated with the bar.

Astrometry beyond 10 kpc is an remaining future issue, which requires improvement of astrometric accuracy by array expansion and/or better calibration. Meanwhile we developed a new method to use 3-D motion measured with VLBI astrometry to determine a kinematic-based distance. Yamauchi *et al.*(2016) applied this technique to G7.47+0.06, an SFR located at a 20 kpc distance, and measured its distance at an accuracy of ~10%. Very recently Sanna *et al.*(2017) successfully measured the parallax for G7.47+0.06 with VLBA and confirmed its location at a distance of 20 kpc, showing the reliability of our 3-D kinematic approach.

3. Results beyond the Galaxy structure

Maser astrometry also provides useful information to investigate individual sources such as AGB stars and star-forming regions. Recently new parallax determinations have been done with VERA for some nearby AGB stars (e.g., Nakagawa *et al.* 2016 and the references therein), which are used to accurately calibrate the period-luminosity (PL) relation of Mira variables. VERA has been conducting survey of tens of AGB parallaxes, and when completed, it is expected to provide an accurate calibration of the absolute magnitude of the Mira's PL relation. Other late-type stars such as OH/IR stars, water-fountain sources and red-super-giant stars are also good targets for maser astrometry to trace mass-loss process in their circumstellar regions through observations of maser spot motions.

SiO masers are quite often observed toward the circumstellar regions of late-type stars, and its radiation mechanism is one of the long-standing question in maser physics. For addressing this issue, a key would be multi-transition observations covering SiO lines at v=1, 2 and 3 transitions. Recently we have developed a wide-band data-acquisition system which is capable of observing the three SiO lines at Q band at the same time, as demonstrated in Oyama *et al.*(2016).

Massive star-forming regions are well-known sites showing strong maser emissions such as those from H_2O and CH_3OH molecules. In addition to distance calibration of

Table 1. Galactic parameter determination

R_0	8.16 ± 0.26 kpc
$\Omega_\odot$	30.49 ± 0.39 km s^{-1} kpc^{-1}
Θ_0	237 ± 8 km s^{-1} (*)

(*) $V_\odot = 12$ km^{-1} is adopted.

star-forming regions, maser astrometry observations provide information on internal motions of maser spots, which are mostly tracing the outflow/jet driven by the central forming star(s). Recently, KaVA (KVN and VERA Array), a joint Korea-Japan VLBI array, started its regular operation, and one of its Large Program is dedicated to survery and monitor of masers in star-forming regions. Kim *et al.*(2018) presented the initial results of KaVA mapping observations for 10 H_2O maser sources. Based on monitoring program starting from 2017 fall, it is anticipated that outflow/jet motions as well as locations of driving sources will be revealed with KaVA.

In the era of ALMA, combination of thermal emissions traced with ALMA and non-thermal maser emissions are essential for further understanding of physical properties of star-forming regions, and ALMA follow-up observations are on-going for the star-forming regions in the KaVA Large Program. In fact, recently Hirota *et al.*(2017) demonstrated the power of such a combined VLBI-ALMA study by successfully tracing the rotation of outflow emanating from the gas disk around Orion source-I.

Time domain astronomy is nowadays getting more and more popular, and the same applies to maser observations. Maser emissions are known to be time-variable, quite often showing a large flare in a specific velocity channel. Therefore, single-dish monitoring of maser sources are important to study such phenomena, and we have been conducting a maser monitoring program of hundreds of sources with VERA single-dish mode, as presented by Sunada *et al.*(2018). During this symposium, there was a discussion on possible global collaborations of maser monitoring, which initiated a coordinated maser monitoring as well as Target-of-Opportunity observations of flaring maser sources with VLBI.

4. Future prospect

In the GAIA era, VLBI astrometry of maser hopefully becomes more important, mainly for two reasons: firstly, comparison between optical and radio astrometry will be the only possibility for consistency check of astrometric results at high accuracy. This should provide firm bases in the future astrometry at 10-μas level. Secondly, the GAIA and maser astrometry traces different species (stars for GAIA and gases for VLBI astrometry). While stars are collisionless components in the Galaxy, the gases are collisional, and hence the motions of these two species are totally different even in the same Galactic potential. Therefore, it is fundamental to trace both collisional and collisionless components for comprehensive understandings of Galactic dynamics.

Future improvement of maser astrometry is two-fold: one way is to extend the distance reach by improving parallax accuracy, and the other is to extend the sky coverage toward the southern hemisphere. To obtain better accuracy, we have been currently extending our VERA array to combine with Korean and Chinese stations, which will be providing better sensitivity as well as better astrometric accuracy. KaVA, Korean-Japan joint array, is already in regular operation, and its astrometry performance is under evaluation. Also, we have initiated a series of test observations of EAVN (East Asian VLBI Network), which consists of VERA in Japan, KVN in Korea, and a few Chinese stations such as Tianma

65m in Shanghai and Nanshan 25m near Urumqi. The inclusion of these Chinese stations is critical for EAVN, because Tianma boosts the array sensitivity and Nanshan station doubles the maximum baseline of the array over ∼5000 km.

Also important is to extend the sky coverage to the souther hemisphere, where currently VERA or VLBA cannot observe. VLBI array in Australia such as LBA and AuScope should be useful for this, and for long-term future, SKA combined with global stations will be the ultimate array for maser astrometry.

References

Boehle, A., *et al.* 2016, *ApJ*, 830, 17

Hirota, T., *et al.* 2007, *PASJ*, 59, 897

Hirota, T., *et al.* 2017, *Nature Astronomy*, 1, 146

Honma, M., *et al.* 2007, *PASJ*, 59, 889

Honma, M., *et al.* 2012, *PASJ*, 64, 136

Kobayashi H., *et al.* 2003, *ASP Conference Series*, Vol. 306, p.367

Kim, J. H., *et al.* 2018, in this volume

Nakagawa, A., *et al.* 2016, *PASJ*, 68, 78

Oyama T., *et al.* 2016, *PASJ*, 68, 105

Reid, M. J. & Honma, M. 2014, *ARA&A*, 52, 339

Reid, M. J.,*et al.* 2014, *ApJ*, 783, 130

Sakai, N., *et al.* 2015, *PASJ*, 67, 69

Sanna, A., *et al.* 2017, *Science*, 358, 227

Sunada, K., *et al.* 2018, in this volume

Yamauchi, A., *et al.* 2016, *PASJ*, 68, 60

Astrophysical Masers:
Unlocking the Mysteries of the Universe
Proceedings IAU Symposium No. 336, 2017
A. Tarchi, M.J. Reid & P. Castangia, eds.

© International Astronomical Union 2018
doi:10.1017/S1743921317010110

Eight new astrometry results of 6.7 GHz CH₃OH and 22 GHz H₂O masers in the Perseus arm

Nobuyuki Sakai, BeSSeL and VERA projects members

Mizusawa VLBI Observatory, National Astronomical Observatory of Japan,
2-21-1 Osawa, Mitaka, Tokyo 181-8588, Japan
email: nobuyuki.sakai@nao.ac.jp

Abstract. We report astrometric results for seven 6.7 GHz CH_3OH and one 22 GHz H_2O masers in the Perseus arm with VLBA and VERA observations. Among the eight sources, we succeeded in obtaining trigonometric parallaxes for all sources, except G098.03+1.44 at 6.7 GHz band. By combining our results with previous astrometry results (Choi *et al.* 2014), we determined an arm width of 0.41 kpc and a pitch angle of 8.2 ± 2.5 deg for the Perseus arm. By using a large sample of the Perseus arm (26 sources), we examined the three-dimensional, non-circular motions (defined as U, V and W) of sources in the Perseus arm as a function of the distance (D) perpendicular to the arm. Interestingly, we found a weighted mean of $< U > = 12.7 \pm 1.2$ km s^{-1} for 14 sources with $D < 0$ kpc (i.e. sources on the interior side of the arm) and $< U > = -0.3 \pm 1.5$ km s^{-1} for 12 sources with $D > 0$ kpc (i.e. sources exterior to the arm). These findings might be the first observational indication of the "damping phase of a spiral arm" suggested by the non-steady spiral arm model of Baba *et al.* (2013). The small pitch angle of the Perseus arm (< 10 deg) also supports the damping phase, based on "pitch angle vs. arm amplitude" relation shown in Grøsbol *et al.* (2004).

Keywords. Galaxy: kinematics and dynamics, ISM: clouds, masers, astrometry.

1. Introduction

Two formation mechanisms for spiral arms have been discussed in the literature:
(1) *the spiral structure rotates nearly uniformly although the material (star and gas) rotate differentially*, or
(2) *the arms are short − lived and reform as open structures*
(Goldreich & Lynden-Bell 1965). Mechanism (1) is known as the density-wave theory (Lin & Shu 1964), while mechanism (2) has often appeared in N-body simulations (e.g. Fujii *et al.* 2011). Measuring 3D velocity fields could potentialy descriminate between these mechanisms, but this is exceedingly difficult to do for external galaxies. In the Milky Way, however, VLBI astrometry studies have allowed us to measure precise distances and 3D velocity fields toward star-forming regions (SFRs) associated with spiral arms (Reid & Honma 2014b). Sakai *et al.* (2012) showed systematic inward motion and slow rotation in the Perseus arm with seven astrometric results and Choi *et al.* (2014) confirmed the same systematic motion, increasing the results to 25 sources.

TGAS in *Gaia* DR1 (with 2.5 million stellar astrometry results) was released in September, 2016, with limited accuracy (systematic error of 0.3 mas)†, but forthcoming catalogs (*Gaia* DR 2 and following) will allow us to study spiral arms beyond several kpc from the Sun including the Perseus arm. We will have precise gas (SFRs) and stellar astrometric data within few years and be able to discriminate between the two mechanisms

† https://gaia.esac.esa.int/documentation/GDR1/pdf/GaiaDR1_documentation_1.1.pdf

discussed above. In other words, we will be able to answer the question "Are spiral arms quasi stationary (like density-waves) or non-stationary (dynamic) structures?" based on observational data.

Here, we report eight new astrometric results from the VLBA and VERA, which better reveal the structure and kinematics of the Perseus arm. Based on the results, we discuss which mechanism (quasi stationary or non-stationary model) is consistent with the data.

2. Observations & Data reduction

We observed 13 CH_3OH ($J_K = 5_1 - 6_0 A^+$) masers, with rest frequency of 6.668519 GHz, under VLBA programs BR149R, S, T and U†. The observation results for seven of these sources are presented here. A more detailed explanation of the observations will be provided in a forthcoming paper (Sakai *et al.* 2017 in preparation). We also observed two H_2O ($J_{K_{-1}K_1} = 6_{16} - 5_{23}$) masers, with rest frequency of 22.235080 GHz with VERA, and we report results here for one of the source. Based on CO $l - v$ emission (Dame *et al.* 2001), the eight sources associated with the Perseus arm.

The VLBA data reduction was done with the NRAO Astronomical Image Processing System (AIPS) and a Parsel tongue script used in previous BeSSeL Survey papers (e.g., Xu *et al.* 2016). Details of the fitting of parallax and proper motion for 6.7 GHz CH_3OH data are described in Reid *et al.* (2017). For the VERA data, AIPS was also used, and the details of the data reduction are summarized in Sakai *et al.* (2015).

3. Results & Discussion

We obtained trigonometric parallaxes and proper motions for seven sources, except for G098.03+1.44. Since G098.03+1.44 was faint and extended, we could not obtain a reliable parallax result. Our results and previous ones obtained from Choi *et al.* (2014) are superposed in a face-on view of the Milky Way in Figure 1.

3.1. *Pitch angle and arm width of the Perseus arm*

We fitted a logarithmic spiral arm model to the locations of the Perseus arm sources using the following equation:

$$\ln(R/R_{\mathrm{ref}}) = -(\beta - \beta_{\mathrm{ref}})\tan\psi, \tag{3.1}$$

where R_{ref} and β_{ref} are a reference Galactocentric distance and azimuth for the arm. β is defined as zero degrees toward the Sun as viewed from the Galactic center and increases with Galactic longitude, and ψ is the pitch angle of the arm.

We obtained $R_{\mathrm{ref}} = 9.97 \pm 0.15$ kpc, $\beta_{\mathrm{ref}} = 13.7$ deg and $\psi = 8.3 \pm 2.5$ deg for the Perseus arm. The results are consistent with those in Reid *et al.* (2014a) within errors. In the fitting, we assumed an arm width 0.43 kpc for the Perseus arm (Reid *et al.* 2014a). Figure 1 shows the result of the fitting.

3.2. *Non-circular motion of the Perseus arm*

Next, we discuss non-circular motions in the Perseus arm derived from the astrometric results and LSR velocities, using a model of the rotation curve and peculiar solar motion $(U_\odot, V_\odot, W_\odot)$. We assume the universal rotation curve (Persic *et al.* 1996) and $(U_\odot, V_\odot, W_\odot) = (10.5 \pm 1.7, 14.4 \pm 6.8, 8.9 \pm 0.9)$ km s^{-1} (Reid *et al.* 2014a). Black arrows in Fig. 1 show residuals (i.e., non-circular motions) to the model.

† Please see the BeSSeL project HP: `http://bessel.vlbi-astrometry.org/observations`

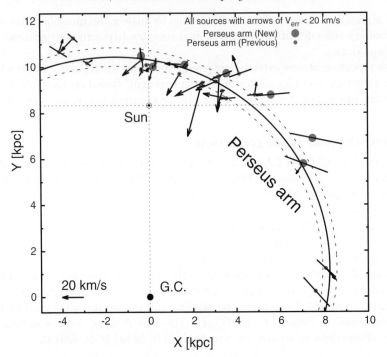

Figure 1. Spatial distribution and non-circular motions of the Perseus arm sources. Arrows indice non-circular motions only for sources with uncertainties less than 20 km s^{-1}. The scale of 20 km s^{-1} is shown at the lower left of the image. The Sun is located at $(X, Y) = (0, 8.34)$ kpc. The universal rotation curve (Persic *et al.* 1996) and $(U_\odot, V_\odot, W_\odot) = (10.5 \pm 1.7, 14.4 \pm 6.8, 8.9 \pm 0.9)$ km s^{-1} (Reid *et al.* 2014a) are assumed.

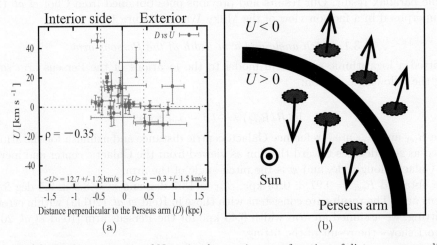

Figure 2. (a) The U component of Non-circular motion as a function of distance perpendicular to the Perseus arm (D). $D < 0$ corresponds to the interior side of the Perseus arm (i.e. closer to the Sun from the center of the arm) while $D > 0$ indicates exterior to the Perseus arm. The symbol $<>$ denoted in each side of the figure shows weighted mean of the non-circular motion while the value ρ superposed on the figure represents the correlation coefficient. Solid line on the figure shows the position of the Perseus arm while vertical dashed lines on the figure represent the arm width ($= 0.43$ kpc, see text). (b) Schematic view of the figure (a). Arrows on the figure show tendency of the non-circular motion of U seen in the Perseus arm.

Figure 2a displays the non-circular motion toward the Galactic center (defined as U) as a function of distance perpendicular to the Perseus arm (D). The data hints at a negative slope with correlation coefficient of -0.35. This suggests that sources on the near side of the Perseus arm (sources with $D < 0$) are moving toward the Galactic center with a weighted mean of $<U> = 12.7 \pm 1.2$ km s^{-1}, while those on the far side (those with $D > 0$) are moving oppositely with $<U> = -0.3 \pm 1.5$ km s^{-1}. This tentative result, if confirmed, would be the first observational indication of the damping phase of the Perseus arm.

Baba *et al.* (2013) conducted three-dimensional N-body simulations to examine the characteristics of stellar spiral arms in disk galaxies. They found that a growing (high density) phase appeared with a large pitch angle >30 deg, and the spiral arm later showed a damping (low density) phase with the samller pitch angle (see Fig. 7 of Baba *et al.* 2013). Observationally, Grøsbol *et al.* (2004) examined structural parameters of 54 normal spiral galaxies and found (1) a typical range of pitch angles between 5-30 deg and (2) a proportional relation of tan(pitch angle) versus relative arm amplitude (see Fig. 8 of Grøsbol *et al.* 2004). If the Milky Way is a normal spiral galaxy, the Grøsbol *et al.* (2004) result suggests the Perseus arm is now in the damping (low density) phase.

We will be able to validate our conclusion regarding the damping phase of the Perseus arm) by adding stellar astrometric results from *Gaia* DR2 (released in April, 2018), which might allow us to obtain a more conclusive answer to a major problem of the Galactic astronomy.

We are grateful to BeSSeL and VERA projects members for the support they offered during observations. We are also grateful to Dr. Mark J. Reid for carefully reading and editing the manuscript. We would like to thank the LOC and SOC of IAUS 336 for organizing the productive conference in Sardegna, Italy.

References

Baba, J., Saitoh, T. R., & Wada, K. 2013, *ApJ*, 763, 46

Choi, Y., Hachisuka, K., Reid, M. J., Brunthaler, A., Menten, K. M., *et al.* 2014, *ApJ*, 790, 99

Dame, T. M., Hartmann, D., & Thaddeus, P. 2001, *ApJ*, 547, 792

Fujii, M., Baba, J., Saitoh, T. R., Makino, J., Kokubo, E., & Wada, K. 2011, *ApJ*, 730, 109

Goldreich, P.,& Lynden-Bell, D. 1965, *MNRAS*, 130, 125

Grøsbol, P., Patsis, P. A., & Pompei, E. 2004, *A&A*, 423, 849

Lin, C. C. & Shu, F. H. 1964, *ApJ*, 140, 646

Persic, M, Salucci, P. & Stel, F. 1996, *MNRAS*, 281, 27

Reid, M. J., Menten, K. M., Brunthaler, A., Zheng, X. W., *et al.* 2014a, *ApJ*, 783, 130

Reid, M. J. & Honma, M. 2014b, *ARA&A*, 52, 339

Reid, M. J., Brunthaler, A., Menten, K. M., Sanna, A., Xu, Y., Li, J. J., *et al.* 2017, *AJ*, 154, 63

Sakai, N., Honma, M., Nakanishi, H., Sakanoue, H., Kurayama, T., *et al.* 2012, *PASJ*, 64, 108

Sakai, N., Nakanishi, H., Matsuo, M., Koide, N., Tezuka, D., *et al.* 2015, *PASJ*, 67, 69

Xu, Y., Reid, M. J., Menten, K., Sakai, N., Li, J., Brunthaker, A., *et al.* 2016, *SciA*, 2, 9

Astrophysical Masers:
Unlocking the Mysteries of the Universe
Proceedings IAU Symposium No. 336, 2017
A. Tarchi, M.J. Reid & P. Castangia, eds.

© International Astronomical Union 2018
doi:10.1017/S1743921317009322

SWAG Water Masers in the Galactic Center

Jürgen Ott[1], Nico Krieger[2], Matthew Rickert[3], David Meier[4], Adam Ginsburg[5], Farhad Yusef-Zadeh[6] and the SWAG team

[1] National Radio Astronomy Observatory,
1003 Lopezville Road, Socorro, NM 87801, USA
email: jott@nrao.edu

[2] Max-Planck-Institut für Astronomie,
Königstuhl 17, 69120 Heidelberg, Germany
email: krieger@mpia.de

[3] Department of Physics and Astronomy and CIERA, Northwestern University,
Evanston, IL 60208, USA; Matthew Rickert is a Reber Fellow at the National Radio
Astronomy Observatory.

[4] New Mexico Institute of Mining and Technology,
801 Leroy Place, Socorro, NM 87801, USA
email David.Meier@nmt.edu; David S. Meier is also an Adjunct Astronomer at the National
Radio Astronomy Observatory.

[5] National Radio Astronomy Observatory,
1003 Lopezville Road, Socorro, NM 87801, USA
email: aginsbur@nrao.edu

[6] Department of Physics and Astronomy and CIERA, Northwestern University,
Evanston, IL 60208, USA
email: zadeh@northwestern.edu

Abstract. The Galactic Center contains large amounts of molecular and ionized gas as well as a plethora of energetic objects. Water masers are an extinction-insensitive probe for star formation and thus ideal for studies of star formation stages in this highly obscured region. With the Australia Telescope Compact Array, we observed 22 GHz water masers in the entire Central Molecular Zone with sub-parsec resolution as part of the large SWAG survey: "Survey of Water and Ammonia in the Galactic Center". We detect of order 600 22 GHz masers with isotropic luminosities down to $\sim 10^{-7}$ L$_\odot$. Masers with luminosities of $\gtrsim 10^{-6}$ L$_\odot$ are likely associated with young stellar objects. They appear to be close to molecular gas streamers and may be due to star formation events that are triggered at pericenter passages near Sgr A*. Weaker masers are more widely distributed and frequently show double line features, a tell-tale sign for an origin in evolved star envelopes.

Keywords. masers, Galaxy: center, radio lines: stars

1. Introduction

The Central Molecular Zone (CMZ; the inner ~ 500 pc) of the Milky Way is the nearest nucleus of any galaxy and allows us to study aspects of star formation processes under extreme conditions, such as high pressure gas, extreme tidal forces, strong radiation fields, and cloud-cloud shock zones. Gas flows from the Milky Way disk to the CMZ, where some of the gas appears to follow specific trajectories also known as orbits or streamers. Kinematic models of the streamers are given, e.g., in Molinari *et al.* (2011), Kruijssen, Dale, & Longmore (2015), or Ridley *et al.* (2017). Kruijssen, Dale, & Longmore (2015), in particular, predict that streamers near the pericenter to Sgr A* (which marks the center of the gravitational potential) may compress the gas and initiate a collapse and trigger star formation. Longmore *et al.* (2013) and Henshaw *et al.* (2016) show that indeed a star formation sequence can be observed in the CMZ, where a streamer after its Sgr A* pericenter passage contains dense gas with no obvious star formation (the "brick"), followed by consecutively more evolved star forming regions, culminating in Sgr B2, the most vigorous star formation site in the Milky Way. Downstream from Sgr B2, stellar clusters appear even more evolved, e.g. in the radio continuum-bright Sgr B1 region and further down in the Arches and Quintuplet stellar clusters. Krieger *et al.* (2017) and Ginsburg *et al.* (2016) also find evidence that gas near pericenter passages exhibit positive temperature gradients, which suggests that gas may indeed be compressed near those spots.

The CMZ is characterized by extreme optical extinction and radio lines are ideal to derive the evolutionary status of star forming regions along the possible star formation sequence. 22 GHz water ($6_{16} - 5_{23}$) masers are not affected by extinction and are produced in extreme environments, typically in shocked envelopes of evolved stars, outflows of young stellar objects (YSOs), or, in their most extreme form, in the accretion disks and jet-gas interfaces of active galactic nuclei (AGN). Maser luminosities from the different sources vary, where AGN related masers are extremely bright megamasers (e.g., Lo 2005; where the term 'megamasers' refers to a comparison with typical maser strengths of individual stellar sources in the Milky Way). Masers near evolved stars and YSOs are much less luminous and the most luminous source in the Milky Way reaches isotropic luminosities of $\sim 0.1\,L_{\odot}$ (W 49N, Liljestrom *et al.* 1989). Palagi *et al.* (1993) compare maser luminosities in the Milky Way and they find that water masers related to YSOs are on the high end of the luminosity function, whereas evolved stars only reach maximum luminosities of $\sim 10^{-4}\,L_{\odot}$. With luminosity cutoffs it is therefore possible to preferentially select YSOs and thus locate the related active zones of star formation.

2. SWAG: Survey of Water and Ammonia in the Galactic Center

The SWAG survey "Survey of Water and Ammonia in the Galactic Center" is ideally suited to obtain a rather comprehensive picture of the molecular gas toward the Galactic Center. This three year survey with the Australia Telescope Compact Array†, covers the entire CMZ in the 21.2-25.4 GHz frequency range with high spectral resolution of targeted 42 specific lines. This includes the 22 GHz water maser line, multiple transitions of the temperature tracer ammonia, photon-dominated and shock-dominated region tracers, and radio recombination lines. The resolution of SWAG is about $\sim 27''$ which corresponds to sub-pc resolution at the distance of the CMZ (8.5 kpc). First results of SWAG are described in Krieger *et al.* (2017), who analyze the temperature properties along the gas streamers based on multiple transitions of ammonia. Here we report on first results of the 22 GHz water maser ($6_{16} - 5_{23}$) transition.

3. Populations of Water Masers toward the CMZ

Maps of ammonia and water masers are shown in Fig. 1. The gas streamers mostly reside in the inner ~ 150 pc, the region that covers the area from Sgr B2 to Sgr C, roughly centered on the supermassive black hole Sgr A*. In Fig. 1, we plot the location and direction of the Kruijssen, Dale, & Longmore (2015) streamers on top of the gas distribution. Starting between Sgr C and Sgr A*, the streamer passes the first pericenter near Sgr A* and continues toward and beyond Sgr B2 (for a 3-dimensional picture, see Kruijssen, Dale, & Longmore 2015).

† The Australia Telescope Compact Array is part of the Australia Telescope National Facility (ATNF), a division of the Commonwealth Scientific and Industrial Research Organisation (CSIRO).

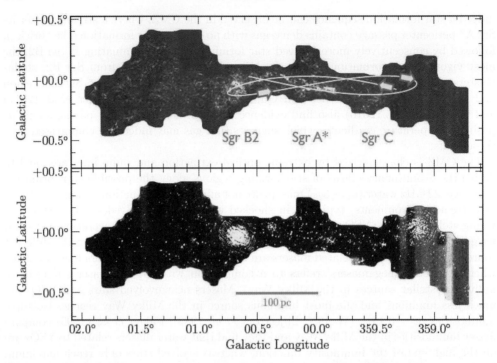

Figure 1. Top: Peak flux map of the ammonia (3,3) emission of the SWAG survey. Prominent features and the gas streamers are marked. The arrows indicate the direction of the flows. **Bottom:** The peak flux map of the 22 GHz water masers in the same region. Note that some extremely bright masers, in particular in the Sgr B2 region show considerable sidelobes.

The bottom of Fig. 1 shows the peak fluxes of the 22 GHz water masers. Note that some masers, especially near Sgr B2 and around $b \sim -0.8°$, are spilling beam sidelobes across many pointings. We consider those regions saturated and of limited use for the current analysis. To first order, the gas and the 22 GHz masers are not particularly correlated and a close inspection of the data shows that even on smaller scales masers and molecular clumps are not necessarily co-spatial.

In Fig. 2, we show masers in the inner region of the CMZ, spanning Sgr B2 to Sgr C, at different luminosity cuts (assuming that all water masers are at 8.5 kpc distance). The most luminous sources with isotropic luminosities exceeding 10^{-6} L$_\odot$ are indeed close to the streamer trajectories. Following the work of Palagi *et al.* (1993), their luminosities are consistent with being related to YSOs and therefore trace current star formation sites. The gas and maser velocities are also frequently separated by less than $\sim \pm 20$ km s^{-1}, which further supports this scenario. Following the streamers, there is only one weak water maser source in the "brick" at $(l, b) \sim (0.253°, 0.016°)$ followed by a very bright $> 10^{-5}$ L$_\odot$ source near $(l, b) \sim (0.38°, 0.04°)$; source "C", see Ginsburg *et al.* 2015) and the extreme Sgr B2 water masers. A few fainter masers but with luminosities still in the YSO regime are observed downstream. This situation corroborates the star formation sequence described by Longmore *et al.* (2013) and Henshaw *et al.* (2016).

At isotropic luminosities $\lesssim 10^{-6}$ L$_\odot$, however, the number of sources increases drastically and in total we observe of order 600 masers. Their distribution does not follow streamers anymore but appear more widely distributed (see also Rickert 2017). Frequently, the masers show double-peaked spectral profiles, which suggests that the majority of them are associated with evolved stars across the entire disk of the Milky Way.

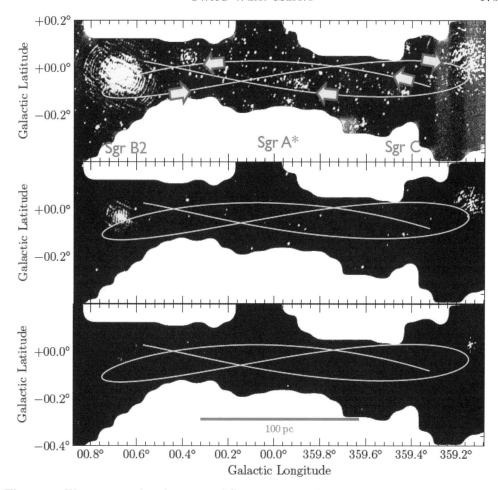

Figure 2. Water maser distribution at different isotropic luminosity cuts, assuming that all masers are at a distance of 8.5 kpc. **Top:** $> 10^{-7}$ L$_\odot$ **Middle:** $> 10^{-6}$ L$_\odot$ **Bottom:** $> 10^{-5}$ L$_\odot$. Overlaid are the Streamers and directional arrows.

References

Ginsburg, A., Walsh, A., Henkel, C., *et al.* 2015, *A&A*, 584, L7

Ginsburg, A., Henkel, C., Ao, Y., *et al.* 2016, *A&A*, 586, A50

Henshaw, J. D., Longmore, S. N., & Kruijssen, J. M. D. 2016, *MNRAS* , 463, L122

Krieger, N., *et al.* 2017, *ApJ*, 850, 77

Kruijssen, J. M. D., Dale, J. E., & Longmore, S. N. 2015, *MNRAS*, 447, 1059

Liljestrom, T., Mattila, K., Toriseva, M., & Anttila, R. 1989, *A&AA*, 79, 19

Lo, K. Y. 2005, *ARA&A*, 43, 625

Longmore, S. N., Kruijssen, J. M. D., Bally, J., *et al.* 2013, *MNRAS*, 433, L15

Molinari, S., Bally, J., Noriega-Crespo, A., *et al.* 2011, *ApJL*, 735, L33

Palagi, F., Cesaroni, R., Comoretto, G., Felli, M., & Natale, V. 1993, *A&AS*, 101, 153

Rickert, M. 2017, Ph.D. Thesis, Northwestern University

Ridley, M. G. L., Sormani, M. C., Treß, R. G., Magorrian, J., & Klessen, R. S. 2017, *MNRAS*, 469, 2251

Yusef-Zadeh, F., Hewitt, J. W., Arendt, R. G., *et al.* 2009, *ApJ*, 702, 178

Astrophysical Masers:
Unlocking the Mysteries of the Universe
Proceedings IAU Symposium No. 336, 2017
A. Tarchi, M.J. Reid & P. Castangia, eds.

© International Astronomical Union 2018
doi:10.1017/S1743921317010213

How maser observations unravel the gas motions in the Galactic Center

**K. Immer[1], M. Reid[2], A. Brunthaler[3], K. Menten[3], Q. Zhang[2],
X. Lu[4], E. A. C. Mills[5], A. Ginsburg[6], J. Henshaw[7], S. Longmore[8],
D. Kruijssen[9] and T. Pillai[3]**

[1] Joint Institute for VLBI ERIC, Postbus 2, 7990 AA Dwingeloo, The Netherlands
email: `immer@jive.eu`
[2] Harvard-Smithsonian Center for Astrophysics, 60 Garden Street, Cambridge, MA 02138, USA
[3] Max-Planck-Institut für Radioastronomie, Auf dem Hügel 69, 53121, Bonn, Germany
[4] National Astronomical Observatory of Japan, 2-21-1 Osawa, Mitaka, Tokyo, JP 181-8588
[5] San Jose State University, 1 Washington Square, San Jose, CA 95192, USA
[6] National Radio Astronomy Observatory, Socorro, NM 87801 USA
[7] Max-Planck-Institut für Astronomie, Königstuhl 17, 69117 Heidelberg, Germany
[8] Astrophysics Research Institute, Liverpool John Moores University, Liverpool, L3 5RF, UK
[9] Astronomisches Rechen-Institut, Zentrum für Astronomie der Universität Heidelberg,
Mönchhofstrasse 12-14, 69120 Heidelberg, Germany

Abstract. The Central Molecular Zone (CMZ), the inner 450 pc of our Galaxy, is an exceptional region where the volume and column densities, gas temperatures, velocity dispersions, etc. are much higher than in the Galactic plane. It has been suggested that the formation of stars and clusters in this area is related to the orbital dynamics of the gas. The complex kinematic structure of the molecular gas was revealed by spectral line observations. However, these results are limited to the line-of-sight-velocities. To fully understand the motions of the gas within the CMZ, we have to know its location in 6D space (3D location + 3D motion). Recent orbital models have tried to explain the inflow of gas towards and its kinematics within this region. With parallax and proper motion measurements of masers in the CMZ we can discriminate among these models and constrain how our Galactic Center is fed with gas.

Keywords. masers, instrumentation: interferometers, astrometry, Galaxy: center, Galaxy: kinematics and dynamics, radio lines: ISM

1. The Central Molecular Zone

The Central Molecular Zone (CMZ), the inner ∼450 pc of the Milky Way, produces 5−10% of the infrared and Lyman continuum luminosity of our Galaxy and 10% of the Galaxy's total molecular gas are located here (Morris & Serabyn 1996 and references therein). The conditions are extreme: high average densities ($>10^4$ cm^{-3}), high gas temperatures (>60 K), large velocity dispersions (> 15 km s^{-1}), etc. (Morris & Serabyn 1996 and references therein for all above quantities).

Several groups have suggested that the formation of stars and clusters is closely linked with the orbital dynamics of the gas (e.g. Molinari *et al.* 2011, Longmore *et al.* 2013, Kruijssen *et al.* 2015). The orbital motion of giant molecular clouds in the CMZ has been studied for decades (e.g. Bally *et al.* 1988, Sofue 1995, Tsuboi *et al.* 1999, Oka *et al.* 1998). Spectral line observations of the molecular gas in this region show a complex kinematic structure in position-position-velocity space. Most of the gas seems to be located in two coherent gas streams, covering a large range of line-of-sight velocities. However, there are several difficulties with this observational approach and its interpretation. Due to our position in the Galactic plane, we see foreground and CMZ clouds in projection and

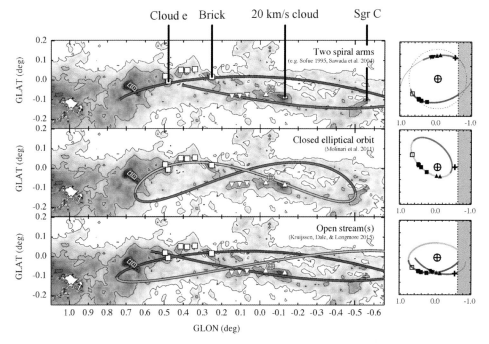

Figure 1. Left: Projected view of the gas orbits predicted by the three models discussed in Section 2 overlaid on a Herschel-derived molecular hydrogen column density map by Battersby *et al.* (in prep.). Right: Top-down view of the same orbits (motion in clockwise direction). The symbols mark important positions along the orbits: Sgr C (plus), 20 and 50 km s^{-1} cloud (upward triangles), "Three Little Pigs" (downward triangles); the Brick and clouds b–f (squares), Sgr B2 (diamonds), Sgr A* (circle with cross). All figures are adapted from Henshaw *et al.* 2016.

thus the relative distance of the clouds along the line-of-sight is difficult to determine. In addition, spectral line observations only yield the line-of-sight velocities of gas clouds but not the motion of the clouds in the plane of the sky. To fully understand the gas motions in the CMZ, we have to know its location in 6D space (position, distance, line-of-sight velocity, velocity in the plane of the sky). With this information, we will be able to test models interpreting the gas kinematics (see section 2).

The only method to develop a well-founded 6D picture of the CMZ is to measure parallaxes and proper motions of masers and to combine them with their positions and line-of-sight velocities. Methanol and water masers in star forming regions are particularly good targets since the motions of the host clouds can be easily inferred from the maser proper motions.

2. Orbital models

Recent models have tried to explain the inflow of gas towards and its kinematics in the CMZ. We will focus on the predictions of three models which interpret the kinematics within $|l| \leqslant 0.7°$: two spiral arms (e.g. Sawada *et al.* 2004, Ridley *et al.* 2017), a closed elliptical orbit (Molinari *et al.* 2011), continuous open streams (Kruijssen *et al.* 2015). Fig. 1 shows the projected and the top-down view of the predicted gas orbits of these three models, adapted from Henshaw *et al.* 2016. Important sources along the orbits are marked.

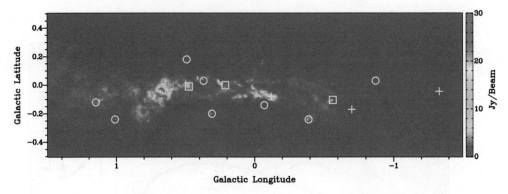

Figure 2. 870 μm emission of the CMZ from the ATLASGAL survey (Schuller *et al.* 2009). Maser sources from the BeSSeL survey are marked with yellow circles. Methanol and water masers from our target list are shown as yellow boxes and crosses, respectively.

At the position of Sgr C, the Molinari model predicts all motion to be along the line-of-sight, implying very small proper motions. The Kruijssen and the Sawada models predict significant proper motions but with opposite signs. For the 20 km/s cloud, the Sawada model predicts proper motion in the direction of decreasing Galactic longitudes, in contrast to the other two models. In all three models, most of the motion at this position should be in the plane of the sky and the proper motions should be large (5−10 mas/yr). At the position of cloud e, all models predict proper motions in the direction of increasing Galactic longitudes but with values differing by factors of several. The velocity structure at the south-west end of the Brick is complex. Proper motion measurements will show if the gas at this position is located at the near or far side of the orbit.

3. Observations

From the parallax observations of the BeSSeL survey (Brunthaler *et al.* 2011) we infer that methanol and water masers need to be stronger than 3 and 8 Jy, respectively, to be used as phase-referencing sources. We searched water and methanol maser catalogues, covering the CMZ, for masers stronger than these limits within $|l| \leqslant 1.5°$ and $|b| \leqslant 0.25°$. We selected three 6.7 GHz methanol and three 22 GHz water masers for parallax and proper motion observations with the Very Long Baseline Array (VLBA, project code: BI042). In addition, four methanol and five water masers from the BeSSeL survey were included in the project. The observations of project BI042 were conducted in 2016/2017. For both the methanol and water masers in BI042, four epochs were observed over the timespan of one year. The location of the masers in the CMZ are shown in Fig. 2.

4. Preliminary results

The analysis of the BI042 data is ongoing while it has recently been concluded for the BeSSeL sources. In the first epoch of BI042, all methanol masers have been detected while two of the water maser candidates do not show maser emission. The parallaxes and proper motion values of the BI042 masers will be determined within the next year.

Here, we will present the preliminary results of two water masers from the BeSSeL survey (G359.61−0.24 and G000.37+0.03, see Fig. 2 for their positions in the CMZ).

G359.61−0.24 To determine the parallax and proper motion of this source, we combined the data of three maser spots and two background quasars (J1748−2907 and

J1752−3001). The parallax is 0.370 ± 0.029 mas, which corresponds to a distance of $2.7^{+0.2}_{-0.2}$ kpc. Although the coordinates of this source position it within the CMZ area, its distance clearly locates it in the Scutum arm. This result shows the importance of combining the proper motion measurements of the CMZ sources with parallax measurements to exclude sources which are not physically part of the CMZ.

G000.37+0.03 For this maser, we combined the parallax measurements of two maser spots, using two different background quasars (J1745−2820 and J1748−2907). This resulted in a parallax of 0.125 ± 0.047 mas which corresponds to a distance of $8.0^{+4.8}_{-2.2}$ kpc. This distance locates the source within the CMZ, however, the uncertainty is very large. Additional epochs from future observations could reduce this uncertainty. The proper motion of this source is $\mu_x = -1.004 \pm 0.108$ mas yr^{-1} and $\mu_y = -2.474 \pm 0.288$ mas yr^{-1} which is consistent with the values that the Kruijssen model predicts at that position ($\mu_x = -0.76^{+0.36}_{-0.78}$ mas yr^{-1}, $\mu_y = -2.88^{+0.84}_{-0.78}$ mas yr^{-1}).

References

Bally, J., Stark, A. A., Wilson, R. W., & Henkel, C. 1988, *ApJ*, 324, 223

Brunthaler, A., Reid, M. J., Menten, K. M., Zheng, X.-W., et al. 2011, *AN*, 332, 461

Henshaw, J. D., Longmore, S. N., Kruijssen, J. M. D., & Davies, B. 2016, *MNRAS*, 457, 2675

Longmore, S. N., Kruijssen, J. M. D., Bally, J., Ott, J., et al. 2013, *MNRAS*, 433, L15

Kruijssen, J. M. D., Dale, J. E., & Longmore, S. N. 2015, *MNRAS*, 447, 1059

Molinari, S., Bally, J., Noriega-Crespo, A., Compiègne, M., et al. 2011, *ApJL*, 735, L33

Morris, M. & Serabyn, E. 1996, *ARA&A*, 34, 645

Oka, T., Hasegawa, T., Sato, F., Tsuboi, M., et al. 1998, *ApJS*, 118, 455

Ridley, M. G. L., Sormani, M. C., Tre, R. G., Magorrian, J., et al. 2017, *MNRAS*, 469, 2251

Sawada, T., Hasegawa, T., Handa, T., & Cohen, R. J. 2004, *MNRAS*, 349, 1167

Schuller, F., Menten, K. M., Contreras, Y., & Wyrowski, F. 2009, *A&A*, 504, 415

Sofue, Y. 1995, *PASJ*, 47, 527

Tsuboi, M., Handa, T., & Ukita, N. 1999, *ApJS*, 120, 1

Astrophysical Masers:
Unlocking the Mysteries of the Universe
Proceedings IAU Symposium No. 336, 2017
A. Tarchi, M.J. Reid & P. Castangia, eds.
© International Astronomical Union 2018
doi:10.1017/S1743921317009292

Stellar SiO masers in the Galaxy: The Bulge Asymmetries and Dynamic Evolution (BAaDE) survey

L. O. Sjouwerman[1], Y. M. Pihlström[2,1], R. M. Rich[3], M. J Claussen[1], M. R. Morris[3] and the BAaDE collaboration

[1] National Radio Astronomy Observatory, Socorro NM 87801 (USA) email: lsjouwer@nrao.edu

[2] Dept. of Physics & Astronomy, University of New Mexico, Albuquerque NM 87131 (USA)

[3] Div. of Astronomy & Astrophysics, University of California, Los Angeles CA 90095 (USA)

Abstract. Circumstellar SiO masers can be observed in red giant evolved stars throughout the Galaxy. Since stellar masers are not affected by non-gravitational forces, they serve as point-mass probes of the gravitational potential and form an excellent sample for studies of the Galactic structure and dynamics. Compared to optical studies, the non-obscured masers are in particular valuable when observed close to the highly obscured Galactic Bulge and Plane. Their line-of-sight velocities can easily be obtained with high accuracy, proper motions can be measured and distances can be estimated. Furthermore, when different mass and metallicity effects can be accounted for, such a large sample will highlight asymmetries and evolutionary traces in the sample. In our Bulge Asymmetries and Dynamic Evolution (BAaDE) survey we have searched 20,000 infrared selected evolved stars for 43 GHz SiO masers with the VLA in the northern Bulge and Plane and are in the process of observing another 10,000 stars for 86 GHz SiO masers with ALMA in the southern Bulge. Our instantaneous detection rate in the Bulge is close to 70%, both at 43 and 86 GHz, with occasionally up to 7 simultaneous SiO transitions observed in a single star. Here we will outline the BAaDE survey, its first results and some of the peculiar maser features we have observed. Furthermore we will discuss the prospects for obtaining proper motions and parallaxes for individual maser stars to reconstruct individual stellar orbits.

Keywords. Masers, Surveys, Infrared: Stars, Stars: Late-type, Circumstellar Matter, Galaxy: Kinematics and Dynamics

1. Introductory Background

The Galactic Bulge formed largely by dynamical evolution, and it provides us with a remarkable laboratory to study the processes that must have taken place in many disk galaxies in the local universe between $z = 1$ and now. The central part of the Galaxy seems strongly dominated by a massive bar, based on analysis of the infrared morphology (e.g., Blitz & Spergel 1991, Dwek *et al.* 1995), spatial distribution of red clump stars (e.g., Stanek *et al.* 1997, Babusiaux & Gilmore 2005), and dynamics of red giants from the Bulge Radial Velocity Assay survey (*BRAVA*, Rich *et al.* 2007, Kunder *et al.* 2012).

Optical surveys of the Bulge are a powerful approach to learn about the populations and dynamics in the less reddened and obscured regions where $|b| > 4°$. To a great extent, however, surveys in the optical or infrared bands have begun to reach impasses that cannot be easily resolved with increases in sample sizes.

Red giant circumstellar maser sources, which can be exploited with the Atacama Large Millimeter/submillimeter Array (ALMA) and the Karl G. Jansky Very Large Array (VLA), offer a bold new approach to address the most pressing problems in the study of the inner Galaxy. Kinematic studies of stellar populations using Galactic masers have

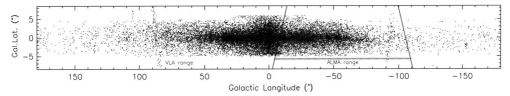

Figure 1. Galactic distribution of ∼ 30,000 color selected MSX sources in the BAaDE survey. Sources south of Declination −35° (roughly −10° < l < −105°) are observed with ALMA at 86 GHz, the others with the VLA at 43 GHz. With a detection rate of about 70% we expect to derive line-of-sight velocities for nearly 20,000 sources, most of which are in the Galactic Bulge.

been proven prosperous once it was recognized that color selection of IR sources detected by the IRAS satellite was quite predictive in finding stellar OH (1612 MHz) masers in asymptotic giant branch (AGB) stars. The instantaneously obtained stellar line-of-sight velocities could be readily used for kinematic studies. About 3000 OH/IR stars have been studied in the Galaxy, mostly in the Galactic Bulge and Center regions (e.g., Lindqvist *et al.* 1992, Deguchi *et al.* 2002, Debattista *et al.* 2002, Habing 1996).

A high chance of SiO maser occurrence associated with objects that may be used as independent dynamical probes can also be achieved by targeting infrared stars. The SiO masers usually reside in thin envelopes with "bluer" colors on the circumstellar envelope sequence (CSE, Sjouwerman *et al.* 2009). The SiO masers may be observed both at J=1–0 transitions at 43 GHz and at J=2–1 transitions at 86 GHz, allowing a complete coverage of the Galactic Plane when combining observations taken with the VLA and ALMA.

Our goal is to produce samples of radio-detected red giants that are comparable to the 10,000 stars of BRAVA, or the 28,000 stars of the Abundances and Radial velocity Galactic Origins Survey (*ARGOS*, Freeman *et al.* 2012). By using SiO masers, which are detectable in red giants spanning a wide range in luminosity and distances, we probe into the highest extinction, most crowded regions of the Milky Way: the Plane and Galactic Center. We have thus begun an observational program (Bulge Asymmetries and Dynamical Evolution, BAaDE) using the VLA and ALMA to undertake a targeted survey that in principle are not reachable with optical surveys.

2. Targeted Observations and Results

We selected candidate targets that had good quality infrared fluxes in the Midcourse Space Experiment survey (MSX) Point Source Catalog version 2.3 (Price *et al.* 2001, Egan *et al.* 2003). Furthermore we required that the candidate targets fulfilled the criteria for MSX region *iiia* – the region corresponding to IRAS region IIIa where we expect a high occurrence of SiO masers (Sjouwerman *et al.* 2009).This selection yielded about 30,000 sources (Figure 1). Dec ⩾ −35° sources were observed at 43 GHz with the VLA during 2012-2017 and the lower Declination sources were reserved for ALMA observations at 86 GHz, which have begun in 2015. A comparison in detection statistics and detection bias between using the two surveys is found elsewhere in this volume (Stroh *et al.* 2018).

Due to the lack of calibrators, we are using a novel self-calibration scheme to use the brightest masers to detect the weaker ones. This method works very well because we can use that the typical detection rate, using our infrared color-selection scheme, is well over 50% for both VLA and ALMA observations, even though we spend less than a minute of observing time per source (Figures 2 and 3). It is promising to show, in Figure 4, that our survey already surpasses the detail obtained with OH/IR stars (see e.g., Habing 1996).

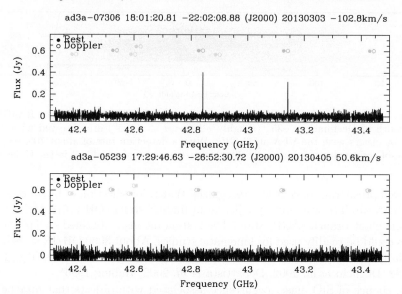

Figure 2. Vector-averaged VLA spectra, ranging from about 42300 to 43500 MHz, labeled with the source name, the position (J2000), the observing date (UT) and the derived LSR velocity. To guide the eye, the filled circles show all potential SiO transitions (plus a carbon line) at $V_{LSR}=0$ and the open circles are those lines shifted to the derived V_{LSR}. Top: a typical VLA detection with two J=1-0 ^{28}SiO maser lines, from left to right (v=2) and (v=1). Bottom: an example of a carbon line detection (HC$_5$N) instead of an SiO maser. Carbon detections also yield a velocity for dynamical purposes and in addition allow for a sub-study toward carbon-rich sources.

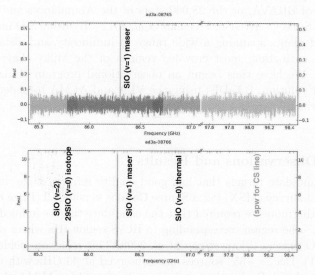

Figure 3. Vector-averaged ALMA spectra, using 3 overlapping spectral windows ranging from 85400 to 87100 MHz where SiO transitions are expected, combined in the same panel with an additional spectral window from about 97550 to 98450 MHz for potential lines of carbon bearing molecules (i.e., CS). Top: a typical ALMA detection with a single J=2-1 (v=1) ^{28}SiO maser line. Bottom: an example of a multi-line detection with all possible SiO transitions detected.

   We are currently compiling our list of detections and selecting candidate sources for VLBI follow-up observations. The extended list of new line-of-sight velocities, in combination with the prospect of deriving proper motions for a subset of sources located right

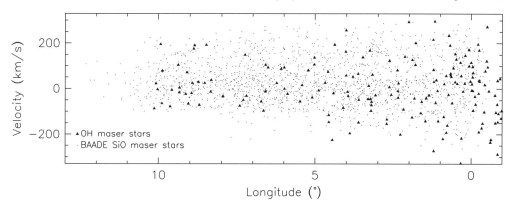

Figure 4. Preliminary longitude-velocity diagram of a subset of masers in the Galactic Bulge and Plane ($|b| \ll 5°$) showing the increased information density obtained with SiO masers in the BAaDE survey (small dots) compared to the known population of OH masers (filled triangles).

in the Galactic Plane and in the Galactic Bulge and Center will contribute significantly to the modeling of Galactic dynamics and the understanding of Galactic evolution.

Acknowledgements

This material is based in part upon work supported by the National Science Foundation under Grant Numbers 1517970 and 1518271. Any opinions, findings, and conclusions or recommendations expressed in this material are those of the author(s) and do not necessarily reflect the views of the National Science Foundation.

This paper makes use of the following VLA data: VLA/11B-091. The Karl G. Jansky Very Large Array (VLA) is operated by The National Radio Astronomy Observatory (NRAO). NRAO is a facility of the National Science Foundation (NSF) operated under cooperative agreement by Associated Universities, Inc. (AUI).

This paper makes use of the following ALMA data: ADS/JAO.ALMA#2013.1.01180.S. ALMA is a partnership of ESO (representing its member states), NSF (USA) and NINS (Japan), together with NRC (Canada), NSC and ASIAA (Taiwan), and KASI (Republic of Korea), in cooperation with the Republic of Chile. The Joint ALMA Observatory is operated by ESO, AUI/NRAO and NAOJ.

References

Babusiaux, C., & Gilmore, G., 2005, *MNRAS*, 358, 1309
Blitz, L., & Spergel, D. N., 1991, *ApJ*, 379, 631
Debattista, V. P., Ortwin, G., & Sevenster, M. N. 2006, *MNRAS*, 334, 355
Deguchi, S., Fujii, T., Miyoshi, M. & Nakashima, J-I. 2002, *PASJ*, 54, 61
Dwek, E., Arendt, R. G., Hauser, M. G., *et al.* 1995, *ApJ*, 445, 716
Egan, M. P., Price, S. D., Kraemer K. E., Mizuno, D. R., Carey, S. J., *et al.* 2003, *Air Force Research Laboratory Technical Report*, AFRL-VS-TR-2003-1589 2819
Freeman, K., Ness, M., Wylie-de-Boer, E., Athanassoula, E., *et al.* 2012, *MNRAS*, 428, 3660
Habing, H. J. 1996, *ARAA*, 7, 97
Kunder, R., Koch, A., Rich, R. M., de Propris, R., *et al.* 2012, *AJ*, 143, 57
Lindqvist, M., Habing, H. J., & Winnberg, A. *A&A*, 259, 118
Price, S. D., Egan, M. P., Carey, S. J., Mizuno, D. R., & Kuchar, T. A. 2001, *AJ*, 121, 2819
Rich, R. M., Reitzel, D. B., Howard, C. D., & Zhao, H-S. 2007, *ApJ*, 658, L29
Sjouwerman, L. O., Capen, S. M., & Claussen, M. J. 2009, *ApJ*, 705, 1554
Stanek, K. Z., Udalski, A., Szymanski, M., *et al.* 1997, *ApJ*, 477, 163
Stroh, M. C., Pihlström, Y. M., & Sjouwerman, L. O., 2018, *This Volume*

Astrophysical Masers:
Unlocking the Mysteries of the Universe
Proceedings IAU Symposium No. 336, 2017
A. Tarchi, M.J. Reid & P. Castangia, eds.

© International Astronomical Union 2018
doi:10.1017/S1743921317010493

Maser, infrared and optical emission for late-type stars in the Galactic plane

L. H. Quiroga-Nuñez[1,2], H. J. van Langevelde[2,1], L. O. Sjouwerman[3], Y. M. Pihlström[4], M. J. Reid[5], A. G. A. Brown[1] and J. A. Green[6]

[1] Leiden Observatory, Leiden University, P.O. Box 9513, Leiden, The Netherlands.
email: quiroganunez@strw.leidenuniv.nl

[2] Joint Institute for VLBI ERIC (JIVE), Postbus 2, Dwingeloo, The Netherlands.

[3] National Radio Astronomy Observatory, P.O. Box 0, Lopezville Road 1001, Socorro, USA.

[4] Department of Physics and Astronomy, University of New Mexico, MSC07 4220, Albuquerque, USA.

[5] Harvard-Smithsonian Center for Astrophysics, 60 Garden Street, Cambridge, USA.

[6] CSIRO Astronomy and Space Science, Australia Telescope National Facility, P.O. Box 76, Epping, Australia.

Abstract. Radio astrometric campaigns using VLBI have provided distances and proper motions for masers associated with young massive stars (BeSSeL survey). The ongoing BAaDE project plans to obtain astrometric information of SiO maser stars located in the inner Galaxy. These stars are associated with evolved, mass-losing stars. By overlapping optical (*Gaia*), infrared (2MASS, MSX and WISE) and radio (BAaDE) sources, we expect to obtain important clues on the intrinsic properties and population distribution of late-type stars. Moreover, a comparison of the Galactic parameters obtained with *Gaia* and VLBI can be done using radio observations on different targets: young massive stars (BeSSeL) and evolved stars (BAaDE).

Keywords. Galaxy: bulge, stars: late-type, masers, astrometry.

1. Context: Maser surveys and *Gaia* counterparts

Molecular masers have been used for decades to study astrophysical environments (see e.g., Elitzur, 1992, Elitzur, 2005 and references therein). Since then, radio campaigns have used masers in order to provide insights into the structure of the Milky Way. In this study, we focus on two specific radio surveys that aim to study different regions of the Galaxy: the spiral structure and the Galactic bulge.

By using data from the Very Large Baseline Array (VLBA) and the European VLBI Network (EVN), the Bar and Spiral Structure Legacy (BeSSeL) survey has obtained accurate astrometric measurements with a resolution up to $\sim 10 \mu$as. The BeSSeL survey targeted high-mass star forming regions (HMSFRs) uniquely associated with water (22 GHz) and methanol maser (6.7 and 12 GHz) emission. By fitting 6D phase-space information of the observed HMSFRs to a spiral Galactic model, BeSSeL has determined the fundamental Galactic parameters and the solar motion with respect to the Local Standard of Rest (Reid *et al.*, 2014). Additionally, simulated data of several Galactic distributions of methanol maser bearing stars has confirmed the accuracy of the Galactic parameter values found by the BeSSeL survey (Quiroga-Nuñez *et al.*, 2017).

The Bulge Asymmetries and Dynamical Evolution (BAaDE) project is a survey for SiO maser bearing stars in the Galactic plane, focusing on the bulge (Sjouwerman *et al.*, 2017). By selecting targets based on their infrared color using MSX data (Sjouwerman *et al.*, 2009), more than 20,000 evolved stars have been searched for SiO maser lines using

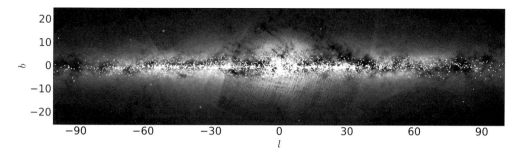

Figure 1. Galactic distribution of the sample obtained by cross-matching BAaDE targets, 2MASS and *Gaia* DR1 over the Gaia first sky map. Credit: ESA/Gaia/DPAC.

the VLA (43 GHz) and ALMA (86 GHz). Thousands of line-of-sight velocities and peak flux densities, along with other line properties, for late-type stars in the inner Galaxy are being compiled.

Although BAaDE is mapping Galactic regions that optical surveys typically do not reach ($|b| \leqslant 5^o$), a refined positional cross-match between *Gaia* DR1, 2MASS and BAaDE sources has shown more than 2,000 coincidences (see Fig. 1), which represents $\sim 8\%$ of the BAaDE sample ($\sim 30,000$ targets). Moreover, around 150 sources of the cross-matched sample are part of the Tycho catalog, and therefore, an estimate of their parallaxes are already available in Tycho-Gaia Astrometric Solution (TGAS).

2. Future studies with the cross-matched sample

1. *Gaia* DR2 will soon provide optical photometry, parallaxes, proper motions and periods that will complement the radio (BAaDE) and infrared (2MASS, MSX and WISE) data known for these 2,000 late-type stars. Using these data, we aim to characterize the stellar population in the Galactic bulge by providing statistics of stellar masses, ages, metallicities, periods and luminosities. It is thought that the stellar population at the bulge can provide clues to understand the dynamical evolution of the Milky Way.

2. VLBI astrometry of some bright BAaDE targets is being defined to provide parallaxes and 3D orbits with an estimated accuracy of $\sim 50 \mu$as. By doing VLBI, we can refine a subset of the orbits and perhaps get the orbit class, and therefore, signatures of past mergers of other stellar systems with the Galaxy.

3. Astrometric VLBI results would also allow a direct comparison of the parallax technique between optical (*Gaia* DR2) and radio data, which will align optical stellar images with SiO maser rings for nearby sources. An initial comparison is ongoing using the parallaxes from TGAS.

4. By getting full phase-space information of a subset of evolved stars in the Galactic bulge, dynamical models for the inner Galaxy can be tested. Orbital motions affected by the bar can be fitted to refine the fundamental parameter values of the Milky Way.

Acknowledgements

This work has made use of data from the ESA mission *Gaia*, processed by DPAC. Funding for DPAC has been provided by national institutions participating in the *Gaia* Multilateral Agreement.

References

Elitzur, M. 1992, *ARA&A*, 30, 75
Elitzur, M. 2005, *Science*, 309, 71
Quiroga-Nuñez, L. H., van Langevelde, H. J., Reid, M. J., & Green, J. A. 2017, *A&A*, 604, A72
Reid, M. J., Menten, K. M., Brunthaler, Zheng, X. W., *et al.* 2014, *ApJ*, 783, 130
Sjouwerman, L. O., Capen, S. M., & Claussen, M. J. 2009, *ApJ*, 705, 1554
Sjouwerman, L. O., Pihlström, Y. M., Rich, R. M., *et al.* 2017, *Proc. IAU Symposium*, 322, 103

Astrophysical Masers:
Unlocking the Mysteries of the Universe
Proceedings IAU Symposium No. 336, 2017
A. Tarchi, M.J. Reid & P. Castangia, eds.

© International Astronomical Union 2018
doi:10.1017/S1743921317010079

Molecular clouds in the Extreme Outer Galaxy

Yan Sun, Ye Xu, Ji Yang, Yang Su, Shao-Bo Zhang, Xue-Peng Chen, Zhi-Bo Jiang and Xin Zhou

Purple Mountain Observatory and Key Laboratory of Radio Astronomy, Chinese Academy of
Sciences, Nanjing 210008, China
email: yansun@pmo.ac.cn

Abstract. More than 200 molecular clouds were newly found distributed beyond the Outer
arm in the extreme outer Galaxy (EOG) region by MWISP. Those MCs roughly following the
HI's distribution well delineate the outermost spiral structure (the Outer Scutum-Centaurus
arm) and warp of our Galaxy. Besides, those MCs show different σ_v-Radius relation and exhibit
higher value of α_{vir} than MCs in the inner Galaxy.

Keywords. Galaxy: structure – ISM: molecules – radio lines: ISM

1. Introduction

Molecular cloud (MC) in the Extreme Outer Galaxy (EOG) not only delineates the
spiral structure and warping of our Galaxy, but it also serves as an excellent laboratory
for studying the star-formation process in a physical environment that is very different
from that of the solar neighborhood. However, because of the far distances involved
and the lack of high-sensitivity observations in the past, the MCs detected in the EOG
regions is less than 30, and even more the star-formation signature is only revealed
in few of these dense cores. The Milky Way Imaging Scroll Painting (MWISP) project
is a high resolution (50″) J=1-0 ^{12}CO, ^{13}CO, and C^{18}O survey of the northern Galactic
Plane, performed with the Purple Mountain Observatory Delingha 13.7 m telescope. The
survey started in 2011, and will cover Galactic longitudes from l=-10.25° to 250.25° and
latitudes from b=-5.25° to 5.25° over a period of ~10 years. The nominal sensitivities in
the survey are set for 0.3 K in ^{13}CO/C^{18}O at the resolution of 0.17 km s^{-1}, and 0.5 K
in CO at the resolution of 0.16 km s^{-1}. The high-sensitivity data from MWISP survey
will provide us a unique opportunity to study the spiral structure and star-formation
activity at the edge of the Milky Way.

2. Overview

The spiral arm traced by the EOG clouds. Using our new CO data of the MWISP, we
have found a total of 72 EOG clouds in Galactic range of 100° < l <150°, which delineate a new segment of a spiral arm between Galactocentric radii of 15 and 19 kpc
in the second Galactic quadrant (Sun *et al.* 2015). The new arm appears to be the
extension of the distant arm discovered by Dame & Thaddeus as well as the Outer
Scutum−Centaurus (OSC) arm into the outer second quadrant. And a total of 168 EOG
clouds were identified in Galactic longitude range of 34.75° ⩽L⩽45.25° (Sun *et al.* 2017),
which in roughly following the OSC arm. All these may provide a robust evidence for
the existence of the OSC arm. Figure 1a shows their locations.

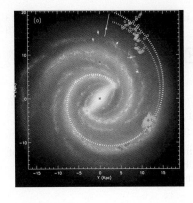
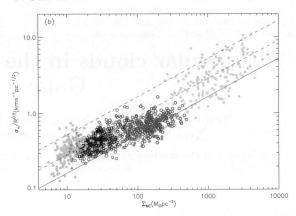

Figure 1. (a) Locations of the EOG clouds superposed on an artist's conception of the Milky Way (R. Hurt: NASA/JPL-Caltech/SSC). Note that the distances are derived from the rotaion curve of Reid *et al.* (2014). (b) Scaling coefficient, $\sigma_v/R^{1/2}$, as a function of the mass surface density for clouds located in EOG region (cyan, from MWISP), Galactic Center (gold, Oka *et al.* 2001), Galactic ring (red and blue, Solomon *et al.* 1987; Dame *et al.* 1986), and outer Galaxy (black, Heyer *et al.* 2001). The solid line and two dashed lines show the loci for $\alpha_{vir}=1$ (lower), 3 (middle) and 10 (upper), respectively. The pluses mark the EOG clouds with ^{13}CO detections.

The warp of the Galactic disk. The Galactic warp structure traced by CO emission of the OSC arm was discussed. Both of the effects of the warp and the tilted plane (because of the Sun's offset from the physical mid-plane) were considered. We find that the slope of the two warp models are very close to that of the observations, especially the gaseous warp model from the HI data. Generally, the Z scale heights of the OSC arm is increased with the increasing of the Galactocentric radii.

Properties of the EOG clouds. Similar to Heyer & Dame (2015), the revised scaling Larson relations were examined and compared across a wide range of Galactic environments in Figure 1b. We find that the EOG clouds are well displaced below the scaling relationship defined by the inner Galaxy MCs. The results are similar to the finding in the outer Galaxy clouds (Heyer & Dame 2015). Interestingly, most of the EOG clouds have a virial ratio $\alpha_{\rm vir} >3$.

Acknowledgements

This work is supported by National Key Research & Development Program of China (2017YFA0402702), National Natural Science Foundation of China (grant nos. 11773077 and 11233007), the Key Laboratory for Radio Astronomy, CAS, and the Youth Innovation Promotion Association, CAS.

References

Dame, T. M., Elmegreen, B. G., Cohen, R. S., & Thaddeus, P. 1986, *ApJ*, 305, 892
Dame, T. M. & Thaddeus, P. 2011, *ApJ*, 734, L24
Heyer, M., Carpenter, J. M., & Snell, R. L. 2001, *ApJ*, 551, 852
Heyer, M. & Dame, T. M. 2015, *ARA&A*, 53, 583
Oka, T., Hasegawa, T., Sato, F., *et al.* 2001, *ApJ* 562, 348
Reid, M. J., Menten, K. M., Brunthaler, A., *et al.* 2014, *ApJ*, 783, 130
Solomon, P. M., Rivolo, A. R., Barrett, J., & Yahil, A. 1987, *ApJ*, 319, 730
Sun, Y., Xu, Y., Yang, J., Li, F. C., Du, X. Y., Zhang, S. B., & Zhou, X. 2015, *ApJ*, 798, L27
Sun, Y., Su, Y. Zhang, S. B., Xu, Y., Chen, X. P., Yang, J., Jiang, Z. B., & Fang, M. 2017, *ApJS*, 230, 17

Top: Mareki Honma
Bottom: Christian Henkel animating the discussion

Photo credits: S. Poppi

Star Formation

Astrophysical Masers:
Unlocking the Mysteries of the Universe
Proceedings IAU Symposium No. 336, 2017
A. Tarchi, M.J. Reid & P. Castangia, eds.

© International Astronomical Union 2018
doi:10.1017/S1743921317010742

Perspectives on star formation: the formation of high-mass stars

Maria T. Beltrán

INAF-Osservatorio Astrofisico di Arcetri,
Largo E. Fermi 5, 50125 Firenze, Italy
email: mbeltran@arcetri.astro.it

Abstract. The formation process of high-mass stars has puzzled the astrophysical community for decades from both a theoretical and an observational point of view. Here, we present an overview of the current theories and status of the observational research on this field, outlining the progress achieved in recent years on our knowledge of the initial phases of massive star formation, the fragmentation of cold, infrared-dark clouds, and the evidence for circumstellar accretion disks around OB stars. The role of masers in helping us to understand the mechanism leading to the formation of a high-mass star are also discussed.

Keywords. stars: formation, ISM: molecules, ISM: kinematics and dynamics

1. Introduction

High-mass stars are defined as those with masses $>8\,M_\odot$ and luminosities $>10^4\,L_\odot$. Following Salpeter's Initial Mass Function (IMF), one sees that high-mass stars are rare objects because for each $30\,M_\odot$ star formed, there are hundred $1\,M_\odot$ formed. These stars have short lifetimes and, because of that, are mainly located in the spiral arms of the galaxies in which they formed (e.g., Urquhart *et al.* 2014). High-mass stars are key elements in galaxies because they dominate their appearance and evolution, are responsible for the production of heavy elements, and influence the interstellar medium through energetic winds and supernovae. They produce enough UV radiation to create HII regions and huge ionized bubbles, and are the most chemically rich sources in the Galaxy, being the best reservoirs of Complex Organic Molecules, in particular of prebiotic ones (building blocks of life) (e.g., G31.41+0.31: Rivilla *et al.* 2017).

2. Challenges

Studying the formation of massive stars represents an observational challenge because high-mass stars are rare objects that are located at large distances. Typical distances of O-type star-forming regions are 5 kpc (Beltrán *et al.* 2006). They have short evolutionary timescales, so it is difficult to trace the earliest phases of their evolution. In addition, massive stars are embedded in rich clusters: only ∼4% of O-field stars might have formed outside clusters (de Wit *et al.* 2005). This makes it very difficult to disentangle their emission from that of the other cluster members and to trace the primordial configuration (initial conditions) of the molecular cloud. Therefore, it is absolutely necessary to carry out high-angular resolution observations, especially at cm and mm wavelengths, to study high-mass protostars.

From a theoretical point of view, massive stars represent a challenge because of their short evolutionary timescales. They form so fast that reach the zero age main sequence while still deeply embedded in the parental cloud and still with active accretion. For stars with masses $>8\,M_{\odot}$, the pre-stellar evolution is dominated by the Kelvin-Helmholtz timescale. This means that the contraction proceeds very fast and the star starts burning hydrogen while still accreting. According to theoretical predictions, in case of spherical accretion, the radiation pressure of the newly formed OB star should stop the accretion preventing further growth (e.g., Kahn 1974; Wolfire & Casinelli 1987). In the 90's, different theoretical scenarios proposed non-spherical accretion as a possible solution for the formation of OB-type stars (Nakano 1989; Jijina & Adams 1996), and in recent years, theoretical ideas and simulations appear to have converged to a disk-mediated accretion scenario (e.g., Krumholz *et al.* 2009; Kuiper *et al.* 2010; Kuiper & Yorke 2013). As a matter of fact, competing theories that propose very different high-mass star-formation mechanisms (monolithic collapse: Mc Kee & Tan 2002, and competitive accretion: Bonnell & Bate 2006) predict the existence of circumstellar accretion disks.

3. Theoretical models: predictions and differences

The two competing theories that discuss the formation of high-mass stars are:

Monolithic collapse. This models predicts that massive stars form via monolithic collapse of a massive turbulent core fragmented from the natal molecular cloud (Mc Kee & Tan 2002, 2003). The forming star gathers its mass from this massive core alone. In the monolithic collapse, massive stars should form in (almost) isolation. The core mass function is similar to the IMF (Krumholz *et al.* 2005). In the turbulent core accretion, the accretion rates are high, of the order of 10^{-2}–$10^{-3}\,M_{\odot}\,\mathrm{yr}^{-1}$, as often observed in high-mass star forming regions (e.g., Beltrán *et al.* 2011). The monolithic collapse process is slow and quasi-static and starts with a strongly peaked density distribution ($n \propto r^{-1.5}$). Given the non-zero angular momentum of the collapsing core, this model predicts the existence of protostellar accretion disks around massive stars. These accretion disks have been recently found around early-B to late-O type stars (see Beltrán & de Witt 2016 for a review).

The existence of truly isolated massive stars is one of the caveats of the model, because as concluded by de Wit *et al.* (2005), only 4% of O-field stars might have an origin outside of a young cluster. Another caveat of the model is the existence of monolithic pre-stellar cores. There have been few claims of their existence (Tan *et al.* 2013; Peretto *et al.* 2014; Wang *et al.* 2014; Sanhueza *et al.* 2017), but in many cases, the cores are not massive or dense enough to form a massive star. Finally, one of the main problems of the core collapse model is how to prevent fragmentation. It has been proposed that protostellar heating could reduce significantly fragmentation because it would increase the Jeans masses (Krumholz *et al.* 2007), or that fragmentation could be controlled by the magnetic field, as suggested by the magneto-hydrodynamics simulations of Hennebelle *et al.* (2011) and Commerçon *et al.* (2011, 2012). In these simulations, highly magnetized cores show a low level of fragmentation, while cores where turbulence dominates over magnetic field show a high fragmentation level.

Competitive accretion. This model predicts the initial fragmentation of a molecular cloud in low-mass cores of Jeans masses, which form stars that compete to accrete unbound gas from the common gas reservoir (Bonnell *et al.* 1997; Bonnell & Bate 2006), that is, from the whole cloud. The cloud experiences global collapse and the stars located near the center of the gravitational potential accrete at a higher accretion rate because of a stronger gravitational pull. A special case of interaction is that causing a merging of low-mass stars, which is predicted for unusually high stellar densities, of the order of 10^8 stars per cubic parsec, although the density could be smaller if the mergers are binary systems, in that case it would be $\sim 10^6$ stars per cubic parsec (Bonnell & Bate 2005). In

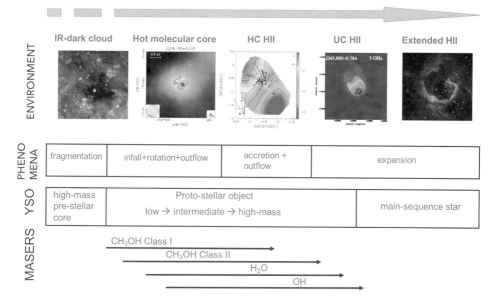

Figure 1. Evolutionary sequence for high-mass stars: from cold IRDCs to chemically rich HMCs to HII regions.

the competitive model, massive stars should always form in clustered environments, as very often observed. The model reproduces the full IMF (Bonnell *et al.* 2004) and the accretion is Bondi-Hoyle. The collapse is a fast, dynamical, and gravity-driven process. The collapse starts with a uniform gas density distribution. In the original competitive accretion model, disks were not predicted, or were very small, because severely affected or destroyed by tidal interactions with the other member of the protocluster (Cesaroni 2006). In more recent updates, the competitive accretion model also predicts the existence of disks around the most massive members in the cluster, although it is not clear how disturbed, large, or transient these disks can be.

Krumholz *et al.* (2005) proposed that the model could not work because the virial parameters and accretion rates were too low. The high velocities of the gas relative to stars and the low accretion rates would not allow the low-mass stars to accrete enough mass to become massive ones. In addition, radiation pressure should stop Bondi-Hoyle accretion above $10\,M_\odot$. Finally, another caveat of the model is that feedback, especially of ionized gas from OB-type stars, was not taken into account. Bonnell & Bate (2006) have answered to most of these criticisms and proposed that initially the relative velocities between gas and stars are low and the turbulence is locally small, which allows the low-mass stars to accrete mass. Regarding the low accretion rates, they should be higher if one considers the local cluster core and not just the global cloud. Finally, regarding feedback from OB-type stars, if accretion proceeds mostly through disks, then all the mass flow is channeled into a relatively small area that intercepts only a small fraction of the incident UV flux. In addition, most of the radiation escapes through the polar regions and does not significantly interact with infalling gas (Krumholz *et al.* 2009; Peters *et al.* 2010) that proceeds unhampered onto the central star (Dale *et al.* 2005).

4. Evolutionary sequence

Figure 1 shows the evolutionary sequence of a high-mass star, with the different phenomena (fragmentation, rotation, infall, outflow) and young stellar (or pre-stellar) objects

(YSOs) associated. Because of the short timescales needed to form a massive star and the difficulty to trace the earliest stages of its formation, to define an evolutionary sequence for high-mass stars is not easy. The formation of a massive star initiates with the fragmentation of a cold cloud, usually dark at infrared wavelengths that is known as infrared-dark cloud (IRDC). Subsequent infall and heating inside each core eventually leads to the formation of chemically rich hot molecular cores (HMCs), which are the cradle of high-mass stars. Accretion continues onto high-mass protostars that gain mass. The UV-radiation from the young embedded star will develop a bubble of ionized gas, known as hypercompact Hɪɪ (HC Hɪɪ) region that will start expanding, becoming first an ultracompact Hɪɪ (UC Hɪɪ) region and later on, an extended or giant Hɪɪ region. Finally, the gas is dissipated by the ionized winds, exposing the newly formed OB star. Note that IRDCs and HMCs are the typical environments associated with the different phenomena, but they not represent evolutionary phases by themselves. Therefore, inside IRDCs we can find from embedded protostellar objects up to more evolved compact Hɪɪ regions. Regarding the masers, the first in appear, already in the earliest phases, are the Class I and Class II methanol masers, then the water masers, and finally the hydroxyl masers. However, it is very difficult to define an evolutionary sequence only with masers because the different types can be found almost in all the evolutionary stages.

4.1. Infrared-dark clouds

The onset of massive star formation takes place inside cold ($T < 20\,\mathrm{K}$), massive (10^3–$10^4\,M_\odot$), parsec-scale (1–5 pc) clouds known as IRDCs. These clouds, which have volume densities of 10^4–$10^5\,\mathrm{cm}^{-3}$, are seen in absorption at infrared wavelengths and in emission at millimeter and sub-millimeter wavelengths (e.g., Pérault et al. 1996; Egan et al. 1998; Carey et al. 1998; Rathborne et al. 2006). Although IRDCs have very different morphologies, they often appear as filaments (e.g., the "snake" nebula: Wang et al. 2014; or the "Nessie" nebula: Jackson et al. 2010; Goodman et al. 2014).

Do massive pre-stellar cores exist? One of the open questions in massive star formation is how *pre-stellar* are the cores embedded in IRDCs. The high sensitivity Spitzer and Herschel observations have shown that many IRDCs present evidence for active star formation. Some of them show enhanced, slightly extended 4.5 μm emission, called "green-fuzzies", and maser emission, all indicators of the presence of molecular outflows, broad SiO emission which indicates shocked gas, bright 24 μm, associated with embedded protostars, or 8 μm, probably from Hɪɪ regions (Chamber et al. 2009; Cyganowski et al. 2009; Jiménez-Serra et al. 2010). Large surveys have been carried out with MSX, Spitzer, and Herschel to search for massive starless cores by looking for cores IR-quiet at 24 or 70 μm and with temperatures below 15 K (e.g., Beltrán et al. 2006; Motte et al. 2007; Chambers et al. 2009; Rathborne et al. 2010; Butler & Tan 2012; Tigé et al. 2017). As a result of these searches many candidates have been proposed as starless (Duarte-Cabra et al. 2013; Tan et al. 2013; Cyganowski et al. 2014; Peretto et al. 2014; Wang et al. 2014; Sanhueza et al. 2017), but in many cases their densities or masses are too low, especially when observed at observed high-angular resolution. Even the best starless candidates identified with ALMA in Cycle 0, like for example G28.37+0.07 C1-S (Tan et al. 2013), appear to be active and associated with molecular outflows when observed at higher sensitivity with ALMA in Cycle 2 (Tan et al. 2016). In conclusion, if massive pre-stellar cores do exist, they are very rare objects. In fact, Motte et al. (2007) have been proposed that starless massive cores might do exist but would be short-lived ($< 10^3$ yr), and therefore difficult to "catch".

What does it control fragmentation? Another open question related to the first stages of evolution of a high-mass star is what controls fragmentation. High-angular resolution sub-millimeter observations of IRDCs show evidence of different levels of fragmentation (Zhang *et al.* 2009; Bontemps *et al.* 2010; Longmore *et al.* 2011; Palau *et al.* 2015; Rathborne *et al.* 2015; Fontani *et al.* 2016; Henshaw *et al.* 2017; see Fig. 2). In many cases, the masses of the fragments are of a few tens of $M_{\odot}$. Since these cores are cold ($T <$ 20 K) and dense ($n \sim 10^4$–10^5 cm^{-3}), their Jeans masses are $M_{\rm Jeans} \sim$ 1–5 $M_{\odot}$. Therefore, these fragments have super-Jeans masses and cannot be only thermally supported. Other support mechanisms such as turbulence of magnetic field are required to account for the large masses observed. 3D radiation-magneto-hydrodynamics simulations of Hennebelle *et al.* (2011) and Commerçon *et al.* (2011, 2012) suggest that magnetic field and radiative transfer strongly interplay at the early stages of star formation. In these simulations, highly magnetized cores, identified thanks to a low mass-to-flux over critical mass-to-flux ratio show a low level of fragmentation, while cores where turbulence dominates over magnetic field show a high fragmentation level. Fontani *et al.* (2016) have confronted real ALMA observations of the fragmented IRDC IRAS 16061−5048c1 with the simulations of Commerçon *et al.* (2011) and concluded that magnetic field indeed plays a crucial role in the fragmentation process even though the cloud fragments. Note that one very important piece is still missing in these comparisons, and this is the strength of the magnetic field. However, dust polarization observations, or even better, Zeeman effect measurements, carried out with ALMA should allow us to estimate it.

4.2. *Hot molecular cores*

High-mass protostars are deeply embedded in chemically rich HMCs. The high temperatures ($T > 100$ K) achieved when the protostar starts heating the surroundings produce the evaporation of the grain mantles giving rise to this rich chemistry (e.g., G31.41+0.31: Rivilla *et al.* 2017). Hot cores have typical sizes of 0.1 pc, luminosities $> 10^4 L_{\odot}$, and high densities of $\sim 10^7$ cm^{-3}. Hot molecular cores are associated with young massive protostars but also with more evolved objects that have already developed a HC or UC H II, like for example G24.78+0.08 (Moscadelli *et al.* 2007; Beltrán *et al.* 2007). One of the hottest topics in the star formation field is how a massive young stellar object gathers its mass. High-mass star-forming theories appear to converge to a disk-mediated accretion scenario, but what do the observations tell us? Do true accretion disks around massive stars really exist?

Disks around early B-type and late O-type (proto)stars. Circumstellar disks have been detected around stars with masses up to 25–30 $M_{\odot}$ by means of NIR and MIR (e.g., IRAS 13481−6214: Kraus *et al.* 2010; CRL 2136: de Wit *et al.* 2011), and (sub)millimeter (e.g., IRAS 20126+4104: Cesaroni *et al.* 2014) interferometric observations. ALMA high-angular resolution observations have allowed us to study in detail the velocity field of these circumstellar disks. The observed velocity gradients have been modeled and are consistent with Keplerian rotation (e.g., G35.20−0.74N: Sánchez-Monge *et al.* 2013; G35.03+0.35: Beltrán *et al.* 2014; AFGL 4176: Johnston *et al.* 2015; G11.92−0.61 MM1: Ilee *et al.* 2016). The radii of these disks are approximately a few 1000 au, although for some sources observed at very high-angular resolution, the radii can be as small as 300–400 au (Beltrán & de Wit 2016, and references therein). Their masses are of a few $M_{\odot}$. For true acretion disks candidates, the mass of the disk is always smaller than (or similar to) the mass of the central star, $M_{\star}$. This suggests that these structures could be rotationally supported.

Disks around early O-type (proto)stars. For higher mass stars, with luminosities $> 10^5 L_{\odot}$ and spectral types earlier than O6–O7, the situation is different. What has been found around these objects are huge and massive rotating structures called toroids. These toroids have masses of a few 100 $M_{\odot}$ and sizes of several 1000 au, which suggests that

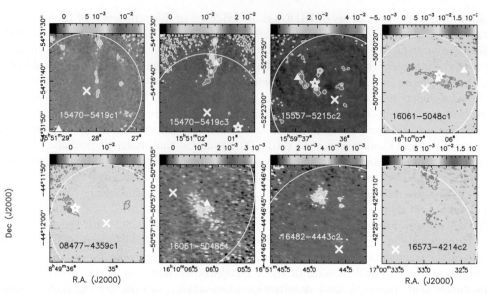

Figure 2. ALMA 278 GHz dust continuum emission maps of a sample of IRDCs showing different levels of fragmentation (Fontani *et al.*, in preparation). The white crosses and circles indicate the ALMA phase center and primary beam, respectively, the stars the position of IR sources, and the triangles the H_2O maser spots.

they are probably surrounding protoclusters. The mass of these toroids is much higher than that of the central star, and therefore, Keplerian rotation is not possible (on scales of 10^4 au) because the gravitational potential of the system is dominated by the toroid and not by the central star. In addition, the mass of the toroid is higher than the dynamical mass, which suggests that these structures might be undergoing fragmentation and collapse.

In a recent study carried out with ALMA, Cesaroni *et al.* (2017) have observed six early O-type star-forming regions looking for circumstellar disks. The 1.4 mm dust continuum emission has revealed that some of the cores fragment in few sources, while others, like G31.41+0.31, do not fragment at all. The CH_3CN observations with an angular resolution of 0.2″ have revealed that i) three of the cores show signatures of Keplerian rotation, with the position-velocity (PV) plots showing the typical butterfly pattern; ii) three of the cores show velocity gradients that suggest rotation, but the PV plots are not consistent with Keplerian motions; and iii) G17.64+0.16 shows no hints of rotation. The luminosity-to-mass ratio as a function of the distance to these sources, including the O-type star AFGL 4176 observed by Johnston *et al.* (2015), shows that true accretion disks are found for sources with an intermediate evolutionary stage, while those with questionable disk evidence are the younger sources (Cesaroni *et al.* 2017). The explanation for the non-detection of Keplerian disks around these young sources could be that their disks are so embedded that their emission is difficult to disentangle from that of the envelopes. Alternatively, disks might start small and grow up with time. On the other hand, for the most evolved source in the sample, G17.64+0.16, the molecular gas might have dispersed and therefore, no disk is found.

Disks versus toroids Disks and toroids are different from a point of view of stability. While accretion disks around Herbig Ae stars and bona-fide Keplerian disks around early B- and late O-type (proto)stars have masses $<0.3\,M_\star$ and a stability Toomre's Q parameter >1, suggesting that they are stable, toroids have all masses $>0.3 M_\star$ and Q$<$1.

Disks and toroids are also dynamically different. Disks around Herbig Ae and those in Keplerian rotation around high-mass stars have dynamical masses higher than those of the central star, while toroids around O-type stars have dynamical masses much smaller, and therefore cannot be centrifugally supported and could be susceptible to gravitational collapse and fragmentation. The ratio of the free-fall time, $t_{\rm ff}$, and the rotational period at the outer radius, $t_{\rm rot}$, is also higher for disks than for toroids. This suggests that if the structure rotates fast, the infalling material has enough time to settle into a centrifugally supported disk, otherwise, the infalling material does not have enough time to reach centrifugal equilibrium and the rotating structure is a transient toroid.

Typical infall rates, $\dot{M}_{\rm inf}$, in intermediate- and high-mass (proto)stars are of the order of 10^{-3}–$10^{-2}\,M_\odot$/yr, while typical accretion rates, $\dot{M}_{\rm acc}$, estimated from the mass loss rate of the associated outflow, are of the order of 10^{-4}–$10^{-3}\,M_\odot$/yr (Beltrán & de Wit 2016). $\dot{M}_{\rm inf}$ are always higher than $\dot{M}_{\rm acc}$, and in some cases up to a factor 1000. This could be a result of stellar multiplicity if the infalling material is not accreted onto the single star driven the molecular outflow but onto a cluster of stars. This explanation seems plausible for the most massive O-type (proto)stars, because as already explained, the sizes and masses of the rotating toroids suggest that they are enshrouding stellar (proto)clusters. However, this explanation cannot solve the problem for the intermediate-mass protostars and probably neither for the B-type (proto)stars. For intermediate-mass protostars with $M_\star \simeq 2$–$3\,M_\odot$, the $\dot{M}_{\rm inf}/\dot{M}_{\rm acc}$ ratio is still 20–300, with mass of the structure of ~ 0.3–$1.4\,M_\odot$. Therefore, although these disks could be circumbinary disks, it seems unlikely that they are circumcluster structures surrounding several members. The apparent implication of this is that the infalling material needs to pile up in the disk and results in disk masses that are tens to hundreds of solar masses given the observed rates. This is massive and suggests a gravitationally unstable disk inducing variable, "FUOri-like" accretion events onto the central object. And this is what has been recently discovered in the high-mass regime by Caratti o Garatti *et al.* (2016). These authors have discovered the first disk-mediated accretion burst from a young stellar object of $\sim 15\,M_\odot$, S255 NIR 3. The NIR photometry reveals an increase in brightness of 2.5–3.5 magnitudes and NIR spectroscopy reveals emission lines typical of accretion bursts in low-mass protostars, but orders of magnitude more luminous, confirming our prediction.

References

Beltrán, M. T., Brand, J., Cesaroni, R. *et al.* 2006, *A&A*, 447, 221

Beltrán, M. T. & de Wit, W. J. 2016, *A&AR*, 24, 6

Beltrán, M. T., Sánchez-Monge, A., Cesaroni, R. *et al.* 2014, *A&A*, 571, A52

Beltrán, M. T., Cesaroni, R., Moscadelli, L., & Codella, C. 2007, *A&A*, 471, L13

Beltrán, M. T., Cesaroni, R., Neri, R., & Codella, C. 2011, *A&A*, 525, A151

Bonnell, I. A. & Bate, M. R. 2005, *MNRAS*, 362, 915

Bonnell, I. A. & Bate, M. R. 2006, *MNRAS*, 370, 488

Bonnell, I. A., Bate, M. R., Clarke, C. J., & Pringle, J. E. 1997, *MNRAS*, 285, 201

Bonnell, I. A., Vine, S. G., & Bate, M. R. 2004, *MNRAS*, 349, 735

Bontemps, S., Motte, F., Csengeri, T., & Schneider, N. 2010, *A&A*, 524, A18

Butler, M. J. & Tan, J. C. 2012, *ApJ*, 754, 5

Caratti o Garatti, A., Stecklum, B., Weigelt, G. *et al.* 2016, *A&A*, 589, L4

Carey, S. J., Clark, F. O., Egan, M. P. *et al.* 1998, *ApJ*, 508, 721

Cesaroni, R. 2006, Galli, D., Lodato, G., Walmsley, C. M., & Zhang, Q. 2006, *Nature*, 444, 703

Cesaroni, R., Galli, D., Neri, R., & Walmsley, C. 2014,*A&A* 566, A73

Cesaroni, R., Sánchez-Monge, Á., Beltrán, M. T. *et al.* 2017, *A&A*, 602, A59

Chambers, E. T., Jackson, J. M., Rathborne, J. M., & Simon, R. 2009, *ApJS*, 181, 360

Commerçon, B., Hennebelle, P., & Henning, Th. 2011, *ApJ*, 742, L9

Commerçon, B., Launhardt, R., Dullemond, C., & Henning, Th. 2012, *A&A*, 545, A98

Cyganowski, C. J., Brogan, C. L., Hunter, T. R., & Churchwell, E. 2009, *ApJ*, 702, 1615

Cyganowski, C. J., Brogan, C. L., Hunter, T. R. *et al.* 2014, *ApJ*, 796, L2

Dale, J. E., Bonnell, I. A., Clarke, C. J., & Bate, M. R. 2005, *MNRAS*, 358, 291

de Wit, W. J., Testi, L.,Palla, F., & Zinnecker, H. 2005, *A&A*, 437, 247

de Wit, W. J., Hoare, M., Oudmaijer, R., Nürnberger, D. *et al.* 2011, *A&A*, 526, L5

Duarte-Cabral, A., Bontemps, S., Motte, F. *et al.* 2013, *A&A*, 558, A125

Egan, M. P., Shipman, R. F., Price, S. D. *et al.* 1998, *ApJ*, 494, L199

Fontani, F., Commerçon, B., Giannetti, A. *et al.* 2016, *A&A*, 593, L14

Goodman, A. A., Alves, J., Beaumont, C. N. *et al.* 2014, *ApJ*, 797, 53

Hennebelle, P., Commerçon, B., Joos, M. *et al.* 2011, *A&A*, 528, A72

Henshaw, J. D., Jiménez-Serra, I., Longmore, S. N. *et al.* 2017, *MNRAS*, 464, L31

Ilee, J. D., Cyganowski, C. J., Nazari, P. *et al.* 2016, *MNRAS*, 462, 4386

Jackson, J. M., Finn, S. C., Chambers, E. T. *et al.* 2010, *ApJ*, 719, L185

Jijina, J. & Adams, F. C. 1996, *ApJ*, 462, 874

Jiménez-Serra, I., Caselli, P., Tan, J. C. *et al.* 2010, *MNRAS*, 406, 187

Johnston, K. G., Robitaille, T. P., Beuther, H. *et al.* 2015, *ApJ*, 813, L19

Kahn, F. D. 1974, *A&A*, 37, 149

Kraus, S., Hofmann, K.-H., Menten, K. *et al.* 2010, *Nature*, 466, 339

Krumholz, M. R., Klein, R. I., McKee, C. F. *et al.* 2009, *Science*, 323, 754

Krumholz, M. R., Klein, R. I., & McKee, C. F. *ApJ*, 656, 959

Krumholz, M. R., McKee, C. F., & Klein, R. I. 2005, *Nature*, 438, 332

Kuiper, R., Klahr, H., Beuther, H., & Henning, T. 2010, *ApJ*, 722, 1556

Kuiper, R. & Yorke, H. W. 2013, *ApJ*, 763, 104

Longmore, S. N., Pillai, T., Keto, E. *et al.* 2011, *ApJ*, 726, 97

McKee, C. F. & Tan, J. C. 2002, *Nature*, 416, 59

Moscadelli, L., Goddi, C.; Cesaroni, R. *et al.* 2007, *A&A*, 472, 867

Motte, F., Bontemps, S., & Schilke, P. 2007, *A&A*, 476, 1243

Nakano, T. 1989, *ApJ*, 345, 464

Palau, A., Ballesteros-Paredes, J., Vázquez-Semadeni, E. *et al.* 2015, *MNRAS*, 453, 3785

Pérault, M., Omont, A., Simon, G. *et al.* 1996, *A&A*, 315, L165

Peretto, N., Fuller, G. A., André, Ph. *et al.* 2014, *A&A*, 561, A83

Peters, T., Banerjee, R., Klessen, R. S. *et al.* 2010, *ApJ*, 711, 1017

Rathborne, J. M., Jackson, J. M., & Simon, R. 2006, *ApJ*, 641, 389

Rathborne, J. M., Longmore, S. N., Jackson, J. M. *et al.* 2015, *ApJ*, 802, 125

Rivilla, V. M., Beltrán, M. T., Cesaroni, R. *et al.* 2017, *A&A*, 598, A59

Sánchez-Monge, A., Beltrán, M. T., Cesaroni, R., Etoka, S. *et al.* 2013, *A&A*, 569, A11

Sanhueza, P., Jackson, J. M., Zhang, Q. *et al.* 2017, *ApJ*, 841, 97

Tan, J. C., Kong, S., Butler, M. J. *et al.* 2013, *ApJ*, 779, 96

Tan, J. C., Kong, S., Zhang, Y. *et al.* 2016, *ApJ*, 821, L3

Tigé, J., Motte, F., Russeil, D. *et al.* 2017, *A&A*, 602, 77

Urquhart, J. S., Figura, C. C., Moore, T. J. T. *et al.* 2014, *MNRAS*, 437, 1791

Wang, K., Zhang, Q., Testi, L., van der Tak, F. *et al.* 2014, *MNRAS*, 439, 3275

Wolfire, M. G. & Cassinelli, J. P. 1987, *ApJ*, 319, 850

Zhang, Q., Wang, Y., Pillai, T., & Rathborne, J. 2009, *ApJ*, 696, 268

Astrophysical Masers:
Unlocking the Mysteries of the Universe
Proceedings IAU Symposium No. 336, 2017
A. Tarchi, M.J. Reid & P. Castangia, eds.

© International Astronomical Union 2018
doi:10.1017/S1743921317009796

Masers as probes of the gas dynamics close to forming high-mass stars

Luca Moscadelli[1], Alberto Sanna[2] and Ciriaco Goddi[3,4]

[1]INAF - Osservatorio Astrofisico di Arcetri
Largo E. Fermi 5, I-50125, Firenze, Italy
email: mosca@arcetri.astro.it

[2]Max-Planck-Institut für Radioastronomie
Auf dem Hügel 69, D-53121, Bonn, Germany
email: asanna@mpifr-bonn.mpg.de

[3]Research Institute for Mathematics, Astrophysics and Particle Physics, Radboud University,
Heyendaalseweg 135, 6525, AJ Nijmegen, The Netherlands
email: C.Goddi@astro.ru.nl

[4]Leiden Observatory,
PO Box 9513, 2300 RA Leiden, The Netherlands

Abstract. Imaging the inner few 1000 AU around massive forming stars, at typical distances of several kpc, requires angular resolutions of better than $0\rlap.{''}1$. Very Long Baseline Interferometry (VLBI) observations of interstellar molecular masers probe scales as small as a few AU, whereas (new-generation) centimeter and millimeter interferometers allow us to map scales of the order of a few 100 AU. Combining these informations all together, it presently provides the most powerful technique to trace the complex gas motions in the proto-stellar environment. In this work, we review a few compelling examples of this technique and summarize our findings.

Keywords. Stars: formation, masers, instrumentation: interferometers

1. Introduction

Low-mass ($\sim 1\ M_\odot$) stars form through disk/jet systems (see, for instance, Lee *et al.* (2017)), whereas accretion disks are a natural consequence of the collapse of a rotating molecular core by angular momentum conservation, and jets help removing the excess of angular momentum, allowing the infalling gas to accrete onto the protostar. The formation route could be different for massive ($\geqslant 8\ M_\odot$) young stellar objects (YSO), since they reach the ZAMS while still accreting and emit energetic UV radiation that could have a strong impact (via photoionization and radiation pressure) on preexisting disk/jet systems (see Beltrán & de Wit (2016), for a recent review). In order to resolve the kinematics of disks/jets around individual YSOs, an angular resolution of $\leqslant 0\rlap.{''}1$ is required, approached only by most recent, millimeter interferometers (ALMA, NOEMA). The demand for very high (~ 1 mas) angular resolution is naturally met by Very Long Baseline Interferometry (VLBI) observations of molecular (water and methanol, in particular) masers, which are commonly observed nearby (at radii of 100–1000 AU from) high-mass YSOs. Multi-epoch VLBI observations allow the measurement of the maser proper motions with high (~ 1 km s^{-1}) accuracy, providing the full 3-D distribution of (masing) gas velocities nearby the forming star. Since a decade, this technique has been successfully used to search for disks and jets around high-mass YSOs (Moscadelli *et al.* 2011; Sanna *et al.* 2010; Goddi *et al.* 2011). In the following, we present recent results from two main programs undertaken by our group.

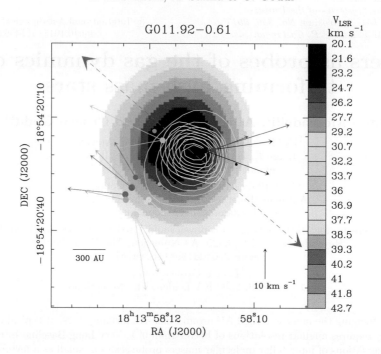

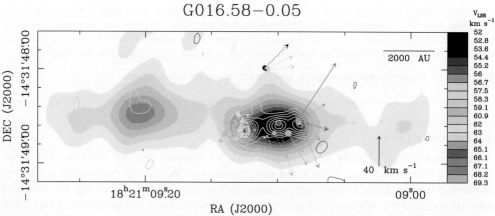

Figure 1. Radio continuum and maser emission towards G011.92−0.61 (upper panel) and G016.58−0.05 (lower panel). In both panels, the JVLA continuum at 6, 13, and 22 GHz is represented by the gray-scale image, cyan and white contours, respectively. Colored dots and arrows give the positions and the proper motions of the 22 GHz water masers, with colors denoting maser "Local Standard of Rest" velocities ($V_{\rm LSR}$). The amplitude scale for the maser velocity is indicated by the *black arrow* in the bottom right of the panels. In the lower panel, colored triangles mark the positions of the 6.7 GHz methanol masers. (See the on-line version of the figure; see Moscadelli *et al.* (2016), Figs. 2 and 4, for further details.)

2. Survey of massive (proto)stellar outflows using water masers

Since a few years, we have started a survey of massive protostellar outflows with the specific goals of both achieving an angular resolution sufficient to resolve individual YSOs and targeting relevant, complementary flow components. We use VLBI of water masers to determine the 3-D velocities of the molecular component of the flow on scales of 10–100 AU, and sensitive (rms noise ≈ 5–10 μJy) JVLA continuum observations to study the ionized emission. Our sample are 40 luminous (B3–O7 ZAMS type) YSOs with accurate distances measured by trigonometric parallaxes (from the BeSSeL† survey). The JVLA observations are performed at 6 and 13 GHz with the A-Array and at 22 GHz with the B-Array, obtaining comparable angular resolutions in the range 0″.2–0″.4.

Our project has been awarded 60 hours of JVLA time during 2012 October and 2014 May, and we have recently presented the first results for a subset (11) of targets (Moscadelli *et al.* 2016). The large majority of the targeted water masers is detected in continuum at one or more frequencies. Most detections are weak ($\sim$100 μJy), compact or slightly resolved sources, likely pinpointing the YSO responsible for the maser excitation. For the subset of sources fully analyzed, the spectral indexes between 13 and 22 GHz are generally positive and consistent with ionized winds/jets. In a few cases, we also find clear negative spectral indexes, between 6 and 13 GHz, hinting at non-thermal emission.

Figure 1, upper panel, presents results of one typical case, G011.92−0.61, where the continuum emission is only slightly resolved and overlays on the cluster of water masers. Although no indication for an ionized jet can be obtained from the continuum morphology, a collimated flow is suggested by the distribution of maser proper motions, which all form small angles with the direction identified by the big arrows in the plot. This direction actually coincides with the axis of a collimated molecular outflow observed with the Submillimeter Array (SMA) at larger angular scales towards this YSO (Cyganowski *et al.* 2011). This is a nice example where the 3-D distribution of water maser velocities can help identifying collimated ejection from an high-mass YSO. Figure 1, lower panel, shows the opposite case, for the source G016.58−0.05, where the continuum emission (both at 6 and 13 GHz) is clearly elongated and witnesses the presence of an ionized jet, and the water maser proper motions are instead quite scattered in direction, and probably trace wide bow-shocks at the radio knots of the jet.

3. Disk/jet system in high-mass YSOs explored with methanol/water masers

G23.01−0.41 is a molecular clump excited by a late O-type YSO with a stellar mass of $\approx$ 20 $M_\odot$. SMA observations by Sanna *et al.* (2014) detect a collimated (^{12}CO(2-1)) molecular outflow at linear scales of $\sim$1 pc, emerging from a molecular core placed at the center of the outflow and elongated perpendicular to the outflow axis (see Fig. 2, left panel). Inside the molecular core, at linear scales of a few 1000 AU, recent JVLA sensitive observations reveal an ionized (thermal) jet, powering the larger scale molecular outflow (see Fig. 2, right panel). Multi-epoch VLBA observations have identified three distinct clusters of water masers, which spread over the radio continuum emission of the jet along a direction almost parallel to the jet and the molecular outflow. Maser spots of the central cluster draw an ark with diverging proper motions, and they are probably tracing a fast bow-shock of the jet impinging against dense molecular material near the YSO. The masers in the cluster to NE move parallel to the jet (see Fig. 2, right panel). The water maser data, taken a few years before the SMA and JVLA observations, have been our first (reliable) evidence of a jet emerging from the high-mass YSO in G23.01−0.41 (Sanna *et al.* 2010).

High-density (thermal) tracers (like the CH$_3$CN emission) reveal that the gas inside the molecular core is undergoing both expansion along and rotation about the outflow axis. (see Fig. 3, left panel). The presence of these two motion components emerges also

---

† The Bar and Spiral Structure Legacy survey is a key Very Long Baseline Array (VLBA) project, which has measured parallaxes and proper motions of hundreds of methanol and water masers distributed across the Galactic disk (Reid *et al.* 2014)

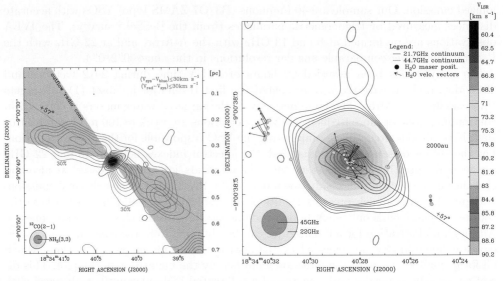

Figure 2. Comparison of the outflow tracers across different linear scales in G23.01−0.41. *Left:* Blue and red contours show SMA maps of the blue- and redshifted ^{12}CO(2-1) emission integrated over a V_{LSR} range up to 30 km s^{-1} from the systemic velocity. Gray contours at the center of the bipolar outflow represent the emission of warm gas traced with NH3 (Codella *et al.* 1997). The gray cone indicates the opening angle of the radio jet. *Right :* Overlay of the JVLA 22 GHz continuum (black contours and gray scale) with the 44 GHz continuum residual map (red contours). Colored dots and black arrows give positions and velocities of the water masers. (See the on-line version of the figure; see Sanna *et al.* (2016), Fig. 2, for further details.)

clearly from position-velocity plots along slices parallel and perpendicular to the outflow axis (see Sanna *et al.* (2014), Fig. 3). The 6.7 GHz masers originate at radii of 100–2000 AU from the forming star (see Fig. 3, right panel) and their 3-D velocities trace a combination of expansion and rotation, consistent with the kinematic pattern from thermal lines. The distribution of the 3-D 6.7 GHz maser velocities, moving radially outward across the equatorial plane and collimating at closer angles with the jet axis at (relatively) larger quotes, reminds that of a rotating disk-wind, qualitatively similar to the disk-wind traced with the SiO masers in Orion-KL Source I (Matthews *et al.* 2010; Greenhill *et al.* 2013).

Towards the ultra-compact (UC) HII region NGC7538 IRS1, we have recently accumulated several pieces of evidence that the 6.7 GHz masers are tracing accretion disks around high-mass YSOs (Moscadelli & Goddi 2014). In this region, the methanol masers show two elongated distributions (labeled in Figure 4 with letters A and B+C, respectively) of size of $\approx$ 500 AU. We used four epochs of European VLBI Network (EVN) observations for: 1) studying the very regular change of maser V_{LSR} with position; 2) measuring the maser proper motions, mostly found to be closely aligned with the maser linear distributions, consistent with planar motion seen (almost) edge-on; 3) deriving the line-of-sight accelerations of many maser features, also varying regularly with maser position. We have complemented the maser data with JVLA B-Array observations of the NH$_3$ line (Goddi *et al.* 2015) from this region, finding that the NH$_3$ V_{LSR} distribution is in good agreement with the maser 3-D velocity pattern. Finally, for each of the two linear maser clusters, we have reproduced our set of observations with a simple kinematic model of a rotating disk in centrifugal equilibrium, constraining the mass of both the disk and the central star. Our hypothesis that the 6.7 GHz masers in NGC7538 IRS1 emerge from

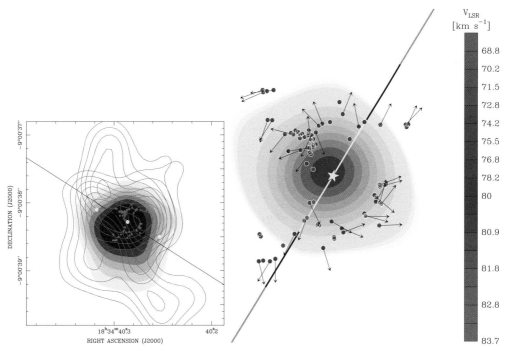

Figure 3. Rotation and expansion in G23.01−0.41. *Left:* The gray-scale image, and the blue and red contours show SMA maps of the CH3CN (12_3-11_3) emission at the systemic, and blue- and redshifted velocities, respectively. Big yellow and small, red and blue dots give the spatial distribution of the 22 GHz water and 6.7 GHz methanol masers, respectively. The NE-SW black line denotes the outflow direction. *Right:* Zoom on the 6.7 GHz methanol masers. The distribution of 6.7 GHz maser V_{LSR} is shown by white-encircled, colored dots, and the proper motion direction by black arrows. The gray scale image represents the JVLA 1.3 cm continuum. The colored axis denotes the disk mid-plane, where each colored segment spans 1000 AU in length. (See the on-line version of the figure; see Sanna *et al.* (2014), Fig. 1, for further details.)

accretion disks around two distinct high-mass YSOs (labeled IRS1a and b in Fig. 4), has been recently confirmed by JVLA A-Array observations of NH_3 and CH_3OH lines by Beuther *et al.* (2017). These authors find a velocity gradient, in (thermal) CH_3OH absorption, in correspondence of each of the two CH_3OH (maser) linear distributions. These velocity gradients, interpreted in terms of accretion disks, show an (approximately) E-W orientation, not very different from the PA of the maser structures observed at smaller scales.

4. Conclusions

Molecular interstellar masers are *reliable* tracers of kinematic structures in high-mass YSOs. 22 GHz water masers invariably arise from (and trace the motion of) fast (20–100 km s^{-1}) shocks in (wide-angle) winds or (collimated) jets ejected from the massive (proto)stars. Depending on the shock properties, water maser velocities are well collimated about the local flow direction in compact, radiative shocks, or present a larger scatter, if emerging from wide-angle, (quasi)adiabatic bow-shocks. In a few well-studied objects, 6.7 GHz methanol masers are found to originate in the flattened, rotating disk/envelope of high-mass YSOs, at radial distances from a few 100 AU to a few 1000 AU. Methanol masers move at typical speeds of 1–20 km s^{-1}, and can show a complex pattern of 3-D velocities. They can trace not only the envelope/disk rotation but also expansion along directions both radial (close to the equatorial plane) and forming

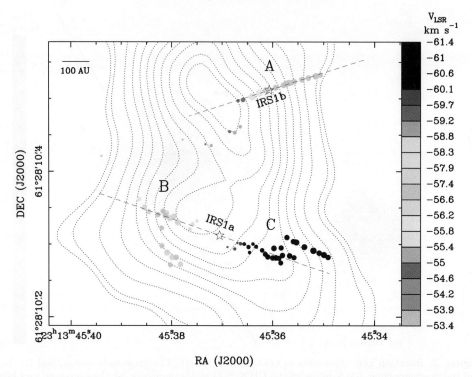

Figure 4. Methanol masers tracing accretion disks in the UC HII region NGC7538 IRS1. Colored dots give positions of the 6.7 GHz methanol masers, with colors denoting maser V_{LSR}. Masers distribute in three elongated clusters labeled with letters "A", "B" and "C". The dashed lines show the linear fits to the spatial distributions of maser features in cluster A, and in the combined clusters B + C. The dotted contours reproduce the VLA A-Array, 22 GHz continuum. The stars labeled IRS1a and IRS1b mark the YSO positions, as determined from fitting a disk model to the maser positions, velocities and line-of-sight accelerations measured via multi-epoch EVN observations. (See the on-line version of the figure; see Moscadelli *et al.* (2014), for further details.)

small angles with the disk axis (at relatively higher quotes). Such a velocity pattern can be interpreted in terms of a (relatively slow) rotating disk-wind.

References

Beltrán, M. T. & de Wit, W. J. 2016, *ARAA*, 24, 6
Beuther, H., Linz, H., Henning, T., Feng, S., & Teague, R. 2017, *A&A*, 605, A61
Cyganowski, C. J., Brogan, C. L., Hunter, *et al.* 2011, *ApJ*, 729, 124
Codella, C., Testi, L., & Cesaroni, R. 1997, *A&A*, 325, 282
Goddi, C., Moscadelli, L., & Sanna, A. 2011, *A&A*, 535, L8
Goddi, C., Zhang, Q., & Moscadelli, L. 2015, *A&A*, 573, A108
Greenhill, L. J., Goddi, C., Chandler, *et al.* 2013, *ApJL*, 770, L32
Lee, C.-F., Ho, P. T. P., Li, Z.-Y., *et al.* 2017, Nature Astronomy, 1, 0152
Matthews, L. D., Greenhill, L. J., Goddi, C., *et al.* 2010, *ApJ*, 708, 80
Moscadelli, L., Cesaroni, R., Rioja, M. J., Dodson, R., & Reid, M. J. 2011, *A&A*, 526, A66+
Moscadelli, L. & Goddi, C. 2014, *A&A*, 566, A150
Moscadelli, L., Sánchez-Monge, Á., Goddi, C., *et al.* 2016, *A&A*, 585, A71
Reid, M. J., Menten, K. M., Brunthaler, A., *et al.* 2014, *ApJ*, 783, 130
Sanna, A., Cesaroni, R., Moscadelli, L., *et al.* 2014, *A&A*, 565, A34
Sanna, A., Moscadelli, L., Cesaroni, R., *et al.* 2016, *A&A*, 596, L2
Sanna, A., Moscadelli, L., Cesaroni, R., *et al.* 2010, *A&A*, 517, A78+

Astrophysical Masers:
Unlocking the Mysteries of the Universe
Proceedings of the IAU Symposium No. 336, 2017
A. Tarchi, M.J. Reid & P. Castangia, eds.

© International Astronomical Union 2018
doi:10.1017/S1743921317010225

ALMA observations of submillimeter H$_2$O and SiO lines in Orion Source I

Tomoya Hirota[1], Masahiro N. Machida[2], Yuko Matsushita[2], Kazuhito Motogi[3], Naoko Matsumoto[1,3], Mikyoung Kim[4,5], Ross A. Burns[6] and Mareki Honma[5]

[1]National Astronomical Observatory of Japan, Mitaka-shi, Tokyo 181-8588, Japan
email: tomoya.hirota@nao.ac.jp

[2]Kyushu University, Fukuoka-shi 819-0395, Japan
[3]Yamaguchi University, Yamaguchi-shi 753-8512, Japan
[4]Korea Astronomy and Space Science Institute, Daejeon 305-348, Republic of Korea
[5]National Astronomical Observatory of Japan, Oshu-shi, Iwate 023-0861, Japan
[6]Joint Institute for VLBI ERIC, Postbus 2, 7990 AA Dwingeloo, The Netherlands

Abstract. We present observational results of the submillimeter H$_2$O and SiO lines toward a candidate high-mass young stellar object Orion Source I using ALMA. The spatial structures of the high excitation lines at lower-state energies of >2500 K show compact structures consistent with the circumstellar disk and/or base of the northeast-southwest bipolar outflow with a 100 au scale. The highest excitation transition, the SiO (v=4) line at band 8, has the most compact structure. In contrast, lower-excitation transitions are more extended than 200 au tracing the outflow. Almost all the line show velocity gradients perpendicular to the outflow axis suggesting rotation motions of the circumstellar disk and outflow. While some of the detected lines show broad line profiles and spatially extended emission components indicative of thermal excitation, the strong H$_2$O lines at 321 GHz, 474 GHz, and 658 GHz with brightness temperatures of >1000 K show clear signatures of maser action.

Keywords. stars: individual (Orion Source I), ISM: jets and outflows, outflow, masers

1. Introduction

Orion Source I is a candidate of a high-mass young stellar object (Menten & Reid 1995) located in the nearest high-mass star-forming region Orion KL at a distance of ~420 pc (Menten *et al.* 2007, Kim *et al.* 2008). It drives a low-velocity bipolar outflow along the northeast-southwest (NE-SW) direction with a 1000 au-scale in edge-on view (Greenhill *et al.* 2013). Source I is known as one of the rare star-forming regions associated with the SiO masers (Menten & Reid 1995). The vibrationally excited SiO masers trace a rotating outflow arising from a circumstellar disk with a ~100 au scale (Kim *et al.* 2008, Matthews *et al.* 2010). The rotation curve of the SiO masers implies an enclosed mass of $(5-7)M_\odot$ under the assumption of Keplerian rotation, which is also confirmed by recent observations with ALMA (Plambeck & Wright 2016). However, the above mass estimate is significantly smaller than that from proper motion measurements with VLA, in which Source I is proposed to be a 20 M$_\odot$ binary system formed by a dynamical encounter event 500 years ago (Goddi *et al.* 2011, Rodríguez *et al.* 2017, Bally *et al.* 2017). Observations of Source I at higher angular resolution would be crucial to investigate detailed physical and dynamical properties of Source I. For this purpose, strong maser lines will be unique probes as they can reveal high density and temperature regions in close vicinity to newly born stars at higher resolution than weaker thermal molecular lines.

208 T. Hirota *et al.*

Table 1. Detected lines (ordered by the lower-state energy)

Molecule	Transition	ν (MHz)	E_l (K)	Beam size (arcsec)	Reference
H_2O	$5_{3,3}$-$4_{4,0}$	474689	702	0.10″	
H_2O	$10_{2,9}$-$9_{3,6}$	321226	1846	0.17″	Hirota *et al.* (2014)
H_2O	$v=1$, $1_{1,0}$-$1_{0,1}$	658007	2329	0.26″	Hirota *et al.* (2016)
H_2O	$v=1$, $4_{2,2}$-$3_{3,1}$	463171	2744	0.10″	Hirota *et al.* (2017)
H_2O	$v=1$, $5_{2,3}$-$6_{1,6}$	336228	2939	0.17″	Hirota *et al.* (2014)
H_2O	$v=1$, $5_{5,0}$-$6_{4,3}$	232687	3451	0.18″	Hirota *et al.* (2012)
H_2O	$v=1$, $7_{4,4}$-$6_{5,1}$	498502	3673	0.09″	
^{29}SiO	10-9	428684	93	0.08″	
SiO	11-10	477505	115	0.10″	
$Si^{18}O$	12-11	484056	128	0.09″	Hirota *et al.* (2017)
^{30}SiO	$v=1$, 11-10	462757	1859	0.10″	
SiO	$v=1$, 11-10	474185	1883	0.10″	
^{29}SiO	$v=2$, 11-10	465014	3611	0.10″	
SiO	$v=2$, 10-9	428087	3614	0.08″	
SiO	$v=4$, 11-10	464245	7085	0.10″	

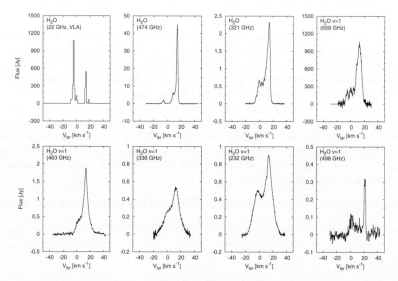

Figure 1. Spectra of the H_2O lines. For the 22 GHz (Gaume *et al.* 1998) and 658 GHz lines, total flux densities are integrated over the primary beam. Flux densities for other lines are integrated over the synthesized beam size (Table 1) around peak positions of each line.

2. Observations

Observations of the submillimeter H_2O and SiO lines were carried out with ALMA in cycles 0, 1, and 2. We also compared the ALMA Science Verification data for Orion KL at band 6. Table 1 summarizes the detected transitions. Details of observations and data analysis are described in previous papers (Hirota *et al.* 2012, 2014, 2016, 2017).

3. Results

Figures 1 and 2 show detected spectra of the H_2O and SiO lines, respectively. Intensive studies with single-dish telescopes have already detected several submillimeter H_2O lines in star-forming regions and late-type stars (Humphreys 2007) while the 463 GHz and 498 GHz lines are detected for the first time with our ALMA observations. All the spectra show broad line profiles and some of them have double peaked structures with the velocity width of $\sim$10-20 km s^{-1}. These results are analogous to previously observed

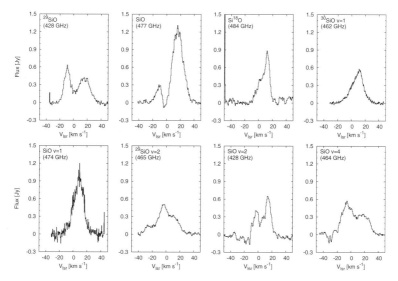

Figure 2. Spectra of the SiO lines observed toward the peak positions of each line. Flux densities are integrated over the beam size (Table 1). Note that the flux scales in all panels are common.

22 GHz H_2O masers (Figure 1; Gaume *et al.* 1998, Greenhill *et al.* 2013) and 43 GHz SiO masers (Menten & Reid 1995, Kim *et al.* 2008, Matthews *et al.* 2010).

Recent ALMA observations have revealed that high excitation molecular lines trace a hot molecular gas disk and/or base of the NE-SW outflow in Source I (Hirota *et al.* 2012, 2014, 2016, 2017, Plambeck & Wright 2016). Figure 3 shows examples of our new results from the ALMA observations. We have found a clear trend that the higher excitation lines of H_2O and SiO at the lower state energy levels of $E_l >2500$ K could trace the compact disk or innermost region of the outflow with a 100 au scale, as can be seen in the vibrationally excited SiO masers at 43 GHz (Kim *et al.* 2008, Matthews *et al.* 2010). The highest excitation transition we have detected is the $v=4$, $J=11$-10 transition of SiO ($E_l=7085$ K), which shows the most compact structure among the detected lines (Figure 3b). On the other hand, the lower excitation lines of $E_l <2500$ K trace the NE-SW outflow. The extended outflow structures can be traced even by the vibrationally excited $v=1$ transition of SiO (Figure 3a) with a size of >200 au. All the maps show velocity gradients perpendicular to the outflow axis, except SiO line at 477 GHz and H_2O line at 474 GHz, which are optically thick and are preferentially excited in the outer part of the outflow, respectively. These velocity structures, indicating rotation motion of the disk and outflow, provide clear evidence of magneto-centrifugal disk winds as a possible launching mechanism of the bipolar outflow from Source I (Matthews *et al.* 2010, Greenhill *et al.* 2013, Hirota *et al.* 2017).

4. Discussions

It is known that Source I is associated with the strong H_2O maser at 22 GHz and SiO masers at 43 GHz (Menten & Reid 1995, Gaume *et al.* 1998, Kim *et al.* 2008, Matthews *et al.* 2010, Greenhill *et al.* 2013). These maser emissions usually show spatially compact structures, higher brightness temperatures compared with gas kinetic temperatures, and narrow spike-like line profiles. Extremely high brightness temperatures of >1000 K of the 321 GHz, 474 GHz, and 658 GHz H_2O lines are clear signatures of maser emissions. It is also likely that a narrow spike-like spectral profile of the 498 GHz H_2O line (Figure 1) could be an evidence of maser action. However, the other lines, in particular for SiO, show

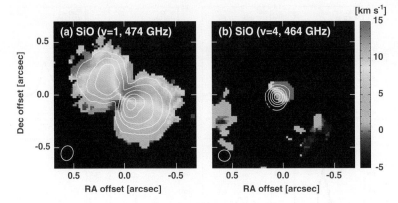

Figure 3. Moment 0 (cotour) maps of (a) SiO (v=1, 474 GHz) and (b) SiO (v=4, 464 GHz) lines superposed on their moment 1 maps. Contours represent 10, 30, 50, 70, and 90% of the peak intensities of (a) 31 Jy beam^{-1} km s^{-1} and (b) 22 Jy beam^{-1} km s^{-1}, respectively.

no clear signature of the above typical characteristics of masers. Flux density ratios of SiO isotopologues are different from those of isotope ratios (Tercero *et al.* 2011) suggesting maser amplification and/or self-absorption in foreground gas. In order to access excitation mechanism of the observed lines, whether they are masers or thermal emissions, further maser pumping and radiative transfer models would be essential.

Acknowledgements

This letter makes use of the following ALMA data: ADS/JAO.ALMA#2011.0.00009.SV, ADS/JAO.ALMA#2011.0.00199.S, and ADS/JAO.ALMA#2013.1.00048.S. ALMA is a partnership of ESO (representing its member states), NSF (USA) and NINS (Japan), together with NRC (Canada), NSC and ASIAA (Taiwan), and KASI (Republic of Korea), in cooperation with the Republic of Chile. The Joint ALMA Observatory is operated by ESO, AUI/NRAO and NAOJ. TH is supported by the MEXT/JSPS KAKENHI Grant Numbers 24684011, 25108005, 15H03646, 16K05293, and 17K05398.

References

Bally, J., Ginsburg, A., Arce, H., *et al.* 2017, *ApJ*, 837, 60
Gaume, R. A., Wilson, T. L., Vrba, F. J., *et al.* 1998, *ApJ*, 493, 940
Goddi, C., Humphreys, E. M. L., Greenhill, L. J., *et al.* 2011, *ApJ*, 728, 15
Greenhill, L. J., Goddi, C., Chandler, C. J., *et al.* 2013, *ApJ*, 770, 32
Hirota, T, Kim, M. K. & Honma, M. 2012, *ApJ* (Letters), 757, L1
Hirota, T., Kim, M. K., & Honma, M. 2016, *ApJ.*, 817, 168
Hirota, T., Machida, M. N., Matsushita, Y., *et al.* 2017, *Nature Aston.*, 1, 146
Hirota, T., Kim, M. K., Kurono, Y., & Honma, M. 2014, *ApJ* (Letters), 782, L28
Humphreys, E. M. L. 2007, in: J. M. Chapman & W. A. Baan (eds.), *Astrophysical Masers and their Environments*, Proc. IAU Symposium No. 242 (Cambridge: CUP), p. 471
Kim, M. K., Hirota, T., Honma, M., *et al.* 2008, *PASJ*, 60, 991
Matthews, L. D., Greenhill, L. J., Goddi, C., *et al.* 2010, *ApJ*, 708, 80
Menten, K. M. & Reid, M. J. 1995, *ApJ*, 445, L157
Menten, K. M., Reid, M. J., Forbrich, J., & Brunthaler, A. 2007, *A&A*, 474, 515
Plambeck, R. L. & Wright, M. C. H. 2016, *ApJ*, 833, 219
Rodríguez, L. F., Dzib, S. A., Loinard, L., *et al.* 2017, *ApJ*, 834, 140
Tercero, B., Vincent, L., Cernicharo, J., *et al.* 2011, *A&A*, 528, A26

Astrophysical Masers:
Unlocking the Mysteries of the Universe
Proceedings IAU Symposium No. 336, 2017
A. Tarchi, M.J. Reid & P. Castangia, eds.

© International Astronomical Union 2018
doi:10.1017/S1743921317010055

Expansion of methanol maser rings

Anna Bartkiewicz[1], Alberto Sanna[2], Marian Szymczak[1], Luca Moscadelli[3] and Huib van Langevelde[4,5]

[1] Centre for Astronomy, Faculty of Physics, Astronomy and Informatics, Nicolaus Copernicus University, Grudziadzka 5, 87-100 Torun, Poland
email: `annan@astro.umk.pl`

[2] Max-Planck-Institut für Radioastronomie, Auf dem Hügel 69, 53121, Bonn, Germany
email: `asanna@mpifr-bonn.mpg.de`

[3] INAF, Osservatorio Astrofisico di Arcetri, Largo E. Fermi 5, 50125, Firenze, Italy
email: `mosca@arcetri.astro.it`

[4] Joint Institute for VLBI ERIC (JIVE), Postbus 2, 7990 AA Dwingeloo, The Netherlands
email: `langevelde@jive.eu`

[5] Sterrewacht Leiden, Leiden University, Postbus 9513, 2300 RA Leiden, The Netherlands

Abstract. Ring−like sources of 6.7 GHz methanol maser emission were discovered a decade ago with the European VLBI Network. In the past years we have been incessantly working to understand the nature of these rings. In general, the methanol rings do not coincide with H II regions nor they show 22 GHz water maser emission. Here, we present a proper motion study over a time baseline up to 10.5 years for the first sub-sample of methanol maser rings. Our findings suggest that in three targets G23.207−00.377, G23.389+00.185, and G23.657−00.127, such rings form in outflows or even in winds close to the central sources, and the masers trace slow proper motions of a few km s^{-1} typically.

Keywords. masers, stars: formation, instrumentation: high angular resolution

1. Introduction

After the discovery of the ring-like structures of the 6.7 GHz methanol maser emission a decade ago (Bartkiewicz *et al.* 2005), we started complementary studies to answer the question "what are those methanol rings?". The morphology suggests a relation with a disc around a massive proto- or young star, but the velocity signature of the maser spots is not consistent with rotation of a disc (Bartkiewicz *et al.* 2009). In general, we found neither water maser emission towards these sources (Bartkiewicz *et al.* 2011) nor radio continuum emission at cm-wavelength range (Bartkiewicz *et al.* 2009). This evidence suggests that the rings are associated with massive young stellar object at early evolution stages. High-angular resolution infrared observations of four methanol maser rings did not support a scenario where the 6.7 GHz maser emission arises from a circumstellar disc (De Buizer *et al.* 2012). Therefore, we started the most direct investigation of methanol maser emission with ring-like morphology using proper motion measurements obtained with the European VLBI Network†. VLBI observations were successfully demonstrated as a powerful tool to trace the 3D kinematics of the masing gas over a time span of a few years (e.g., Moscadelli *et al.* 2005, 2006; Sanna *et al.* 2010a, 2010b; Goddi *et al.* 2011). Here, we report preliminary results for three methanol maser rings, which are a part of a larger survey towards 12 targets.

† The European VLBI Network is a joint facility of independent European, African, Asian, and North American radio astronomy institutes. Scientific results from data presented in this publication are derived from the following EVN project codes: EN003, EB052.

2. Observations

The first epoch data of three targets, G23.207−00.377, G23.389+00.185 and G23.657 −00.127 were observed on 11 November 2004 (Bartkiewicz *et al.* 2009). The second epoch data were observed on 15 March 2015. The observations were phase−referenced with a switching cycle of 175 s+95 s (maser + phase-calibrator) resulting in a total on-source time of ca. 1.5 hr. The EVN Mk IV Data Processor at JIVE was used in the first epoch data and the SFXC software correlator (Keimpema *et al.* 2015) in the second one; spectral resolution on the maser lines was 0.1 km s^{-1}, i.e. a bandwidth of 2 MHz was divided into 1024 spectral channels. In the project EB052, in order to increase the signal-to-noise ratio on the phase-reference source, we used eight BBCs per polarization for a second correlator pass with 128 channels per BBC (each 2 MHz wide). The data reduction was carried out in AIPS with standard procedures for spectral line observations. Phase calibration was performed on the strongest maser channel from the first epoch observations (Bartkiewicz *et al.* 2009). Finally, we searched the maser emission using the task SAD of AIPS and a cutoff of 7σ for each channel map.

In order to study the displacements of maser spots in time, the following procedures were used: first we selected single maser spots that were visible at both epochs, next we fitted the flux−weighted ellipses to the overall maser spots distributions seen at both epochs using the code by Fitzgibbon *et al.* (1999), and then we aligned the centres of best fitted ellipses. This approach removed any bulk motion of the ring in the plane of the sky. Finally, for each group of maser spots that were clearly separated each one from the other, we constructed the averaged proper motion vector.

3. Results and Discussion

The barycentre of each maser group and its motion vector for three targets G23.207−00.377, G23.389+00.185, and G23.657−00.127 are shown in Fig. 1.

In G23.207−00.377 we detected tangential displacements of 0.13–5.6 mas between 2004 and 2015. For a distance of 4.17 kpc, obtained from a parallax measurement within the BeSSeL Survey† (Reid priv. comm.), these offsets correspond to velocities in the sky plane from 0.24 to 10.6 km s^{-1}. In the second source, G23.389+00.185, displacements were from 0.5 to 7.5 mas in the same period. For a distance of 5 kpc (Reid priv. comm.) this range corresponds to velocities from 1.1 to 17.0 km s^{-1}. The methanol maser spots in the third target, G23.657−00.127, are the most circularly distributed; the ellipticity of the best fitted ellipse is 0.39 (Fig.1). The registered shifts are from 0.6 to 3.8 mas corresponding to the velocities from 0.9 to 5.5 km s^{-1} for a distance of 3.19 kpc (Bartkiewicz *et al.* 2008). The proper motion vectors of 33 maser groups in G23.657−00.127 do not show any obvious sign of expansion or rotation, furthermore there is neither a particular direction where the motion would be greater or smaller relatively to the mean value of 2.7 km s^{-1} (Bartkiewicz *et al.* 2014). The width of "the shell" of methanol maser emission was established as 29 mas, corresponding to 92 AU (Bartkiewicz *et al.* 2005). The proper motion of each maser group do not seem to depend on a radius; the further from the centre of the best-fitted ellipse the shift is larger (Fig. 1). This points the possible expansion. De Buzier *et al.* (2012) detected a source towards this target at near- and mid-infrared. The peak emission coincided with the centre of the methanol ring to within 2.5σ. However, the 2.12 μm morphology was fan-shaped; this evidence argues in favor of scattered and/or reflected emission from the walls of the outflow cavity. Both morphologies, of methanol

† http://bessel.vlbi-astrometry.org

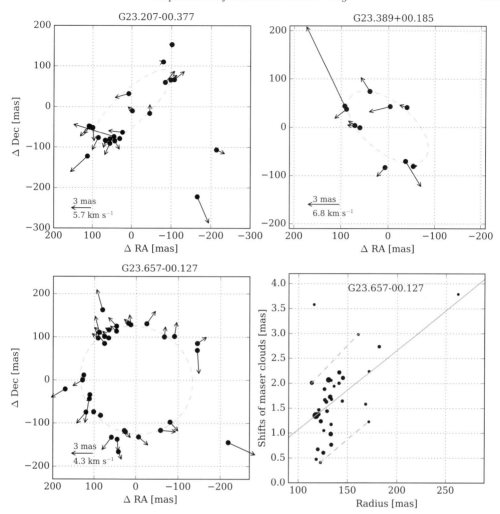

Figure 1. In the first three panels proper motion vectors of the 6.7 GHz methanol maser groups in G23.207−00.377, G23.389+00.185, and G23.657−00.127 as derived from two epoch observations (2004 and 2015), are presented. The (0,0) points correspond to the centres of the best fitted ellipses to the data from 2004. **In the bottom right panel**, for G23.657-00.127 between 2004 and 2015, we show maser shifts versus their radii from the center of the ellipse. The sizes of dots are proportional to the intensities of the brightest spots within each maser groups (a logarithmic scale). The solid line shows the least square fit with following coefficients: a=−0.52, b=0.0159. The dashed lines connect the most outer data points at both side of the fitted line, but in close position to the majority of data.

maser and near-infrared, were not consistent with a scenario where the maser ring arises from a face-on circumstellar disc.

One can see that in all three sources the proper motions are not consistent with radial motions of maser spots from the centre of the ellipse.. That is opposite to the result for the well−known high-mas star-forming region Cep A, where a combination of infall and rotation were reported (Sugiyama *et al.* 2014, Torstensson *et al.* 2011, Sanna *et al.* 2017). We rather observe an expansion and possible rotation as reported for G23.01−0.41 (Sanna *et al.* 2010b). We also note a similarity to the proper motions of SiO masers in Orion Source I, where they arise from a wide-angle bipolar wind emanating from a rotating, edge-on disc (Matthews *et al.* 2010). The extended disc wind has been recently seen by ALMA using CO lines in the TMC1A low-mass protostellar system (Bjerkeli *et al.* 2016). We therefore want to explore the possibility that methanol maser rings may be a part of a disc wind close to the central sources (rings are relatively small in sizes 400–1000 AU in diameter). In order to locate the position of the central objects we are planning high-angular resolution observations with the most sensitive interferometers such as the Jansky VLA and ALMA.

Acknowledgements

AB and MS acknowledge support from the National Science Centre, Poland through grant 2016/21/B/ST9/01455. The research leading to these results has received funding from the European Commission Seventh Framework Programme (FP/2007-2013) under grant agreement No. 283393 (RadioNet3).

References

Bartkiewicz, A., Szymczak, M., & van Langevelde, H. J. 2005, *A&A*, 442, L61

Bartkiewicz, A., Brunthaler, A., Szymczak, M., van Langevelde H. J., & Reid, M. J. 2008, *A&A*, 490, 787

Bartkiewicz, A., Szymczak, M., van Langevelde, H. J., Richards, A. M. S., & Pihlström, Y. M. 2009, *A&A*, 502, 155

Bartkiewicz, A., Szymczak, M., Pihlström, Y. M., van Langevelde H. J., Brunthaler, A., & Reid, M. J. 2011, *A&A*, 525, A120

Bartkiewicz, A., Sanna, A., Szymczak, M., & Moscadelli, L. 2014, Proceedings of 12th European VLBI Network Symposium and Users Meeting (EVN 2014) at http://pos.sissa.it/, 039

Bjerkeli, P., van der Wiel, M. H. D., Harsono, D., Ramsey J. P., & Jørgensen J. K. 2016, *Nature*, 540, 406

De Buizer, J. M., Bartkiewicz, A., & Szymczak, M. 2012, *ApJ*, 754, 149

Fitzgibbon, A., Pilu, M., & Fisher, R. B. 1999, *IEEE Transactions on Pattern Analysis and Machine Intelligence*, 21, 476

Goddi, C., Moscadelli, L., & Sanna, A. 2011, *A&A*, 535, L8

Keimpema, A., Kettenis, M. M., & Pogrebenko, S. V. *et al.* 2015, *ExA*, 39, 259

Matthews, L. D., Greenhill, L. J., Goddi, C., Chandler, C. J., Humphreys, E. M. L., & Kunz, M. W. 2010, *ApJ*, 708, 80

Moscadelli, L., Cesaroni, R., & Rioja, M. J. 2005, *Astronomy and Astrophysics*, 438, 889

Moscadelli, L., Testi, L., Furuya, R. S., Goddi, C., Claussen, M., Kitamura, Y., & Wootten, A. 2006, *A&A*, 446, 985

Sanna, A., Moscadelli, L., Cesaroni, R., Tarchi, A., Furuya, R. S., & Goddi, C. 2010a, *A&A*, 517, 71

Sanna, A., Moscadelli, L., Cesaroni, R., Tarchi, A., Furuya, R. S., & Goddi, C. 2010b, *A&A*, 517, 78

Sanna, A., Moscadelli, L., Surcis, G., van Langevelde, H. J., Torstensson, K. J. E., & Sobolev, A. M. 2017, *A&A*, 603, 94

Sugiyama, K., Fujisawa, K., Doi, A., Honma, M., Kobayashi, H., *et al.* 2014, *A&A*, 562, 82

Torstensson, K. J. E., van Langevelde, H. J., Vlemmings, W. H. T., & Bourke, S. 2011, *A&A*, 526, 38

Astrophysical Masers:
Unlocking the Mysteries of the Universe
Proceedings IAU Symposium No. 336, 2017
A. Tarchi, M.J. Reid & P. Castangia, eds.

© International Astronomical Union 2018
doi:10.1017/S1743921318000017

Measuring Magnetic Fields from Water Masers Associated with a Synchrotron Protostellar Jet

Ciriaco Goddi[1,2] and Gabriele Surcis[3]

[1]Department of Astrophysics/IMAPP, Radboud University,
P.O. Box 9010, 6500 GL Nijmegen, The Netherlands

[2]ALLEGRO/Leiden Observatory, Leiden University,
PO Box 9513, NL-2300 RA Leiden, the Netherlands
email: cgoddi@strw.leidenuniv.nl

[3] INAF - Osservatorio Astronomico di Cagliari
Via della Scienza 5 - I-09047 Selargius, Italy
email: surcis@oa-cagliari.inaf.it

Abstract. The Turner-Welch Object in the W3(OH) high-mass star forming complex drives a synchrotron jet, which is quite exceptional for a high-mass protostar, and is associated with a strongly polarized water maser source, $W3(H_2O)$, making it an optimal target to investigate the role of magnetic fields on the innermost scales of protostellar disk-jet systems. We report here full polarimetric VLBA observations of water masers. The linearly polarized emission from water masers provides clues on the orientation of the local magnetic field, while the measurement of the Zeeman splitting from circular polarization provides its strength. By combining the information on the measured orientation and strength of the magnetic field with the knowledge of the maser velocities, we infer that the magnetic field evolves from having a dominant component parallel to the outflow velocity in the pre-shock gas (with field strengths of the order of a few tens of mG), to being mainly dominated by the perpendicular component (of order of a few hundred of mG) in the post-shock gas where the water masers are excited. The general implication is that in the undisturbed (i.e. not-shocked) circumstellar gas, the flow velocities would follow closely the magnetic field lines, while in the shocked gas the magnetic field would be re-configured to be parallel to the shock front as a consequence of gas compression.

Keywords. stars: formation, ISM: jets and outflows, ISM: magnetic fields

1. Introduction

One relevant open question in high-mass star formation (HMSF) is the relation between the gas dynamics and the magnetic field in regulating mass-accretion and mass-loss. VLBI measurements of molecular masers can provide a detailed description of gas kinematics (Goddi *et al.* 2005, 2006, 2011, Matthews *et al.* 2010, Greenhill *et al.* 2013, Moscadelli & Goddi 2014, Moscadelli *et al.* 2016, Issaoun *et al.* 2017) and magnetic field structure (Surcis *et al.* 2014, 2015, Sanna *et al.* 2015) on scales from tens to hundreds AU, which are the smallest accessible scales in studies of HMSF, and have therefore the potential to address such an open question. $W3(H_2O)$ contains the best known (archetypical) case of synchrotron jet driven by an embedded high-mass young stellar object or YSO (Reid *et al.* 1995) and associated with H_2O masers, and therefore provides a good target for investigating this issue.

Here we summarize the main results from a full polarimetric study of H_2O masers in $W3(H_2O)$. A full description of results is reported in Goddi *et al.* (2017).

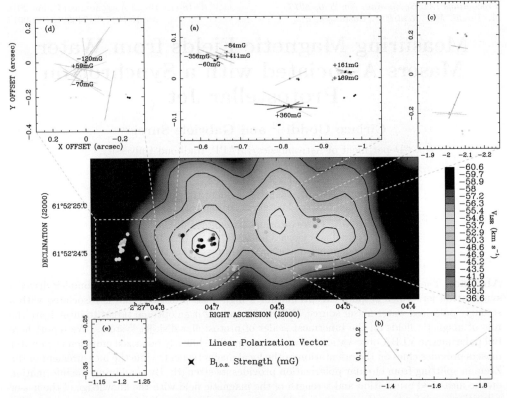

Figure 1. Overlay of the water masers detected with the VLBA in W3(H_2O) onto the 1.4 mm continuum emission mapped with the PdBI by Wyrowski *et al.* (1999) (gray scale and black contours). The circles show positions of the H_2O masers, while the colors denote their l.o.s. velocity in km s^{-1} (color scale on the right-hand side). The three 1.4 mm continuum peaks are labelled "A", "B", and "C", from east to west. The insets show the linear polarization vectors of individual maser features in different clusters (from "a" to "e"), where the length of the line segments scales logarithmically with the polarization fraction (in the range $P_l = 0.9\% - 42\%$). We also report the magnetic field strengths (in mG) along the l.o.s. ($B_{l.o.s.}$) in the maser features for which we measured the Zeeman splitting. The positions are relative to the reference maser feature.

2. Results

We identified a total of 148 individual maser features and we measured their physical properties, including positions, flux densities, l.o.s. velocities (V_{LSR}), and (when polarized) their fraction of linear and circular polarizations, as well as the corresponding linear polarization angles and magnetic field strengths along the l.o.s.. The methodology adopted to derive the physical properties of individual maser features is described in Appendix A in Goddi *et al.* (2017).

Out of the 148 features detected, we measured linear polarization in 34 maser features, with a fractional percentage varying in the range $P_l = 0.9\% - 42\%$. In Figure 1, line segments indicate the linear polarization vectors of the maser features, whose length scales logarithmically with the polarization fraction, P_l. We established an *empirical threshold* of linear polarization fraction ($P_l = 5\%$) above which the water masers enter into the saturation regime, and their polarized signal does not trace the magnetic field anymore (see Goddi *et al.* 2017 for details). Figure 2 shows the resulting (sky-projected) magnetic field vectors for the maser features with $P_l < 5\%$, overplotted on the 1.4 mm

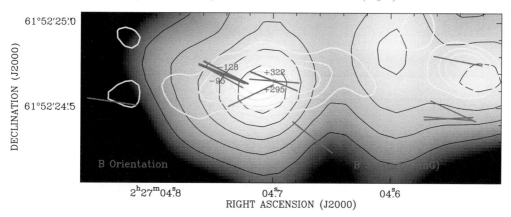

Figure 2. Magnetic field orientation (in the plane of the sky) for 17 individual masers with $P_l < 5\%$ (red segment) and strength for 4 (non-saturated) masers for which the Zeeman splitting was measured. The 8.4 GHz emission imaged with the VLA (beamsize~$0''.2$) by Wilner *et al.* (1999) (yellow contours) is overploted onto the 1.4 mm continuum emission mapped with the PdBI (beamsize~$0''.5$) by Wyrowski *et al.* (1999) (gray scale and black contours: same as in Fig. 1).

continuum map tracing the dust emission (same as in Figure 1), as well as the 8.4 GHz continuum imaged with the VLA (Wilner *et al.* 1999), tracing the synchrotron emission. Besides linear polarization, we also detected circularly polarized emission toward ten maser features, varying in the range $P_V = 0.2 - 1.6\%$. Considering only masers with $P_l < 5\%$, we estimated magnetic field strengths of ~100-300 mG (see Figure 2).

3. Discussion

The magnetic field inferred from the H₂O masers is on average oriented along E-W (Figure 2), well aligned with the axis of the synchrotron jet, suggesting that the molecular masers may probe the magnetic field in the protostellar jet. Figure 3 shows an overlay of the dust emission (greyscale) from the hot-core (Wyrowski *et al.* 1999), synchrotron emission (white contours) from the radio jet (Wilner *et al.* 1999), water maser proper motions (arrows) measured by Hachisuka *et al.* 2006, and the direction of magnetic field vectors as determined from our polarization measurements (purple segments). This overlay reveals a misalignment between the magnetic field and the velocity vectors, which can be explained with an origin in magnetically supported shocks. In particular, the shock passage alters the initial magnetic field configuration in the circumstellar gas, by compressing and enhancing the component of the magnetic field perpendicular to the shock velocity with respect to the parallel component (by a factor equal to the ratio between the post- and pre-shock densities: typically a 100). In the gas shocked by the synchrotron jet, we estimate a total field strength in the range ~100-300 mG (at densities of 10^9 cm^{-3}). Although the water maser polarization measurements alone cannot provide a *direct* measurement of the magnetic field properties in the *quiescent* (pre-shock) circumstellar gas, by combining the information on the orientation and strength of the magnetic field (in the post-shock gas) with the knowledge of the maser velocities, we are able to constrain the magnetic field strength in the pre-shock circumstellar gas (at densities of 10^7 cm^{-3}) to 10-20 mG (see Goddi *et al.* 2017 for details).

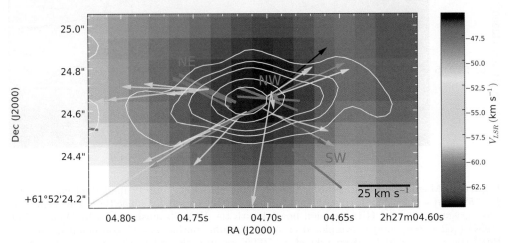

Figure 3. Comparison between the magnetic field orientations of H$_2$O masers (purple segments) and proper motions of H$_2$O masers (arrows). The 8.4 GHz emission imaged with the VLA by Wilner *et al.* (1999) (white contours) is overploted onto the 1.4 mm continuum emission mapped with the PdBI by Wyrowski *et al.* (1999) (gray scale). Colors denote l.o.s. velocity (color scales on the right-hand side) and the scale for the proper motion amplitude is given in the lower right corner (both in km s^{-1}). There are four main knots of masers, located towards NE, NW, SE, and SW, with respect to the radio continuum peak . The proper motions identify a biconical, bipolar molecular outflow. Note the misalignment between magnetic field and velocity vectors, particularly in the NE and NW knots.

4. Conclusions

Our results in W3(H$_2$O) suggest the presence of a local coupling between the magnetic field and the gas kinematics, indicating that magnetic fields can be dynamically important in driving the gas outflowing from a high-mass protostar.

References

Goddi, C., Moscadelli, L., Alef, W., *et al.* 2005, *A&A*, 432, 161
Goddi, C., Moscadelli, L., Torrelles, J. M., Uscanga, L., & Cesaroni, R. 2006, *A&A*, 447, L9
Goddi, C., Moscadelli, L., & Sanna, A. 2011, *A&A*, 535, L8
Goddi, C., Surcis, G., Moscadelli, L., *et al.* 2017, *A&A*, 597, A43
Greenhill, L. J., Goddi, C., Chandler, *et al.* 2013, *ApJ* (Letters), 770, L32
Hachisuka, K., Brunthaler, A., Menten, K. M., *et al.* 2006, *ApJ*, 645, 337
Issaoun, S., Goddi, C., Matthews, L. D., *et al.* 2017, *A&A*, 606, 126
Matthews, L. D., Greenhill, L. J., Goddi, C., *et al.* 2010, *ApJ*, 708, 80
Moscadelli, L. & Goddi, C. 2014, *A&A*, 566, A150
Moscadelli, L., Sánchez-Monge, Á., Goddi, C., *et al.* 2016, *A&A*, 585, A71
Reid, M. J., Argon, A. L., Masson, C. R., Menten, K. M., & Moran, J. M. 1995, *ApJ*, 443, 238
Sanna, A., Surcis, G., Moscadelli, L., *et al.* 2015, *A&A*, 583, L3
Surcis, G., Vlemmings, W. H. T., van Langevelde, H. J., *et al.* 2014, *A&A*, 565, L8
Surcis, G., Vlemmings, W. H. T., van Langevelde, H. J., *et al.* 2015, *A&A*, 578, A102
Wilner, D. J., Reid, M. J., & Menten, K. M. 1999, *ApJ*, 513, 775
Wyrowski, F., Schilke, P., Walmsley, C. M., & Menten, K. M. 1999, *ApJ* (Letters), 514, L43

Astrophysical Masers:
Unlocking the Mysteries of the Universe
Proceedings IAU Symposium No. 336, 2017
A. Tarchi, M.J. Reid & P. Castangia, eds.

© International Astronomical Union 2018
doi:10.1017/S174392131701050X

A golden age for maser surveys

Shari L. Breen

Sydney Institute for Astronomy (SIfA), School of Physics, University of Sydney, NSW 2006,
Australia
email: `shari.breen@sydney.edu.au`

Abstract. Masers are becoming increasingly important probes of high-mass star formation, revealing details about the kinematics and physical conditions at the elusive, early stages of formation. Over the last decade significant investment has been made in a number of large-scale, sensitive maser surveys targeting transitions found in the vicinity of young, high-mass stars. Individually, these searches have led to valuable insights into maser populations, their associated star formation regions, and often revealed further details such as Galactic structure. In combination, they become even more powerful, especially when considered together with complementary multi-wavelength data. Another consequence of large maser surveys has been the identification of a number of especially interesting sources that have been the subject of subsequent detailed studies. I summarize the recent plethora of maser surveys, their results, and how they are contributing to our understanding of star formation. Ongoing searches will ensure a bright future of maser surveys in the decade to come.

Keywords. masers, stars: formation, ISM: molecules

1. Introduction

In recent years, maser surveys have been able to capitalise on increased sensitivity of radio telescopes, allowing for large-scale, sensitive searches for a number of maser transitions that are commonly associated with star formation regions. We are now able to routinely cover large sections of the Galaxy and achieve high sensitivities and spatial resolutions, either in the initial survey or in follow up observations. Such surveys have allowed for meaningful population statistics and revealed new 'special' sources that have become the subjects of intensive targeted observations. Large-scale systematic searches are essential to our understanding of star formation and the masers themselves.

Here I summarise five significant surveys at various stages of completion, conducted in the southern hemisphere. They target a number of different maser transitions, including the ground-state OH masers, 6035 MHz excited-state OH masers, 6.7 GHz methanol masers, 22 GHz water masers and 44 GHz methanol masers. Particularly in the case of the Methanol Multibeam (MMB) survey, a plethora of targeted observations have also been conducted towards the large sample of 6.7-GHz methanol masers and these are mentioned in Section 2.1.3.

2. The surveys

2.1. The Methanol Multibeam (MMB) survey and follow-up observations

The Methanol Multibeam (MMB) survey was an unbiased, systematic search for both 6.7 GHz methanol masers and 6035 MHz excited-state OH masers within the Southern Galactic plane. The initial survey observations were made with the Parkes 64 m radio telescope, allowing 65% of the Galactic plane to be searched, covering a longitude range of 186°, through the Galactic Centre (GC), to 60° and latitudes of ±2°, as well as both the Small and Large Magellanic Clouds. The survey was completed in scanning mode with a purpose-built 7-beam receiver which dramatically reduced the time required to complete the observations. In the end it took 120 days of observing time to conduct the initial survey which was chiefly allocated between 2006 and 2009. All detections without previously derived precise positions, of both the methanol and excited-state OH maser transitions, were subsequently followed up with high-resolution observations using either the Australia Telescope Compact Array or the Multi-Element Radio Linked Interferometer Network (MERLIN) resulting in positions accurate to 0.4 arcsec. The MMB survey is the largest, sensitive, systematic search for 6.7 GHz methanol and 6035 MHz excited-state OH masers ever conducted in the Galactic plane. A full description of the survey parameters can be found in Green *et al.* (2009).

2.1.1. *6.7-GHz methanol masers*

Methanol masers at 6.7 GHz are exclusively associated with sites of high-mass star formation (e.g. Minier *et al.* 2003, Xu *et al.* 2008, Breen *et al.* 2013) making them especially useful probes of these elusive regions. A total of 972 methanol masers were detected across the southern portion of the Galaxy (see Fig. 1) to a 3-σ detection limit of $\sim$0.51 Jy and are presented in a series of five catalogue papers (Caswell *et al.* 2010, Green *et al.* 2010, Caswell *et al.* 2011, Green *et al.* 2012a, Breen *et al.* 2015). Approximately 40% of the detections were discovered in the survey, and, combined with the known sources, have proved to be important targets for other types of masers, parallax observations and detailed multi-wavelength studies.

Recently, Green *et al.* (2017) presented the overall statistics of the MMB maser population, focusing on the 6.7 GHz methanol detections. They found that:

- The 972 detections implied a total Galactic 6.7 GHz methanol maser population of $\sim$1290 sources above the 3-σ detection limit of $\sim$0.51 Jy;
- 51 sources surpassed 100 Jy, 330 surpassed 10 Jy and the strongest detection was G 9.621+0.196 at 5200 Jy. 118 sources had peak flux densities less than 1 Jy;
- Velocity extents of the detected emission ranged from 0.3 km s^{-1} for G 37.767$-$0.214 and 28.5 km s^{-1} for G 305.475$-$0.096 and had a median value of 6.0 km s^{-1};
- 7% of the MMB sample shows variability of a factor of two, and weaker sources show the greatest levels of variability;
- Lower luminosity sources have smaller velocity ranges compared to higher-luminosity sources;
- There was evidence that brighter sources were seen towards arm origins.

2.1.2. *Excited-state OH masers*

The MMB survey also included the 6035 MHz excited-state OH maser transition, detecting a total of 127 sources, 47 of which were new to the survey (see Avison *et al.* 2016 for full details). All detections without previously derived precise positions were observed with the ATCA, and, combined with the 6.7 GHz methanol maser observations, allowed Avison *et al.* (2016) to determine that 52% of their 6035 MHz excited-state OH maser detections fell within 2 arcsec of a 6.7 GHz methanol maser detected in the MMB survey. Follow up spectra taken with the Parkes telescope also allowed for sensitive observations of both the 6035 and 6030 MHz transitions, resulting in the detection of 32 6030 MHz OH masers.

Since a number of the 6035 MHz excited-state OH masers had been presented previously in the literature, Avison *et al.* (2016) was able to assess the temporal variability of these sources. Through comparisons with observations made by Caswell & Vaile (1995) and Caswell (2003), Avison *et al.* (2016) were able to determine that 20% of the detected

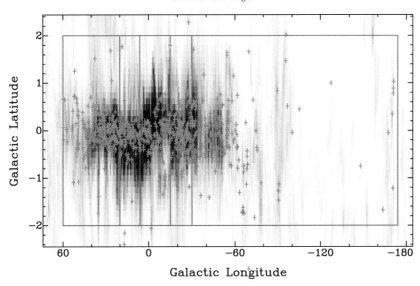

Figure 1. All 972 6.7 GHz methanol masers (crosses) detected in the MMB survey overlaid on CO emission from Dame, Hartmann, & Thaddeus (2001). The large rectangle indicates the full extent of the survey and the internal vertical lines indicate the boundaries of each of the survey catalogues.

6035 MHz OH masers had varied by more than a factor of two on timescales of seven or more years. A total of 12 sources that were previously presented in the literature where unable to be confidently detected in the MMB survey.

Full polarisation data of the 6035 MHz sources detected in the MMB survey, allowing the consideration of Zeeman pairs and the inferred magnetic field strengths and directions of sources across the Galaxy, will be presented in a forthcoming paper (Avison *et al.* in prep).

2.1.3. *Follow-up observations at 12.2-GHz methanol, 22-GHz water and ground-state OH*

Each of the 6.7 GHz methanol masers detected in the MMB survey were subsequently targeted for accompanying 12.2 GHz methanol maser emission using the Parkes radio telescope over four seperate observing sessions. A total of 431 12.2 GHz methanol masers were detected (and presented in four seperate catalogue papers; Breen *et al.* 2012a,b, 2014, 2016), equating to a detection rate of 45.3 %. The flux density distribution of the 12.2 GHz sources ranged from 0.3 to 976 Jy and only 11 sources surpassed peak flux densities of 100 Jy. From this work it was proposed that 12.2 GHz methanol masers were present in the second half of the 6.7 GHz methanol maser lifetime and hence trace slightly more evolved star formation regions.

In other follow-up work, Titmarsh *et al.* (2014, 2016) used the ATCA to target all of the 6.7 GHz MMB sources in the Galactic longitude range 341° (through the GC) to 20°. With a 5-σ detection limit of $\sim$250 mJy or better, 156 of the 323 6.7 GHz methanol masers in this longitude range were found to have associated water maser emission, equating to a detection rate of 47%. They further found that the water maser detection rate was slightly higher towards MMB sources that had accompanying 12.2 GHz methanol maser emission. A simple interpretation of this might suggest that water masers are therefore also seen at a slightly later stage of high-mass star formation than 6.7-GHz methanol masers, but both Breen *et al.* (2016) and Titmarsh *et al.* (2016) argue that water masers do not follow a simple evolutionary scheme, and the fact that they are collisionally pumped means that their presence is more often dependant on the interaction of the star formation region with the surrounding environment.

Follow up ground-state OH maser observations have also been conducted towards the MMB sources as part of the MAGMO project (the project to study the Magnetic fields of the Milky Way through OH masers), using the ATCA. This project simultaneously observes 1612, 1665, 1667 and 1720 MHz OH masers and the results of pilot observations are presented in Green *et al.* (2012b).

2.2. *H_2O Southern Galactic Plane Survey (HOPS)*

HOPS surveyed a large section of the southern Galactic Plane ($290° <$ longitude $< 30°$ and latitudes $\pm 0.5°$) for a number of spectral lines near 20 GHz, including the 22 GHz water maser line (Walsh *et al.* 2011). The initial survey was conducted with Mopra and identified 540 maser sites but had a slightly variable RMS across the survey region, resulting in a detection limit that was around 2 Jy, but up to 10 Jy in some areas. All water maser detections were followed up with high-resolution, high-sensitivity ATCA observations and resulted in the identification of 631 maser sites (Walsh *et al.* 2014). Comparison between the water maser positions with complementary data revealed that 433 of the 631 maser sites were associated with star formation (69% of sites). Data products from the HOPS survey are available at https://research.science.mq.edu.au/hops/public/index.php.

Walsh *et al.* (2014) found, as was the case in the MMB survey, that the highly variable water masers tended to be weaker. They also found that the water maser sites that were associated with star formation had fewer spots than those associated with evolved stars, but that the spots within a site were distributed further in the case of the star formation masers. They also found that a small number of sources showed linear spot distributions, arranged both parallel and perpendicular to the orientation of associated outflows.

Forthcoming high-sensitivity water maser surveys such as SWAG (Survey of Water and Ammonia in the Galactic Center; see Ott *et al.* in these proceedings) and RAMPS (Radio Ammonia Mid-Plane Survey; Hogge *et al.* in prep) will reveal the weaker water maser population that fell below the detection limit of HOPS.

2.3. *Southern Parkes Large-Area Survey in Hydroxyl (SPLASH)*

SPLASH is an unbiased, fully-sampled, large-scale OH and 1.6 - 1.7 GHz radio continuum survey of the Milky Way (Dawson *et al.* 2014). While the primary science driver is the diffuse OH emission, the sensitivity and the Galactic coverage of this survey make it one of the most extensive, sensitive maser surveys for all four ground state transitions of OH (1612, 1665, 1667, 1720 MHz). The survey covers 156 square degrees, extending from Galactic longitude $332°$, through the GC to $8°$ with a latitude coverage of $\pm 2°$.

The initial Parkes survey has low spatial resolution (HPBW of 15 arcmin) but high sensitivity (rms of 65 mJy in a 0.18 km s^{-1} channel). All detections were re-observed with the ATCA in order to derive precise positions and achieve full polarisation observations. The high-resolution maser results from the SPLASH pilot region ($334° < l < 344°$, latitudes $\pm 2°$) detected a total of 215 OH maser sites, 111 of which were new detections and 64 sites (or 30%) were associated with star formation (Qiao *et al.* 2016). Future publications will detail the detections in the remaining longitude regions and present the full polarisation data, including the analysis of Zeeman splitting.

2.4. *MALT45*

Using an innovative survey strategy, MALT45 (Millimetre Astronomers Legacy Team 45 GHz) systematically surveyed five square degrees of the Galactic plane for a number

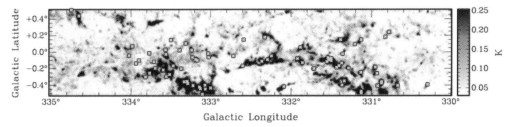

Figure 2. CS (1−0) map overlaid with the positions of 77 44 GHz class I methanol masers (crosses) from Jordan *et al.* (2015)

of 7 mm spectral lines, including the 44 GHz class I methanol maser, as well as SiO masers (Jordan *et al.* 2015). This initial survey observations were conducted with the ATCA, but in autocorrelation mode which resulted in a HPBW of ∼1 arcmin. From these observations, Jordan *et al.* (2015) identified 77 44 GHz class I methanol masers which are overlaid on a CS map (made in the same survey) in Fig. 2. Of the 77 class I methanol masers detected, 58 were new detections and 42 were located at sites devoid of other maser species. This survey achieved an rms noise of 0.9 Jy in a 0.2 km s^{-1} channel, and covered Galactic longitudes of 330° to 335° and latitudes ±0.5°.

Jordan *et al.* (2017) conducted further observations, made with the ATCA in a standard interferometric mode, towards all of the detected masers. In these observations they were able to identify 238 maser spots across 77 maser sites, and by comparing their cross-correlation data to auto-correlation data they were able to investigate the spatial scales of the emission regions and identify quasi-thermal contributions in the 44 GHz class I methanol maser line. They present comparisons between their masers with other star formation tracer such as CS (1−0), SiO v = 0 and the H53α radio-recombination line, along with dust continuum emission from the ATLASGAL survey (Contreras *et al.* 2013) and other maser species. Among other things, they find that the 44 GHz methanol masers without accompanying OH masers or recombination lines have lower luminosities.

2.5. *Dense Gas Across the Milky Way - The Full-Strength MALT45*

Building on the success of the MALT45 survey, we have begun a 2700h ATCA legacy survey to map 90 square degrees of the southern Galactic Plane at 7 mm. Like the initial MALT45 survey, this expanded survey will target a host of dense gas, shock tracers, recombination lines, 44 GHz class I methanol masers and SiO masers. The observations are conducted in auto-correlation mode and use on-the-fly mapping to allow us to cover a large portion of the Southern Galactic plane. Combined with existing MALT45 data, we will provide full coverage of the longitude range 270°, through the GC to 5° and latitudes of ±0.5°. Observations began in May 2017 and have already revealed many new sites of class I methanol maser emission which we hope to publish by mid-2018.

3. Summary

In the last five to 10 years there has been a wealth of large-scale, sensitive maser surveys conducted within the Galactic plane, specifically targeting masers that are associated with star formation regions. Together with surveys that are ongoing, they represent a huge resource for studies of maser emission and the star formation regions that they are associated with. While the surveys I have described are those which I have been involved with (and so are biased towards those conducted from the southern hemisphere) there are a number of other exciting maser surveys. In particular I note the water maser surveys SWAG and RAMPS, while limited to smaller regions of the Galaxy compared to HOPS, are much more sensitive and will provide very complementary information, particularly revealing the weaker population of water masers within our Galaxy. Other, large surveys such as THOR (The Hi/OH/Recombination line survey of the Milky Way), GLOSTAR (Global view of star formation in the Milky Way) and KuGARS (Ku-band GAlactic Reconnaissance Survey) will nicely complement the southern surveys that I have described and will ensure that the golden age for maser surveys extends well into the future.

The success of the maser surveys I have described is largely due to the wisdom of Jim Caswell, who will continue to inspire large-scale maser surveys, and those conducting them, for many years to come.

References

Avison A., Quinn L., Fuller G. A., *et al.*, 2016, *MNRAS*, 461, 134
Breen S. L., Ellingsen S. P., Caswell J. L., Green J. A., Voronkov M. A., Fuller G. A., Quinn L. J., & Avison A., 2012a, *MNRAS*, 421, 1703
Breen S. L., Ellingsen S. P., Caswell J. L., Green J. A., Voronkov M. A., Fuller G. A., Quinn L. J., & Avison A., 2012b, *MNRAS*, 426, 2189
Breen, S. L., Ellingsen S. P., Contreras Y., Green J. A., Caswell J. L., Stevens J. B., Dawson J. R., & Voronkov M. A., 2013, *MNRAS*, 435, 524
Breen, S. L., Ellingsen S. P., Contreras Y., Green J. A., Caswell J. L., Stevens J. B., Dawson J. R., & Voronkov M. A., 2013, *MNRAS*, 435, 524
Breen S. L., Ellingsen S. P., Caswell J. L., Green J. A., Avison A., Voronkov M. A., Fuller G. A., Quinn L. J., & Titmarsh, A., 2014, *MNRAS*, 438, 3368
Breen S. L., Fuller G. A., Caswell J. L., *et al.*, 2015, *MNRAS*, 450, 4109
Breen S. L., Ellingsen S. P., Caswell J. L., *et al.*, 2016, *MNRAS*, 459, 4066
Caswell J. L., Vaile R. A., 1995, *MNRAS*, 273, 328
Caswell J. L., 2003, *MNRAS*, 341, 551
Caswell J. L., Fuller G. A., Green J. A., *et al.*, 2010, *MNRAS*, 404, 1029
Caswell J. L., Fuller G. A., Green J. A., *et al.*, 2011, *MNRAS*, 417, 1964
Contreras Y., Schuller F., Urquhart J. S., *et al.*, 2013, *A&A*, 549, A45
Dame T. M., Hartmann D., & Thaddeus P., 2001, *ApJ*, 547, 792
Dawson J. R., Walsh A. J., Jones P. A., *et al.*, 2014, *MNRAS*, 438, 1596
Green J. A., Caswell J. L., Fuller G. A., *et al.*, 2009, *MNRAS*, 392, 783
Green J. A., Caswell J. L., Fuller G. A., *et al.*, 2010, *MNRAS*, 409, 913
Green J. A., Caswell J. L., Fuller G. A., *et al.*, 2012a, *MNRAS*, 420, 3108
Green J. A., McClure-Griffiths N. M., Caswell J. L., Robishaw T., & Harvey-Smith L., 2012b, *MNRAS*, 425, 2530
Green J. A., Breen S. L., Fuller G. A., *et al.*, 2017, *MNRAS*, 469, 1383
Jordan C. H., Walsh A. J., Lowe V., *et al.*, 2015, *MNRAS*, 448, 2344
Jordan C. H., Walsh A. J., Breen S. L., *et al.*, 2017, *MNRAS*, 471, 3915
Minier V., Ellingsen S. P., Norris R. P., Booth R. S., 2003, *A&A*, 403, 1095
Qiao H.-H., Walsh A. J., Green J. A., *et al.* 2016, *ApJS*, 227, 26
Titmarsh A. M., Ellingsen S. P., Breen S. L., Caswell J. L., Voronkov M. A., 2014, *MNRAS*, 443, 2923
Titmarsh A. M., Ellingsen S. P., Breen S. L., Caswell J. L., Voronkov M. A., 2016, *MNRAS*, 459, 157
Walsh A. J., Breen S. L., Britton T., *et al.*, 2011, *MNRAS*, 416, 1764
Walsh A. J., Purcell C. R., Longmore S. N., *et al.*, 2014, *MNRAS*, 442, 2240
Xu Y., Li J. J., Hachisuka K., Pandian J. D., Menten K. M., & Henkel C., 2008, *A&A*, 485, 729

Astrophysical Masers:
Unlocking the Mysteries of the Universe
Proceedings IAU Symposium No. 336, 2017
A. Tarchi, M.J. Reid & P. Castangia, eds.

© International Astronomical Union 2018
doi:10.1017/S174392131800011X

Periodic masers in massive star forming regions

S. Goedhart[1], R. van Rooyen[1], D. J. van der Walt[2],
J. P. Maswanaganye[2], G. C. MacCleod[3] and A. Sanna[4]

[1]SKA SA, The Park, Park Rd, Pinelands, Cape Town, 7405, South Africa
email: sharmila@ska.ac.za

[2]Space Research Unit, Physics Department, North-West University, Potchefstroom, South Africa

[3] Hartebeesthoek Radio Astronomy Observatory, PO Box 443, Krugersdorp, 1740, South Africa

[4]Max-Planck-Institut für Radioastonomie, Auf dem Hügel 69, D-53121 Bonn, Germany

Abstract. The first periodic Class II methanol maser was reported on in 2003. Since that time, a number of different monitoring programmes have found periodic masers, as well as other modes of variability. In a few cases, periodicity has been found in other maser species such as formaldehyde and water. Several distinct characteristics of light curves have been noted, possibly pointing to different underlying mechanisms for periodicity if one assumes a linear response to incoming radiation. I will give a brief overview of the known periodic sources, discuss current theories, and present new results obtained from monitoring mainline hydroxyl masers using the seven-element Karoo Array Telescope (KAT-7) during its science verification phase.

Keywords. masers, stars: formation, ISM: molecules, radio lines: ISM

1. Introduction

The first periodic masers were found by Goedhart, Gaylard & van der Walt (2004) while comducting a large scale monitoring programme of 6.7 GHz methanol masers. An extension of the programme found correlated variations at 12.2 GHz, in cases where the masers were detected (Goedhart *et al.* 2014). In other studies, correlated periodic variability in other maser molecules was found in a single source in formaldehyde (Araya *et al.* 2010), Green & Caswell (2012) found a faint hint that the hydroxyl maser in G12.89-0.49 may undergo a dip in flux density at the onset of the methanol maser flare, while anti-correlated variability in a water maser compared to methanol in G107.298+5.639 was recently reported by Szymczak *et al.* (2016). Two more large scale monitoring programmes have been reported on during this conference (Sugiyama *et al.*, Szymczak *et al.*, this volume), bringing the number of periodic masers now known to exceed 50. Thus periodic variability may not be as unusual as initially thought.

The sample of periodic masers show a range of profiles, many of which are similar to each other. By analogy with optical variability studies, one could argue that the shape of the light curves could hint to the origin of the periodic modulation, if there is a direct relation of the maser output to the incoming radiation. A number of mechanisms have been proposed to explain the periodic masers. This includes

• Colliding-wind binary system producing shocks which leads to increased ionisation of the background HII region (van der Walt, 2011).

• Dust temperature variations in an accretion disk around a binary system, modulating the pump rate (Parfenov & Sobolev, 2014).

- Pulsational instability in a bloated protostar during rapid accretion (Inayoshi *et al.* 2013).
- Eclipsing binary system modulating the infrared radiation (Maswanganye *et al.* 2015).

Thus far, no conclusive way has been found to verify any of these hypotheses, but monitoring at other wavelengths, particularly of the radio continuum or the infrared, may help solve this question.

Hydroxyl and methanol masers are thought to have a common pump mechanism (Cragg *et al.* 2002) but, up to now, no conclusive evidence has been found of correlated variability in methanol and hydroxyl. A sample of seven periodic methanol maser sources were monitored in the hydroxyl main lines using the KAT-7 telescope, while the methanol and water transitions were monitored using the HartRAO 26m telescope. Thus far only two of the sources have been analysed and have been confirmed to show periodic variability in the hydroxyl masers. Here we present the results on G9.62+0.20E, which was the first periodic source discovered. The masers towards G9.62+0.20E have been accurately mapped relative to each other and the background HII region (Sanna *et al.* 2015), giving us a unique opportunity to study the relation of the various maser species given high cadence monitoring of as many transitions as possible.

2. Observations

The 7-dish Karoo Array Telescope (KAT-7) (Foley *et al.* 2016), built as an engineering prototype for the 64-dish MeerKAT Array, consists of 12m prime focus dishes equiped with L-band receivers covering 12.2 to 1.95 GHz. The shortest and longest baselines are 26m and 186 m, respectively. The system temperature, when the cryostats were functional, was approximately 30K, and the apperture efficiency is on average 65%. The OH maser observations were done as part of the KAT-7 science verification programme. We used the narrowest correlator mode, which gives a velocity resolution of 68 m s^{-1}. Observations were run at both 1665 and 1667 MHz, from February 2013 to June 2015. The typical rms noise achieved ranged from 0.15 to 0.2 Jy. The typical beam size (not all antennas were always available) was $\sim$ 3 arcmin, thus the masers are unresolved and relative positions cannot be measured since all spots are located within a single beam.

The 26m dish at Hartebeesthoek Radio Astronomy Observatory was used to concurrently monitor the associated methanol masers at 6.7 and 12.2 GHz, and the water masers at 22 GHz.

Observations were typically done on a weekly basis, with increased cadence when flares were expected.

3. Results

Figure 1 shows the spectra for each of the transitions monitored, with an indication of the range of variation by plotting the minimum, maximum and mean of the timeseries in each channel. The methanol and hydroxyl masers show periodic flaring behaviour in the same velocity ranges, while the water masers cover a different, slightly higher velocity range and, while variable, appear to be uncorrelated with the other two species.

Figure 2 shows the time-series for the strongest maser channel in each transition. Note that these results are for Stokes I only due to problems encountered in polarisation calibration of KAT-7. The most notable feature in the hydroxyl masers is the drop in flux density, which occurs at the same time as the start of the flare in methanol. The hydroxyl masers show a drop in intensity over a period of about 5 days, after which they continue

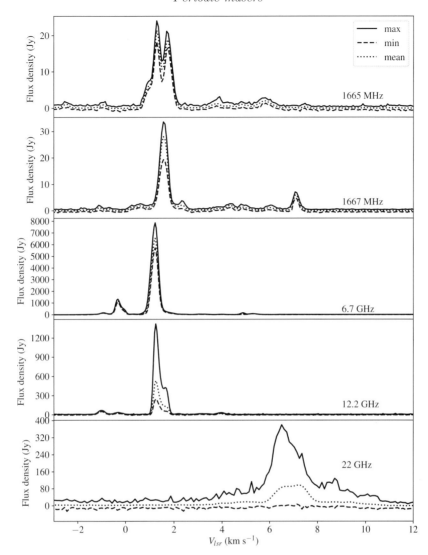

Figure 1. Spectra of the maser transitions monitored, showing the range of variation over the monitoring period, in each channel.

to increase in flux density, with the 1665 MHz transition peaking between $\sim$ 7–10 days after the methanol masers, which flare and peak simultaneously in both transitions, while the 1667 MHz masers peak $\sim$ 20 days later. The amplitude of the variation is higher in the 1667 MHz transition than for 1665 MHz. The 1667 MHz masers also show a steady rise in the baseline flux density over the course of the monitoring programme. Not all of the hydroxyl maser features behave in the same way. Some velocity features show very pronounced dips while others show stronger flaring behaviour and not as much of a dip (see Figure 3). Some features, which are spatially offset from the HII regions, do not show variations. The periodic variations occur only in the velocity ranges 1.2 to 1.8 km s^{-1} and 2.12 to 2.4 km s^{-1}. Examination of the VLBI spot maps and spectra indicates that the periodically flaring hydroxyl masers are situated to the north-east of the methanol masers, close to the HII region component E2 (as designated by Sanna *et al.* 2015). The

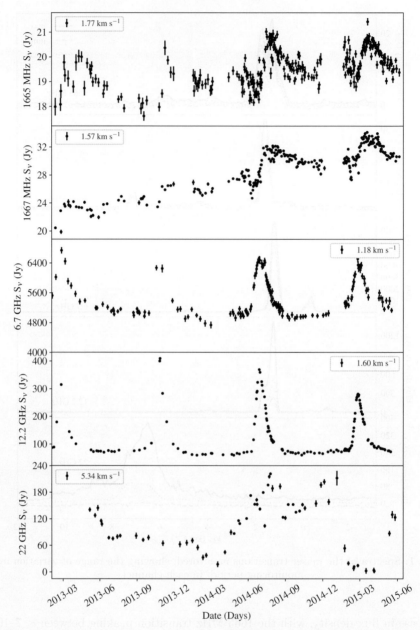

Figure 2. Comparison of the dominant velocity channel flux density as a function of time for each transition monitored.

hydroxyl masers to the west and south of the methanol maser group do not show periodic variability.

4. Discussion

The monitoring programme using KAT-7 covered four cycles of periodic flaring of G9.62+0.20. The first two cycles were not very well sampled but showed clear variability. The subsequent two cycles were then followed at a higher cadence, and a pronounced

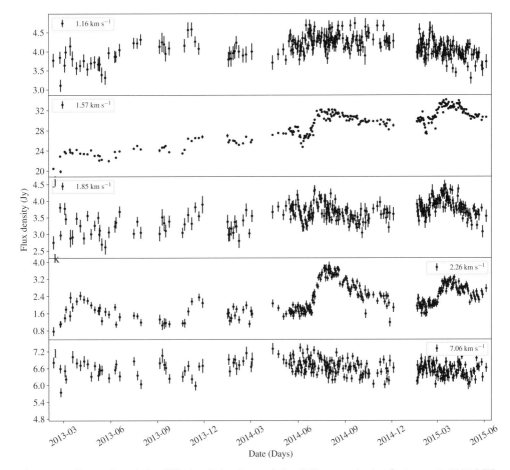

Figure 3. Examples of the differing behaviour of the different velocity features at 1667 MHz.

drop in the hydroxyl maser flux density is seen over a week, coinciding with the start of the flare in methanol. The hydroxyl masers are located approximately 300 mas from the methanol masers, which at a distance of 5.2 kpc corresponds to a projected distance of 1560 AU or 9 light days. For the masers to see an effect simultanously the source of the periodic phenomenon would have to be situated midway between the two locations, or the methanol masers are closer to us along the line of sight, as would be seen on a sphere. In any case, it is clear that the hydroxyl and methanol do not respond in the same way to the incoming impulse, whether it be in the background continuum flux or pump photons. The water masers, while they are variable, do not show any obvious correlated variability.

Maswanganye *et al.* (this volume, and PhD thesis) show that hydroxyl masers can react very differently to the same thermal profile, depending on the column density of the particular maser. Thus these very detailed pulse profiles could potentially be used to infer specific conditions in the maser locales.

5. Conclusions and recommendations

These results show that hydroxyl masers associated with periodic methanol maser sources can also show periodic variability, but there may not necessarily be a one-to-one correspondence in pulse profiles. Much more work needs to be done in simulating the radiative transfer in the different proposed mechanisms. It will also be helpful to increase the sample of masers monitored at multiple transitions, while having high angular resolution maps of their relative positions, in order to build up a global picture of the conditions around the young stellar object.

References

Araya, E., Hofner, P., Goss, W. M., Kurtz, S., Richards, a. M. S., Linz, H., Olmi, L., & Sewi,
    lo M., 2010, *ApJ*,717, L133
Cragg, D. M., Sobolev, A. M., & Godrey, P. D., 2002, *MNRAS*, 331, 52
Foley, A. R., Alberts, T., Armstrong, R. *et al.* 2016, *MNRAS*, 460, 1664
Goedhart, S., Gaylard, M. J. & van der Walt, D. J., 2004, *MNRAS*, 355, 553
Goedhart, S., Maswanganye, J. P., Gaylard, M. J., & van der Walt, D. J., 2014, *MNRAS*, 437,
    1808
Green, J. A. & Caswell, J. L., 2012, *MNRAS*, 425, 1504
Inayoshi, K., Sugiyama, K., Hosokawa, T., Motogi, K., & Tanaka, K. E. I., 2013, *ApJ*, 769, L20
Maswanganye, J. P., Gaylard, M. J., Goedhart, S., van der Walt, D. J., & Booth, R. S., 2015,
    MNRAS, 446, 2730
Parfenov, S. Y. & Sobolev, A. M., 2014, *MNRAS*, 444, 620
Sanna, A., Menten, K. M., Carrasco-González, C., Reid, M. J., Ellingsen, S. P., Brunthaler, A.,
    Moscadelli, L., Cesaroni, R., & Krishnan V., *ApJ*, 804, L2
Szymczak, M., Olech, M., Wolak, P., Bartkiewicz, A., & Gawroński M., 2016, *MNRAS*, 459, L56
van der Walt, D. J., 2011, *AJ*, 141, 152

Astrophysical Masers:
Unlocking the Mysteries of the Universe
Proceedings IAU Symposium No. 336, 2017
A. Tarchi, M.J. Reid & P. Castangia, eds.

© International Astronomical Union 2018
doi:10.1017/S1743921318000121

The CepHeus-A Star formation and proper Motions (CHASM) Survey

Alberto Sanna

Max-Planck-Institut für Radioastronomie,
Auf dem Hügel 69, 53121 Bonn, Germany
email: asanna@mpifr-bonn.mpg.de

Abstract. The "CepHeus-A Star formation and proper Motions" (CHASM) survey is a large project consisting of a combination of astrometric Very Long Baseline Array (VLBA) and Jansky Very Large Array (VLA) observations, to map both the stellar and dense molecular gas components in the star-forming region Cepheus A. With the VLBA, we make use of the CH_3OH and H_2O maser emission in the vicinity of Cepheus A HW2, in order to measure accurate proper motions and parallax distances to both T Tauri stars and massive young stellar objects (YSOs) belonging to the same star-forming region. With the Jansky VLA, we make use of the interstellar thermometer NH_3, in order to image the molecular clump surrounding Cepheus A HW2 and to determine its physical conditions. By combining these informations all together, we can provide, for instance, a direct measurement of the Bondi-Hoyle accretion radius for a massive young star, namely, HW2.

Keywords. surveys, astrometry, ISM: clouds, ISM: kinematics and dynamics, stars: individual (Cepheus A HW2)

Overview

Models of star formation predict that the interplay between the clump gas, and the young stellar objects (YSOs) which form inside the clump, might be fundamental to aid the formation of the most massive objects and to set the stellar initial mass function (e.g., Krumholz *et al.* 2014). With the aim to provide a detailed analysis of how young stars influence their environment (and vice versa), we have initiated the "CepHeus-A Star formation and proper Motions" (CHASM) survey, which makes use of a combination of Jansky Very Large Array (VLA) and astrometric Very Long Baseline Array (VLBA) observations. The target of this study is the massive star-forming region Cepheus A (Figure 1), which, at a trigonometric distance of 700 pc from the Sun (Moscadelli *et al.* 2009; Dzib *et al.* 2011), is the second nearest after Orion. In the following, we highlight the objectives of the VLBA observations.

Astrometric Very Long Baseline Interferometry (VLBI) measurements of both, compact spectral line emission (maser) in the vicinity of massive YSOs, and radio continuum emission from chromospherically active YSOs (low-mass, T Tauri stars), provide us with parallax measurements typically accurate to about $\pm 10 \, \mu$as, and yield measurements of secular proper motions with accuracies as good as a few $0.1 \, \mathrm{km \, s^{-1}}$, for sources within a few kpc (e.g., Reid & Honma 2014; Loinard *et al.* 2011). When applied to stars distributed across the same region, these observations allow us to measure *relative* distances and proper motions of gravitationally bound YSOs, and their 3D distribution and the internal kinematics of the region can be reconstructed.

With this in mind, we are conducting multi-epoch, phase referencing, VLBA observations of three distinct tracers towards Cepheus A: the 6 GHz radio continuum emission from T Tauri stars surrounding HW2 (see below); the 6.7 GHz CH_3OH maser emission

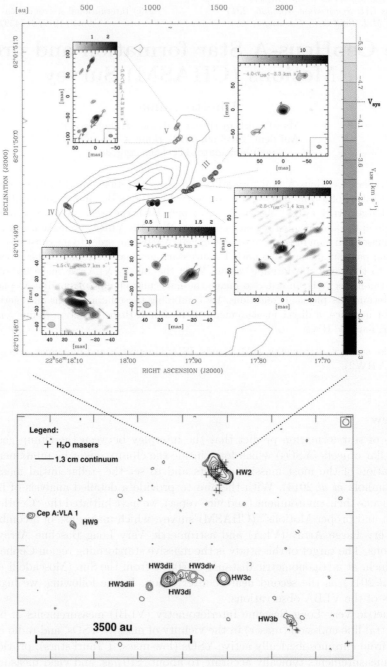

Figure 1. Lower panel: map of the 1.3 cm continuum emission (contours) of the central region of Cepheus A from Torrelles *et al.* (1998). Main sources are labeled according to the literature. Crosses indicate the positions of the 22.2 GHz H_2O masers detected in the region. Liner scale in the bottom left corner **Upper panel:** 6.7 GHz CH_3OH maser distribution (circles) in the vicinity of the HW2 object (star) from Sanna *et al.* (2017). Maser colors give the local LSR velocity according to the right-hand scale (see the on-line version). Linear scale on the upper axis. Individual insets zoom in onto each maser group and show the morphology of the local maser emission. Red arrows trace the proper motions vectors of the maser cloudlets. More details can be found in Figure 2 of Sanna *et al.* (2017).

within 1000 au of HW2 (Figure 1, upper); the 22.2 GHz H_2O maser emission clustered within 5000 au of HW2 (Figure 1, lower). HW2 is the most massive YSO which contributes half the bolometric luminosity of the region of 2–3×10^4 $L_\odot$ (e.g. De Buizer *et al.* 2017).

At C band, we are targeting simultaneously the T Tauri stars and the CH_3OH masers; the former will be searched for blindly, across a large field of radius >0.2 pc centered on HW2, whereas the latest are associated with HW2 solely (e.g., Torstensson *et al.* 2011; Sugiyama *et al.* 2014; Sanna *et al.* 2017). Based on the YSOs population density in a low-mass star-forming region such as Ophiuchus, and taking into account that massive stars grow up in a much more crowded environment, we expect to find more than 40 YSOs spread over an area of $(0.2 \,\text{pc})^2$. Note that, Cepheus A is the closest star-forming region showing bright 6.7 GHz CH_3OH maser emission, and this property can be used, in particular, to improve on the detection of faint ($<$ mJy) radio continuum emission coming from young T Tauri stars. On the other hand, H_2O masers are the most suited target to undertake a systematic measurement of distances to high-mass components in Cepheus A. In the literature, tens of maser spots, with intensities as high as several 10 Jy, have been observed to cluster around 5 distinct YSOs (Figure 1, lower): HW2–main, and sub-regions R 4 and R 5 (e.g., Torrelles *et al.* 2001; Torrelles *et al.* 2011), and radio components HW3d (e.g., Chibueze *et al.* 2012) and HW3b (e.g., Torrelles *et al.* 1998).

The immediate goal of the VLBA survey is to measure both accurate proper motions and (absolute and relative to HW2) parallaxes to low- and high-mass YSOs surrounding HW2, as well as to map the young low-mass population in the region. Relative distances allow us to constrain the position of the young stars with respect to HW2 within a few pc (1 pc $=$ 2 μas at 700 pc), in order to prove that these YSOs belong to the same clump (a few pc in size). Having on hand the proper motion measurements of tens of YSOs belonging to Cepheus A, we can study the magnitude and direction of their (3D) velocity vectors with respect to HW2, and vice versa (e.g., Rivera *et al.* 2015). Note that the line-of-sight velocity component will be inferred from complementary observations of the local gas emission (through VLA observations). The full-space motion of HW2 inside the clump, combined with the kinematics of the clump gas, can be used to measure the Bondi-Hoyle accretion radius (r_{BH}) of HW2 directly. The Bondi-Hoyle accretion radius sets the outer radius of the mass reservoir for HW2, and depends on the relative velocity (v_{rel}) of the star with respect to the clump gas ($r_{BH} \propto v_{rel}^{-2}$). Knowing r_{BH} would eventually tell us whether HW2 can accrete mass (mainly) from its immediate surroundings (i.e., the pre-stellar core, $\ll 0.1$ pc), or its final mass can be clump-fed (i.e., the mass reservoir is supplied outside a radius $\geqslant 0.1$ pc from the star).

References

Chibueze, J. O., Imai, H., Tafoya, D., *et al.* 2012, *ApJ*, 748, 146

De Buizer, J. M., Liu, M., Tan, J. C., *et al.* 2017, *ApJ*, 843, 33

Dzib, S., Loinard, L., Rodríguez, L. F., Mioduszewski, A. J., & Torres, R. M. 2011, *ApJ*, 733, 71

Krumholz, M. R., Bate, M. R., Arce, H. G., *et al.* 2014, *Protostars and Planets VI*, 243

Loinard, L., Mioduszewski, A. J., Torres, R. M., *et al.* 2011, *Revista Mexicana de Astronomia y Astrofisica Conference Series*, 40, 205

Moscadelli, L., Reid, M. J., Menten, K. M., *et al.* 2009, *ApJ*, 693, 406

Reid, M. J. & Honma, M. 2014, *ARA&A*, 52, 339

Rivera, J. L., Loinard, L., Dzib, S. A., *et al.* 2015, *ApJ*, 807, 119

Sanna, A., Moscadelli, L., Surcis, G., *et al.* 2017, *A&A*, 603, A94

Sugiyama, K., Fujisawa, K., Doi, A., *et al.* 2014, *A&A*, 562, A82

Torrelles, J. M., Patel, N. A., Gómez, J. F., *et al.* 2001, *ApJ*, 560, 853

Torrelles, J. M., Patel, N. A., Curiel, S., *et al.* 2011, *MNRAS*, 410, 627

Torrelles, J. M., Gómez, J. F., Garay, G., *et al.* 1998,*ApJ*, 509, 262

Torstensson, K. J. E., van Langevelde, H. J., Vlemmings, W. H. T., & Bourke, S. 2011, *A&A*, 526, A38

Astrophysical Masers:
Unlocking the Mysteries of the Universe
Proceedings IAU Symposium No. 336, 2017
A. Tarchi, M.J. Reid & P. Castangia, eds.

© International Astronomical Union 2018
doi:10.1017/S1743921317010535

The Structure of the Radio Recombination Line Maser Emission in the Envelope of MWC349A

James M. Moran, Qizhou Zhang and Deanna L. Emery

Harvard-Smithsonian Center for Astrophysics,
Mail Stop 42, 60 Garden St., Cambridge, MA 02138, USA
email: `jmoran@cfa.harvard.edu`

Abstract. The Submillimeter Array (SMA) has been used to image the emission from radio recombination lines of hydrogen at subarcsecond angular resolution from the young high-mass star MWC349A in the H26α, H30α, and H31α transitions at 353, 232, and 211 GHz, respectively. Emission was seen over a range of 80 km s^{-1} in velocity and 50 mas (corresponding to 60 AU for a distance of 1200 pc). The emission at each frequency has two distinct components, one from gas in a nearly edge-on annular disk structure in Keplerian motion, and another from gas lifted off the disk at distances of up to about 25 AU from the star. The slopes of the position-velocity (PV) curves for the disk emission show a monotonic progression of the emission radius with frequency with relative radii of 0.85 ± 0.04, 1, and 1.02 ± 0.01 for the H26α, H30α, and H31α transitions, respectively. This trend is consistent with theoretical excitation models of maser emission from a region where the density decreases with radius and the lower transitions are preferentially excited at higher densities. The mass is difficult to estimate from the PV diagrams because the wind components dominate the emission at the disk edges. The mass estimate is constrained to be only in the range of 10–30 solar masses. The distribution of the wind emission among the transitions is surprisingly different, which reflects its sensitivity to excitation conditions. The wind probably extracts significant angular momentum from the system.

Keywords. masers – stars: emission-line, Be – stars: winds, outflows – radio lines: stars

1. Introduction

MWC349A is a highly unusual object that has intrigued and confused astronomers since its identification as a peculiar emission line star in the Mount Wilson Catalogue of Be stars by Merrill & Burwell (1933). The evolutionary state of the star, its relation to its apparent companion, MWC349B, and the nature of its circumstellar neutral and ionized envelopes are subjects of lively debate. For a comprehensive description of its characteristics, see, for example, Gvaramadze & Menten (2012). Its distance is usually taken to be 1.2 kpc (Cohen *et al.* 1985), which is derived from the spectral type (B0III) of MWC349B. On the other hand, its distance could be as high as 1.7 kpc, if it is within the Cygnus OB2 association. Its bolometric luminosity based on the nearer distance is about 40,000 $L_\odot$, which corresponds to a ZAMS mass of about 40 $M_\odot$. MWC349A is among the brightest radio stars known because of emission from a biconical region of ionized gas, which has a nearly perfect $\nu^{0.6}$ power law spectrum from below 1 GHz to above 1 THz. This appears to be a classic case of a constant velocity ionized outflow in which the electron density decreases as the square of the distance from the source.

Of particular interest to this conference is that MWC349A hosts one of only three known examples of radio recombination line (RRL) maser emission (Martín-Pintado *et al.* 1989). At large principal quantum numbers, e.g., $n > 41$, the spectral profile is a simple

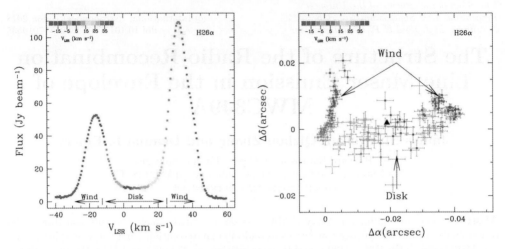

Figure 1. Observations of the H26α RRL in MWC349A made on Oct. 12, 2015, with the SMA. (left) Spectrum with 0.35 km s^{-1} resolution. The velocity ranges of the origin of the dominant emission are shown. (right) The positions of the emission as a function of velocity determined by centroid analysis. The stellar position is shown as a black triangle. At a distance of 1200 pc, $1'' = 1200$ AU. Adapted from Zhang *et al.* (2017).

Gaussian function, whereas at smaller values of n, down to at least 15, the profile develops a prominent double structure (see Fig. 1) suggestive of a rotating disk viewed edge-on with limb brightening. It is interesting that RRL masers are apparently so rare, since the structure of the hydrogen energy levels seem particularly suited to generate maser action. In the 1930s, Menzel (e.g., Menzel 1937) made extensive theoretical calculations of the expected departures of the hydrogen level populations from their expected LTE values. Menzel, in the 1937 paper, noticed that the effects he calculated could lead to population inversion but said, "The condition may conceivably arise when the [excitation temperature] turns out to be negative. ... The process merely puts energy back into the original beam, as if the atmosphere had a negative opacity. This extreme will probably never occur in practice." Menzel's calculations became important in interpreting the RRLs from HII regions starting in the mid-1960s because the kinetic temperatures inferred from the line-to-continuum ratios gave kinetic temperatures that were much lower than measured by other means (Dupree & Goldberg 1970). In these cases, the non-LTE effects were significant but did not produce outright population inversion.

The proclivity of the hydrogen atom to natural maser action can be understood from a simple model in which hydrogen atoms in the envelope of a massive star are ionized by the strong ultraviolet radiation field and then recombine and cascade down the principal energy levels to the ground state. The Einstein A coefficient, which gives the rate of spontaneous emissions, varies approximately as n^{-5}. If collisions are negligible, then the population levels will be largely controlled by the spontaneous emissions. Hence, for a given level, the transition rate into it from the level above is slightly less than the transition rate to the next lower level. This "bottleneck" effect means that all levels will be inverted. Collisions disrupt this inversion, and higher densities are required to quench the inversion as n decreases (see, e.g., Strelnitski *et al.* 1996). This situation is the opposite of that found in molecules undergoing rotational level transitions for which $A \sim J^3$, where J is the rotational quantum number. Of course, detectable maser emission requires high negative optical depth in addition to population inversion.

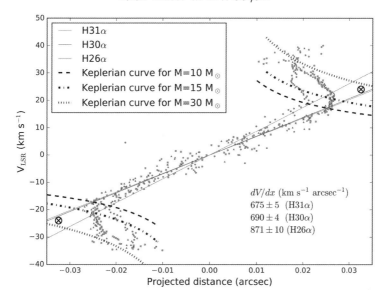

Figure 2. The PV diagram for the maser emission centroids for MWC349A for the H26α (green), H30α (blue), and H31α (red) transitions. The increasing slope of the PV curves with decreasing n is expected for a masing disk with decreasing density with increasing radius. The Keplerian curves for masses of 10, 15, and 30 $M_\odot$ are shown as dashed lines. The two-point measurement of Planesas *et al.* (1992) is shown by circled crosses. Adapted from Emery (2017).

The critical observational development in the study of the RRL line in MWC349A came with the measurement of the spatial separation of the two most prominent components of the spectrum with the OVRO interferometer in the H30α line by Planesas *et al.* (1992), which gave a separation of 65 mas for the components separated by 48 km s^{-1}. Interpreting this in a Keplerian disk model gave a mass $M = v^2 R/G$ of $25 \pm 6 \ M_\odot$. This work was followed by SMA observations of the detailed distribution of masers, which showed the features following a disk structure (Weintroub *et al.* 2008; Zhang *et al.* 2017). Martín-Pintado *et al.* (2011) used the PdBI to show that the velocity components greater than 15 km s^{-1} from the systemic velocity were positioned off the disk and probably attributable to gas in a wind created by photoionization of the disk.

We report here on observations of the H26α, H30α, and H31α lines. Figure 1 shows the spectra and spatial distribution for the H26α transition. The position of the emission at each velocity was determined by a centroiding analysis, which is strictly appropriate only for an unresolved distribution. The formal position accuracy can exceed the resolution of the interferometer by a factor of about the signal-to-noise ratio. If multiple components are present, the centroiding analysis gives some sort of flux-weighted mean position whose interpretation can be unclear and misleading. The elongated distribution corresponding to features within 15 km s^{-1} for the systemic velocity corresponds to a nearly edge-on disk. The position-velocity (PV) diagrams for emission in all three transitions are shown in Figure 2. The emission is expected to peak in an annular ring, where the density is optimal for maser action. The radii can be estimated from the slope of the PV plot since $V = \sqrt{GM/R}\sin\phi = \sqrt{GM/R^3}x$, where G is the gravitational constant, R is the radius, ϕ is the azimuthal angle in the disk measured from the line-of-sight, and x is the projected distance. The ratios of the radii can be determined with respect to the radius of the H30α emission without knowledge of the stellar mass. The relative radii of the H26α, H30α, and H31α emission peaks are 0.85 ± 0.04, 1 and 1.02 ± 0.01, respectively,

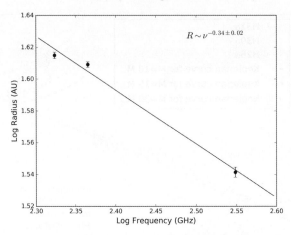

Figure 3. The annular radius of each RRL maser as a function of transition frequency derived from the slope of the PV diagram. A power-law fit is also plotted. From Emery (2017).

and for a mass of 25 $M_\odot$, the radii are 35, 40, and 41 AU. These results are shown in Figure 3. The radii scale as $\nu^{-0.34\pm0.02}$ or $n^{1.0\pm0.06}$. This dependence is in reasonable agreement with theoretical models (e.g., Strelnitski *et al.* 1996; Thum *et al.* 1994).

The mass is rather difficult to estimate from the data. The two-point Keplerian-model analysis (Planesas *et al.* 1992) gives 25 ± 6 $M_\odot$ (see Fig. 2). If the emission in the straight line portion of the PV diagrams shown in Figure 2 is assumed to trace the whole range of disk azimuth angles ($\pm90°$), e.g., limb-to-limb, then the mass estimate would be only 10 $M_\odot$. However, a more reasonable interpretation of the data is that a centroiding process obscures the disk emission at $|\phi| > 50°$ because of the dominance of the wind emission. In this case, the true mass could be easily as large as 30 $M_\odot$.

Finally, we note that the wind emission seems to be excited along rather different trajectories in three transitions. This is probably due to a particular sensitivity of the maser emission in any transition to density distribution. The origin of this wind is discussed by Báez-Rubio *et al.* (2014).

We thank Vladimir Strelnitski for pedagogical discussions about the population distribution in the hydrogen atom and Alejandro Báez-Rubio for helpful discussions about radiative transfer modeling.

References

Báez-Rubio, A., Martín-Pintado, J., Thum, C., Planesas, P., & Torres-Redondo, J. 2014 *A&A*, 571, L4
Cohen, M., Bieging, J. H., Dreher, J. W., & Welch, W. J. 1985 *ApJ*, 292, 249
Dupree, A. K., & Goldberg, L. 1970 *ARAA*, 8, 231
Emery, D. 2017 BS thesis, Harvard College
Gvaramadze, V. V., & Menten, K. M. 2012 *A&A*, 541, A7
Martín-Pintado, J., Bachiller, R., Thum, C., & Walmsley, M. 1989 *A&A*, 215, L13
Martín-Pintado, J., Thum, C., Planesas, P., & Báez-Rubio, A. 2011 *A&A*, 530, L15
Menzel, D. H. 1937 *ApJ*, 85, 330
Merrill, P. W., & Burwell, C. G. 1933 *ApJ*, 78, 87
Planesas, P., Martín-Pintado, J., & Serabyn, E. 1992 *ApJ* (Letters), 386, L23
Strelnitski, V. S., Smith, H. A., & Ponomarev, V. O. 1996 *ApJ*, 470, 1134
Thum, C., Matthews, H. E., Martín-Pintado, J., *et al.* 1994 *A&A*, 283, 582
Weintroub, J., Moran, J. M., Wilner, D. J., *et al.* 2008 *ApJ*, 677, 1140
Zhang, Q., Claus, B., Watson, L., & Moran, J. 2017 *ApJ*, 837, 53

Astrophysical Masers:
Unlocking the Mysteries of the Universe
Proceedings IAU Symposium No. 336, 2017
A. Tarchi, M.J. Reid & P. Castangia, eds.

© International Astronomical Union 2018
doi:10.1017/S1743921317011504

Interferometric and single-dish observations of 44, 84 and 95 GHz Class I methanol masers

Carolina B. Rodríguez-Garza[1], **Stanley E. Kurtz**[1],
Arturo I. Gómez-Ruiz[2], **Peter Hofner**[3,4], **Esteban D. Araya**[5]
and Sergei V. Kalenskii[6]

[1]Instituto de Radioastronomía y Astrofísica, Universidad Nacional Autónoma de México,
Morelia, Michoacán, México, C. P. 58089
email: ca.rodriguez@irya.unam.mx

[2]Instituto Nacional de Astrofísica, Óptica y Electrónica,
Luis E. Erro 1, Tonantzintla, Puebla, C.P. 72840, México

[3]Physics Department, New Mexico Tech
801 Leroy Pl., Socorro, NM 87801, USA
[4]Adjunct Astronomer at the National Radio Astronomy Observatory
1003 Lopezville Road, Socorro, NM 87801, USA

[5]Physics Department, Western Illinois University
1 University Circle, Macomb, IL 61455, USA

[6]Astro Space Center, Lebedev Physical Institute
Profsoyuznaya 84/32, Moscow, 117997, Russia

Abstract. We present observations of massive star-forming regions selected from the IRAS Point Source Catalog. The observations were made with the Very Large Array and the Large Millimeter Telescope to search for Class I methanol masers. We made interferometric observations of 125 massive star-forming regions in the 44 GHz methanol maser transition; 53 of the 125 fields showed emission. The data allow us to demonstrate associations, at arcsecond precision, of the Class I maser emission with outflows, HII regions and shocks traced by 4.5 μm emission. We made single-dish observations toward 38 of the 53 regions with 44 GHz masers detected to search for the methanol transitions at 84.5, 95.1, 96.7, 107.0, and 108.8 GHz. We find detection rates of 74, 55, 100, 3, and 45%, respectively. We used a wide-band receiver which revealed many other spectral lines that are common in star-forming regions.

Keywords. stars: formation — stars: massive — stars: protostars — ISM: masers — ISM: molecules

1. Introduction

Methanol masers are empirically divided into two classes based on their environments and pumping mechanisms (Batrla *et al.* 1987; Menten 1991). In particular, the Class I methanol masers arise in molecular gas shocked by outflows in high-mass protostellar objects (HMPOs). These masers were first associated with protostellar outflows by Plambeck & Menten (1990) and more recently by Cyganowski *et al.* (2009) by their coincidence with the so-called extended green objects (EGOs), which are regions of shocked gas seen in the 4.5 μm band of the InfraRed Array Camera (IRAC) on the Spitzer Space Telescope.

Two well-known samples of HMPOs are reported in the literature: the samples of Molinari *et al.* (1998; hereafter M98) and of Sridharan *et al.* (2002; hereafter S02). Each sample consists of a collection of 69 HMPOs selected systematically to satisfy specific selection criteria. Half of the sample of M98 contains sources with colors similar to UCHII regions and the other half with colors of deeply embedded sources; 35 of the 69 objects have molecular outflows detected by Zhang *et al.* (2001; 2005). The S02 sample consists of HMPOs with colors similar to UCHII regions but with 5 GHz flux densities lower than 25 mJy; almost all sources have CO line wings indicative of high velocity gas from molecular outflows (S02, Beuther *et al.* 2002).

Here, we present interferometric and single-dish observations of both samples to search for Class I methanol maser emision.

2. VLA surveys of methanol masers

We made observations of Class I 44 GHz methanol masers with the NRAO† Karl G. Jansky Very Large Array (VLA) in D configuration toward 69 and 56 HMPOs from the samples of M98 and S02, respectively. The data were observed in Q band (7 mm) using the dual IF mode and fast switching method. One IF of 3.125 MHz bandwith was centered at the 44 GHz methanol transition. The bandwidth was divided in 127 channels providing a spectral resolution of 0.16 km s^{-1} and a velocity coverage of 21 km s^{-1}. The observations have an angular resolution of about 2″ and a sensitivity of 0.15 Jy. The results of these maser surveys toward the M98 and S02 samples are presented in Gómez-Ruiz *et al.* (2016; Survey I) and Rodríguez-Garza *et al.* (2017; Survey II), respectively.

We find a similar detection rate of 43% for Class I 44 GHz methanol masers in Surveys I and II. Multiple sources are present in each of the fields from both samples. In general, we note that when an EGO is in the field, the 44 GHz masers seem to favor it, as shown in Figure 1. The spatial coincidence of the 44 GHz masers and the shocked molecular gas supports the idea that these masers may arise from molecular outflows and, presumably, the youngest sources in the field.

Class I methanol masers are thought to be saturated with little time variation over timescales of several years, although this field has not been extensively studied (Leurini *et al.* 2016). Our observations suggest that variability on these timescales may occur. IRAS 20126+4104 was part of VLA observations of 44 GHz masers made in March 1999 (Kurtz *et al.* 2004), March 2007 (Survey I) and August 2008 (Survey II). Four 44 GHz maser components were found clustered toward the NW of the IRAS source and one isolated feature toward the SE of the IRAS central source. Interestingly, the flux density of the isolated feature remains nearly the same during these years while the flux densities of the NW maser group are similar between 1999 and 2007 but are amplified by a common factor of 2 during the observations of 2008. This implies a flux variation on timescales of less than 15 months.

3. LMT survey of methanol emission

We observed 38 HMPOs selected from the VLA 44 GHz methanol maser surveys presented in Section 2; 19 HMPOs were taken from Survey I and another 19 from Survey II. This sub-sample was observed during the Early Science Phase of the Large Millimeter Telescope (LMT) in 2016. The observations were made with the Redshift Search Receiver (RSR; Erickson *et al.* 2007), a broad bandwidth spectrometer that covers the frequency range from 73 to 111 GHz, at 31 MHz (100 km s^{-1}) spectral resolution. At the time of

† The National Radio Astronomy Observatory (NRAO) is operated by Associated Universities, Inc., under a cooperative agreement with National Science Foundation.

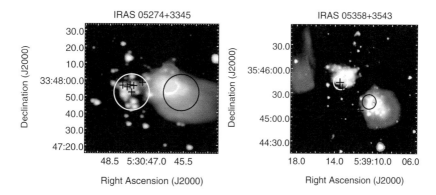

Figure 1. Left panel: IRAS 05274+3345 from Survey I. Right panel: IRAS 05358+3543 from Survey II. Spitzer images of the IRAC bands at 3.6, 4.5, and 8 μm. The crosses are the 44 GHz methanol masers detected in the VLA surveys. The white circles represent the LMT primary beam (18″ at 111 GHz) pointed toward the position of the brightest 44 GHz maser (related to shocked gas traced by EGOs) and the black circles represent the LMT primary beam toward the IRAS source.

the observations, the LMT had a usable surface 32-m in diameter, providing a beam size of $\sim 18''$ at 111 GHz.

The methanol lines within the RSR bandpass are at 84.5, 95.1, 96.7, 107.0, and 108.8 GHz. We find detection rates of 74, 55, 100, 3 and 45%, respectively.

Many molecular lines are blended in a single LMT channel but in most cases, the methanol emission dominates the channel emission which allows us to identify the spectral line. The low spectral resolution precludes a definitive interpretation of the methanol emission (maser or thermal) but even so, these data serve to identify promising methanol maser candidates for follow-up studies at high spectral resolution.

The 73-111 GHz band includes many molecular lines commonly found in massive star-forming regions. In order to compare the molecular composition of the shocked regions (EGOs) traced by the 44 GHz masers with the central protostars (IRAS sources), we made observations at both positions. For example, IRAS 05274+3345 and IRAS 05358+3543 show two clusters of star formation in different evolutionary stages (see Figure 1). In both cases, the 44 GHz masers are associated with the youngest cluster, separated by $\sim 30''$ from the IRAS source. We made LMT pointings toward each region to compare their chemical composition. We find that the clusters related to shocked gas are more chemically active than the cluster associated to the IRAS source (see Figure 2).

4. Conclusions

Our VLA surveys support the relation of Class I methanol masers with shocked regions around massive protostars. In one case, we found evidence of maser variability, with a time variation of less than 15 months.

The spectral line survey performed with the LMT in the 3 mm band revealed a rich chemical composition of the shocked gas related to the methanol masers. In the few cases tested, we also found that shocked regions have a greater number of spectral lines than the central regions where the massive protostars are located.

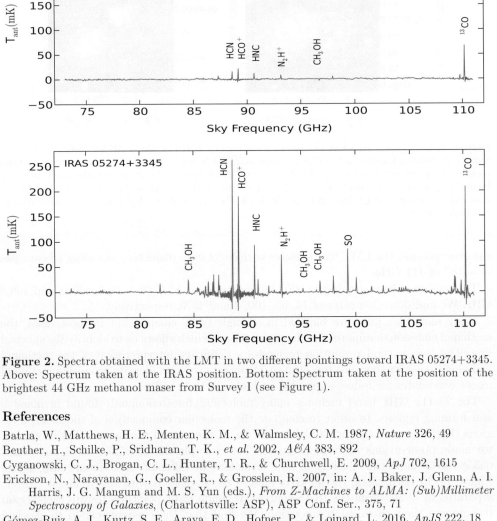

Figure 2. Spectra obtained with the LMT in two different pointings toward IRAS 05274+3345. Above: Spectrum taken at the IRAS position. Bottom: Spectrum taken at the position of the brightest 44 GHz methanol maser from Survey I (see Figure 1).

References

Batrla, W., Matthews, H. E., Menten, K. M., & Walmsley, C. M. 1987, *Nature* 326, 49

Beuther, H., Schilke, P., Sridharan, T. K., *et al.* 2002, *A&A* 383, 892

Cyganowski, C. J., Brogan, C. L., Hunter, T. R., & Churchwell, E. 2009, *ApJ* 702, 1615

Erickson, N., Narayanan, G., Goeller, R., & Grosslein, R. 2007, in: A. J. Baker, J. Glenn, A. I. Harris, J. G. Mangum and M. S. Yun (eds.), *From Z-Machines to ALMA: (Sub)Millimeter Spectroscopy of Galaxies*, (Charlottsville: ASP), ASP Conf. Ser., 375, 71

Gómez-Ruiz, A. I., Kurtz, S. E., Araya, E. D., Hofner, P., & Loinard, L. 2016, *ApJS* 222, 18

Kurtz, S., Hofner, P., & Álvarez, C. V. 2004, *ApJS* 155, 149

Leurini, S., Menten, K. M., & Walmsley, C. M. 2016, *A&A* 592, A31

Menten, K. M. 1991, *ApJ (Letters)* 380, L75

Molinari, S., Brand, J., Cesaroni, R., Palla, F., & Palumbo, G. G. C. 1998, *A&A* 336, 339

Plambeck, R. L. & Menten, K. M. 1990, *ApJ* 364, 555

Rodríguez-Garza, C. B., Kurtz, S. E., Gómez-Ruiz, A. I., *et al.* 2017, *ApJS* 233, 4

Sridharan, T. K., Beuther, H., Schilke, P., Menten, K. M., & Wyrowski, F. 2002, *ApJ*, 566, 931

Zhang, Q., Hunter, T. R., Brand, J., *et al.* 2001, *ApJ (Letters)* 552, L167

Zhang, Q., Hunter, T. R., Brand, J., *et al.* 2005, *ApJ* 625, 864

Astrophysical Masers:
Unlocking the Mysteries of the Universe
Proceedings IAU Symposium No. 336, 2017
A. Tarchi, M.J. Reid & P. Castangia, eds.

© International Astronomical Union 2018
doi:10.1017/S1743921317010195

Linear polarisation
of Class I methanol masers
in massive star formation regions

Ji-hyun Kang[1], Do-Young Byun[1], Kee-Tae Kim[1], Aran Lyo[1],
Jongsoo Kim[1], Mi-kyoung Kim[1], Wouter Vlemmings[2],
Boy Lankhaar[2] and Gabriele Surcis[3]

[1]Korea Astronomy and Space Science Institute 776, Daedeokdae-ro, Yuseong-gu, Daejeon,
Republic of Korea
email: jkang.kasi.re.kr

[2]Department of Earth and Space Sciences, Chalmers University of Technology, Onsala Space
Observatory, SE-439 92 Onsala, Sweden

[3] INAF-Osservatorio Astronomico di Cagliari, Via della Scienza 5, 09047 Selargius, Italy

Abstract. We present the results of the linear polarisation observations of methanol masers at 44 and 95 GHz towards 39 massive star forming regions (Kang *et al.* 2016). These two lines are observed simultaneously with the 21-m Korean VLBI Network (KVN) telescope in single dish mode. About 60% of the observed showed fractional polarisation of a few percents at least at one of the two transition lines. We note that the linear polarisation of the 44 GHz methanol maser is first detected in this study including single dish and interferometer observations. We find the polarisation properties of these two lines are similar as expected, since they trace similar regions. As a follow-up study, we have carried out the VLBI polarisation observations toward some 44 GHz maser targets using the KVN telescope. We present preliminary VLBI polarisation results of G10.34-0.14, which show consistent polarisation properties in multiple epoch observations.

Keywords. masers, polarisation, stars: formation, magnetic fields

1. Introduction

Masers (OH, H2O, SiO, and CH3OH) are the tracers that can be used to study the physical conditions of dense regions ($n_H = 10^{5-11} \mathrm{cm}^{-3}$), close to protostellar disks or outflows, that are embedded in thick dust envelopes. Class I methanol masers, including 44 and 95 GHz transitions, are known to trace the regions where the outflow of a massive young stellar object is interacting with the surrounding medium, while Class II methanol masers are tightly correlated with the central (proto)stars (Plambeck & Menten 1990).

The polarimetric observations of Class II methanol masers have increased during the last decade (e.g., Vlemmings 2012) and have been able to provide information about the magnetic fields near the central objects. However, the polarisation studies of Class I methanol masers have been very limited. For example, the detection of linear polarisation at 95 GHz has been reported only for two sources before our study (Wiesemeyer *et al.* 2004). No linear polarisation has been reported for 44 GHz methanol maser sources. Circular polarisation at 44GHz, detected with the Very Large Array (VLA), has been presented only for one object, OMC2 (Sarma & Momjian 2011).

In this paper, we report on single-dish measurements of linear polarisation toward 44 and 95 GHz methanol maser sources. The full results are published in Kang *et al.* (2016). We also report on the preliminary results of follow-up VLBI observations, especially on the G10.34-0.14.

2. Observations and Calibration

We observed the Class I methanol $7_0 - 6_1 A+$ (44.06943 GHz) and $8_0 - 7_1 A +$ (95.169463 GHz) maser transitions in full polarisation spectral mode towards 39 massive star forming regions. The observations were conducted simultaneously in both transitions using the KVN 21 m telescope at the Yonsei station in single-dish mode from August to December in 2013. The beam sizes of data are $65''$ and $30''$, at 44 and 95 GHz, respectively. The typical rms levels (1σ) of final spectra are 0.5 Jy and 1.2 Jy at 44 and 95 GHz at a velocity resolution of 0.2 km s^{-1} , respectively .

The instrumental cross talk and phase offset were corrected using planets (Jupiter, Venus, or Mars) and the Crab nebula, which were observed at least once a day as calibrators. The polarized intensity (PI) given here is $(Q^2 + U^2)^{1/2}$, and the given error is the standard deviation of the measurement sets. Detection criterion is that $PI > 3\sigma$. The measured position angle was derived by $\chi = \frac{1}{2}\arctan(\frac{U}{Q}) + 152°$. Here the latter term is the absolute position angle of the Crab nebula, which is known to be nearly constant near the brightest region from millimeter to X-ray wavelengths. (Aumont *et al.* 2010)

3. Results

We detected fractional linear polarisation toward 23 (59%) of the 39 Class I methanol maser sources at 44 and/or 95 GHz. Figure 1 shows the spectral profiles of the total flux (I), the polarized intensity (PI), the polarisation degree (P_L), and the polarisation position angle measured counterclockwise from north (χ) of G10.34-0.14 and G18.34+1.78SW at both transitions. The profiles of the total flux and the polarized intensity tend to peak at similar velocities. At 44 and 95 GHz, 21 (54%) and 17 (44%) sources show linear polarisation, respectively. We emphasize that this is the first detection of linear polarisation of the 44 GHz methanol masers. Fifteen (38%) sources were detected at both frequencies.

The rms weighted means of the fractional linear polarisation detected sources are $2.7 \pm 0.3\%$ and $4.8 \pm 0.1\%$ at 44 and 95 GHz, respectively. All sources have $P_L < 11\%$ except W33Met (24.6%), which has a large error of 6.9% (1σ). The ranges of polarisation fractions are $1.1\% - 9.5\%$ and $2.0\% - 24.6\%$ at 44 and 95 GHz, respectively. The polarisation detection rate tends to increase with the total flux at both transitions. The polarisation fraction and the total flux do not show clear correlation, whereas the error of P_L increases as I decreases, because $P_L \propto I^{-1}$ while the observational noise is relatively constant for all sources.

Since we observed the two transition lines at the same time, our data are ideal to compare their polarisation properties. The polarisation degrees of the two transitions appear to have a positive linear correlation, although the correlation is not very tight. The polarisation fractions of the 95 GHz masers tend to be greater than those of the 44 GHz masers. The polarisation angles of the 44 and the 95 GHz maser transitions are well correlated. In general, the polarisation properties of the 44 and 95 GHz transitions are similar, indicating that the masers at these two transition lines are indeed experiencing magnetic fields of similar regions.

4. Discussion

We have not detected any source with linear polarisation above 30% that Wiesemeyer *et al.* (2004) has reported at 132 GHz Class I and 157 GHz Class II maser transitions. Linear polarisation above 33% is expected to be rare (Elitzur 2002). Nedoluha & Watson (1990) indicated that about 30% would be the highest for an angular momentum $J = 2-1$ and higher transitions, unless significant anisotropic pumping is present. Weisemeyer *et al.*(2004) found that a large fractional linear polarisation $(P_L > 33\%)$ is not rare (2 out

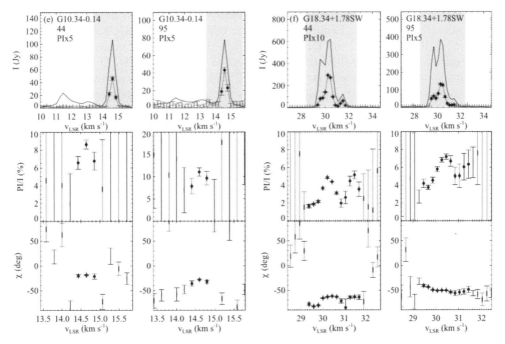

Figure 1. Polarisation-detected maser profiles of G10.34-0.14 and G18.34+1.78SW at 44 GHz and 95 GHz. For each source, the polarized intensity (PI) in grey line with errors (1σ) and the total flux intensity (I) in black solid line (top), and the polarisation fraction P_L (middle) and angle χ (bottom) are presented. The multiplication factor of PI is presented under the line frequency. The points with $PI > 3\sigma$ are indicated with filled circles. The gray-shaded area in the top panel indicates the velocity range for the middle and bottom panels.

of 10 for Class I and 1 out of 3 for Class II), giving an impression that anisotropic pumping or loss may be commonly achievable due to an unequal population of the magnetic substates of the maser levels (Nedoluha & Watson 1990). Since transitions at higher frequencies tend to have a higher fraction of polarisation, we cannot directly compare our results with those of Wiesemeyer *et al.* (2004). However, our observational results, which lacked high fractional polarisation, out of 23 polarisation-detected sources, still suggest that the anisotropic pumping or loss mechanism may not be common for 44 and 95 GHz maser transitions.

We have found that the degree of linear polarisation at 95 GHz tends to be greater than that at 44 GHz. The reduced fractional polarisation at the lower frequency transition compared to higher frequency transitions seems to be general trend in maser polarisation observations (e.g., McIntosh & Predmore 1993, Wiesemeyer *et al.* 2004). The depolarisation due to the Faraday rotation and the different beam size effects are negligable, because the internal/external Faraday rotation at these frequencies are generally smaller than the angle measurement errors and maser features of our targets are mostly confined within the beam of the 95 GHz transition in the previous VLA observations (Kogan & Slysh 1998, Kurtz *et al.* 2004). The fractional linear polarisation of the 44 and 95 GHz transitions could be intrinsically different. Pérez-Sánchez & Vlemmings (2013) and Nedoluha & Watson (1990) showed that the linear polarisation fraction of masers depends on intrinsic properties of individual transitions, such as the degree of maser saturation R/Γ and the ratio between the Zeeman splitting rate and the stimulated emission rate, etc. These are not well known and needs future studies.

We have tried to investigate the association between the orientations of linear polarisation and outflows. We have found 7 sources, i.e., OMC2, S255N, NGC2264, G49.49−0.39 (W51 e2), DR21W, DR21, and DR21(OH), where the direction of maser-associated outflow is rather simple in the literature when searched using the SIMBAD website (Wenger *et al.* 2000). Although the sample size is small, we tried the Kolmogorov-Smirnov (K-S) statistics to see whether the angle differences are similar to or different from the projected angles of aligned, perpendicular, or randomly aligned samples, by comparing the observed angle difference to the results from the Monte Carlo simulations as discussed in Hull *et al.* (2014). The K-S tests rule out the scenario where the outflows and polarisation angles are tightly aligned ($P < 0.01$). The probability of them being perpendicular ($P = 0.7$) appears to be higher than that of them being random ($P = 0.4$).

DR21(OH) and G82.58+0.20 appear to be interesting targets because they show the $90°$ polarisation angle flip in the 44 GHz polarisation profiles, while it is not visible in their 95 GHz polarisation profiles. The high angular resolution polarimetry observations in both frequencies and the theoretical studies will reveal whether these polarisation properties are due to the van Vleck angle crossing or change of maser saturation level, providing more information on the not-well-known methanol maser polarisation physics.

As a follow-up, we have carried out the VLBI observations of 7 targets in multiple epochs with a resolution of a few mas with full polarisation modes. In case of G10.34-0.14, observations of 3 epochs (2014-2017) showed consistent flux and polarized intensity profiles at $v_{LSR} \sim +14.6$ km s^{-1} with a fractional polarisation of about 13%, indicating successful performance of the KVN polarisation system. More VLBI observations have been performed, whose results will be soon published.

5. Summary

We have been investigating the linear polarisation properties of a significant number of Class I methanol masers at 44 and 95 GHz using the KVN telescope in single dish and VLBI modes, which enables us to understand the polarisation properties of the Class I methanol masers in a statistical sense. Follow-up observations of the VLA and the ALMA telescopes would help to understand the magnetic fields of individual maser features. We have obtained the ALMA polarisation data in the 95 GHz methanol maser transition line for G10.34-0.14, which shows maser features at the tip of outflows and also near the central source. This will help us to improve our understanding on the magnetic field environment of G10.34-0.14.

References

Aumont, J., Conversi, L., Thum, C., *et al.* 2010, *A&A*, 514, A70
Elitzur, M. 2002, *in Astrophysical Spectropolarimetry*, ed. J. Trujillo-Bueno, F. Moreno-Insertis, & F. Sánchez, 225–264
Hull, C. L. H., Plambeck, R. L., Kwon, W., *et al.* 2014, *ApJS*, 213, 13
Kang, J., Byun, D., Kim, K., Kim, J., Lyo, A., & Vlemmings, W. H. T. 2016, *ApJ*, 227, 17
Kogan, L. & Slysh, V. 1998, *ApJ*, 497, 800
Kurtz, S., Hofner, P., & Álvarez, C. V. 2004, *ApJS*, 155, 149
McIntosh, G. C. & Predmore, C. R. 1993, *ApJL*, 404, L71
Nedoluha, G. E. & Watson, W. D. 1990, *ApJ*, 354, 660
Pérez-Sánchez, A. F. & Vlemmings, W. H. T. 2013, *A& A*, 551, A15
Plambeck, R. L. & Menten, K. M. 1990, *ApJ*, 364, 555
Sarma, A. P. & Momjian, E. 2011, *ApJL*, 730, L5
Vlemmings, W. H. T. 2012, *in IAU Symposium, Vol. 287, IAU Symposium*, ed. R. S. Booth, W. H. T. Vlemmings, & E. M. L. Humphreys, 31–40
Wiesemeyer, H., Thum, C., & Walmsley, C. M. 2004, *A& A*, 428, 479

Astrophysical Masers:
Unlocking the Mysteries of the Universe
Proceedings IAU Symposium No. 336, 2017
A. Tarchi, M.J. Reid & P. Castangia, eds.

© International Astronomical Union 2018
doi:10.1017/S1743921317011413

Class II 6.7 GHz Methanol Maser Association with Young Massive Cores Revealed by ALMA

James O. Chibueze[1,2,3], Timea Csengeri[4], Ken'ichi Tatematsu[5,6], Tetsuo Hasegawa[5], Satoru Iguchi[5,6], Jibrin A. Alhassan[2], Aya E. Higuchi[7], Sylvain Bontemps[6] and Karl M. Menten[4]

[1]SKA Africa, 3rd Floor, The Park, Park Road, Pinelands, Cape Town, 7405, South Africa
email: jchibueze@ska.ac.za

[2]Department of Physics and Astronomy, Faculty of Physical Sciences,
University of Nigeria, Carver Building, 1 University Road, Nsukka, Nigeria

[3]Space Research Unit, Physics Department, NorthWest University, Private Bag X6001,
Potchefstroom, 2520, South Africa

[4]Max Planck Institute for Radioastronomy, Auf dem Hügel 69, 53121 Bonn, Germany
email: tcsengeri@mpifr-bonn.mpg.de

[5]National Astronomical Observatory of Japan, National Institutes of Natural Sciences, 2-21-1
Osawa, Mitaka, Tokyo 181-8588, Japan

[6]The Graduate University for Advanced Studies), 2-21-1 Osawa, Mitaka, Tokyo 181-8588,
Japan

[7]The Institute of Physical and Chemical Research (RIKEN), 2-1, Hirosawa, Wako-shi, Saitama
351-0198, Japan

[8]OASU/LAB-UMR5804, CNRS, Université Bordeaux 1, 33270 Floirac, France

Abstract. The association of 6.7 GHz class II methanol (CH_3OH) masers with ATLASGAL/ ALMA 0.9 mm massive dense cores is presented in this work from a statistical viewpoint. 42 of the 112 cores (37.5%) detected with the Atacama Compact Array (ACA) excite 6.7 GHz CH_3OH masers. ACA cores have offsets $0\rlap.{''}17$ to $4\rlap.{''}79$ from the methanol multibeam survey (MMB), with a median of 2."19. Approximately 90% of the MMB-associated cores are of masses > 40 M$_\odot$. Because all the cores show evidence of outflow activity, and only a fraction of the cores excited CH_3OH masers, we suggest that outflows precede the emergence of maser emission. This first ALMA survey of massive dense cores combined with the MMB survey along with other maser specie surveys is a promising tool to trace the evolutionary sequence of high-mass stars.

Keywords. stars: formation – stars: winds, outflows – ISM: H II regions – surveys

1. Introduction

Large surveys of massive young cores at various wavelengths are crucial for understanding the evolutionary sequence of high-mass stars, especially at their early formation stages. A number of surveys both in dust continuum and different maser species toward massive dense cores have been carried out (Breen 2010, 2011, 2013; Gerner *et al.* 2014; Urquhart *et al.* 2013a, 2013b; Codella *et al.* 2004; de Villiers *et al.* 2015), and could be statistically analysed to filter out information about high-mass star formation.

While 6.7 GHz CH_3OH are known to be exclusively associated with massive stars, $\sim$22 GHz H_2O masers as well as other maser species can also trace sites of star formation. The formative stages at which massive cores excite this masers remain an unsettled issue with Reid 2007 suggesting that H_2O maser excitation precede CH_3OH masers which precedes OH masers with some level of overlap in time.

de Villiers *et al.* (2015) suggested that outflows are launched prior to the excitation of 6.7 GH and this is supported by the results of Bayandina *et al.* (2012). To confirm the above results a statistical approach is required and this is the aim of this paper.

2. ALMA and MMB Data

The data used in this work was taken from Atacama Compact Array (ACA) Cycle 2 0.9 mm observations (Csengeri *et al.* 2017a contains the details of the observing setup and data reduction procedure; Chibueze *et al.* 2017). The primary beam was 28.″9, while the synthesized (geometric mean of the major and minor axes) beam was 3.″5 to 4.″6.

The 6.7 GHz CH_3OH maser information was taken from the methanol multibeam (MMB) catalog of Green *et al.* (2012) [covering Galactic longitude 186° – 330°], and Caswell *et al.* (2010, 2011) [covering Galactic longitude 330° – 6°]. The Australia Telescope Compact Array (ATCA) astrometric accuracy for the MMB is 0.″4 (an order of magnitude better angular resolution than the ACA observations).

To check for the presence of 44 GHz class I CH_3OH masers, we have used the class I methanol maser catalog of Bayandina *et al.* (2012), which contains 206 sources selected from the literature up to the end of 2011 (see Bayandina *et al.* 2012 and references therein). Maser information extracted from individual publications lack completeness but we found 13% of the ALMA sources to be associated with 44 GHz class I CH_3OH masers, and in addition 4 with large offsets ($> 5''$) from the ALMA cores.

H_2O maser information was taken from the H_2O Southern Galactic Plane Survey (HOPS) by Walsh *et al.* (2011). G351.4441+0.6579 [NGC 6334 I(N)] H_2O maser details was taken from Chibueze *et al.* (2014).

3. Association and Selection Criteria

With 6.7 GHz class II CH_3OH maser (MMB) association with the ACA cores as the primary focus, we crossed matched and determined association with the following conditions:

(1) distance to the core is < 5 kpc;

(2) angular offset of the ACA core peak and the MMB peak is <5″.

(3) difference between the V_{LSR} of the MMB peak and that of its associated ACA core is within ± 8 km s^{-1}.

4. Results and Discussions

Of the 125 ACA core, we confirmed 6.7 GHz observations toward 112 cores of the 42 ATLASGAL clumps (Csengeri *et al.* 2017b). 31 of the 42 ATLASGAL clumps (73.8%) were associated with one or more 6.7 GHz CH_3OH masers. 42 of the 112 (37.5%) ACA cores were found to be exciting 6.7 GHz CH_3OH masers. The ACA-MMB core with the lowest mass has a mass of ~ 12 $M_\odot$. Figure 1 shows the 0.9 mm continuum emission of G329.1835 with markers to indicate the positions of different maser species (see Chibueze *et al.* 2017).

We compare ACA and ATLASGAL dust continuum peak flux densities and found a good correlation of coefficient 0.56 with a significance value of 2×10^{-10}.

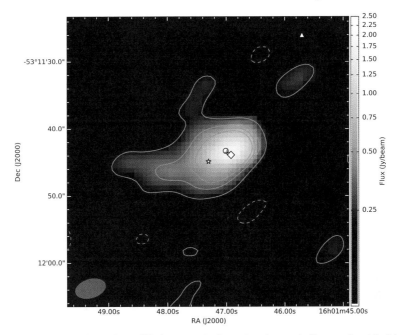

Figure 1. G329.1835. The white filled star, circle, triangle, and diamond with black edges represent the peak position of ATLASGAL clumps, 6.7 GHz CH$_3$OH masers, H$_2$O masers and 44 GHz CH$_3$OH masers, respectively (see Appendix of Chibueze *et al.* (2017).

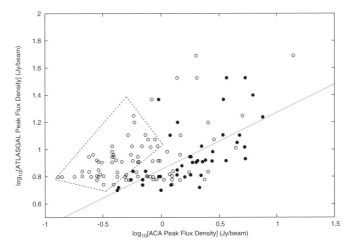

Figure 2. Plot of ATLASGAL clump peak flux densities against ACA core peak flux densities. Open circles represent sources without 6.7 GHz CH$_3$OH masers, while the filled circles are those associated with 6.7 GHz CH$_3$OH masers (*unflagged ACA-MMB*). The dotted line represents the best fit of the ATLASGAL-ACA flux relation of the 6.7 GHz CH$_3$OH maser associated cores only. The skewed square represent the ACA weak-continuum region which lack CH$_3$OH maser emissions.

Figure 2 is a plot of peak flux densities of the ATLASGAL clumps against those of their associated ACA cores. The filled circles represent cores/clumps associated with CH$_3$OH masers while the open circles are those with no CH$_3$OH maser association. The skewed box (drawn with dashed lines) in Figure 2 encloses cores/clumps that are massive, driving outflows but not associated with CH$_3$OH masers. This is likely an indication that

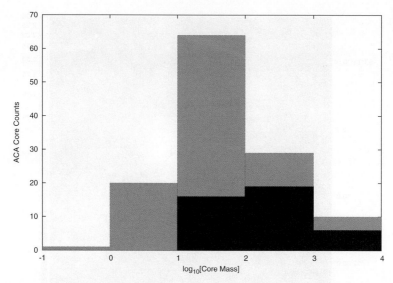

Figure 3. Mass distribution of all the ACA cores/clumps with 10 $M_\odot$ binning (gray: all core, black: 6.7 GHz CH_3OH maser associated cores).

the cores within the enclosed clump-to-core flux density relation represent the earlier phase of protostellar evolution. While these cores are driving outflows, absence of maser association could suggest that the cores do not have sufficient radiative power to pump CH_3OH masers.

Using the core masses of Csengeri *et al.* (2017a) derived with dust temperature of 25 K, we compared the mass distribution of ACA cores with 6.7 GHz CH_3OH masers (ACA-MMB). The mass range of the entire sample is 11.8 – 3876.3 $M_\odot$, and 11.8 – 403.4 $M_\odot$ for the more strictly selected sample of 27 cores. Figure 3 shows the distribution of the core masses.

Our statistical analysis supports the notion that outflows precede 6.7 GHz CH_3OH maser excitation in massive dense cores.

References

Bayandina, O. S., Val'tts, I. E., & Larionov, G. M. 2012, *Astron. Rep.*, 56, 553
Breen, S. L., Ellingsen, S. P., Caswell, J. L., & Lewis, B. E. 2010, *MNRAS*, 401, 2219
Breen, S. L., Ellingsen, S. P., Caswell, J. L., *et al.* 2011, *ApJ*, 733, 80
Breen, S. L., Ellingsen, S. P., Contreras, Y., *et al.* 2013, *MNRAS*, 435, 524
Caswell, J. L., Fuller, G. A., Green, J. A., *et al.* 2010, *MNRAS*, 404, 1029
Caswell, J. L., Fuller, G. A., Green, J. A., *et al.* 2011, *MNRAS*, 417, 1964
Chibueze, J. O., Omodaka, T., Handa, T., *et al.* 2014, *ApJ*, 784, 114
Chibueze, J. O., Csengeri, T., Tatematsu, K., *et al.* 2017, *ApJ*, 836, 59
Csengeri, T., Bontemps, S., Wyrowski, F., *et al.* 2017b, *A&A*, 601, A60
Csengeri, T., Bontemps, S., Wyrowski, F., *et al.* 2017a, *A&A*, 600, L10
Codella, C., Lorenzani, A., Gallego, A. T., Cesaroni, R., & Moscadelli, L. 2004, *A&A*, 417, 615
de Villiers, H. M., Chrysostomou, A., Thompson, M. A., *et al.* 2015, *MNRAS*, 449, 119
Gerner, T., Beuther, H., Semenov, D., *et al.* 2014, *A&A*, 563, A97
Green, J. A., Caswell, J. L., Fuller, G. A., *et al.* 2012, *MNRAS*, 420, 3108
Reid, M. J. 2007, Astrophysical Masers and their Environments, 242, 522
Urquhart, J. S., Moore, T. J. T., Schuller, F., *et al.* 2013, *MNRAS*, 431, 1752
Urquhart, J. S., Thompson, M. A., Moore, T. J. T., *et al.* 2013, *MNRAS*, 435, 400

Astrophysical Masers:
Unlocking the Mysteries of the Universe
Proceedings IAU Symposium No. 336, 2017
A. Tarchi, M.J. Reid & P. Castangia, eds.

© International Astronomical Union 2018
doi:10.1017/S1743921317010250

The extraordinary outburst in NGC6334I-MM1: the rise of dust and emergence of 6.7 GHz methanol masers

Todd R. Hunter[1], Crystal L. Brogan[1], James O. Chibueze[2], Claudia J. Cyganowski[3], Tomoya Hirota[4] and Gordon C. MacLeod[5]

[1] National Radio Astronomy Observatory,
520 Edgemont Rd, Charlottesville, VA 22903, USA
email: `thunter@nrao.edu`

[2] SKA South Africa,
3rd Floor, The Park, Park Road, Pinelands, Cape Town, 7405, South Africa

[3] SUPA, School of Physics and Astronomy, University of St. Andrews,
North Haugh, St. Andrews KY16 9SS, UK

[4] Mizusawa VLBI Observatory, National Astronomical Observatory of Japan,
Osawa 2-21-1, Mitaka-shi, Tokyo 181-8588, Japan

[5] Hartebeesthoek Radio Astronomy Observatory,
PO Box 443, Krugersdorp 1740, South Africa

Abstract. Our 2015-2016 ALMA 1.3 to 0.87 mm observations (resolution $\sim$ 200 au) of the massive protocluster NGC6334I revealed that an extraordinary outburst had occurred in the dominant millimeter dust core MM1 (luminosity increase of 70×) when compared with earlier SMA data. The outburst was accompanied by the flaring of ten maser transitions of three species. We present new results from our recent JVLA observations of Class II 6.7 GHz methanol masers and 6 GHz excited OH masers in this region. Class II masers had not previously been detected toward MM1 in any interferometric observations recorded over the past 30 years that targeted the bright masers toward other members of the protocluster (MM2 and MM3=NGC6334F). Methanol masers now appear both toward and adjacent to MM1 with the strongest spots located in a dust cavity $\sim$1 arcsec (1300 au) north of the MM1B hypercompact HII region. In addition, new excited OH masers appear on the non-thermal source CM2. These data reveal the dramatic effects of episodic accretion onto a deeply-embedded high mass protostar and demonstrate its ongoing impact on the surrounding protocluster.

Keywords. Massive protostars, accretion, masers

1. Introduction

Episodic accretion in protostars is increasingly recognized as being an essential phenomenon in star formation (Evans *et al.* 2009). Outbursts from deeply-embedded low-mass protostars have recently been detected via large increases in their mid-infrared or submillimeter emission, including a Class 0 source HOPS-383 (Safron *et al.* 2015) and a Class I source EC53 (Yoo *et al.* 2017). It has also become clear that massive protostars also exhibit outbursts as evidenced by the 4000 $L_\odot$ erratic variable V723 Carinae (Tapia *et al.* 2015), the far-infrared flare from the 20 $M_\odot$ protostar powering S255IR-NIRS3 (Caratti o Garatti *et al.* 2016), and the ongoing (sub)millimeter flare of the deeply-embedded source MM1 in the massive protocluster NGC6334I (Hunter *et al.* 2017). The large increases in bolometric luminosity observed in these events provide evidence for accretion outbursts similar to those predicted by hydrodynamic simulations of massive star formation (Meyer *et al.* 2017). Identifying additional phenomena associated with these events will help explore the mechanism of the outbursts, particularly if spectral line tracers can be identified, as they can potentially trace gas motions at high angular resolution. As the brightest line emission emerging from regions of massive star formation, masers offer an important avenue of study, especially since the recent methanol maser flare in S255IR (Moscadelli *et al.* 2017) and the multi-species flares in NGC6334I-MM1 (Brogan *et al.* 2018; MacLeod *et al.* 2018) have provided a direct link between protostellar accretion outbursts and maser flares.

2. Observations

Observations of the massive protocluster NGC6334I were performed with the Karl G. Jansky Very Large Array (VLA)† in A-configuration in Oct.-Nov. 2016 (mean epoch 2016.9). In addition to the coarse resolution spectral windows used to image the continuum emission in C-band (4-8 GHz), we observed several maser transitions with the 8-bit digitizers and high spectral resolution, dual-polarization correlator windows. The 6.66852 GHz (hereafter 6.7 GHz) CH_3OH 5(1)-6(0) A+ line (E_{Lower}=49 K) was observed with a channel spacing of 1.953125 kHz (0.0878 km s^{-1}) over a span of 90 km s^{-1} centered at -7 km s^{-1}. Three lines of excited-state OH maser emission were also observed: J=1/2-1/2, F=0-1 at 4.66024 GHz (E_{Lower}=182 K), and J=5/2-5/2, F=2-2 at 6.03075 GHz and F=3-3 at 6.03509 GHz (both having E_{Lower}=120 K). The data were calibrated using the VLA pipeline with some additional flagging required. The calibrated data were Hanning smoothed to reduce ringing from strong spectral features and then iteratively self-calibrated using the bright continuum emission as the initial model. Multiscale clean was used with scales of 0, 5, and 15 times the image pixel size to produce Stokes I continuum images with a beamsize of 0.63″ × 0.14″. The bright UCHII region (MM3 = NGC 6334F, Hunter *et al.* 2006) limits the dynamic range of the continuum images. Stokes I cubes of the maser emission were made with a channel spacing (and effective spectral resolution) of 0.15 km s^{-1}. We also made cubes of right and left circular polarization for the strongly-polarized high frequency OH lines. The 4.66 GHz OH line was undetected at an rms of 3.3 mJy beam^{-1}.

3. Results

We fit each channel of the maser cubes that had significant emission with an appropriate number of Gaussian sources, using the pixel position of each emission peak and the beam shape as the initial guesses for the fitted parameters. Their positions and relative intensities are shown in Fig. 1. Many maser features have now appeared within the continuum source MM1 and along its eastern and northern peripheries. These are the first detections of Class II methanol maser emission toward this location in nearly 30 years of interferometric observations, all of which were sufficiently sensitive to detect masers of this strength. Previous epochs at 6.7 GHz include Jan. 1992 (Norris *et al.* 1993), May 1992 (Ellingsen *et al.* 1996), July 1994 (Walsh *et al.* 1998), July 1995 (Caswell *et al.* 1997), Sep. 1999 (Dodson *et al.* 2012), Mar. 2005 (Krishnan *et al.* 2013), May 2011 (Brogan *et al.* 2016), and Aug. 2011 (Green *et al.* 2015). In all previous epochs, masers originate only from the UCHII region (NGC6334F) and MM2. In addition, the original VLBI imaging of the Class II 12.2 GHz methanol maser did not detect any features toward

† The National Radio Astronomy Observatory is a facility of the National Science Foundation operated under agreement by the Associated Universities, Inc.

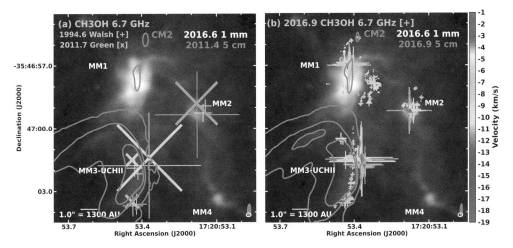

Figure 1. (a) The methanol maser positions (+ symbols) observed by the Australia Tele-scope Compact Array (ATCA) on July 31, 1994 (Walsh *et al.* 1998) and September 2, 2011 (Green *et al.* 2015) are overlaid on an epoch 2016.6 ALMA 1 mm continuum image in greyscale (Hunter *et al.* 2017). Epoch 2011.5 VLA 5 cm continuum contours are also overlaid with levels of $2.2 \times 10^{-2} \times (4, 260, 600)$ mJy beam^{-1} (Brogan *et al.* 2016). Maser intensity is indicated by the symbol diameter (proportional to the square root of the intensity). The millimeter continuum sources are labeled for reference. (b) Same as (a) but the VLA epoch 2016.9 masers are overlaid.

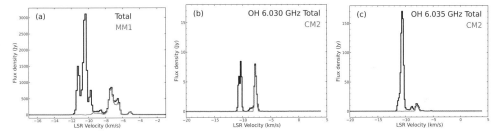

Figure 2. (Panel a) Integrated spectra from the epoch 2016.9 6.7 GHz methanol maser cube taken over all the maser features (upper profile) and from an elliptical region surrounding only the maser features associated with MM1 (lower profile). (Panels b-c) Integrated spectra of the excited-state OH masers constructed by summing the flux from all directions showing emission (upper profiles) and from CM2 only (lower profiles): (b) 6.030 GHz line; (c) 6.035 GHz line.

MM1 (Norris *et al.* 1988), nor are any features toward MM1 reported in the two epochs of the Methanol MultiBeam survey of 12.2 GHz masers (Breen *et al.* 2012).

The brightest features in MM1 (peak = 545 Jy at -7.25 km s^{-1}) lie in the valley of emission in the northern part, between the 1.3 mm continuum sources MM1F and MM1G. This velocity matches that of the thermal hot core emission (-7.3 km s^{-1}, Zernickel *et al.* 2012). The integrated spectrum of MM1 compared to the total field (Figure 2) further demonstrates that the dominant emission arises from -6 to -8 km s^{-1}. Additional strong features appear on MM1F and weaker features near the LSR velocity extend northward from MM1F, with the northernmost spot coincident with the non-thermal radio source CM2. Weak redshifted features lie just south of MM1F. Another set of near-LSR features coincide with the millimeter source MM1C, however the three brightest and most central millimeter sources (MM1A, 1B, and 1D) are notably lacking in any maser emission. A large number of moderate strength, primarily blueshifted masers lie in the cavities of dust emission located west and southwest of MM1A. Finally, a number of weak, moderately redshifted features lie in the depression between MM1C and MM1E.

The tendency of the new masers to lie along the heated surfaces of dust cavities while avoiding the regions of highest density are consistent with pumping schemes for 6.7 GHz masers. This transition is radiatively pumped by mid-infrared photons (Sobolev & Deguchi 1994), requiring dust temperatures above ~120 K and gas densities below about 10^8 cm^{-3} (Cragg *et al.* 2005). Following the millimeter continuum outburst, the dust temperature of MM1 exceeds 250 K toward the central components, and was above 150 K over a several square arcsec region (Brogan *et al.* 2016). Prior to the outburst, when the dust and gas temperatures were presumably in better equilibrium, the most compact organic species exhibited excitation temperatures of 100 K (CH$_3$OH) to 150 K (CH$_3$CN) (Zernickel *et al.* 2012). Thus, the rapid heating of dust grains by more than ~100 K by the recent accretion outburst can plausibly explain the appearance of 6.7 GHz masers in the gas in the vicinity of the powering source MM1B. Furthermore, the group of masers to the west and southwest of MM1 coincide with an area of warm dense gas where methanol is abundant, as traced by the 279.3519 GHz thermal transition ($E_{lower} = 177$ K) in our post-outburst ALMA observations.

In the 6 GHz OH masers, our analysis of Zeeman pairs confirms the north/south dichotomy in the sign of the line-of-sight magnetic field across the UCHII region previously reported by Caswell *et al.* (2011). As shown in Fig. 2b-c, we find a new velocity component at -7 to -8 km s^{-1} that originates from the non-thermal radio continuum source CM2 (Brogan *et al.* 2016). It shows an abnormally comparable intensity ratio between the 6.035 and 6.030 GHz lines and has likely been excited by the recent outburst in MM1 and shows a smaller magnetic field (+0.5 to +3.7 mG) compared to the UCHII region.

References

Breen, S. L., Ellingsen, S. P., Caswell, J. L., *et al.* 2012, *MNRAS* 421, 1703

Brogan, C. L., Hunter, T. R., *et al.* 2018, these proceedings

Brogan, C. L., Hunter, T. R., Cyganowski, C. J., *et al.* 2016, *ApJ* 832, 187

Caratti o Garatti, A., Stecklum, B., Garcia Lopez, R., *et al.* 2016, *Nature* (Physics) 13, 276

Caswell, J. L., Kramer, B. H., & Reynolds, J. E. 2011, *MNRAS* 414, 1914

Caswell, J. L. 1997, *MNRAS* 289, 203

Cragg, D. M., Sobolev, A. M., & Godfrey, P. D. 2005, *MNRAS* 360, 533

Dodson, R. & Moriarty, C. D. 2012, *MNRAS* 421, 2395

Ellingsen, S. 1996, Ph.D. Thesis, University of Tasmania

Evans, N. J., II, Dunham, M. M., Jørgensen, J. K., *et al.* 2009, *ApJ* (Supplement) 181, 321-350

Green, J. A., Caswell, J. L., & McClure-Griffiths, N. M. 2015, *MNRAS* 451, 74

Hunter, T. R., Brogan, C. L., MacLeod, G., *et al.* 2017, *ApJ* (Letters) 837, L29

Hunter, T. R., Brogan, C. L., Megeath, *et al.*, 2006, *ApJ* 649, 888

Krishnan, V., Ellingsen, S. P., Voronkov, M. A., & Breen, S. L. 2013, *MNRAS* 433, 3346

MacLeod, G., *et al.*, 2018, these proceedings

Meyer, D. M.-A., Vorobyov, E. I., Kuiper, R., & Kley, W. 2017, *MNRAS* 464, L90

Moscadelli, L., Sanna, A., Goddi, C., *et al.* 2017, *A&A* 600, L8

Norris, R. P., Caswell, J. L., Wellington, K. J., *et al.* 1988, *Nature* 335, 149

Norris, R. P., Whiteoak, J. B., Caswell, J. L., *et al.* 1993, *ApJ* 412, 222

Safron, E. J., Fischer, W. J., Megeath, S. T., *et al.* 2015, *ApJ* (Letters) 800, L5

Sobolev, A. M. & Deguchi, S. 1994, *A&A* 291, 569

Tapia, M., Roth, M., & Persi, P. 2015, *MNRAS* 446, 4088

Walsh, A. J., Burton, M. G., Hyland, A. R., & Robinson, G. 1998, *MNRAS* 301, 640

Yoo, H., Lee, J.-E., Mairs, S., *et al.* 2017, *ApJ* 849, 69

Zernickel, A., Schilke, P., Schmiedeke, A., *et al.* 2012, *A&A* 546, A87

Astrophysical Masers:
Unlocking the Mysteries of the Universe
Proceedings IAU Symposium No. 336, 2017
A. Tarchi, M.J. Reid & P. Castangia, eds.

© International Astronomical Union 2018
doi:10.1017/S1743921317010237

The extraordinary outburst in NGC6334I-MM1: dimming of the hypercompact HII region and destruction of water masers

Crystal L. Brogan[1], Todd R. Hunter[1], Gordon MacLeod[2], James O. Chibueze[3] and Claudia J. Cyganowski[4]

[1]National Radio Astronomy Observatory,
520 Edgemont Rd, Charlottesville, VA 22903, USA
email: thunter@nrao.edu

[2]Hartebeesthoek Radio Astronomy Observatory,
PO Box 443, Krugersdorp 1740, South Africa

[3]SKA South Africa,
3rd Floor, The Park, Park Road, Pinelands, Cape Town, 7405, South Africa

[4]SUPA, School of Physics and Astronomy, University of St. Andrews,
North Haugh, St. Andrews KY16 9SS, UK

Abstract. We present subarcsecond resolution pre- and post-outburst JVLA continuum and water maser observations of the massive protostellar outburst source NGC6334I-MM1. The continuum data at 5 and 1.4 cm reveal that the free-free emission powered by MM1B, modeled as a hypercompact HII region from our 2011 JVLA data, has dropped by a factor of 5.4. Additionally, the water maser emission toward MM1, which had previously been strong (500 Jy) has dramatically reduced. In contrast, the water masers in other locations in the protocluster have flared, with the strongest spots associated with CM2, a non-thermal radio source that appears to mark a shock in a jet emanating 2″ (2600 au) northward from MM1. The observed quenching of the HCHII region suggests a reduction in uv photon production due to bloating of the protostar in response to the episodic accretion event.

Keywords. Massive protostars, accretion, masers

1. Introduction

Episodic accretion in protostars is increasingly recognized as being an essential phenomenon in star formation (Kenyon *et al.* 1990, Evans *et al.* 2009). The total luminosity of a protostar scales with the instantaneous accretion rate, so variations in that rate will lead to observable brightness changes (e.g., Offner & McKee 2011). Direct evidence for episodic accretion events towards massive protostars have recently emerged via dramatic increases in the brightness, and hence luminosity observed in the near-IR towards S255IR-NIRS3 (Caratti o Garatti *et al.* 2016), as well as in the millimeter toward NGC6334I-MM1 (Hunter *et al.* 2017a). The impact of such an accretion event on the free-free emission from a massive protostar is as yet relatively unexplored, though Hosokawa & Omukai (2009) (for example) suggest that significant suppression of uv-photons is expected.

Another eruptive phenomenon associated with sites of high mass star formation are maser flares, such as the three past events observed in the water masers in the vicinity of Orion KL (Abraham *et al.* 1981, Omodaka *et al.* 1999, Hirota *et al.* 2014). The repeating nature of the Orion maser outbursts (Tolmachev 2011) as well as the periodic

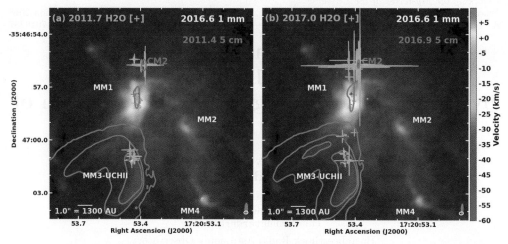

Figure 1. (a) Water maser positions (+ symbols) from epoch 2011.7 (Brogan *et al.* 2016) are overlaid on an epoch 2016.6 ALMA 1 mm continuum image (Hunter *et al.* 2017a). Epoch 2011.5 JVLA 5 cm continuum contours are also overlaid with contour levels of $3.7 \times 10^{-5} \times$ (4, 260, and 600). Maser intensity is indicated by the size of the symbol, while velocity is indicated by the color. Continuum sources are labeled for reference; sources with millimeter emission are labeled in white. (b) Same as (a) but the epoch 2016.9 5 cm emission contours are shown (contour levels $2.2 \times 10^{-5} \times$ (4, 260, and 600)), along with the 2017.0 water maser positions. In both panels, the size of the symbol is $\propto$ flux density$^{0.5}$, and the synthesized beams are shown in the lower right.

features seen toward many high mass protostellar objects (HMPOs) in the 6.7 GHz methanol maser line (e.g. Goedhart *et al.* 2004, MacLeod & Gaylard 1996), have suggested that variations in the underlying protostellar activity could be responsible for the maser flares. Indeed, the recent methanol maser flare associated with the infrared outburst in S255IR-NIRS3 (Moscadelli *et al.* 2017), and the emergence of methanol masers toward the (sub)millimeter outburst in NGC6334I-MM1 (Hunter *et al.* 2017b) have provided the first direct link between protostellar accretion outbursts and maser flares. Detailed study of additional maser species is of critical importance to understanding accretion process in high-mass protostars.

In this proceeding, we describe multi-epoch results from subarcsecond resolution Karl G. Jansky Very Large Array (VLA)† observations of the centimeter-λ emission and 22.235 GHz H_2O maser emission toward the millimeter outburst source NGC6334I-MM1 (Hunter *et al.* 2017a). NGC6334I is a protocluster forming massive stars (Hunter *et al.* 2006) at a distance of 1.3 ± 0.1 kpc (Chibueze *et al.* 2014, Reid *et al.* 2014). Its massive constituents include a more evolved UCHII region (MM3), two millimeter bright cores, MM1 and MM2, each of which contain multiple massive protostars as well as copious hot core line emission, and a line poor but millimeter-bright massive dust source MM4 (Brogan *et al.* 2016). One of the massive protostars in MM1, MM1B has already formed a hypercompact HII region (Brogan *et al.* 2016), and in 2015 appears to have undergone a significant episodic accretion event resulting in a large (factor of ~70) increase in its luminosity (Hunter *et al.* 2017a).

† The National Radio Astronomy Observatory is a facility of the National Science Foundation operated under agreement by the Associated Universities, Inc.

2. Results

Figure 1 shows the pre-outburst (epoch 2011.4) and post-outburst (epoch 2016.9) 5 cm continuum emission, as well as the locations of the 2011.7 and 2017.0 epoch H_2O masers superposed on an ALMA 2016.6 1 mm image of the dust continuum emission toward NGC6334-I. We find that the 5 cm emission is not significantly changed between the pre- and post-outburst epochs (the most apparent differences are due to the lower image fidelity of the 2011 data). Of particular note is the persistence of the non-thermal source CM2, located $\sim 2''$ (2600 au) north of the origin of the millimeter outburst MM1-B. The nature of this source, which appears to be completely devoid of dust emission from our sensitive 1 mm ALMA data, is a mystery. Indeed, this source has not been detected shortward of 5 cm. Its location along the same axis as a 5 cm jet emanating from MM1-B suggests that it is a related shock structure.

In contrast, the H_2O maser data shows considerable differences: (i) the maser emission has significantly flared toward the mysterious non-thermal source CM2; (ii) the emission toward MM1 has significantly decreased in both flux density and spatial extent; (iii) a number of new maser regions are detected toward the northern part of the UCHII region MM3; and (iv) the weak masers detected toward MM4 in the 2011.4 epoch are no longer detectable in the 2017.0 epoch. Based on single dish monitoring of the water maser emission in this source from the HartRAO 26m telescope over the last decade (MacLeod *et al.* 2018), a major water maser flaring event began in early 2015 (presumably marking the start of the accretion event). The JVLA data presented here localize the flare to the location of CM2 for the first time.

Figure 2 shows a close-up view of the pre- and post-outburst H_2O maser activity toward MM1, as well as contours of pre- and post-outburst 1.5/1.35 cm continuum emission. This figure demonstrates both the significant decrease in water maser activity toward MM1 post-outburst, and a dramatic drop in free-free continuum emission from the hypercompact HII region MM1B, while the emission from MM1D has remained constant. We suggest that changes in the physical conditions in the vicinity of MM1B due to the accretion event have disrupted much of the water maser activity. A potential culprit is the increase in dust temperature that accompanied the outburst (Hunter *et al.* 2017a).

To further assess 1.5/1.35 cm dimming, we fit two-dimensional Gaussian models to each source, and after interpolating the measurements to a common frequency using the spectral index of each object, we find that the flux density of MM1D at 1.5 cm has remained constant to within 10%, giving confidence in the relative flux scaling between the 2011.4 and 2017.0 datasets. In contrast, the flux density at 1.5 cm of the hypercompact HII region MM1B has dropped by a factor of ~ 5.4 during this interval. The pre-outburst 1.5 cm emission from MM1B (epoch 2011.4, 1.78$\pm$0.11 mJy) is consistent with an ionizing photon rate (Q) of $10^{43.94}$ s^{-1} (using Eq. 6 of Turner & Matthews 1984). This value of Q is consistent with a zero age main sequence (ZAMS) stellar surface temperature of 19500 K (Diaz-Miller *et al.* 1998), which in turn corresponds to a ZAMS spectral type of B2.7, mass of 5.6 $M_\odot$ and radius of 3.2 $R_\odot$ (Hanson *et al.* 1997). The observed dimming of the 1.35 cm emission in epoch 2017.0 requires a lower stellar temperature, while the luminosity increase requires a larger radius by a factor of 10 (i.e. up to 32 $R_\odot$), a combination that is equivalent to a B3.4 supergiant star. This level of stellar bloating in response to a rapid increase in accretion rate is predicted by Hosokawa & Omukai (2009) (also see Inayoshi *et al.* 2013) for an accretion rate of 10^{-3} $M_\odot$ yr^{-1}. Future subarcsecond JVLA monitoring of the centimeter wavelength emission at 1.35 and 0.7 cm will help to further characterize how this emission will evolve with time and, consequently, how the underlying protostar's properties continue to respond to the accretion event.

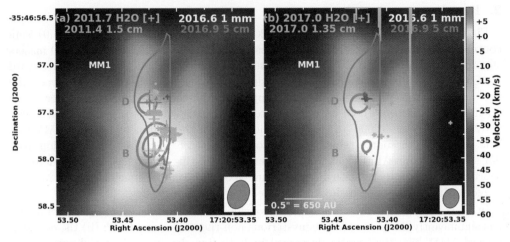

Figure 2. Similar to Fig. 1, except showing a smaller field of view toward MM1, note both panels show the 5 cm 2016.9 continuum contours. In (a) 1.5 cm epoch 2011.4 continuum contours are also overlaid (thicker contours) at levels of $7.2 \times 10^{-5} \times (4, 10)$ Jy beam^{-1} (the beam is shown in the lower right). In (b) 1.35 cm epoch 2017.0 continuum contours are overlaid at $4.4 \times 10^{-5} \times (4)$ Jy beam^{-1} (the beam is shown in the lower right). In both panels, the size of the symbol is $\propto$ flux density$^{0.5}$. For reference the LSR velocity of this source is about -7 km s^{-1} (McGuire *et al.* 2017, Zernickel *et al.* 2012).

References

Abraham, Z., Cohen, N. L., Opher, R., Raffaelli, J. C., & Zisk, S. H. 1981, *A&A* 100, L10
Brogan, C. L., Hunter, T. R., Cyganowski, C. J., *et al.* 2016, *ApJ* 832, 187
Caratti o Garatti, A., Stecklum, B., Garcia Lopez, R., *et al.* 2016, *Nature* (Physics) 13, 276
Chibueze, J. O., Omodaka, T., Handa, T., *et al.* 2014, *ApJ* 784, 114
Diaz-Miller, R. I., Franco, J., & Shore, S. N. 1998, *ApJ* 501, 192
Evans, N. J., II, Dunham, M. M., Jørgensen, J. K., *et al.* 2009, *ApJ* (Supplement) 181, 321-350
Goedhart, S., Gaylard, M. J., & van der Walt, D. J. 2004, *MNRAS* 355, 553
Hanson, M. M., Howarth, I. D., & Conti, P. S. 1997, *ApJ* 489, 698
Hirota, T., Tsuboi, M., Kurono, Y., *et al.* 2014, *PASJ* 66, 106
Hosokawa, T. & Omukai, K. 2009, *ApJ* 691, 823
Hunter, T. R., *et al.* 2017b, *ApJ*, submitted
Hunter, T. R., Brogan, C. L., MacLeod, G., *et al.* 2017a, *ApJ* 837, L29
Hunter, T. R., Brogan, C. L., Megeath, S. T., *et al.*, 2006, *ApJ* 649, 888
Inayoshi, K., Sugiyama, K., Hosokawa, T., Motogi, K., & Tanaka, K. E. I. 2013, *ApJ* 769, L20
Kenyon, S. J., Hartmann, L. W., Strom, K. M., & Strom, S. E. 1990, *AJ* 99, 869
MacLeod G., *et al.*, 2018, *MNRAS*, accepted
MacLeod, G. C. & Gaylard, M. J. 1996, *MNRAS* 280, 868
McGuire, B. A. 2017, *ApJ* 851, L46
Moscadelli, L., Sanna, A., Goddi, C., *et al.* 2017, *A&A* 600, L8
Offner, S. S. R. & McKee, C. F. 2011, *ApJ* 736, 53
Omodaka, T., Maeda, T., Miyoshi, M., *et al.* 1999, *PASJ* 51, 333
Reid, M. J., Menten, K. M., Brunthaler, A., *et al.* 2014, *ApJ* 783, 130
Tolmachev, A. 2011, *The Astronomer's Telegram* 3177
Turner, B. E. & Matthews, H. E. 1984, *ApJ* 277, 164
Zernickel, A., Schilke, P., Schmiedeke, A., *et al.* 2012, *A&A* 546, A87

Astrophysical Masers:
Unlocking the Mysteries of the Universe
Proceedings IAU Symposium No. 336, 2017
A. Tarchi, M.J. Reid & P. Castangia, eds.

© International Astronomical Union 2018
doi:10.1017/S1743921317011681

Understanding high-mass star formation through KaVA observations of water and methanol masers

Kee-Tae Kim[1], Tomoya Hirota[2], Koichiro Sugiyama[2], Jungha Kim[2], Do-Young Byun[1], James Chibueze[3], Kazuya Hachisuka[2], Bo Hu[4], Eodam Hwang[1], Ji-Hyun Kang[1], Jeong-Sook Kim[1], Mikyoung Kim[2], Tie Liu[1], Naoko Matsumoto[2], Kazuhito Motogi[5], Chung Sik Oh[1], Kazuyoshi Sunada[2], Yuanwei Wu[2] and KaVA star formation group

[1] Korea Astronomy & Space Science Institute, 776 Daedeokdae-ro, Yuseong-gu, Daejeon 34055, Republic of KOREA

[2] Mizusawa VLBI Observatory, National Astronomical Observatory of Japan, 2-21-1 Osawa, Mitaka, Tokyo 181-8588, Japan

[3] Department of Physics and Astronomy, Faculty of Physical Sciences, University of Nigeria, Carver Building, 1 University Road, 410001 Nsukka, Nigeria

[4] School of Astronomy and Space Science, Nanjing University, 22 Hankou Rd., Nanjing, Jiangsu 210093, China

[5] Graduate School of Sciences and Technology for Innovation, Yamaguchi University, Yoshida 1677-1, Yamaguchi 753-8512, Japan

Abstract. Despite their importance in the formation and evolution of stellar clusters and galaxies, the formation of high-mass stars remains poorly understood. We recently started a systematic observational study of the 22 GHz water and 44 GHz class I methanol masers in high-mass star-forming regions as a four-year KaVA large program. Our sample consists of 87 high-mass young stellar objects (HM-YSOs) in various evolutionary phases, many of which are associated with two or more different maser species. The primary scientific goals are to measure the spatial distributions and 3-dimensional velocity fields of multiple maser species, and understand the dynamical evolution of HM-YSOs and their circumstellar structures, in conjunction with follow-up observations with JVN/EAVN (6.7 GHz class II methanol masers), VERA, and ALMA. In this paper we present details of our KaVA large program, including the first-year results and observing/data analysis plans for the second year and beyond.

Keywords. ISM: molecules — masers: ISM — stars: formation

1. Introduction

Although high-mass stars are fundamental in the formation and evolution of clusters and galaxies, their formation process of is still poorly understood (Zinnecker & Yorke 2007). This is mainly due to their fast and clustering formation and evolution in heavily obscured regions at large distances. Despite significant progress in the recent years, there are at least two competing models: turbulent core accretion vs. competitive accretion. Therefore, direct observations of the circumstellar structures (e.g., jets/outflows, accretions disks, infalling envelopes) of individual high-mass young stellar objects (HM-YSOs) are essential for understanding the formation and evolution of high-mass stars.

Various masers are associated with HM-YSOs in a wide range of evolutionary phases, and they have been utilized to probe high-mass star formation processes. Water (H_2O) masers at 22 GHz trace shocked gas regions associated with various kinds of dynamical

259

structures including jets/outflows, disks, and HII regions. Methanol (CH$_3$OH) masers are divided into two classes (I & II) on the basis of their relationship with the central (proto)stars (Menten 1991). The brightest class II masers at 6.7 GHz are located close to HM-YSOs (e.g., Fujisawa *et al.* 2014), while class I masers at 44 GHz are usually located farther from HM-YSOs (e.g., Kurtz *et al.* 2004). Because of extremely high intensities, 22 GHz water masers and 6.7 GHz methanol masers have been employed as powerful tools to investigate the circumstellar spatial and velocity structures of HM-YSOs using VLBIs at resolutions of ∼1 milli-arcsecond (mas). In addition, we have recently reported the first VLBI imaging of 44 GHz class I methanol masers (Matsumoto *et al.* 2014), demonstrating the unique capability of the KVN and VERA array (KaVA) providing relatively short and dense *uv* coverage. These three maser species are complementary with each other for investigating overall 3-dimensional (3D) structures and dynamics around HM-YSOs by multi-epoch and multi-species VLBI studies. Such well-compiled and time-resolved VLBI datasets are quite unique in the ALMA era, allowing us to quantitatively understand the evolution of HM-YSOs and their circumstellar structures.

We started a KaVA large program in late 2015 to conduct systematic monitoring observations of 22 GHz water and 44 GHz class I methanol masers: *Understanding high-mass star formation through KaVA observations of water and methanol masers* (co-PIs: Tomoya Hirota & Kee-Tae Kim). Our program will provide the proper motion data of two maser species. We have been also conducting similar monitoring observations of 6.7 GHz class II methanol masers with the Japanese VLBI Network (JVN)/East Asian VLBI Network (EAVN) (PI: Koichiro Sugiyama). Our sample consists of 87 HM-YSOs in various evolutionary stages. The sources were selected mainly from the KVN single-dish and fringe surveys of water and class I methanol masers (e.g., Kang *et al.* 2015; Kang *et al.* 2016; Kim *et al.* 2018; Kim, Kee-Tae *et al.* in prep.). Many of them are associated with two or more different maser species and/or high-velocity water maser features.

2. Immediate Objectives

The primary scientific goal of our KaVA large program is to understand the dynamical evolution of HM-YSOs and their circumstellar structures by measuring the spatial distributions and 3D velocity fields of the three maser species. These studies can be done only by the KaVA, which is capable of imaging all three representative masers. Through the KaVA program, we will address key issues in high-mass star formation. First, we will establish an evolutionary sequence of different maser species with statistical samples. Although various maser species are known to be associated with HM-YSOs, their relationship with the central (proto)stars is still controversial (see, e.g., Ellingsen at al. 2007 and Reid 2007). This is mainly due to a lack of high-resolution observations. We will investigate the 3D velocity structures of three maser species at a resolution of ∼1 mas to identify their possible powering sources at a scale of ∼1000 AU (e.g., Fujisawa *et al.* 2014), by combining ancillary follow-up observational data.

In addition, our program will provide powerful tools to reveal the driving mechanism of jets/outflows from HM-YSOs. Jets/outflows are one of the most important processes in star formation because of their significant role to extract angular momentum. Beuther & Shepherd (2005) proposed that initially well-collimated jets/outflows from HM-YSOs evolve into wide opening-angle outflows. However, Seifried *et al.* (2012) suggested an opposite scenario from their 3D MHD simulations, in which poorly-collimated outflows evolve into well-collimated ones with the development of Keplerian disks. It is thus worth establishing a scenario explaining when and how jets/outflows from HM-YSOs with different morphology are forming and evolving.

Several previous studies suggested that high-velocity water maser sources are a good subsample of high-mass protostellar jets. Especially, dominant blue-shifted maser sources are the best candidates of very compact and young sources (Caswell & Phillips 2008). High-velocity water maser sources are usually characterized by quite short lifetimes of the individual maser features and hence, are studied only by well organized dense monitoring as in our KaVA program. Such a time-variability can be an important clue to time-dependent nature of accretion and outflow processes (Motogi *et al.* 2016). Moreover, we will be able to estimate inclination angles of disk-jet/outflow systems, which is essential for quantitative spectral energy distribution modeling, from the 3D velocity field and spatial distribution of high-velocity maser features. Thus, the results will provide key parameters for the theoretical works to understand physical properties of HM-YSO themselves.

3. Observing Strategy and Timeline

First year (finished): We have first conducted snap-shot imaging observations of selected targets (25 water and 19 methanol maser sources) for which no previous VLBI data were available, to check detectability of maser features. We found a variety of maser morphology and selected appropriate sources for further monitoring observations in the second year (see the next section).

Second year: We will start monitoring observations toward the selected targets (16 water and 3 methanol maser sources) to measure the internal proper motions of maser features. The number of observing epochs will be 5 for each source, and the total observing time will be 160 hours and 40 hours for K- and Q-bands, respectively.

Third year and beyond: We will continue monitoring observations to extend target sources not observed in the second year. Because non-detection (∼50 sources) could be due to time variability, we will conduct fringe-check and single-dish monitoring in the second year to find additional good targets not observed in the first and second years. In addition, we will carry out intensive monitoring for specific highly variable sources reflecting episodic accretion events.

4. First-year Results

In the first year, we undertook snap-shot imaging observations of 25 water and 19 methanol maser sources without available previous VLBI data. Water masers were detected toward 21 (84%) sources. They show various distributions of water maser features, including elongated and arc-like ones (see Kim, Jungha *et al.* in this volume for details). Both blue- and red-shifted maser features were detected in 16 of the 21. We will investigate the physical and dynamical properties of the jets/outflows in combination with follow-up ALMA observations, e.g., 2015.1.01571.S (PI: Mi Kyoung Kim).

None of the observed methanol maser sources have been observed with any VLBI except G18.34+1.78SW, which was already imaged with the KaVA (Matsumoto *et al.* 2014). Among the 19, correlated data were available for 17 sources so far and 16 sources were succeeded in VLBI imaging. Only single spectral features were detected in most sources, while 2–4 maser features were detected in four sources: G357.967-0.163, G18.34+1.78SW, G28.37+0.07MM1, and G49.49-0.39 (see Figure 1). In that case, individual features are typically separated by 50–500 AU with each other and sometimes more than 1,000 AU at the largest scale. The peak flux density ratios of CLEAN component to auto-correlated spectra range from 2% to 30% (Figure 1).

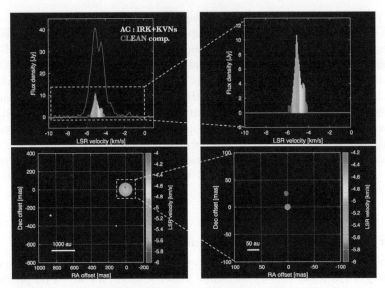

Figure 1. 44 GHz class I methanol masers in G357.967-0.163. (Upper panel) Auto-correlated spectrum (white) and sum of CLEAN components (color). (Lower panel) Spatial distribution of detected maser features.

We selected the 16 water maser sources with both blue- and red-shifted maser features and 3 methanol maser sources with three or more maser features for the second-year monitoring observations. We will be able to measure relative proper motions with respect to the brightest feature in each source, because there are sufficient number of features (>10 for the water maser and $\geqslant 3$ for the methanol maser sources).

5. Follow-up Observations

In addition to the JVN/EAVN observations of 6.7 GHz class II methanol masers, we have been performing follow-up observations using VERA and ALMA, as well. The VERA observations (PI: Tomoya Hirota) aim at determining the distances by measuring the annual parallaxes using water masers. The ALMA observations (PI: Mi Kyoung Kim) aim at investigating the physical and dynamical properties of molecular outflows associated with 44 GHz methanol masers and to identify the outflow driving sources.

References

Beuther, H. & Shepherd, D. 2005, *in Cores to Clusters (ASSL Volume 324)*, 105
Caswell, J. & Phillips, C. 2008, *MNRAS*, 386, 1521
Ellingsen, S. P., Voronkov, M. A., Cragg, D. M., *et al.* 2007, *in IAU Symposium 242*, 213
Fujisawa, K., Aoki, N., Nagadomi, Y., *et al.* 2014, *PASJ*, 66, 31
Kang, H.-W., Kim, K.-T., & Byun, D.-Y., *et al.* 2015, *ApJS*, 221, 6
Kang, J.-H., Byun, D.-Y., & Kim, K.-T., *et al.* 2016, *ApJS*, 227, 17
Kim, C.-H., Kim, K.-T., & Park, Y.-S. 2018, *ApJS*, in press
Kurtz, S., Hofner, P., & Álvarez, C. V. 2004, *ApJS*, 155, 149
Matsumoto, N., Hirota, T., Sugiyama, K., *et al.* 2014, *ApJ*, 789, L1
Menten, K. M. 1991, *in Atoms, ions and molecules (ASP-CS Volume 16)*, 119
Motogi, K., Sorai, K., Honma, M., *et al.* 2016, *PASJ*, 68, 69
Reid, M. J. 2007, *in IAU Symposium 242*, 522
Seifried, D. Pudritz, R. E., Banerjee, R., *et al.* 2012, *MNRAS*, 422, 347
Zinnecker, H. & Yorke, H. W. 2007, *ARAA*, 45, 481

Astrophysical Masers:
Unlocking the Mysteries of the Universe
Proceedings IAU Symposium No. 336, 2017
A. Tarchi, M.J. Reid & P. Castangia, eds.

© International Astronomical Union 2018
doi:10.1017/S1743921317010584

Water masers in bowshocks: Addressing the radiation pressure problem of massive star formation

Ross A. Burns

Joint Institute for VLBI ERIC (JIVE),
Postbus 2, 7990 AA, Dwingeloo, The Netherlands
email: burns@jive.nl

Abstract. Ejection activities in S255IR-SMA1 and AFGL 5142 were investigated by multi-epoch VLBI observations of 22 GHz water masers, tracing bowshocks leading collimated jets. The history of ejections, revealed by the 3D maser motions and supplemented by the literature, suggests that these massive stars formed by episodic accretion, inferred via the accretion-ejection connection. This contribution centers on the role of episodic accretion in overcoming the radiation pressure problem of massive star formation - with maser VLBI and single-dish observations providing essential observational tools.

Keywords. Masers, Stars: formation, ISM: jets and outflows, etc.

1. Background

The formation of massive stars persists as one of the many exciting branches of modern astronomy. One frustration of the massive star formation community is overcoming the 'radiation pressure problem' in which harsh radiation from the embedded star counteracts accretion - a problem which would limit spherical accretion models to producing stars of maximum 8 $M_\odot$. While non-spherical disk accretion circumvents this issue to some extent, accreting material re-encounters the radiation pressure problem at smaller radii - where gas (ionized by stellar radiation) accretes onto the star (review given in Zinnecker & Yorke 2007). Such regions cannot be resolved by today's instruments. These proceedings draw on various sources for the literature to advocate episodic accretion (EA) as a means of overcoming radiation to form massive stars, and discuss observational tests of EA using VLBI observations of masers. The recipe begins with a brief introduction to its three main ingredients: the accretion-ejection relation, EA in low-mass stars, and EA in high-mass stars.

[1] The accretion-ejection relation in young stellar objects (YSOs) is continuous over several orders of magnitude in mass and luminosity; bright accretion tracers correlate with bright ejection tracers (Caratti o Garatti *et al.* 2015). Its unbroken extension into the regime of massive star formation suggesting some degree of continuity in the physical processes governing low- and high-mass star formation, while also advocating the correlation between accretion activity and ejection activity. Via the hypothesis that each accretion event induces an ejection event, the accretion history of a young star can be inferred from its history of ejections - traced as symmetric, bipolar jet-shocks extending from the accreting object at ever increasing distances.

[2] EA in low-mass YSOs is exemplified observationally by the FUori and EXori classes of protostars (see review by Audard *et al.* 2014). One mechanism by which low-mass stars might accrete episodically has been explored by Stamatellos *et al.* (2011), invoking

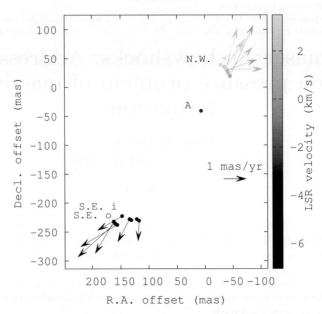

Figure 1. Vector map of water masers in AFGL5142 MM1 from Burns *et al.* (2017). North West, South East inner and South East outer bowshocks (labelled in abbreviation) extend from an MYSO located near the maser feature labelled 'A'. Bowshocks indicate a junction in physical conditions generated by the most recent ejection events.

a magnetic barrier which is episodically disrupted in a 'magneto-rotational instability' (MRI) - leading to an accretion burst; long periods of apparent quiescence perforated by short accretion bursts, a behaviour that lends a solution to the 'luminosity problem' of protostars. While single-event bursts are commonly reported in low- and high-mass stars (Contreras Peña *et al.* 2017; Forbrich *et al.* 2017), establishing the episodicity of such bursts would require monitoring for long timescales; up to thousands of years. An alternative approach is therefore desirable.

[3] *EA in massive stars* may provide a means of suppressing the intense radiation pressure thought to curtail accretion on to the central object. The mechanism, applicable to current-day stars but discussed in the context of primordial massive stars, is described in detail in Hosokawa *et al.* (2016). It is summarised as follows: Accretion of material on to the protostar causes it to 'bloat'. The subsequent increase in radius leads to a drop in effective temperature. Consequently, the peak of spectral radiation (blackbody) migrates to lower frequencies, thus reducing emissivity at UV wavelengths - thereby permitting further accretion. In the absence of accretion the bloated star contracts (Kelvin-Helmholtz) toward its compact, high-temperature state, taking $\sim 10^4$ yrs. However, contraction can be stopped by further accretion events, repeatedly bloating the star and permitting further mass accumulation. This requires that accretion events occur at least every $\sim 10^4$ yrs. Accretion events become less frequent as the reservoir of material in the envelope and disk deplete, allowing contraction to set in. UV emissivity eventually increases until accretion is finally halted and the star reaches ZAMS.

Combining the topics discussed above - EA enables disk mediated accretion to form very massive stars by periodically suppressing stellar radiation (see [3]). Some (if not all) low-mass stars undergo a phase of episodic accretion (see [2]). If high-mass star formation resembles a scaled up version of low-mass star formation, as suggested by the accretion-ejection relation (see [1]), then a class of periodically accreting massive stars should be recognised - and their accretion histories can be investigated by their jet shocks. Water masers trace such shocks, and multi-epoch observations provide their 3D motion. By combination with a parallax measurement, masers can therefore be used to accurately reveal the dynamic timescales of ejection events - and by association, accretion events - occurring in deeply embedded massive young stellar objects (MYSOs).

2. Observational evidence

Episodic jets operating on timescales of 10^{3-4} yrs were inferred from proper motion and parallax observations of 22 GHz water masers in S255IR-SMA1 and AFGL5142. In both works VLBI water maser observations trace the youngest collimated ejections, while shock tracers at larger scales, tracing older ejections, were sourced from the literature - details given in Burns *et al.* (2016) and Burns *et al.* (2017), respectively, with references therein. Both cases are examples of jets where the maser distributions trace clear bowshocks propagating symmetrically from the central MYSO (see Fig 1); the junction in physical conditions characteristic of the onset of a new ejection.

Reports of episodic ejection in S255IR-SMA1 (Burns *et al.* 2016) were shortly followed by an accretion burst in the same source, leading to a several magnitude increase in infrared continuum emission (Caratti o Garatti *et al.* 2017), followed by a maser burst in the radiatively pumped 6.7 GHz methanol maser line (Moscadelli *et al.* 2017). These works demonstrate the commutation of low-mass star formation principles into the high-mass regime, namely accretion bursts (as opposed to steady accretion) and episodic jets. Furthermore, these results promote the aforementioned objects as prime candidates of MYSOs undergoing episodic accretion. Follow-up observations of masers S255IR and AFGL5142, and other eruptive MYSOs, are underway by several groups.

3. Conclusions

While EA has long been discussed in the framework of low-mass stars it has only recently been pursued in the context of massive star formation. This contribution highlights two examples of EA in MYSOs, where the inferred accretion episodes operate on periods shorter than the 10^4 yrs required to outpace contraction, thus consistent with the mechanism described in Hosokawa *et al.* (2016). EA can be explored by the history of ejections from YSOs and MYSOs, providing an alternative to long-term monitoring for accretion events. Recent simulations exploring EA in massive stars were conducted by Meyer *et al.* (2017) who also find accretion events to occur on timescales shorter than 10^4 yr - their paper serves as a useful source of information on the topic of EA in low-and high-mass star formation.

Previously, EA has primarily been evinced by enhancements in continuum emission. The detection of such events typically requires interferometric observations. Maser super burst events may provide an alternative approach to detecting accretion bursts and such events can be readily identified as part of maser monitoring observations conducted by single-dish radio observatories, covering large samples of sources (for example Szymczak *et al.* 2018). Two such maser super burst events were detected and announced during this Symposium in G25.65+1.05 and W49N (Volvach *et al.* 2017, and private communication) and data from rapid follow-up VLBI observations are now being analysed. Maser observations - VLBI for investigating episodic jets, and single-dish for detecting burst events - will therefore be crucial to the integration of EA into the framework of massive star formation.

References

Audard, M., Ábrahám, P., Dunham, M. M., *et al.* 2014, *Protostars and Planets VI*, Univ. Arizona
      Press, Tucson, AZ, 387
Burns, R. A., Handa, T., Nagayama, T., Sunada, K., & Omodaka, T. 2016, *MNRAS*, 460, 283
Burns, R. A., Handa, T., Imai, H., *et al.* 2017, *MNRAS*, 467, 2367
Caratti o Garatti, A., Stecklum, B., Linz, H., Garcia Lopez, R., & Sanna, A. 2015, *A&A*, 573,
      A82
Caratti o Garatti, A., Stecklum, B., Garcia Lopez, R., *et al.* 2017, *Nature Physics*, 13, 276
Contreras Peña, C., Lucas, P. W., Kurtev, R., *et al.* 2017, *MNRAS*, 465, 3039
Forbrich, J., Reid, M. J., Menten, K. M., *et al.* 2017, *ApJ*, 844, 109
Hosokawa, T., Hirano, S., Kuiper, R., *et al.* 2016, *ApJ*, 824, 119
Meyer, D. M.-A., Vorobyov, E. I., Kuiper, R., & Kley, W. 2017, *MNRAS*, 464, L90
Moscadelli, L., Sanna, A., Goddi, C., *et al.* 2017, *A&A*, 600, L8
Stamatellos, D., Whitworth, A. P., & Hubber, D. A. 2011, *ApJ*, 730, 32
Szymczak, M., Olech, M., Sarniak, R., Wolak, P., & Bartkiewicz, A. 2018, *MNRAS*, 474, 219
Volvach, A. E., Volvach, L. N., MacLeod, G., *et al.* 2017, The Astronomer's Telegram, 728
Zinnecker, H. & Yorke, H. W. 2007, *ARAA*, 45, 481

Astrophysical Masers:
Unlocking the Mysteries of the Universe
Proceedings IAU Symposium No. 336, 2017
A. Tarchi, M.J. Reid & P. Castangia, eds.

© International Astronomical Union 2018
doi:10.1017/S1743921317010973

A Face-on Accretion System in High Mass Star-Formation: Possible Dusty Infall Streams within 100 Astronomical Unit

Kazuhito Motogi[1], **Tomoya Hirota**[2,3], **Kazuo Sorai**[4,5],
Yoshinori Yonekura[6], **Koichiro Sugiyama**[2], **Mareki Honma**[7],
Kotaro Niinuma[1], **Kazuya Hachisuka**[7], **Kenta Fujisawa**[8] and
Andrew J. Walsh[9]

[1] Graduate School of Sciences and Technology for Innovation, Yamaguchi University,
Yoshida 1677-1, Yamaguchi 753-8512, Japan
email: kmotogi@yamaguchi-u.ac.jp

[2] Mizusawa VLBI Observatory, National Astronomical Observatory of Japan,
Osawa 2-21-1, Mitaka, Tokyo 181-8588, Japan

[3] Department of Astronomical Sciences, SOKENDAI
(The Graduate University for Advanced Studies),
Osawa 2-21-1, Mitaka, Tokyo 181-8588, Japan

[4] Department of Physics, Faculty of Science, Hokkaido University,
Sapporo 060-0810, Japan

[5] Department of Cosmosciences, Graduate School of Science, Hokkaido University,
Sapporo 060-0810, Japan

[6] Center for Astronomy, Ibaraki University,
2-1-1 Bunkyo, Mito, Ibaraki 310-8512, Japan

[7] Mizusawa VLBI Observatory, National Astronomical Observatory of Japan,
Hoshigaoka 2-12, Mizusawa, Oshu, Iwate 023-0861, Japan

[8] The Research Institute for Time Studies, Yamaguchi University,
Yoshida 1677-1, Yamaguchi 753-8511, Japan

[9] International Centre for Radio Astronomy Research, Curtin University,
Bentley, WA 6102, Australia

Abstract. We report on interferometric observations of a face-on accretion system around the high mass young stellar object, G353.273+0.641. The innermost accretion system of 100-au radius was resolved in a 45-GHz continuum image taken with the Jansky Very Large Array. Our SED analysis indicated that the continuum could be explained by optically-thick dust emission. 6.7 GHz CH_3OH masers associated with the same system were also observed with the Australia Telescope Compact Array. The masers showed a spiral-like, non-axisymmetric distribution with a systematic velocity gradient. The line-of-sight velocity field is explained by an infall motion along a parabolic streamline that falls onto the equatorial plane of the face-on system. The streamline is quasi-radial and reaches the equatorial plane at a radius of 16 au. The physical origin of such a streamline is still an open question and will be constrained by the higher-resolution thermal continuum and line observations with ALMA long baselines.

Keywords. ISM: individual objects (G353.273+0.641) – molecules – masers – radio continuum: ISM – stars: formation

---

1. Introduction

G353.273+0.641 (hereafter G353), is a relatively nearby (1.7 kpc) high mass young stellar object (HMYSO) in the southern sky (Motogi *et al.* 2016). The bolometric luminosity of $\sim 5 \times 10^3$ $L_\odot$ corresponds a B1-type ZAMS, implying a stellar mass of ~ 10 $M_\odot$ (Motogi *et al.* 2017). G353 is also known as a Dominant Blue-Shifted Maser (DBSM) source that is a class of 22 GHz H_2O masers showing a highly blue-shift dominated spectrum (Caswell & Phillips 2008). The line-of-sight (LOS) velocities of the maser typically range from -120 to -45 km s^{-1} and the systemic velocity (V_{sys}) is $\sim$-5.0 km s^{-1} in the case of G353 (e.g., Motogi *et al.* 2016). Caswell & Phillips (2008) proposed that DBSMs are candidates of HMYSOs with a face-on protostellar jet. The inclination angle of the jet, that was estimated by maser proper motions, is $8 - 17$deg from the LOS (Motogi *et al.* 2016). G353 is, at present, the best candidate of a face-on HMYSO in the active accretion phase.

2. Observation

We have searched for a face-on accretion system associated with G353, via the highest-resolution (A-configuration) continuum imaging using the Jansky Very Large Array (JVLA). In addition to the 45 GHz data reported in Motogi *et al.* (2017), new 20 – 30 GHz continuum data were added. We also performed mapping observations of the associated 6.7 GHz class II CH_3OH maser (e.g., Caswell & Phillips 2008) by the Australia Telescope Compact Array (ATCA) in 6A configuration (see Motogi *et al.* 2017 for details).

3. Results

3.1. *Centimeter Continuum*

Figure 1 shows a 45-GHz continuum image taken by the JVLA in A-configuration (Motogi *et al.* (2017)). A very compact continuum source was detected at the center of the bipolar H_2O maser jet. The source was resolved along the minor axis of the synthesized beam (E-W direction). Total flux of 3.5 mJy and the size of the deconvolved Gaussian (FWHM: $0''.123 \times 0''.073$) indicate the averaged brightness temperature T_b of 235 K.

Figure 2 shows a spectral energy distribution at centimeter wavelength. The SED contains both of the low-resolution ATCA data tracing the extended ($\sim 1''$) radio jet (Motogi *et al.* 2013) and high-resolution JVLA data tracing the newly detected compact emission. We found that the former could be modeled by a typical radio jet emission with a positive spectral index (α; $S(\nu) \propto \nu^\alpha$) of 0.5. The JVLA data, on the other hand, clearly showed α of ~ 2, suggesting the optically thick emission.

The relatively low T_b and slightly resolved source structure can exclude any possibility of a very compact HII region. For example, the observed flux of 3.5 mJy at 45 GHz corresponds a source size of only 18 milli-arcsecond in the case of optically thick HII region with the electron temperature of 10^4 K. Alternatively, such a low T_b and spectral index of 2 are naturally explained by the optically thick dust emission. The association of the 6.7 GHz CH_3OH maser, which is excited by warm dust emission (Cragg *et al.* 2005), is also consistent in this case. We, thus, suggest that the continuum traces warm dust in the innermost circumstellar system.

3.2. *6.7 GHz CH_3OH maser*

Figure 1 also presents the internal distribution of 30 bright CH_3OH maser spots (signal-to-noise ratio > 30) that have accurate relative positions compared to the scale of the maser distribution. Since it is difficult to directly compare the 45-GHz continuum position and maser distribution, due to the significant astrometric error in ATCA data ($\sim$400 mas), we superposed the maser distribution on the continuum image.

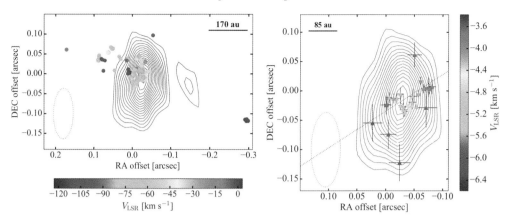

Figure 1. Left: The contours present JVLA 45-GHz continuum image, which are from 19 (3 σ) to 99% with step of 5% of the image peak flux (1.39 mJy beam^{-1}). Filled circles show VLBI map of the H$_2$O maser jet in Motogi *et al.* (2016) with a color indicating the LOS velocity of each maser spot (See the online color version). The coordinate origin is the phase tracking center, $17^h 26^m 01^s.59$, $-34°15'14''.90$ (J2000.0). Right: Internal distribution of the 6.7 GHz CH$_3$OH masers. Filled triangle shows a position of each maser spot with a color indicating the LOS velocity (See the online color version). The maser map is superposed on the continuum image, assuming that the peak position of the 6-GHz continuum is identical to that of the 45-GHz continuum. The synthesized beam of JVLA is shown in lower left corner in both figures.

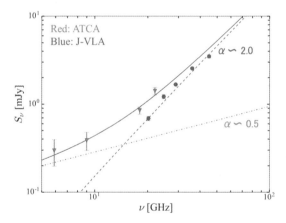

Figure 2. The centimeter SED between 6 – 45 GHz. Filled circles and triangles indicate high--resolution JVLA data and low-resolution ATCA data, respectively. The dotted and dash-dot lines show the best-fit model of the optically thick compact dust emission ($\alpha \sim 2.0$) and the extended radio jet ($\alpha \sim 0.5$), respectively.

The maser distribution in Figure 1 appears spiral-like rather than ring or linear shape, accompanied by systematic velocity gradient ($\pm$ 1.5 km s^{-1} over 0.15 arcsec). We have found that this velocity structure is well explained by infall streams which fall down to the equatorial plane of the inner accretion disk, along the point symmetric trajectory. Figure 3 shows the schematic view of our parabolic infall model with the best-fit result (see Motogi *et al.* 2017 for details). We note that recent ALMA observations towards a low mass class 0 object have found that similar non-axisymmetric streams fall down onto the edge-on accretion disk along a parabolic orbit (Yen *et al.* 2014).

The infall streams reach the equatorial plane at the landing radius (R_0) of 16 au in the best-fit case. This is clearly smaller than that of typical accretion disks in high mass

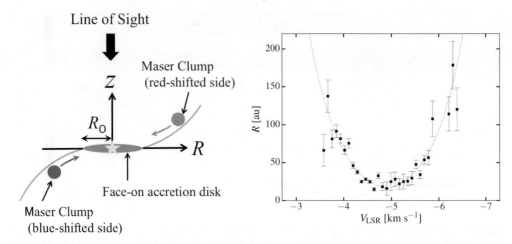

Figure 3. Left: Schematic view of the parabolic infall model. Two filled circles indicates a blue-shifted and red-shifted infalling CH_3OH maser clumps, respectively. The LOS is along the Z direction. Right: Distance-velocity diagram of the CH_3OH masers. A dotted line indicates the best-fit result of our parabolic infall model (see Motogi *et al.* 2017 for details). The x and y axes show LOS velocities and a projected distances of maser spots with respect to the dynamical center that was determined by the model fitting, respectively.

star-formation (Beltrán & de Wit 2016, and references therein). This fact indicates that the initial angular momentum in G353 is very small, or the CH_3OH masers selectively trace accreting material that has a small angular momentum.

Our model also expects that the streamline is quasi-radial ($Z/R < 10$ per cent) even at the outer region ($R \sim 200$ au). This fact indicates that centrifugal force is basically negligible, although the head-tail like distribution of the blue-shifted masers in Figure 1 may imply non-zero angular momentum. Such an infall-dominated motion of the class II CH_3OH maser at 100-au scale has actually been detected in a VLBI study by Goddi, Moscadelli & Sanna (2011).

3.3. *Conclusion*

Our results suggested the presence of the innermost (~ 100 au) dusty and infall-dominated accretion system in G353. However, the origin of such a specific structure is still an open question. The infall model will be verified by thermal continuum and line observations in the higher-resolution (~ 10 mas), resolving a face-on disk, infall streams, etc. This will be done by our ALMA long baseline project. In addition, infall motion of the CH_3OH maser itself will be directly examined by on-going proper motion measurements by VLBA. 3D velocity field will also constrain an accretion rate within 100 au, combined with mass information obtained by ALMA.

References

Beltrán, M. T. & de Wit, W. J. 2016, *A&A Rv*, 24, 6
Caswell, J. L. & Phillips, C. J. 2008, *MNRAS*, 386, 1521
Cragg, D. M., Sobolev, A. M., & Godfrey, P. D. 2005, *MNRAS*, 360, 533
Goddi, C., Moscadelli, L., & Sanna, A. 2011, *A&A*, 535, L8
Motogi, K., Sorai, K., Niinuma, K., *et al.* 2013, *MNRAS*, 428, 349
Motogi, K., Sorai, K., Honma, M., *et al.* 2016, *PASJ*, 68, 69
Motogi, K., Hirota, T., Sorai, K., *et al.* 2017, *ApJ*, 849, 23
Yen, H.-W., Takakuwa, S., Ohashi, N., *et al.* 2014, *ApJ*, 793, 1

Astrophysical Masers:
Unlocking the Mysteries of the Universe
Proceedings IAU Symposium No. 336, 2017
A. Tarchi, M.J. Reid & P. Castangia, eds.

© International Astronomical Union 2018
doi:10.1017/S1743921317010766

Maser Effects in Recombination Lines: the case of Eta Carinae

Zulema Abraham[1], Pedro P. B. Beaklini[1] and Diego Falceta-Gonçalves[2]

[1]Instituto de Astronomia, Geofísica e Ciências Atmosféricas, Universidade de São Paulo, Rua do Matão 1226, 05508-090, São Paulo, SP, Brazil.
email: `zulema.abraham@iag.usp.br`

[2]Escola de Artes, Ciências e Humanidades, Universidade de Saõ Paulo, R. Arlindo Bettio 1000, 03828-000, São Paulo, SP, Brazil

Abstract. Population of high quantum number states can differ from their LTE values at high densities (Ne $\sim 10^6 - 10^8$ cm^{-3}) and temperatures of the order of 10^4 K. In this case, the intensity of recombination lines can be strongly amplified. The amount of amplification depends on density and temperature, and it is different for different quantum numbers, allowing the determination of the physical and kinematic conditions of the emitting region through the observation of recombination lines of different quantum numbers. This was the case of the massive binary system η Carinae. This system was observed with ALMA in the recombination lines H21α, H28α, H30α, H40α and H42α and the continuum at the frequencies of the corresponding lines. The continuum spectrum was characteristic of a compact HII region, becoming optically thin at around 300 GHz. From the intensity and width of the recombination lines we concluded that the not-resolved emission region, assumed spherically symmetric, is a shell of 40 AU radius and 4 AU width, expanding at velocities between 20 and 60 km s^{-1}, with density of 10^7 cm^{-3} and temperature of 17000 K.

Keywords. masers, mass loss, stellar winds, η Carinae

1. Introduction

η Carinae is one of the most massive stars in our Galaxy. It is possibly an LBV (Luminous Blue Varaible) star that presented several episodes of mass ejection. The largest occurred in 1840, when it ejected ~ 12 M$_\odot$ that formed the Homunculus Nebula, now expanding at about 600 km s^{-1}. At that time, its optical magnitude reached -1, and after that its brightness decreased as the cloud expanded and dust formed. Another smaller episode of mass ejection occurred in 1890, and maybe a third one in 1942 (Fernández-Lajús *et al.* 2009).

η Carinae is not a single star, but a binary system in a very eccentric orbit. The 5.52 yr periodicity in the X-ray light curve, observed by the RXTE (Corcoran *et al.* 2001), and also in the intensity of lines of highly ionized elements (Damineli 1996) and in continuum flux density at radio frequencies (Abraham *et al.* 2005), are proof of the binary nature of the system. The companion star has not been detected, because of absorption by the Homunculus and the strong wind of η Carinae. The secondary star must also have a wind, and the winds' collision produces a shock that emits at X-rays.

At radio frequencies it was mapped by Duncan, White & Lim (1997) at 5 GHz with the Australian Compact Array (ATCA). The maps resemble an edge-on disk, with an arc-like structure, which shrinks to a point-like source close to the dips in the X-ray light curve. Both regions are probably ionized by the companion star.

271

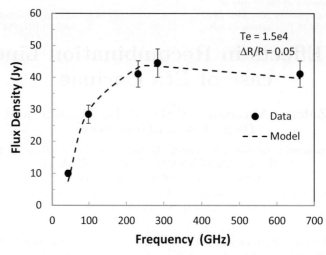

Figure 1. Interferometric continuum observations of η Carinae from ALMA and single dish (43GHz) from Itapetinga.

η Carinae was also observed with SEST in the continuum at 1.3 and 3 mm, and at the recombination lines H29α, H40α and H50β (Cox *et al.* 1995a,b). The source was not resolved in the continuum and showed a spectrum that increased with frequency as $S(\nu) \propto \nu^{0.6}$; its flux density also depended on the orbital phase. The recombination lines were asymmetric and strong, affected by NLTE effects. The continuum emission was attributed to the free-free process in the η Carinae wind. Further observations were made with SEST, between 1999 and 2003, in the continuum at 100, 150 and 230 GHz, as well as in the recombination lines H30α, H35α and H40α (Abraham, Damineli & Durouchoux 2002; Abraham *et al.* 2005). The observations required long integration times, especially for the continuum, and had uncertainties due to the lack of calibration sources and pointing corrections.

The continuum emission was also observed at 7 mm (43 GHz) with the Itapetinga radiotelescope, in Brazil, from 2003 to 2014 (Abraham *et al.* 2005).

2. Observations and Results

η Carinae was observed with ALMA in Cycle 0, at four continuum bands, centered at 92, 225, 291 and 672 GHz, with resolutions of 2.8, 1.5, 0.8 and 0.45 arcsec, and in the recombination lines H42α, H40α, H30α, H28α and H21α (Abraham, Falceta-Gonçalves & Beaklini 2014). The source was not resolved, neither in the continuum nor in the recombination lines.

The continuum spectrum is shown in Fig. 1. It includes 43 GHz single dish data from the Itapetinga radiotelescope in Brazil. As we can see, it is a typical spectrum of a compact HII region that becomes optically thin at a frequency of about 300 GHz.

In Fig. 2 we present the recombination line spectra. Notice the high intensity of the H lines and the presence of corresponding He lines.

In Fig. 3 we show the position vs. velocity diagram for the H28α line, in the direction of the clean beam major axis. This diagram is typical of an expanding region and for that reason we modeled the source as a spherically symmetric expanding shell and solved the equations of radiative transfer for the continuum and lines, using the NLTE line emission coefficients calculated by Storey & Hummer (1995). The model parameters are: electron density N_e and temperature T_e, size R and width ΔR of the shell, and bulk expansion

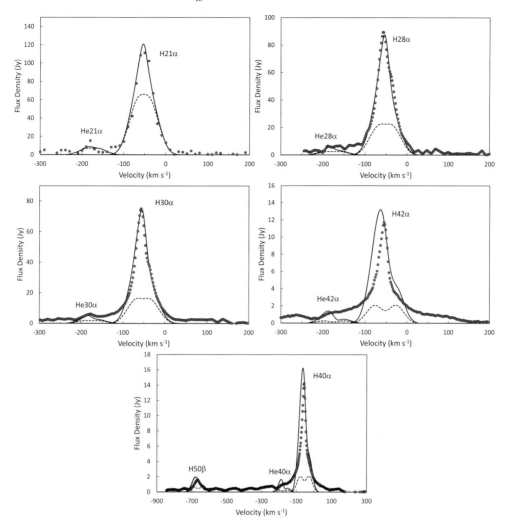

Figure 2. Recombination lines from η Carinae observed with ALMA (points). The continuum line represents the best fit to a NLTE model and the broken line the same model but considering LTE conditions

velocity v_0 and its gradient across the shell (internal v_i and external v_e velocities). The parameters for which the best fit of the continuum and the line profiles are obtained are: $N_e = 1.5 \times 10^7 \text{ cm}^{-3}$, $T_e = 1.7 \times 10^4 \text{ K}$, $R = 40 \text{ AU}$, $\Delta R = 0.1R$, $v_0 = -52 \text{ km s}^{-1}$, $v_i = 60 \text{ km s}^{-1}$, $v_e = 20 \text{ m s}^{-1}$.

The model for the continuum emission is presented in Fig. 1 as a broken line and for the line profiles in Fig. 2 as continuous lines. We also modeled the corresponding He lines, the H50β line and the line profiles using LTE emission coefficients, which are shown as broken lines.

Since η Carinae is probably an LBV star, the compact shell can be the result of a recent mass ejection accompanied by a brightness variation, as that observed in the optical light curve in 1942 (Fernández-Lajús *et al.* 2009).

We are grateful to the Brazilian research agencies FAPESP and CNPq for financial support (FAPESP Projects: 2008/11382-3 and 2014/07460-0).

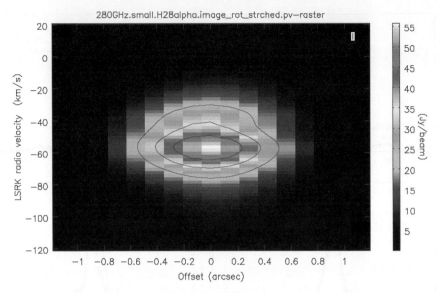

Figure 3. Position vs. velocity diagram for the H28α line in the direction of the clean beam major axis. The contours are 0.2, 0.4, 0.6 and 0.8 of the maximum flux density, which is 55.9 Jy bem^{-1}.

References

Abraham, Z., Damineli, A., Durouchoux, P., in Cosmic MASERS, from Protostars to Black Holes, 2002, IAU Symposium 206, 234, ed. V. Migenes & M. J. Reid

Abraham, Z., Falceta-Gonçalves, D., Dominici, T. M. *et al.*, 2005, *A&A*, 437, 997

Abraham, Z., Falceta-Gon calves, D., & Beaklini, P. P. B., 2014, *ApJ*, 791, 95

Corcoran, M. F., Ishibashi. K., Swank, J. H., & Petre, R., 2001, *ApJ*, 547, 1034

Cox, P. G., Martin-Pintado, J., Bachiller, R. *et al.*, 1995a, *A&A*, 295, L39

Cox, P. G., Mezger, P. G., Sievers, A. *et al.*, 1995b, *A&A*, 297, 168

Damineli, A., 1996, *ApJ*, 460, L49

Duncan, R. A., White, S. M., & Lim, J., 1997, *MNRAS*, 290, 680

Fernández-Lajús, E., Fariña, C., Torres, A. F., *et al.*, 2009, *A&A*, 493, 1093

Storey, P. J. & Hammer, D. G., 1995, *MNRAS*, 272, 41

Astrophysical Masers:
Unlocking the Mysteries of the Universe
Proceedings IAU Symposium No. 336, 2017   © International Astronomical Union 2018
A. Tarchi, M.J. Reid & P. Castangia, eds.   doi:10.1017/S1743921317009565

Revealing the kinematics and origin of ionized winds using RRL masers

Alejandro Báez-Rubio

Instituto de Astronomía, Universidad Nacional Autónoma de México,
Apartado Postal 70-264, 04510, México, CDMX, Mexico
email: abaez@astro.unam.mx

Abstract. The detection of hydrogen radio-recombination maser lines (RRLs) toward MWC349A in the year 1989 opened the chance to place constraints on the kinematics of an ionized circumstellar disk around a massive star. Since then, a significant number of observations have allowed improving our understanding of this source to the point that we have established that its ionized wind launching occurs at a distance of ~24 au as claimed by disk wind models. On the other hand, this field of study has undergone considerable development over the last six years with the detection of new RRL maser sources. Here, we present a brief summary of all these recent advances and the promising future prospects.

Keywords. masers, stars: winds, outflows, HII regions, accretion, accretion disks

---

1. Historical background

Since the first detection of hydrogen radio-recombination lines (RRLs), they have allowed us to study the physical conditions and kinematics of ionized (HII) regions which are highly extincted by dust. This is the case of some post-main sequence stars (e.g. pre-planetary nebulae) and pre-main sequence massive stars. Thus, it was found that while many of these stars have low-velocity ionized winds moving at the thermal speed, ~ 20 km s^{-1}, those with the most compact and densest HII regions reach higher velocities (Jaffe & Martín-Pintado 1999). However, the weak intensity of the detected RRLs did not allow to study key issues such as where and how their ionized winds are launched.

In the year 1989, the detection of the first RRLs affected by maser amplification toward MWC349A, a high-mass B[e] star with an edge-on circumstellar disk (Danchi *et al.* 2001), opened a new panorama. The huge intensity of its double-peak mm RRL profiles allowed to study, for the first time, not only the kinematics of its ionized wind but also its ionized circumstellar disk (Planesas *et al.* 1992). Thus at the time of the previous maser conference, after 13 years of the first RRL maser detection, it had already been well established that the ionized layers of its circumstellar disk rotate, apparently following a Keplerian law, like its ionized wind (Martín-Pintado *et al.* 2011). Further detailed constraints on the kinematics and physical conditions were obtained comparing observations with the predictions of a 3D radiative transfer model (Báez-Rubio *et al.* 2013).

As mentioned, MWC349A has been extensively studied after finding that it emits RRL masers. However, with the exception of the detection of these lines toward Eta Carinae (Cox *et al.* 1995), it was not until more than twenty years later when new RRL maser sources were revealed. In particular, the detection of these RRL masers toward Cepheus A HW2 (Jiménez-Serra *et al.* 2011) and Monoceros R2-IRS2 (Jiménez-Serra *et al.* 2013) showed the need for performing observations with high spatial resolution or broad wideband respectively for detecting a larger sample of these objects.

2. Progress in the last five years

Constraint of the launching point of the ionized wind around MWC349A

Herschel/HIFI observations revealed the abrupt appear of two spectral components, blueshited with respect to the maser spikes, when the principal quantum number of the Hnα RRL changes from n = 22 to n = 21 (Báez-Rubio *et al.* 2014). These components are observed in all the RRLs with $17 \leqslant n \leqslant 21$. Since RRLs with lower principal quantum number trace inner regions, it might indicate that these RRLs trace the region where the wind is launched from the circumstellar disk. In fact, the predictions of the MORELI radiative transfer model confirmed that they might be tracing the ejection of the ionized wind at a distance of $\sim$ 24 au. This is also consistent with recent SMA observations (see Zhang *et al.* 2017 or the contribution by J. Moran in this book).

Discovery of new RRL maser sources

Since the last maser conference, RRL masers have been discovered toward G5.89-0.39A and MWC922, making a total of five known RRL maser sources. In the case of G5.89-0.39A, ALMA observations revealed that the H26α RRL emission is masering. Even if its kinematical structure has not been constrained yet, this finding is really interesting because it is the first pre-main sequence massive star where RRL masers have been revealed exclusively based on the measured Hmβ-to-Hnα flux ratio, in particular with spectroscopic measurements of the H32β and H26α RRLs.

On the other hand, the detection toward MWC922 using the IRAM-30m is quite interesting (Sánchez-Contreras *et al.* 2017). This source is thought to be a pre-planetary nebula, although an evolved nature has not been ruled out. Both MWC922 and MWC349A have the same spectral type, B[e] (Lamers *et al.* 1998), and their RRL profiles are strongly similar. These RRL change from a single peak for the H41α to a double peak for the H30α, which is clearly masering. The profiles of MWC922 can be reproduced if this source is modelled as an ionized wind and an edge-on rotating ionized circumstellar disk like in the case of MWC349A. Thus, MWC922 is a very promising source for understanding how the wind is launched from its disk. This, together with the possible future detection of new RRL maser sources, as done in the last few years, will allow us statistically claim if disk wind models explain how winds are launched or if other models are also valid.

References

Báez-Rubio, A., Martín-Pintado, J., Thum, C., & Planesas, P. 2013, *A&A*, 553, A45

Báez-Rubio, A., Martín-Pintado, J., Thum, C., Planesas, P., & Torres-Redondo, J. 2014, *A&A*, 571, L4

Cox, P., Martín-Pintado, J., Bachiller, R., Bronfman, L., Cernicharo, J., Nyman, L.-A & Roelfsema, P. R. 1995, *A&A*, 295, L39

Danchi, W. C., Tuthill, P. G., & Monnier, J. D. 2001, *ApJ*, 562, 440

Jaffe, D. T. & Martín-Pintado, J. 1999, *A&A*, 520, 162

Jiménez-Serra, I., Martín-Pintado, J., Báez-Rubio, A., Patel, N., & Thum, C. 2011, *ApJ*, 732, L27

Jiménez-Serra, I., Báez-Rubio, A., Rivilla, V. M., Martín-Pintado, J., Zhang, Q., & Patel, N. 2013, *A&A*, 764, L4

Lamers, H. J. G. L. M., Zickgraf, F.-J., de Winter, D., Houziaux, L., & Zorec, J. 1998, *A&A*, 340, 117

Martín-Pintado, J., Thum, C., Planesas, P., & Báez-Rubio, A. 2011, *A&A*, 530, L15

Planesas, P., Martín-Pintado, J., & Serabyn, E. 1999, *ApJ*, 386, L23

Sánchez Contreras, C., Báez-Rubio, A., Alcolea, J., Bujarrabal, V., & Martín-Pintado, J. 2017, *A&A*, 603, A67

Zhang, Q., Claus, B., Watson, L., & Moran, J. 2017, *A&A*, 837, 53

Astrophysical Masers:
Unlocking the Mysteries of the Universe
Proceedings IAU Symposium No. 336, 2017
A. Tarchi, M.J. Reid & P. Castangia, eds.

© International Astronomical Union 2018
doi:10.1017/S1743921317009437

Long term 6.7 GHz methanol maser monitoring program

Artis Aberfelds, Ivar Shmeld and Karlis Berzins

Ventspils University College, Engineering Research Institute Ventspils International Radio
Astronomy Centre (VIRAC),
Inzenieru 101,LV-3601, Ventspils, Latvia
email: `artis.aberfelds@venta.lv`

Abstract. The first long-term maser (mainly methanol) monitoring program is under way with the radio telescopes of Ventspils International Radio Astronomy Center. The first activity of this program was to develop an observations methodology and data registration and reduction software for the Ventspils telescopes. The developed routines are to be used for maser variability monitoring, investigating short bursts of intensity and a search for new, previously unknown, maser sources. Currently the program consists of 41 methanol masers observed at 6.7 GHz, while new ones are periodically added. The maser sources are observed at 3 – 5 day intervals. It was found that most the sources display a significant level of variability with time, ranging from a few days, up to several months and, perhaps, years. In addition to non-varying masers, several types of maser variability behavior were observed, including: monotonic increases or decreases, un-periodical, quasi-periodic and periodic variations.

Keywords. masers, methanol, radio astronomy, circumstellar matter.

1. INSTRUMENTATION

After a modernization program, the capabilities of Ventspils International Radio Astronomy Center (VIRAC) 32-m and 16-m telescopes have been greatly improved. A methanol maser monitoring program has been started, employing the two radio telescopes. Digital backend consisting from DBBC-2 (*Digital Base Band Convector* developed by HAT-LAB, Italy) and *Flexbuff* (data storage system based on commercially available server system) is used for data digitalization and registration. The spectra were obtained by autocorrelation using program package *mark5access* (tool *m5spec*). We attempted to ensure 3–5 day observation intervals for all selected sources. For bright masers (Flux >200 Jy), the recording time was taken to be 3 min, while for weaker ones – 7 min.

More details about instrumentation and data reduction are given by Shmeld *et al.* (2018).

2. SOURCE SELECTION

The Torun methanol source catalogue (Szymczak *et al.* 2012) was used as the initial list of the monitored methanol masers. The first task after the recent telescope renovation was telescope testing and calibration. The main initial criteria for sources to be included in our list were: (i) visibility from Northern hemisphere, (ii) Flux >3 Jy, (iii) easy to observe and well-known masers. Initially, 13 brightest and well-studied 6.7 GHz methanol masers were selected. However, the main task of our maser monitoring program is researching the properties of their variability and now our list is expanded by adding of 29 sources, mostly known as being variable.

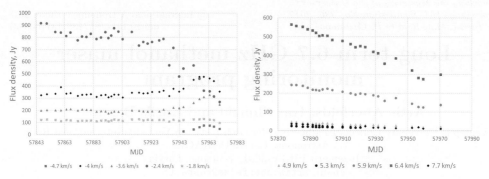

Figure 1. Cepheus A: intensity changes of spectral features.

Figure 2. S255: dimming of 6.4 and 5.3 km s^{-1} spectral features.

3. FIRST RESULTS AND RECENT HEADLINES

Currently we have accumulated up to 9 months of intensity variability data for 41 maser sources. This is not sufficient for a thorough analysis, so the program is to be continued. However, more or less clear results can be seen towards constantly dimming or rapidly bursting sources Berzins *et al.* (2017). The 3–5 day interval seems to be sufficient for noticing a short-time maser bursts or other rapid brightness changes with high probability. Also, it is possible to increase the rate of observations during the period of rapid intensity changes even up to multiple times per day.

Despite of the relatively short duration of the monitoring, some notable intensity changes for our program sources can be already noticed. The most notable maser spectrum changes were observed towards Cepheus A and S255. For Cepheus A, the previously relative stable -2.4 km s^{-1} spectral feature has lost 2/3 of its amplitude in about 40 days and has been overtaken by two other features, which have witnessed 30% gains in the same period. For S255, monotone dimming in two strongest spectral features has continued for the last 5 months.

Acknowledgements The authors would like to thank all members of staff at VIRAC. This work was financed by ERDF project "Physical and chemical processes in the interstellar medium", No. 1.1.1.1/16/A/213, being implemented in Ventspils University College. Invaluable assistance for solving calibration tasks was provided by Torun Radio Astronomy Observatory staff.

References

Shmeld, I., Aberfelds, A., & Berzins, K. 2018, these proceedings
Szymczak, M., Wolak P., BartKiewicz A., & BorKowski K. 2012, *Astron Nach.*, 333, 634-639
Berzins, K., Aberfelds, A., & Shmeld, I., 2018, these proceedings

Astrophysical Masers:
Unlocking the Mysteries of the Universe
Proceedings IAU Symposium No. 336, 2017
A. Tarchi, M.J. Reid & P. Castangia, eds.

© International Astronomical Union 2018
doi:10.1017/S1743921317011437

Variability of Water Masers in W49N: Results from Effelsberg Long-term Monitoring Programme

Busaba H. Kramer[1,2], Karl M. Menten[1] and Alex Kraus[1]

[1] Max-Plank-Institut für Radioastronomie, Auf dem Hügel 69, D-53121 Bonn, Germany
email: bkramer@mpifr-bonn.mpg.de

[2] National Astronomical Research Institute of Thailand, 260 Moo 4, T. Donkaew, Amphur Maerim, Chiang Mai, 50180, Thailand

Abstract. We present the results from an ongoing long-term monitoring of the 22 GHz H_2O maser in W49N with the 100-m Effelsberg radio telescope from February 2014 to September 2017. The unique Effelsberg's spectral line observation capability provides a broad velocity range coverage from -500 to $+500$ km s^{-1} with a spectral resolution better than 0.1 km/s. Following the strong major outburst in W49N in late 2013, we have started a long-term monitoring programme at Effelsberg. The major outburst feature (up to 80,000 Jy at $V_{LSR} - 98$ km s^{-1}) faded away by June 2014. However, we found that the site is still active with several high velocity outbursts (both blue and redshifted). Some features appear at extremely high velocities (up to ± 280 km s^{-1}) and show rapid flux variations within a 1-2 month period. This sub-year scale variability implies that the water masers could be excited by episodic shock propagation caused by a high-velocity protostellar jet.

Keywords. masers, stars: formation, ISM: molecules, radio lines: ISM

1. Summary of the Results

Spectra of the 22 GHz H_2O masers in W49N during 2014-2017 are shown in Figure 1 where selected observations are plotted. Using the VLBA, we identified the location of the 2013 strong outburst to be near the centre of the north-south arc-like structure similar to the previous strong outburst observed with VERA in 2003 (Honma *et al.* 2004) but at different V_{LSR} (-30.7 km s^{-1}). Even though the 2013 outburst faded away by June 2014, we found that the site is still active with several high velocity outbursts (both blue and red shifted). Full detailed analysis will be reported in Kramer *et al.* (in prep.).

On September 7, 2017, we detected a new strong outburst (up to 34,000 Jy at V_{LSR} -82 km s^{-1}) which have led to further follow-up single dish monitoring and VLBI observations, which are clearly needed.

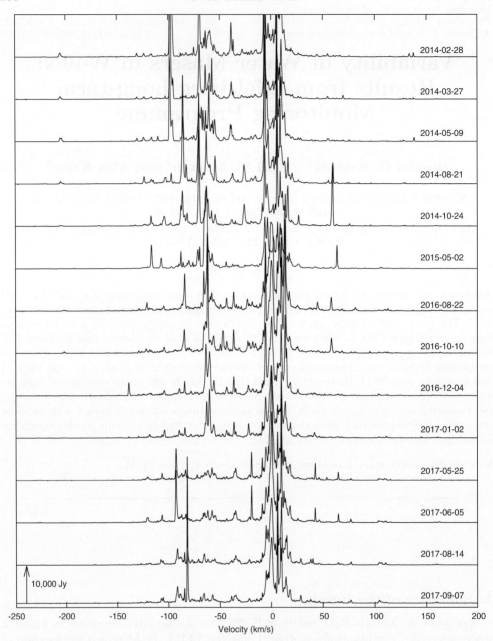

Figure 1. Effelsberg Spectra of the 22-GHz H_2O masers in W49N.

References

Honma, M., Choi Y. K., Bushimata, T., *et al.* 2004, *PASJ*, 56, L15

Kramer, B. H., *et al.* (in prep).

Astrophysical Masers:
Unlocking the Mysteries of the Universe
Proceedings IAU Symposium No. 336, 2017
A. Tarchi, M.J. Reid & P. Castangia, eds.

© International Astronomical Union 2018
doi:10.1017/S1743921317010717

Ubiquitous millimeter-wavelength Class I methanol masers associated with massive (proto)stellar outflows: ALMA and SMA results

C. J. Cyganowski[1], D. Hannaway[1], C. L. Brogan[2], T. R. Hunter[2] and Q. Zhang[3]

[1] Scottish Universities Physics Alliance (SUPA), School of Physics and Astronomy
University of St. Andrews, North Haugh, St. Andrews, Fife KY16 9SS, UK
email: cc243@st-andrews.ac.uk

[2] NRAO, 520 Edgemont Rd, Charlottesville, VA 22903, USA

[3] Harvard-Smithsonian Center for Astrophysics, 60 Garden Street, Cambridge, MA 02138, USA

Abstract. We report the discovery of widespread millimeter-wavelength Class I methanol maser emission associated with protostellar molecular outflows in the massive (proto)cluster G11.92−0.61. Our ∼0.5″-resolution SMA and ALMA observations of the 229 GHz and 278 GHz Class I transitions reveal seven and twelve candidate masers, respectively: all 229 GHz masers have 278 GHz counterparts, and five are also coincident with 44 GHz Class I masers previously detected with the VLA. For paired masers, the peak intensities at 229 GHz and 278 GHz are correlated. We also find tentative evidence for a correlation between the strength of millimeter-wavelength Class I maser emission and the energy of the associated molecular outflow.

Keywords. masers, ISM: individual objects (G11.92-0.61), ISM: molecules, ISM: jets and outflows, stars: formation, techniques: interferometric

1. Introduction

Numerous observational studies, spanning nearly three decades, have shown that Class I CH_3OH masers at 36, 44 and 95 GHz are associated with the outflows driven by high-mass protostars (e.g. Plambeck & Menten 1990, Kurtz *et al.* 2004, Cyganowski *et al.* 2009, Voronkov *et al.* 2014). More recently, Kalenskii *et al.* (2013) detected Class I CH_3OH masers (at 44, 84, and 95 GHz) towards outflows in low-mass star-forming regions, and found that Class I maser luminosities scaled with the luminosity of the driving protostar. The Class I maser series extend into the (sub)millimeter regime (e.g. Voronkov *et al.* 2012), but higher-frequency transitions were relatively unstudied prior to the commissioning of the Atacama Large Millimeter/submillimeter Array (ALMA). Here, we report observations of two millimeter-wavelength Class I CH_3OH maser transitions towards the massive protocluster G11.92−0.61. Located at a distance of $3.37^{+0.39}_{-0.32}$ kpc (Sato *et al.* 2014), G11.92−0.61 is notable for containing both high- and low-mass (proto)stars that are actively accreting and driving outflows (Cyganowski *et al.* 2017).

2. Observations

We have observed G11.92−0.61 with subarcsecond resolution in three Class I CH_3OH maser transitions: 44.069 GHz (with the Very Large Array (VLA)), 229.759 GHz (with

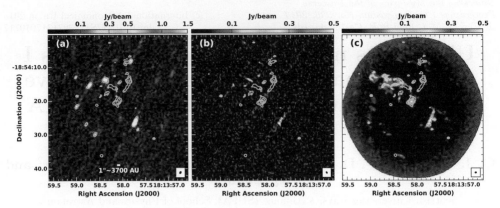

Figure 1. Peak intensity maps, in colorscale, of (a) VLA 44 GHz, (b) SMA 229 GHz, and (c) ALMA 278 GHz CH_3OH emission, overlaid with ALMA 1.05 mm continuum contours from Cyganowski *et al.* (2017) (levels: [5,15,100] $\times$ 0.35 mJy beam^{-1}).

the Submillimeter Array (SMA)), and 278.305 GHz (with ALMA). Details of the observations are given in Cyganowski *et al.* (2009), Cyganowski *et al.* (2014), and Cyganowski *et al.* (2017), respectively. The key parameters of the maser image cubes are: $\theta_{\rm syn}$ ~0.6″ and $\sigma_{\rm line}$ ~ 23 mJy beam^{-1} (VLA 44.069 GHz), $\theta_{\rm syn}$ ~0.5″ and $\sigma_{\rm line}$ ~ 23 mJy beam^{-1} (SMA 229.759 GHz), and $\theta_{\rm syn}$ ~0.4″ and $\sigma_{\rm line}$ ~ 3 mJy beam^{-1} (ALMA 278.305 GHz).

3. Results

Peak intensity maps of the 44.069 GHz, 229.759 GHz, and 278.305 GHz CH_3OH emission observed towards G11.92−0.61 are shown in Figure 1. From the image cubes, we identified 12 candidate 278.305 GHz and 7 candidate 229.759 GHz masers (based on T_B, line width, and line ratios). All of the 229.759 GHz candidate masers have 278.305 GHz counterparts, and five are also coincident with 44.069 GHz masers. Both the 229.759 and 278.305 GHz transitions are in the 36 GHz maser series (Voronkov et al. 2012); for 229/278 GHz maser pairs, the peak intensities of the two transitions are correlated.

The millimeter-wavelength Class I CH_3OH masers are associated with shocked gas at outflow-cloud interfaces (c.f. Figs. 2 and 3 of Cyganowski *et al.* 2017). The new masers are found in association with outflows from high-mass (MM1) and low-mass (MM7/9) protostars within the G11.92−0.61 protocluster; based on the limited data available, the maser strengths appear correlated with the outflow energy, as measured from ^{12}CO (Cyganowski *et al.* 2011, Cyganowski *et al.* 2017).

References

Cyganowski, C. J., Brogan, C. L., Hunter, T. R., *et al.* 2017, *MNRAS*, 468, 3694
Cyganowski, C. J., Brogan, C. L., Hunter, T. R., *et al.* 2014, *Ap. Lett.*, 796, L2
Cyganowski, C. J., Brogan, C. L., Hunter, T. R., *et al.* 2011, *ApJ*, 729, 124
Cyganowski, C. J., Brogan, C. L., Hunter, T. R., & Churchwell, E. 2009, *ApJ*, 702, 1615
Kalenskii, S. V., Kurtz, S., & Bergman, P. 2013, *Astron. Rep.*, 57, 120
Kurtz, S., Hofner, P., & Álvarez, C. V. 2004, *ApJS*, 155, 149
Plambeck, R. L. & Menten, K. M. 1990, *ApJ*, 364, 555
Sato, M., Wu, Y. W., Immer, K., *et al.* 2014, *ApJ*, 793, 72
Voronkov, M. A., Caswell, J. L., Ellingsen, S. P., *et al.* 2014,*MNRAS*, 439, 2584
Voronkov, M. A., Caswell, J. L., Ellingsen, S. P., *et al.* 2012, Cosmic Masers - from OH to H0, 287, 433

Astrophysical Masers:
Unlocking the Mysteries of the Universe
Proceedings IAU Symposium No. 336, 2017
A. Tarchi, M.J. Reid & P. Castangia, eds.

© International Astronomical Union 2018
doi:10.1017/S1743921318000091

VLBI astrometry of a water maser source in the Sgr B2 complex with VERA

Daisuke Sakai[1,2]**, Tomoaki Oyama**[2]**, Takumi Nagayama**[2]**,**
Mareki Honma[2,3,4] **and Hideyuki Kobayashi**[1,3]

[1]Department of Astronomy, Graduate School of Science, The University of Tokyo, 7-3-1
Hongo, Bunkyo-ku, Tokyo 113-0033, Japan
[2]Mizusawa VLBI Observatory, National Astronomical Observatory of Japan, 2-12
Hoshi-ga-oka, Mizusawa-ku, Oshu-shi, Iwate 023-0861, Japan
[3]Mizusawa VLBI Observatory, National Astronomical Observatory of Japan, 2-21-1, Osawa,
Mitaka, Tokyo 181-8588, Japan
[4]Department of Astronomical Sciences, Graduate University for Advanced Studies, 2-21-1,
Osawa, Mitaka, Tokyo 181-8588, Japan

Abstract. We have conducted astrometric observations toward a 22 GHz water maser source associated with the Sgr B2 complex in the Galactic center region with VERA (VLBI exploration of Radio Astrometry). We measured a trigonometric parallax and absolute proper motion of the Sgr B2 complex with respect to an extra-galactic source by observing the water maser source at 10 epochs from 2014 to 2017. The measured distance was $7.52^{+3.01}_{-1.67}$ kpc for the Sgr B2M region.

We also succeeded to measure internal motions of maser spots in Sgr B2M, and N region. The number of spots which we could measure the internal motions is about 400. The distribution of the maser spots shows that the maser spots are associated with envelope of HII region seen in radio continuum image obtained with VLA and ALMA. We discuss relative motions between Sgr B2M, and N by using the internal motion.

Keywords. Galaxy: kinematics and dynamics, masers, astrometry

1. Overview

We conducted astrometric observations toward a 22 GHz water maser source associated with the Sgr B2 region, which is one of the most intense star-forming region in our Galaxy. Trigonometric parallax of this source was measured by Reid *et al.* (2009) with VLBA. Their results suggested that the distance was $7.8^{+0.8}_{-0.7}$ kpc. The precise position and motion of Sgr B2 region on the central molecular zone (CMZ) is important to understand the dynamics of the Galactic center (Sawada *et al.* 2004, Molinari *et al.* 2011). Then, it is important to measure the proper motion of the source accurately. We measured internal motions of maser spots in Sgr B2 region, and corrected the proper motion obtained by phase-reference observations with VERA. Observations have been conducted at 10 epochs from 2014 to 2016. In phase reference observations, we observed an extragalactic source J1745-2820 as a phase and position reference source, simultaneously.

2. Results

Figure 1 depicts the internal proper motions for each maser spot. In Sgr B2 Main region, maser spots extend over 2 arcsec, and the line-of-sight velocities range from 30 to $110 \, \mathrm{km \, s^{-1}}$. Redder spots located at east side of the map have larger internal motions of $5 \, \mathrm{mas \, yr^{-1}}$ than bluer spots whose internal motions is about $2 \, \mathrm{mas \, yr^{-1}}$. In Sgr B2 North, maser spots distribute on two different sites about 1 arcsec separated with each

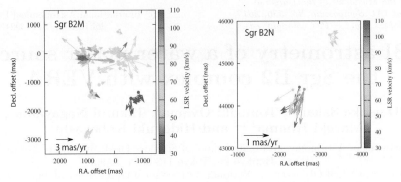

Figure 1. Internal motion of water maser in Sgr B2 region for (Left) Sgr B2M region and (Right) Sgr B2N region.

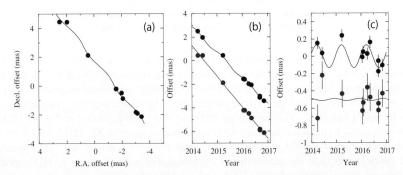

Figure 2. (a) Absolute proper motions of the maser feature at the line-of-sight velocity of $59.58\,\mathrm{km\,s^{-1}}$ Filled circles show the observed points from phase reference observations. (b) Motions toward R.A. and Dec. as a function of time. Black circles show the motion in the R.A. direction, and gray circles show the motion in the Dec. direction. (c) Result of parallax fitting. Error bars are evaluated so that a χ^2 value in the model fitting becomes unity.

other. The line-of-sight velocities of these spots are $80\,\mathrm{km\,s^{-1}}$ for the northern spot, and $50\,\mathrm{km\,s^{-1}}$ for the southern spots.

Figure 2 shows the parallax and absolute proper motion measured by phase-referencing observations for a maser spot at Sgr B2M. Fitting results indicate that the absolute proper motions are $-2.17\pm0.03\,\mathrm{mas\,yr^{-1}}$ and $-2.63\pm0.06\,\mathrm{mas}^{-1}$ in the direction of R.A. and Dec., respectively. The proper motions are consistent with those obtained by Reid *et al.* (2009) within errors. The measured parallax is $0.133\pm0.038\,\mathrm{mas}$, corresponding to $d = 7.52^{+3.01}_{-1.67}\,\mathrm{kpc}$ in the distance domain. The errors on the figure are derived by setting χ^2 per degree to be unity for both the R.A. and Dec. data.

After the correction of internal motion for the spot and the conversion of the coordinate system from the equatorial heliocentric flame (μ_α, μ_δ) to the Galactic heliocentric frame (μ_l, μ_b), we obtained the systematic proper motion of Sgr B2M relative to Sgr A* as $(\mu_l \cos b, \mu_b) = (2.66, -0.29)\,\mathrm{mas\,yr^{-1}}$. This suggests that Sgr B2M is located at nearside of Sgr A* when we assume a low eccentric orbit of the CMZ.

References

Molinari, S., Bally, J., Noriega-Crespo, A., *et al.* 2011, *ApJL*, 735, L33
Reid, M. J., Menten, K. M., Zheng, X. W., Brunthaler, A., & Xu, Y. 2009, *ApJ*, 705, 1548
Sawada, T., Hasegawa, T., Handa, T., & Cohen, R. J. 2004, *MNRAS*, 349, 1167

Astrophysical Masers:
Unlocking the Mysteries of the Universe
Proceedings IAU Symposium No. 336, 2017
A. Tarchi, M.J. Reid & P. Castangia, eds.

© International Astronomical Union 2018
doi:10.1017/S174392131701016X

Methanol masers and magnetic field in IRAS18089-1732

Daria Dall'Olio[1], W. H. T. Vlemmings[1], G. Surcis[2], H. Beuther[3], B. Lankhaar[1], M. V. Persson[1], A. M. S. Richards[4] and E. Varenius[1]

[1]Department of Space, Earth and Environment, Chalmers University of Technology, Onsala
Space Observatory, Observatorievägen 90, 43992 Onsala, Sweden
email: daria.dallolio@chalmers.se
[2]INAF–Osservatorio Astronomico di Cagliari, Via della Scienza 5, 09047 Selargius, Italy
[3]Max-Planck-Institute for Astronomy, Königstuhl 17, 69117 Heidelberg, Germany
[4]Jodrell Bank Centre for Astrophysics, Department of Physics and Astronomy, University of
Manchester, M139PL Manchester, UK

Abstract. Theoretical simulations have shown that magnetic fields play an important role in massive star formation: they can suppress fragmentation in the star forming cloud, enhance accretion via disc and regulate outflows and jets. However, models require specific magnetic configurations and need more observational constraints to properly test the impact of magnetic fields. We investigate the magnetic field structure of the massive protostar IRAS18089-1732, analysing 6.7 GHz CH_3OH maser MERLIN observations. IRAS18089-1732 is a well studied high mass protostar, showing a hot core chemistry, an accretion disc and a bipolar outflow. An ordered magnetic field oriented around its disc has been detected from previous observations of polarised dust. This gives us the chance to investigate how the magnetic field at the small scale probed by masers relates to the large scale field probed by the dust.

Keywords. stars: formation, magnetic field, masers, polarization

1. Introduction

The importance of the magnetic field in high mass star formation (HMSF) is not yet fully clear and there are still many open questions concerning its role in the accretion processes and generation of jets and outflows. In the past few years, masers have been successfully used to study the magnetic field at few AU scales around massive protostars. Thanks to their narrow and strong spectral lines and through their polarized emissions, it is possible to reconstruct the morphology and the strength along the line of sight of the magnetic field, by measuring linear polarization angles and Zeeman splitting (Vlemmings *et al.* 2010; Surcis *et al.* 2015). This can be done on scale comparable to circumstellar discs ($\sim$ 100 au). In order to improve our models and to build a complete picture of the magnetic field evolution, we need more observational constraints; in particular we need to verify that the magnetic field at small scales probed by masers traces the field at larger scales, probed, for example, by the dust, and not small-scale fluctuations. Currently, few observations of both masers and dust polarisation exist towards the same regions (e.g. Surcis *et al.* 2014). IRAS18089-1732 is a well studied high mass star forming region, showing a hot core chemistry and a disk-outflow system. Previous observations of polarized dust made with SMA revealed an ordered magnetic field oriented around the disk at large scales ($\sim$ 5000 au, Beuther *et al.* 2010).

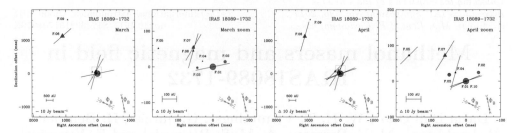

Figure 1. Masers identified in March (left) and April (right). The zoom is made on the region marked by the dashed grey boxes. Triangles and circles represent masers of the blue and of red group respectively. The different sizes of the triangles and the circles represent the intensity. Line segments mark the direction of the polarisation angle for the maser features that show linear polarisation. The average direction of the resulting magnetic field Φ_B obtained for two groups of masers is indicated in the bottom right corners of each panel.

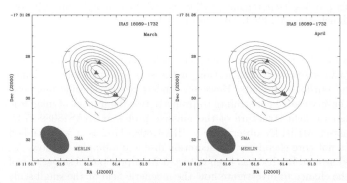

Figure 2. The figure shows that the magnetic field in the maser regions (triangles and long dashed segments) presents the same orientation as observed from the dust tracing the disc (contours and isolated short segments). The orientation is preserved in the two epochs March (left panel) and April (right panel).

2. Masers and dust probe the same large-scale magnetic field

From the analysis of a three-epoch MERLIN observations of the 6.7 GHz CH_3OH masers, Dall'Olio *et al.* 2017 identified two groups separated in velocity and polarisation angles as showed in Fig. 1. The blue group tracing the disc (triangles) show a velocity range between 30.0–36.4 km s^{-1}, and an orientation of the magnetic field $62° \pm 3°$. The red group that is tracing a cloud of gas close to the outflow (circles) shows a velocity range between 37.7–39.2 km s^{-1}, and an orientation of the magnetic field $14° \pm 4°$. From the analysis of the group of masers generated in the same region of the disc, Dall'Olio *et al.* 2017 obtained the first polarised map of the masers for IRAS 18089-1732 and showed that the small-scale magnetic field probed by the masers is consistent with the large-scale magnetic field traced by the dust. This confirms that methanol masers trace the large scale field, and that the large scale field component, even at the AU scale of the masers, dominates over any small scale field fluctuations Fig. 2.

We acknowledge funding from the ERC under the European Union's Seventh Framework Programme (FP/2007-2013) / ERC Grant Agreement n. 614264.

References

Beuther, H., Vlemmings, W. H. T., Rao, R., & van der Tak, F. F. S. 2010, *ApJ*, 724, L113

Dall'Olio, D., Vlemmings, W. H. T., Surcis, G., *et al.* 2017, *A&A*, 607, 111

Surcis, G., Vlemmings, W. H. T., van Langevelde, H. J., *et al.* 2015, *A&A*, 578, A102

Surcis, G., Vlemmings, W. H. T., van Langevelde, H. J., Moscadelli, L., & Hutawarakorn Kramer, B. 2014, *A&A*, 563, A30

Vlemmings, W. H. T., Surcis, G., Torstensson, K. J. E., & van Langevelde, H. J. 2010, *MNRAS*, 404, 134

Astrophysical Masers:
Unlocking the Mysteries of the Universe
Proceedings IAU Symposium No. 336, 2017
A. Tarchi, M.J. Reid & P. Castangia, eds.

© International Astronomical Union 2018
doi:10.1017/S1743921318000297

Long-term photometric observations in the field of the star formation region NGC7129

Evgeni Semkov[1], Stoyanka Peneva[1], Sunay Ibryamov[1,2] and Asen Mutafov[1]

[1]Institute of Astronomy and National Astronomical Observatory, Bulgarian Academy of Sciences, Sofia, Bulgaria
email: esemkov@astro.bas.bg

[2]Department of Theoretical and Applied Physics, University of Shumen, Shumen, Bulgaria

Abstract. We present results from long-term optical photometric observations of the Pre-Main Sequence (PMS) stars, located in the star formation region around the bright nebula NGC 7129. Using the long-term light curves and spectroscopic data, we tried to classify the PMS objects in the field and to define the reasons for the observed brightness variations. Our main goal is to explore the known PMS stars and discover new, young, variable stars. The new variable PMS star 2MASS J21403576+6635000 exhibits unusual brightness variations for very short time intervals (few minutes or hours) with comparatively large amplitudes ($\Delta I = 2.65$ mag).

Keywords. Stars: pre-main-sequence, stars: variables: T Tauri, ISM: Herbig-Haro objects

1. Introduction

The region NGC 7129 is a part of a larger structure, called Cepheus Bubble, and its represents a region with active star formation. The presence of a large number of Herbig-Haro objects, collimated jets, Herbig's Ae/Be and T Tauri stars, water masers, molecular outflow and other young objects have been reported in previous studies in the region (Hartigan & Lada 1985; Miranda *et al.* 1995, Magakian & Movsessian 1997; Kun *et al.* 2009). Using recent data from photometric monitoring and data from the photographic plate archives we aim at studing, the long-term photometric behavior of PMS objects in the field.

Our main goal is to study in more detail the known PMS objects in the field of NGC 7129 and to search for new, young, variable stars. The variability of PMS stars manifests itself as transient increases or temporary drops in brightness or as large amplitude irregular or regular variations. During our photometric monitoring, several PMS stars were found that could be classified as T Tauri stars (Semkov 2006). Such a long-term study of the fields of star formation has shown its effectiveness in discovering new phenomena important for astronomy (Semkov *et al.* 2010).

2. Results and Discussion

Our optical photometric observations were performed with the 2-m RCC, 50/70-cm Schmidt and 60-cm Cassegrain telescopes of the National Astronomical Observatory Rozhen (Bulgaria) and the 1.3-m RC telescope of the Skinakas Observatory (Crete, Greece). The observations were performed with eight different types of CCD cameras. The technical parameters for the cameras used, observational procedure and data reduction process are described in Ibryamov *et al.* (2015). Most suitable for long-term photometric study are the plate archives of the big Schmidt telescopes that have a large field of view, as

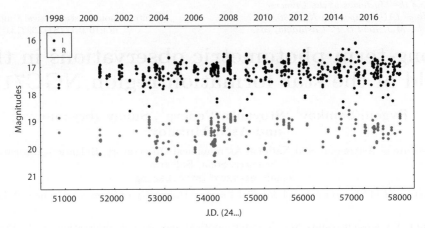

Figure 1. *RI* light curves of V4 for the period February 1998−November 2016.

the 105/150 cm Schmidt telescope at Kiso Observatory, the 67/92 cm Schmidt telescope at Asiago Observatory, the Palomar Schmidt telescope and others (Semkov *et al.* 2013).

The two most extensively studied PMS objects in the field of NGC 7129 are the stars V350 Cep and V391 Cep. Both stars show strong photometric variability and a spectrum rich of emission lines typical of classical T Tauri stars. The historical *B*-light curve of V350 Cep is very similar to eruptive PMS stars from FU Orionis type, but its spectrum is similar to another type of eruptive stars - EX Lupi. It is very likely that V350 Cep is an intermediate object between the two types of PMS stars showing large amplitude outbursts. A characteristic features of V391 Cep are periods of high amplitude brightness variability, followed by periods of smaller amplitude variability.

The star 2MASS J21403576+6635000 (hereafter V4) was recently discovered as a PMS variable. V4 shows very strong and fast photometric variability during short time periods (several minutes or hours) with large amplitude. The brightness of V4 during the period of our photometric observations 1998−2016 varies in the range 16.14−18.79 mag for the *I*-band and 18.40−20.48 mag for the *R*-band. Evidences of periodicity in the brightness variability of V4 are not detected. It can be seen from Fig. 1 that during our study the brightness of V4 vary around some intermediate level. The presence of V4 in the field of star formation and the irregular variability with large amplitude suggest a PMS nature of the star. We suggest that the V4 is a T Tauri star and the observed fast variability with large amplitudes probably is caused from a strong irregular accretion rate from circumstellar disk onto the stellar surface.

References

Hartigan, P. & Lada, C. J. 1985, *ApJS*, 59, 383

Ibryamov, S., Semkov, E., & Peneva, S. 2015, *PASA*, 32, e021

Kun, M., Balog, Z., Kenyon, S. J., Mamajek, E. E., & Gutermuth, R. A. 2009, *ApJS*, 185, 451

Magakian, T. Yu. & Movsessian, T. A. 1997, *Astr. Rep.*, 41, 483

Miranda, L. F., Eiroa, C., & Gomez de Castro, A. I. 1993, *A&A*, 271, 564

Semkov, E. H., Peneva, S. P., Munari, U., Milani, A., & Valisa, P. 2010, *A&A*, 523, L3

Semkov, E. H., Peneva, S. P., Munari, U. *et al.* 2013, *A&A*, 556, A60

Semkov, E. H. 2006, *Bulg. Astr. J.*, 8, 81

Astrophysical Masers:
Unlocking the Mysteries of the Universe
Proceedings IAU Symposium No. 336, 2017
A. Tarchi, M.J. Reid & P. Castangia, eds.

© International Astronomical Union 2018
doi:10.1017/S1743921317011644

The innermost regions of massive protostars traced by masers, high-resolution radio continuum, and near-infrared imaging

Fabrizio Massi[1], Luca Moscadelli[1], Carmelo Arcidiacono[2] and Francesca Bacciotti[1]

[1] INAF - Osservatorio Astrofisico di Arcetri,
Largo E. Fermi 5, I–50125, Firenze, Italy
email: `fmassi, mosca, fran@arcetri.astro.it`

[2] INAF - Osservatorio Astronomico di Bologna,
Via Piero Gobetti, 93/3, I–40129 Bologna, Italy
email: `carmelo.arcidiacono@oabo.inaf.it`

Abstract. Whether high-mass stars ($M > 7M_\odot$) emerge from a scaled-up version of the low-mass star formation scenario, i. e. through disk-mediated accretion, is still debated. We present the first results of an observational programme aimed to map the innermost regions of high-mass stellar objects by combining together high-spatial resolution maser and radio continuum observations, and near-infrared imaging.

Keywords. masers, stars: formation, infrared: stars

Disks and jets play a major role in the early evolution of low-mass stars, but their role in high-mass ($M_* > 7M_\odot$) star formation, if any, is still unclear. This is because high-mass young stellar objects (HMYSOs) are rare and difficult to observe. Recent calculations indicate that magnetic fields could be less efficient than in low-mass YSOs in collimating outflows, since magnetic collimation could be weakened by gravitational fragmentation of the accretion disk and by thermal pressure of the ionised gas at the base of the jet (Peters *et al.* 2011). In addition, outflows from HMYSOs could be intrinsically less collimated if driven by radiation pressure rather than by coherently rotating magnetic fields (Vaidya *et al.* 2011). Finally, OB-type stars emit powerful stellar winds, and if these were already present during the accretion phase they would contribute to the associated outflows.

Although mm interferometric observations in typical molecular outflow tracers (^{12}CO, ^{13}CO, SiO, HCO$^+$) have resolved a few individual high-mass protostellar outflows, the sampled scales are tens to hundreds times larger than model-predicted disk sizes (~ 100 AU) and do not allow investigating the launching regions of the flows. Much higher spatial resolution is then needed to determine the physical nature (e.g., stellar- vs. disk-wind, ionised vs. neutral) and the geometry (wide-angle vs. collimated) of the mechanism powering the associated molecular outflow. To tackle these limitations and better constrain models of high-mass star formation, we have recently started an observational programme of massive protostellar outflows. We have been investigating both the molecular and ionised components in these outflows with unprecedented angular resolution and sensitivity, by complementing multi-epoch VLBI observations of water masers, with high-angular resolution, multi-frequency (6, 13, and 22 GHz), deep imaging (rms noise 6–$10 \sim \mu$Jy) of radio continuum emission with the JVLA. A sample of 40 HMYSOs was selected from the BeSSeL (Bar and Spiral Structure Legacy; Reid *et al.* 2014) survey of water masers over the Galactic disk, which met all the following criteria: 1) strong

Figure 1. Continuum subtracted, pure line emission H_2 image of G111.25–0.77, overlaid with *purple* and *yellow contours* (see the on-line version) representing the JVLA A-Array continuum at 13 and 22 GHz, respectively (Moscadelli *et al.* 2016). A few H_2 knots SE and NW of the compact radio source clearly delineate a collimated jet with the same orientation as the innermost radio emission (observed at higher angular resolution). The black area is an artifact of continuum subtraction and indicates that the NIR counterpart of the radio source exhibits an SED rising with wavelength, a landmark of young stellar objects.

water masers; 2) associated sources with bolometric luminosity corresponding to ZAMS B3–O7 stars ($M_* > 7M_\odot$); 3) objects that, in previous surveys, had gone undetected or only showed compact (size $\leqslant 1$ arcsec) and weak (flux $\leqslant 650$ mJy) radio continuum emission (to exclude extended and more evolved HII regions). This sample was observed with JVLA in three different runs (for details see: Moscadelli *et al.* 2016). A comparison of radio continuum morphologies and water maser spatial and velocity distributions shows that 5 out of a sub-sample of 11 HMYSOs drive a collimated (conical) outflow. The derived outflow momentum rates are in the range $10^{-3} - 10M_\odot$ yr^{-1} km s^{-1}, among the highest values reported in the literature. Radio continuum morphology and maser velocity patterns are more difficult to understand in the remaining 6 sources.

However, critical information on jet collimation can ultimately only be obtained by linking the sub-arcsec structure (closer than 100 AU for the examined targets) to the outflowing gas pattern on a scale of $10^2 - 10^5$ AU from the driving source (i. e., up to 0.5–1 arcmin), a region which is still heavily embedded in dust and gas. To investigate this region we carried out very deep observations of a sub-sample of 4 HMYSOs from the BeSSeL survey, located at distances $d < 4$ kpc, through the H2 narrow-band filter by exploiting one of or both LUCI NIR cameras in parallel mode at the Large Binocular Telescope in Arizona. H_2 line emission at 2.12 μm has been routinely used to trace collimated gas outflows from YSOs. We detected H_2 line emission apparently associated with all the targeted compact radio sources. The most remarkable case, G111.25–0.77, is shown in Fig. 1. Our results confirm the presence of collimated jets in the vicinity of newly formed high-mass stars, pointing to a scaled-up version of low-mass star formation.

References

Moscadelli, L., Sánchez-Monge, A., Goddi, C., *et al.* 2016, *A&A*, 585, A71
Peters, T., Banerjee, R., Klessen, R. S., & Mac Low, M.-M. 2011, *ApJ*, 729, 72
Reid, M., Menten, K. M., Brunthaler, A, *et al.* 2014, *ApJ*, 783, 130
Vaidya, B., Fendt, C., Beuther, H., & Porth, O. 2011, *ApJ*, 742, 56

Astrophysical Masers:
Unlocking the Mysteries of the Universe
Proceedings IAU Symposium No. 336, 2017
A. Tarchi, M.J. Reid & P. Castangia, eds.

© International Astronomical Union 2018
doi:10.1017/S174392131700895X

SMA, VLA and VLBA observations in a 10^5 $L_\odot$ high mass star formation region IRAS 18360-0537

Gang Wu[1,2,3,4], Keping Qiu[2], Jarken Esimbek[1,4] and Xingwu Zheng[2]

[1]Xinjiang Astronomical Observatory, Chinese Acadmy of Sciences, Science 1-street 150, Beijing Road, Urumuqi, P. R. China, email: wug@xao.ac.cn
[2]School of Astronomy and Space Science, Nanjing University, Nanjing 210093, P. R. China
[3]Key Laboratory of Radio Astronomy, Chinese Academy of Sciences, Urumqi 830011, P. R. China
[4]University of the Chinese Academy of Sciences, Beijing 100080, P. R. China

Abstract. We have observed a young stellar object, IRAS 18360-0537, with a far-infrared luminosity of 1.2×10^5 $L_\odot$. It is perhaps the most promising candidate of a high-mass protostar associated with a Keplerian disk and a jet/outflow system in the regime of $L > 10^5 L_\odot$. We are conducting the SMA, VLA, and VLBA studies to provide a comprehensive understanding of this interesting high mass star formation scenario.

1, Introduction

In star formation studies, whether high-mass stars form mediated by disk/outflow systems as their low-mass counterparts is a key question under debate. Outflows have been proved to be omnipresent in high-mass star forming regions. Until now, there are only ~40 candidates observed to harbor rotating disks or toroids in high-mass star forming regions (Beltrán & de Wit 2016). Furthermore, most of these candidates are limited to objects with masses up to 25-30 $M_\odot$ or $L < 10^5 L_\odot$. The low number of disk detections, especially in $L > 10^5 L_\odot$ star formation regions, might be an observational bias or might be a real effect to be explained by detailed models of high mass star formation, involving turbulent core or competitive accretion models (e.g. McKee & Tan 2003, Bonnell & Bate 2002). We have observed a young stellar object, IRAS 18360-0537, with a far-infrared luminosity of 1.2×10^5 $L_\odot$ (Qiu *et al.* 2012). It is perhaps the most promising candidate of a high-mass protostar associated with a Keplerian disk and a jet/outflow system in the regime of $L > 10^5 L_\odot$. We are conducting the SMA, VLA, and VLBA studies to provide a comprehensive understanding of this interesting high mass star formation scenario.

2, Results

SMA: In IRAS 18360-0537, the SMA 1.3 mm continuum map shows two condensations, MM1 and MM2. Meanwhile the SMA CO and SiO indicate a northeast-southwest bipolar outflow centered at MM1 while CH_3OH and CH_3CN trace a northwest-southeast rotation gradient perpendicular to the outflow axis. Furthermore, CN spectra also from the SMA, present typical inverse P-Cygni profiles which demonstrate infall motions (see the panels in the first row of Fig. 1, and also Qiu *et al.* 2012).

VLA: To constrain the ionized gas, we carried out VLA 3.6 cm, 1.3 cm, and 7 mm radio continuum observations. The lower flux at 3.6 cm indicates that IRAS 18360-0537 is presently in a very early evolutionary stage, e.g in a stage prior to the formation of an HII region. The existing VLA observations are not well confining the parameters of free-free emission. We are proposing 2 cm and 6 cm observations with the JVLA to further

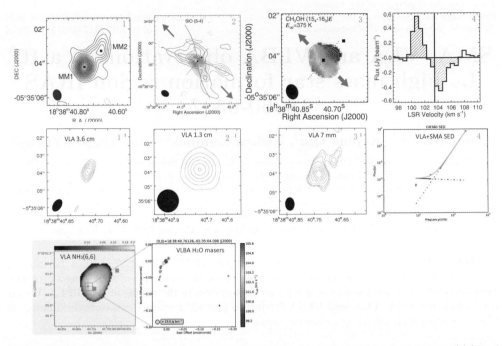

Figure 1. *First row:* SMA 1.3 mm continuum (1), SiO (2), CH₃OH (the first moment)(3) and the inverse P-Cygni profiles of CN (4) (adapted from Qiu *et al.* 2012). *Second row:* VLA 3.6 cm (1), 1.3 cm (2), 7 mm (3) continuum and SED of VLA and SMA continuum data (4). *Third row:* VLA NH₃ (6,6)(color) emission (Left) and BeSSeL H₂O masers (Right).

constrain the properties of the ionized gas in IRAS 18360-0537. We are also proposing an A configuration observation at 7 mm to reveal the spatial morphology with a resolution of ∼ 0.065" (see the panels in the second row of Fig. 1).

VLBA: For a better understanding of IRAS18360-0537, we are conducting the OH, H₂O, and CH₃OH maser studies in IRAS 18360-0537 with VLBA to investigate the immediate vicinity of the central (proto) star. The panels in the third row of Fig. 1 present the H₂O masers in the region obtained from the BeSSeL Survey. According to our previous identified morphologies of outflow and 'disk', H₂O masers are likely associated with the outflow. CH₃OH and OH masers were observed with the VLBA in August 2017. We will use these maser spots to constrain the kinematics with a millisecond (10 AU) resolution and explore the B field along line of sight with the Zeeman splitting of OH masers.

Acknowledgements
This work was funded by the Program of the Light in China's Western Region under grant 2015-XBQN-B-03, the National Natural Science foundation of China under grant 11603063, 11433008.

References
Beltrán, M. T. & de Wit, W. J. 2016, *A&AR*, 24, 6
Bonnell, I. A. & Bate, M. R. 2002, *MNRAS*, 336, 659
McKee, C. F. & Tan, J. C. 2003, *ApJ*, 585, 850
Qiu, K., Zhang, Q., Beuther, H., & Fallscheer, C. 2012, *ApJ*, 756, 170

Astrophysical Masers:
Unlocking the Mysteries of the Universe
Proceedings IAU Symposium No. 336, 2017
A. Tarchi, M.J. Reid & P. Castangia, eds.

ⓒ International Astronomical Union 2018
doi:10.1017/S1743921317010171

A Masing Event in the Cat's Paw

G. MacLeod[1], D. Smits[2], S. Goedhart[3], S. Ellingsen[4], T. Hunter[5] and C. Brogan[5]

[1] Hartebeesthoek Radio Astronomy Observatory,
P.O. Box 443, Krugersdorp, RSA 1740
email: gord@hartrao.ac.za

[2] Dept. of Mathematics, University of South Africa, RSA [3] SKA SA, RSA [4] School of Physical Sciences, U. of Tasmania, Australia [5] NRAO, USA

Abstract. We present Kitty, an unprecedented and near simultaneous flaring event in ten transitions (6 hydroxyl, 1 water and 3 methanol), that began on 1 January 2015 in the massive star-forming region NGC6334F located in the Cat's Paw Nebula. The brightest components in each transition increased by factors of 20 to 70 in line with a factor of ∼70 increase in dust emission luminosity for the source MM1. We also report the detection of only the fifth known 4.660 GHz hydroxyl maser and that it varied in a correlated fashion with 1.720, 6.031, and 6.035 GHz hydroxyl counterparts. We postulate that if Kitty, and two historical flares in 1965 & 1999, are accretion events and are caused by the successive passages of a secondary star disrupting the accretion disk, where the frequency of occurrence is cycling down at a rate of ∼2.2, it is possible another event will occur in 2022.

Keywords. masers, stars:formation, radio lines:molecular:interstellar.

1. Introduction

NGC6334F is a well-studied massive star-formation region with many associated species of masers (Cohen *et al.* 1995 and references therein). Weaver *et al.* (1968) first reported a flare of the 1665 and 1667 MHz hydroxyl masers of NGC6334 occurred in 1965. Variability studies of the 6.7 and 12.2 GHz methanol masers have been presented by MacLeod *et al.* (1993) and Goedhart *et al.* (1994); the former reported significant variation in a single 12.2 GHz maser feature while the latter reports a single, but different, feature also underwent a large increase during their monitoring period. We report a significant flaring event in ten maser transitions, including 6 of OH, 1 of H_2O, and 3 of CH_3OH, associated with NGC6334F and identify a third flare event in 1999 through re-analysis of Goedhart *et al.* (1994) data.

2. Results & Prediction

We were alerted to an unprecedented flaring event in NGC6334F when we discovered new 6.031 & 6.035 GHz excited-OH maser features. We determined the magnetic field strength and orientation, both were B ∼ +1.4 and +1.9 mG for 6.031 and 6.035 GHz respectively, which were not the same as previously reported (Caswell *et al.* 2011). This source was included in the maser monitoring programme at Hartebeesthoek Radio Astronomy (HartRAO) as a calibrator. In Fig. 1 we present the dynamic spectrum of the 6.7 GHz methanol masers associated with NGC6334F. A flaring event is clearly visible; it began on 1 January 2015 and peaked on 15 August 2015, at least in several features. The source is presently undergoing a secondary flare. Analysis of the associated 1.665

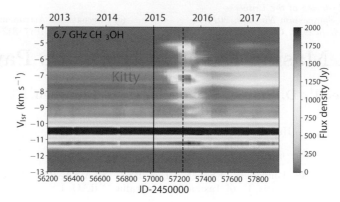

Figure 1. Dynamic spectrum of the 6.7 GHz methanol masers associated with NGC6334F.

GHz Hydroxyl, 12.2 GHz methanol, and 22.2 GHz water maser transitions showed correlated flaring. We refer to all features associated with this flare in all transitions as Kitty because NGC6334F is located in the Cat's Paw Nebula. Kitty also had flaring features in 1.720 & 4.660 GHz hydroxyl and 23.1 GHz methanol transitions. The 4.660 GHz OH maser is only the fifth discovered and varied in a correlated fashion with the other hydroxyl maser features in particular the 1.720 GHz maser; contrary to theory in Cragg *et al.* (2002) and Gray *et al.* (1992) where 4.765 GHz OH masers are predicted.

Previously detected features in NGC6334F (Ellingsen *et al.* 1996) did not change in a correlated way to features of Kitty and polarisation results suggested Kitty resides in a distinctly different region. Very long baseline interferometry (unpublished) and Very Large Array observations (Hunter *et al.* these proceedings) confirm this for the 6.7 GHz masers; Kitty is associated with MM1 where the total luminosity traced by the dust emission increased by a factor of ∼70 (Hunter *et al.* 2017).

Finally, re-analysis of historical data has been used to identify two other flares associated with NGC6334F. We postulate that if these are accretion events and are caused by successive periastron passages of a secondary star disrupting the accretion disk (e.g. Bally 2002) occurring in 1965, 1999, & 2015, then the frequency of occurrence is cycling down at a rate of ∼2.2 per event, and it is possible that another event will occur in 2022.

References

Bally, J. 2002, *ASP Conference Series*, vol. 267, p. 219

Caswell, J. L., Kramer, B. H., & Reynolds, J. E. 1995, *MNRAS*, 414, 1914

Cragg, D. M., Sobolev, A. M., & Godfrey, P. D. 2002, *MNRAS*, 331, 521

Cohen, R. J., Masheder, M. R. W., & Caswell, J. L. 1995, *MNRAS*, 274, 808

Ellingsen, S. P., Norris, R. P., Diamond, P. J., McCulloch, P. M., Amy, S. W., Beasley, A. J., Ferris, R. H., Gough, R. G., King, E. A., Lovell, J. E. J., Reynolds, J. E., Tzioumis, A. K., Troup, E. R., Wark, R. M., & Wieringa, M. H. 1996, *Astrophysics e-prints*, astroph/9604024

Goedhart, S., Gaylard, M. J., & van der Walt, D. J. 2004, *MNRAS*, 355, 553

Gray, M. D. & Field, D., Doel R. C. 1992, *A&A*, 262, 555

Hunter, T. R., Brogan, C. L., MacLeod, G., Cyganowski, C. J., Chandler, C. J., Chibueze, J. O., Friesen, R., Indebetouw, R., Thesner, C., & Young, K. H. 2017, *ApJ*, 837, L29

MacLeod, G. C., Gaylard, M. J., & Kemball, A. J. 1993, *MNRAS*, 262, 343

Weaver, H., Dieter, N. H., & Williams, D. R. W. 1968, *ApJS*, 16, 219

Astrophysical Masers:
Unlocking the Mysteries of the Universe
Proceedings IAU Symposium No. 336, 2017
A. Tarchi, M.J. Reid & P. Castangia, eds.
© International Astronomical Union 2018
doi:10.1017/S1743921317009747

Current stage of the ATCA follow-up for SPLASH

Hai-Hua Qiao[1,2] Andrew J. Walsh[3] and Zhi-Qiang Shen[2,4]

[1]National Time Service Center, Chinese Academy of Sciences, Xi'An, Shaanxi, China, 710600
email: qiaohh@shao.ac.cn

[2]Shanghai Astronomical Observatory, Chinese Academy of Sciences,
80 Nandan Road, Shanghai, China, 20003

[3]International Centre for Radio Astronomy Research, Curtin University,
GPO Box U1987, Perth WA 6845, Australia

[4]Key Laboratory of Radio Astronomy, Chinese Academy of Sciences, China

Abstract. Four ground-state OH transitions were detected in emission, absorption and maser emission in the Southern Parkes Large-Area Survey in Hydroxyl (SPLASH). We re-observed these OH masers with the Australia Telescope Compact Array to obtain positions with high accuracy (∼1 arcsec). According to the positions, we categorised these OH masers into different classes, i.e. star formation, evolved stars, supernova remnants and unknown origin. We found one interesting OH maser source (G336.644-0.695) in the pilot region, which has been studied in detail in Qiao *et al.* (2016a). In this paper, we present the current stage of the ATCA follow-up for SPLASH and discuss the potential future researches derived from the ATCA data.

Keywords. masers, stars: AGB and post-AGB, stars: formation

1. Introduction

Hydroxyl (OH) was the first molecule detected at radio wavelength in the interstellar medium (ISM; Weinreb *et al.* 1963) and OH masers were also the first astrophysical maser species detected in the ISM (Weaver *et al.* 1965). Ground-state OH masers occur at 18 cm, with frequencies at 1612, 1665, 1667 and 1720 MHz. The 1612 MHz OH masers are usually detected toward the circumstellar envelopes of evolved stars (stellar OH masers). The 1665/1667 MHz OH masers are mainly associated with high-mass star forming regions (interstellar OH masers). The 1720 MHz OH masers are probes for shocks and exist in various environments, such as supernova remnants (SNRs), star forming regions (SFRs) and occasionally evolved stars.

SPLASH (the Southern Parkes Large-Area Survey in Hydroxyl) simultaneously observed four ground-state OH transitions in an unbiased way (Dawson *et al.* 2014). The survey region of SPLASH is 176 square degrees including the Galactic Centre region. The spatial resolution of the Parkes radio telescope at 18 cm is about 13 arcmin, which is insufficient to provide positional accuracies to identify OH masers reliably. Our motivation is to get accurate positions for SPLASH OH masers using the Australia Telescope Compact Array (ATCA). The ATCA observations have been completed, which took about 330 hours. In the SPLASH pilot region (40 square degrees), we detected 215 OH maser sites, most of which are associated with evolved stars (Qiao *et al.* 2016b). The preliminary results in the Galactic Centre region (40 square degrees) showed an increase in the number of stellar OH masers and a decrease in the number of interstellar OH masers compared with the pilot region (Qiao *et al.* 2017). In the pilot region, we detected the 1720 MHz OH maser emission toward a planetary nebula (G336.644-0.695;

OH-maser-emitting PN; OHPN) and the magnetic fields were also measured based on the Zeeman splitting of the 1720 MHz OH masers (Qiao *et al.* 2016a). The data reduction excluding the pilot region is still under way.

2. Future plans

Polarization study. Based on Qiao *et al.* (2016a), ATCA data have the ability to study the Zeeman splitting of OH masers. Thus we can use these data to study the magnetic fields in both evolved star and star formation categories. Similar to the "MAGMO" project (Green *et al.* 2012), our aim is also to check whether Galactic magnetic fields can be traced with Zeeman splitting of OH masers associated with SFRs, especially for the OH masers not associated with 6.7 GHz methanol masers. Moreover, we can also investigate the in situ magnetic fields of SNR 1720 MHz OH masers. Zeeman splitting of evolved star OH masers will also be obtained.

Maser time line in SFRs. We will select the star formation OH maser sites from the SPLASH survey region and compare these OH masers with other maser species, such as 6.7 GHz methanol masers from the Methanol Multibeam survey and 22 GHz water masers from the H_2O southern Galactic Plane Survey. Refer to the methods in Breen *et al.* (2010), we will construct a maser time line for star forming regions.

Searching for OHPNe. In the pilot region, 122 OH maser sites are associated with evolved stars. According to the definition of Qiao *et al.* (2016b), the 1612 MHz spectra of 29 evolved star OH maser sites are asymmetric. Deviation from the typical double-horned profile suggests that these evolved sources may be post-AGB stars or planetary nebulae. Therefore, we can select evolved star OH maser sites with asymmetric 1612 MHz spectra to check their continuum emission in order to determine whether these maser sites are associated with post-AGB stars or planetary nebulae.

Acknowledgements: We acknowledge the full SPLASH OH maser team who have contributed to the work presented here, i.e. Green, James A., Breen, Shari L., Dawson, J. R., Ellingsen, Simon P., Gómez, José F., Jordan, Christopher H., Lowe, Vicki; Jones, Paul A.. H.-H.Q. is partially supported by the Special Funding for Advanced Users, budgeted and administrated by Center for Astronomical Mega-Science, Chinese Academy of Sciences (CAMS-CAS) and CAS "Light of West China" Program. This work was supported in part by the Major Program of the National Natural Science Foundation of China (Grant No. 11590780, 11590784) and the Earth rotation measurement using giant fiber-optic gyroscope program.

References

Breen, S. L., Ellingsen, S. P., Caswell, J. L., *et al.* 2010, *MNRAS*, 401, 2219
Dawson, J. R., Walsh, A. J., Jones, P. A., *et al.* 2014, *MNRAS*, 439, 1596
Green, J. A., McClure-Griffiths, N. M., Caswell, J. L., *et al.* 2012, *MNRAS*, 425, 2530
Qiao, H.-H., Walsh, A. J., Gómez, J. F., *et al.* 2016a, *ApJ*, 817, 37
Qiao, H.-H., Walsh, A. J., Green, J. A., *et al.* 2016b, *ApJS*, 227, 26
Qiao, H.-H., Walsh, A. J., Shen, Z.-Q., *et al.* 2017, *The Multi-Messenger Astrophysics of the Galactic Centre*, Proc. IAU Symposium No. 322, p. 141
Weaver, H., Williams, D. R. W., Dieter, N. H., *et al.* 1965, *Nature*, 208, 29
Weinreb, S., Barrett, A. H., Meeks, M. L., *et al.* 1963, *Nature*, 200, 829

Astrophysical Masers:
Unlocking the Mysteries of the Universe
Proceedings IAU Symposium No. 336, 2017
A. Tarchi, M.J. Reid & P. Castangia, eds.

ⓒ International Astronomical Union 2018
doi:10.1017/S1743921317010067

New water maser source near HW3d in the massive star-forming region Cepheus A

Jeong-Sook Kim[1] and Soon-Wook Kim[1,2]

[1] Korea Astronomy and Space Science Institute, Daejeon 34055, Republic of Korea
email: evony@kasi.re.kr, skim@kasi.re.kr

[2] Korea University of Science and Technology, Daejeon 34113, Republic of Korea

Abstract. Cepheus A is the second nearest high mass star-forming region after Orion. It is characterized by the presence of several phenomena, such as a complex molecular outflow, and multiple radio continuum sources, known as HW sources. The radio continuum and water maser emission have been detected toward HW2, HW3b and HW3d regions, and all of them are considered harboring young stellar objects. In 2014, we performed KaVA observations and detected a new bright maser feature, ∼700 mas apart from HW3d, which has not been detected with previous VLBI observations. The relative proper motion of the new maser feature is faster than other regions. It can be a clue for a newly forming star. Alternatively, it may be caused by outflow shock from the star-forming regions such as HW3d or HW3c.

Keywords. masers − stars: formation − ISM: individual objects: Cepheus A − ISM: jets and outflows

1. Introduction

Cepheus A is the second nearest massive star forming region, located at a distance of 725 pc, after Orion (Moscadelli *et al.*2009). Several radio continuum sources such as HW2, HW3c and HW3d have been reported and water masers are detected in HW2, HW3b and HW3d (Torrelles *et al.*1998). The HW2 region is associated with a powerful radio thermal jet, and a complicated water maser structure perpendicular to the HW2 radio jet traces a circumstellar disk of ∼300 AU in radius (Curiel *et al.* 2006). On the other hand, HW3d is thought to harbor protostars, based on the spectral indices and bipolar outflow motion (Garay *et al.* 1996 and Chibueze *et al.* 2012). However, the origin of the thermal continuum outflow in HW3d is still unclear (Zapata *et al.* 2013).

2. New water maser source detected in 2014

We performed KaVA observations at 22 GHz in 2014 to investigate the time-evolving feature of the water maser distribution in Cepheus A. The maser features detected in HW2, HW3d and the new source nearby HW3diii are presented in Fig. 1. We compare the water maser distribution of HW2 observed in 2014 with the VLA observation in 1995 and the VLBA observations in 1996, 2001 and 2002 (Torrelles *et al.* 2011). It is hard to distinguish an appreciable change over 20 years, although no VLBI results have been reported since 2003.

In HW3d, we detected two maser-emitting regions and a new maser feature close to HW3diii. The brightest maser spots were detected in the new region, compared to other regions. The mean proper motion in the new region is the fastest among the values of mean proper motion estimated using the relative proper motion for maser spots detected in all regions including HW2 and HW3d.

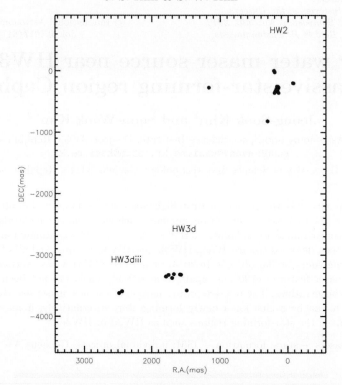

Figure 1. Water maser distribution of Cephues A in 2014 KaVA observations.

The new maser feature can be a signature of a newly forming star. Alternatively, it may be caused by outflow shock from the star-forming region HW3c, or from HW3d which is thought to be a star-forming region. Sudden brightening of water masers in Cep A was reported based on observations of the RT-22 radio telescope in 2012 (Tolmachev 2012). It can be speculated that the observed maser flare in 2012 is related to the new maser feature we detected in 2014. If it is the case, the new maser feature near HW3diii could be a new star-forming region, and is not formed by the outflow shock from other continuum sources such as HW3d or HW3c. Thirdly, it can be possible that the new maser feature is part of HW3diii.

References

Chibueze, J. O., Imai, H., Tafoya, D., Omodaka, T., Kameya, O., Hirota, T., Chong, S.-N., & Torrelles, J. M., 2012, *ApJ*, 748, 146

Curiel, S., Ho, P. T. P., Patel, N. A., Torrelles, J. M., Rodríguez, L. F., Trinidad, M. A., Cantó, J., Hernández, L., Gómez, J. F., Garay, G., & Anglada, G., 2006, *ApJ*, 638, 878

Garay, G., Ramirez, S., Rodriguez, L. F., Curiel, S., & Torrelles, J. M., 1996, *ApJ*, 459, 193

Moscadelli, L., Reid M. J., Menten, K. M., Brunthaler, A., Zheng, X. W., & Xu, Y., 2009, *ApJ*, 693, 406

Tolmachev, A., 2012, *The Astronomer's Telegram*, 4245

Torrelles, J. M., Gómez, J. F., Garay, G., Rodríguez, L. F., Curiel, S., Cohen, R. J., & Ho, P. T. P., 1998, *ApJ*, 509, 262

Torrelles, J. M., Patel, N. A., Curiel, S. *et al.*, 2011, *MNRAS*, 410, 627

Zapata, L. A., Fernandez-Lopez, M., Curiel, S., Patel, N., & Rodriguez, L. F., 2013, *arXiv:1305.4084*

Astrophysical Masers:
Unlocking the Mysteries of the Universe
Proceedings IAU Symposium No. 336, 2017
A. Tarchi, M.J. Reid & P. Castangia, eds.

© International Astronomical Union 2018
doi:10.1017/S1743921317009954

Filamentary Flows and Clump-fed High-mass Star Formation in G22

J. Yuan[1], J.-Z. Li[1] and Y. Wu [2]

[1] National Astronomical Observatories, Chinese Academy of Sciences, 20A Datun Road, Chaoyang District, Beijing 100012, China; email: `jhyuan@nao.cas.cn`

[2] Department of Astronomy, Peking University, 100871 Beijing, China;

Abstract. G22 is a hub-filament system composed of four supercritical filaments. Velocity gradients are detected along three filaments. A total mass infall rate of 700 $M_\odot$ Myr^{-1} would double the hub mass in about three free-fall times. The most massive clump C1 would be in global collapse with an infall velocity of 0.26 km s^{-1} and a mass infall rate of 5×10^{-4} $M_\odot$ yr^{-1}, which is supported by the prevalent HCO$^+$ (3-2) and ^{13}CO (3-2) blue profiles. A hot molecular core (SMA1) was revealed in C1. At the SMA1 center, there is a massive protostar (MIR1) driving multipolar outflows which are associated with clusters of class I methanol masers. MIR1 may be still growing with an accretion rate of 7×10^{-5} $M_\odot$ yr^{-1}. Filamentary flows, clump-scale collapse, core-scale accretion coexist in G22, suggesting that high-mass starless cores may not be prerequisite to form high-mass stars. In the high-mass star formation process, the central protostar, the core, and the clump can grow in mass simultaneously.

Keywords. ISM: clouds – ISM: kinematics and dynamics – stars: formation – stars: massive

1. Introduction

The two extensive debated high-mass star formation scenarios have plotted largely different pictures of mass accumulation process. Extensive investigations show that the prevalent filaments are the most important engine of forming stars, especially for the high-mass ones (André *et al.* 2014). How the gas flows detected in filaments help individual cores grow in mass is still a key open question. In this work, a filamentary cloud G22 is extensively investigated to reveal a promising mass accumulation scenario.

2. Mass accumulation process in G22

G22: a collapsing hub-filament system. With a distance of 3.51 kpc, the G22 cloud contains ten *Spitzer* infrared dark clouds (IRDCs). These IRDCs are mainly distributed in in a hub-filament system. As shown in Figure 1 (b), systematic velocity changes are detected along filaments F1, F1, and F3 based on ^{13}CO (1-0) observations. The differences between the velocities of the filaments and the junction as a function distance to the center shows monolithically increasing profiles for F1, F2, and F3 (see Figure 1 (c)). This suggests that gas is transfered to the hub region along these filaments with an estimated total mass infall rate of 700 $M_\odot$ Myr^{-1}.

G22-C1: a collapsing high-mass clump. Located at the hub region, C1 is the most massive clump with a mass of 466 $M_\odot$. Prevalent blue profiles are detected toward C1 (see Figure 2 (b)), suggestive of clump-scale global collapse. The estimated mass infall rate is 5.2×10^{-4} $M_\odot$ yr^{-1}

G22-C1-SMA1: A collapsing hot molecular core. At the center of C1, a hot molecular core SMA1 with a gas temperature higher than 220 K is detected. The spectrum of ^{13}CO

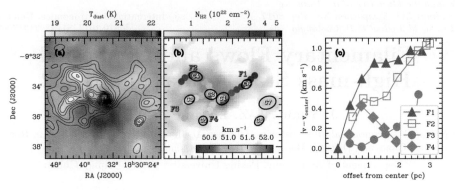

Figure 1. (a) Dust temperature map from SED fitting with column density overlaid as contours. (b) Velocity centroids of ^{13}CO (1-0) spectra extracted along filaments overlaid on top of the N_{H_2} map. The eight clumps, designated as C1 to C8, are shown as open ellipses. (c) LOS velocity of ^{13}CO as a function of distance from the potential well centers, i.e., clump C1 for F1, F2, and F4, and clump C2 for F3

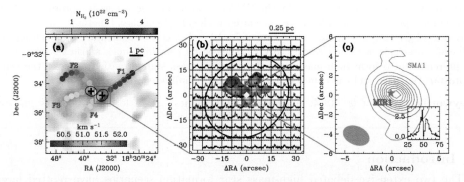

Figure 2. (a) Velocity centroids of ^{13}CO (1-0) extracted along filaments overlaid on the N_{H_2} map. (b) Spectra of JCMT/ ^{13}CO (3-2) overlaid on SMA/CO (2-1) outflows. The large ellipse delineates clump C1. (c) A close-up view of the SMA 1.3 mm continuum. The mono-core is designated as SMA1. A filled star shows the MIR source SSTGLMC G022.0387+00.2222 (MIR1) from the GLIMPSE survey. The insert plot shows the ^{13}CO (2-1) spectrum at the SMA1 peak.

(2-1) and C^{18}O (2-1) show blue profiles (see Figure 2 (c)), indicating infall motions in SMA1. The estimated mass accretion rate is about 7×10^{-5} $M_\odot$ yr^{-1}.

3. Conclusions

Inward motions have been detected along filaments, in the center clump and dense core. The continuous mass growth from large to small scales suggests that high-mass starless cores might not be prerequisite to form high-mass stars. The deeply embedded protostar, the core, and the clump can simultaneously grow in mass.

Reference

André, P., Di Francesco, J., Ward-Thompson, D., Inutsuka, S.-I., Pudritz, R. E., & Pineda, J. E. 2014, in: Henrik Beuther, Ralf S. Klessen, Cornelis P. Dullemond, and Thomas Henning (eds.), *Protostars and Planets VI* (University of Arizona Press, Tucson), p. 27

Astrophysical Masers:
Unlocking the Mysteries of the Universe
Proceedings IAU Symposium No. 336, 2017
A. Tarchi, M.J. Reid & P. Castangia, eds.

© International Astronomical Union 2018
doi:10.1017/S1743921317010699

Sub-mm observations of periodic methanol masers

D. J. van der Walt[1], J.-M. Morgan[1], J. O. Chibueze[2,1,3] and Q. Zhang[4]

[1] Centre for Space Research, NWU, Potchefstroom, South Africa
email: johan.vanderwalt@nwu.ac.za, jeanmariemorgan0@gmail.com,
jchibueze@ska.ac.za, qzhang@cfa.harvard.edu

[2] SKA-SA, Pinelands, South Africa;
[3] Dept. Physics & Astronomy, University of Nigeria, Nsukka, Nigeria; [4] Harvard-Smithsonian CfA, USA

Abstract. We present the results of sub-millimetre observations on three periodic methanol maser sources. Our results indicate that there are geometric differences between some periodic methanol masers which have different variability profiles.

Keywords. Stars:formation, ISM:molecules, submillimetre

1. Introduction

One of the fundamental questions about the periodic methanol masers concerns the driving mechanism underlying the periodic behaviour. The light curves of the periodic masers must in some way reflect the origin of the underlying physical mechanism (see eg. van der Walt *et al.* 2016). We observed G22.357+0.066 with a maser light curve similar to that of G9.62+0.20E, and G25.411+0.105 which has a maser light curve resembling a $|\sin(x)|$ curve to investigate whether there are obvious differences between these sources.

2. SMA observations and ALMA archival data

The target sources G22.357+0.066 and G25.411+00.105 were observed on June 27, 2016, during a 5-hour track with the Submillimeter Array (SMA) in a compact array configuration. The SMA $\sim$ 230 GHz broad band covered the dust continuum and the J = 2-1 spectral lines of ^{12}CO, ^{13}CO, $C^{18}O$, using 7 of the 8 antennas. The data calibration and imaging were done in CASA. We also obtained ALMA band 6 archival data on G9.62+0.20E (Project ID: 2013.1.00957.S). More details of these observations can be found in Liu, *et al.* (2017). The respective maps are shown in Figs. 1 & 2.

3. Results and Discussion

G22.357+0.066 (Fig. 1a): Comparison of the ^{13}CO emission (black contours) and the dust continuum (white contours) very strongly suggests that the detected ^{13}CO emission is not associated with the dust continuum emission. The masers (filled circle) also seem to be associated with the dust continuum rather than with the CO emission. Separate NIR counterparts (black triangles) for the CO and dust emission could be found from UKIDSS. The absence of ^{12}CO and ^{13}CO emission with the dust continuum is likely to be a density effect, suggesting that the masers are associated with a very young object.

G25.411+0.105 (Fig. 1b): The ^{13}CO line profile show strong blue- and redshifted wings, suggesting the presence of an outflow oriented in the direction of the line of

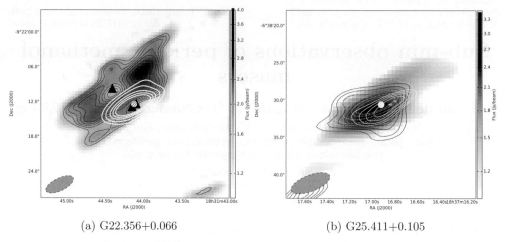

(a) G22.356+0.066   (b) G25.411+0.105

Figure 1. SMA maps of G22.357+0.066 and G25.411+0.105

sight. Strong ^{13}CO emission not associated with the outflow has also been detected (gray scale). The ^{13}CO emission indicates the presence of lower density gas which might suggest a somewhat later evolutionary phase than for G22.357+0.066. The masers are seen to be projected against the outflows. The maser spot distribution is elliptical and has been interpreted by Bartkiewicz *et al.* (2009) as being to due to a disk viewed almost face on. This is consistent with the direction of the outflows as shown in Fig. 1b.

G9.6+0.20E (Fig. 2): No clear ^{12}CO emission directly associated with the dust continuum (white contours) and with the 5_{-1}-4_{-2} E-CH$_3$OH (grayscale) emission, has been detected. A well defined bi-polar outflow has been detected in CO (black contours: red shifted, gray contours: blue shifted). The thermal E-CH$_3$OH emission is a high density tracer ($n_{cr} \sim 10^6$ cm^{-3}) and forms a thick structure with the dust continuum projected on the center of this structure. As in the case of G22.357+0.066, the absence of ^{12}CO emission might be a density effect.

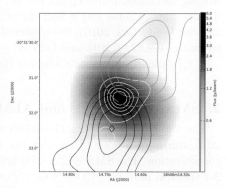

Figure 2. G9.62+0.20E

The first obvious difference between G9.62+0.20E and G25.411+0.105 (with maser light curves that are different) is clearly geometric, with G9.62+0.20E having outflows in the plane of the sky, while for G25.411+0.105 the line of sight is almost parallel to the outflow axis. In the case of G25.411+0.105 the periodic masers are projected onto the outflow; it is not clear whether it affects the maser light curve. For G22.357+0.066 and G9.62+0.20E the absence of ^{12}CO and ^{13}CO emission associated with the dust emission might indicate an earlier evolutionary phase than for G25.411+0.105.

References

Bartkiewicz, A., Szymczak, M., van Langevelde, H. J., *et al.* 2009, *A&A*, 502, 155

Liu, T., Lacy, J., Li, P. S., *et al.* 2017, *ApJ*, 849, 25

van der Walt, D. J., Maswanganye, J. P., Etoka, S. Goedhart, S., & van den Heever, S. P. 2016, *A&A*, 588, 47

Astrophysical Masers:
Unlocking the Mysteries of the Universe
Proceedings IAU Symposium No. 336, 2017
A. Tarchi, M.J. Reid & P. Castangia, eds.

© International Astronomical Union 2018
doi:10.1017/S174392131701167X

Dynamics of jet/outflow driven by high-mass young stellar object revealed by KaVA 22 GHz water maser observations

Jungha Kim[1,2], Tomoya Hirota[1,2], Kee-Tae Kim[3], Koichiro Sugiyama[2] and KaVA Science Working Group for Star-formation

[1]Department of Astronomical Science, SOKENDAI (The Graduate University for Advanced Studies), 2-21-1 Osawa, Mitaka, Tokyo 181-8588, Japan
email: jungha.kim@nao.ac.jp

[2]Mizusawa VLBI Observatory, National Astronomical Observatory of Japan, 2-21-1 Osawa, Mitaka, Tokyo 181-8588, Japan

[3]Korea Astronomy and Space Science Institute, 776 Daedeokdaero, Yuseong, Daejeon 34055, Korea

Abstract. We have started survey observations of the 22 GHz water maser sources associated with high-mass young stellar objects (HM-YSOs) as a part of the KaVA (KVN and VERA Array) large program (LP). The aim of our LP is to understand dynamical evolution of jets/outflows from HM-YSOs by analyzing 3D velocity structures of water maser features. In the first year (2016-2017), an imaging survey toward 25 HM-YSOs has been conducted and the 22 GHz water masers are detected toward 21 sources. Spatial distributions of maser features for individual sources are mapped. To complement physical properties in the vicinity of HM-YSOs, we have carried out ALMA cycle 3 observations of thermal molecular lines and continuum emissions toward 11 selected samples. Summary of the KaVA first year observations and the initial results from the ALMA toward one of our targets, G25.82-0.17, are reported.

Keywords. ISM: kinematics and dynamics, ISM: jets and outflows

1. Introduction

High mass star formation is far from understanding observationally because of its distant location and high surface density of HM-YSOs. The primary scientific goal of KaVA LP is understanding dynamical evolution of HM-YSOs using measured 3D velocity field and spatial structure of 22 GHz water and 44 GHz Class I methanol masers which could trace different evolutionary stages and different dynamical structures. Especially, water maser at 22 GHz is a good tracer of jet/outflow which is one of the most important signposts of star formation. Driving mechanisms of jet/outflow can be investigated to understand the key role of jet/outflow in massive star formation.

2. Observations and Results

We have carried out VLBI observations at 22 GHz toward 25 sources in total within the first year of project since 2016 with the KaVA. Sources associated with the water masers but with no previous VLBI data at the beginning of the KaVA LP (2014) were selected from the source list (87 HM-YSOs) for the first year observations. The highest angular resolution achieved with the longest base line between Mizusawa and Ishigaki station is 1.2 mas at 22 GHz.

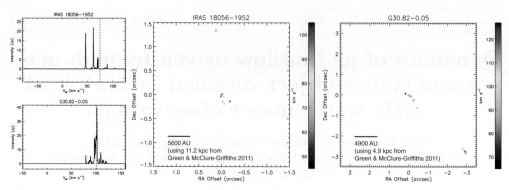

Figure 1. The 22 GHz water maser spectra (*left*) and spatial distribution maps of IRAS 18056-1952 (*center*) and G30.82–0.05 (*right*), respectively. Vertical dashed lines indicate the systemic velocities of each source adopted from the previous thermal molecular line surveys (Shirley *et al.* 2013, Urquhart *et al.* 2011, etc).

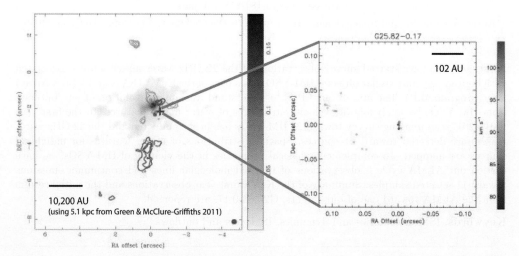

Figure 2. The integrated intensity map of SiO 5–4 overlaid onto dust continuum emission (*left*) and spatial distribution map (*right*) of G25.82–0.17. Bold contours represent the intensities over blueshifted range ($\leqslant$91 km s^{-1}) while normal contours represent the intensities over redshifted range ($\geqslant$93 km s^{-1}). A cross on the left panel denotes the absolute position of water maser emission derived using AIPS task FRMAP.

Water masers were detected toward 21 sources among the observed 25 samples with the detection rate of 84 %. Distributions of water maser features show source-to-source variation; such as elongated structures like G30.82-0.05, arc-like structures as seen in IRAS 18056-1952, etc (Figure 1). Physical properties of the jets/outflows and their driving sources will be investigated by the follow-up studies with ALMA . Preliminary results from ALMA cycle 3 observations (PI:Mikyoung Kim, 2015.1.01571.S) at band 6 (239 GHz) are presented in Figure 2. The SiO 5-4 line and water maser emission are tracing totally different scale of structure (Figure 2). The inner most part near HM-YSOs can be investigated by 3D velocity structure of water maser emission obtained with the KaVA.

References

Green, J. A. & McClure-Griffiths, N. M. 2011, *MNRAS*, 417, 2500
Shirley, Y. L., Ellsworth-Bowers, T. P., Svoboda, B., *et al.* 2013, *ApJS*, 209, 2
Urquhart, J. S., Morgan, L. K., Figura, C. C., *et al.* 2011, *MNRAS*, 418, 1689

Astrophysical Masers:
Unlocking the Mysteries of the Universe
Proceedings IAU Symposium No. 336, 2017
A. Tarchi, M.J. Reid & P. Castangia, eds.

© International Astronomical Union 2018
doi:10.1017/S1743921317010936

6.7 GHz Methanol Masers Observation with Phased Hitachi and Takahagi

Kazuhiro Takefuji[1], Koichiro Sugiyama[2], Yoshinori Yonekura[3], Tagiru Saito[4], Kenta Fujisawa[5] and Tetsuro Kondo[1]

[1] National Institute of Information and Communications Technology, 893-1 Hirai, Kashima, Ibaraki 314-8501, Japan. email: takefuji@nict.go.jp

[2] Mizusawa VLBI Observatory, National Astronomical Observatory of Japan, 2-21-1 Osawa, Mitaka, Tokyo 181-8588, Japan

[3] Center for Astronomy, Ibaraki University, 2-1-1 Bunkyo, Mito, Ibaraki 310-8512, Japan

[4] College of Science, Ibaraki University, 2-1-1 Bunkyo, Mito, Ibaraki 310-8512, Japan

[5] The Research Institute for Time Studies, Yamaguchi University, 1677-1 Yoshida, Yamaguchi, Yamaguchi 753-8511, Japan

Abstract. For the high-sensitivity 6.7 GHz methanol maser observations, we developed a new technology for coherently combining the two signals from the Hitachi 32 m radio telescope and the Takahagi 32 m radio telescope of the Japanese VLBI Network. Furthermore, we compared the SNRs of the 6.7 GHz maser spectra for two methods. One is a VLBI method and the other is the newly developed digital position switching, which is a similar technology to that used in noise-cancelling headphones. We report the phase-up technique and the observation.

Keywords. instrumentation: spectrographs techniques: interferometric

1. Introduction

The establishment of a synthesis technology using two signals from two stations with a maser observation as a reference is our first goal. Once this technology is realized, the phase difference of the synthesis parameters can be determined by the maser observation with two antennas. As a result, the observation efficiency is expected to improve. Moreover, in the case of spectral observation, the position switching of a single dish is performed by moving the antenna physically. Then, the system noise and other background noises can be removed and only the maser signal will remain. Once two antennas are phased, it is expected to create a virtual-off source by changing the phase of the synthesis parameters similarly to noise-cancelling headphones. Without physically moving the antenna, the digital position switching observation is considered to be performed. This is the second goal of our research.

2. Parameters for phasing two radio telescopes by maser observation

If we define P_a as the signal power of antenna A and P_b as the signal power of antenna B, then P_{sum}, the signal power after combining antennas A and B with phase difference $\Delta\theta$, can be expressed as

$$P_{sum} = |P_a + P_b| = \sqrt{P_a^2 + P_b^2 + 2P_a P_b \cos \Delta\theta}. \qquad (2.1)$$

The maximum value of the combined signal is $P_a + P_b$ and the minimum value is $P_a - P_b$ (when $P_a \geqslant P_b$). If the signal powers of the two antennas are equal, the power is doubled. However, depending on the phase difference $\Delta\theta$, the signal will decrease and disappear in the worst case. The extinction state can be used as an off-source observation. By adding an offset of $180°$ to the perfectly matched angle $\Delta\theta$, it can be applied for digital position switching. Figure 1 shows the sample spectra which made by the coherently combined signals of Hitachi 32m and Takahagi 32m. We confirmed the sensitivity (SEFD) by comparison of the daily monitoring result became about 85 Jy, where the SEFD of each telescope was 170 Jy.

To determine the phase difference $\Delta\theta$, two methods can be considered,

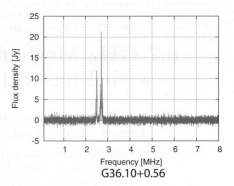

Figure 1. Sample Spectra of phased two telescope of Hitachi 32 m and Takahagi 32 m after digital position switching performed in 60 s integration and 1 kHz resolution.

- Performing the cross-correlation (VLBI) to the maser emission between two antennas and obtain the phase difference [Takefuji *et al.* (2017)]..
- Combining the two signals by adding an offset of $120°$, $-120°$, and $0°$ and compare the amplitude of the maser emission by performing the cross-correlation spectrometry [Takefuji *et al.* (2016)]. Figure 2 shows the spectra of Hitachi 32 m and Takahagi 32 m by the cross-correlation spectrometry.

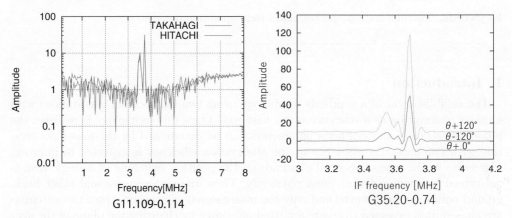

Figure 2. Left: Sample cross-spectra of the observed maser sources with Hitachi 32m and Takahagi 32m by the cross-correlation spectrometry. Right: The phase difference of two telescopes are obtained after comparing the XCS amplitudes of three cases of the phased result by adding an offset of $120°$, $-120°$, and $0°$. We added offsets to the plots for a clarity.

References

Takefuji, K., Imai, H., & Sekido, M., 2016, *PASJ*, 68, 86

Takefuji, K., Sugiyama, K., Yonekura, Y., Saito, T., Fujisawa, K., & Kondo, T., 2017, *PASP*, 129, 981

Astrophysical Masers:
Unlocking the Mysteries of the Universe
Proceedings IAU Symposium No. 336, 2017
A. Tarchi, M.J. Reid & P. Castangia, eds.

© International Astronomical Union 2018
doi:10.1017/S174392131701105X

VERA Single Dish Observations

Kazuyoshi Sunada[1], **Takumi Nagayama**[1], **Aya Yamauchi**[1],
Tomoya Hirota[2], **Katsunori M. Shibata**[2] **and Mareki Honma**[1]

[1]Mizusawa VLBI observatory, NAOJ
2-12, Hoshigaoka, Mizusawa, Oshu, Iwate 023-0861, Japan
email: `kazu.sunada@nao.ac.jp`

[2]Mizusawa VLBI observatory, NAOJ,
2-21-1 Osawa, Mitaka, Tokyo 181-8588 , Japan

Abstract. We will report the activities of the VERA single-dish observations. We are carrying out single-dish observations with two purposes. The first purpose is the monitoring of known H_2O maser sources. At present, we are carrying out monitoring observations for 312 H_2O maser sources at intervals of two months. The second purpose is the search for new water maser sources. We selected 901 target sources from the AKARI FIS Bright Source Catalogue. We found 61 new H_2O maser sources.

Keywords. masers, surveys

1. Monitoring Obsrvations

Because water maser emission shows time variability, it's important to monitor its intensity to progress with the high resolution observations by the VLBI Exploration of Radio Astrometry (VERA) effectively. Periodicity and a flare of the maser emission can be detected only through long-term monitoring observations.

Monitoring observations have been started since December 2015 at intervals of two months. We observed the maser line of H_2O ($6_{16} \rightarrow 5_{23}$, 22.23508 GHz). We used three stations of the VERA, the 20 m telescopes Mizusawa, Ogasawara, and Ishigaki-jima. The beam width (HPBW) and the aperture efficiency of the telescopes at the observation frequency were 145" and 0.45, respectively. The pointing accuracy was better than 10". The spectra were obtained using the digital spectrometer. This provided a bandwidth of 32 MHz and a frequency resolution of 31.25 kHz. This corresponds to a velocity coverage of $\pm$ 215 km s^{-1} and a velocity resolution of 0.4 km s^{-1} at 22.2 GHz. To get the same velocity resolution, data obtained at Mizusawa was smoothed to the same frequency resolution of the data obtained at the other telescopes. The integration time of each observation was 6 minutes. All the data were analyzed by the automatic pipeline. We considered the H_2O maser emission detected when the peak intensity was higher than the 4σ noise level.

We update the monitoring list at August every year. More than 300 sources were replaced because their H_2O maser emission was not detected for more than one year. The present monitoring list consists of 312 H_2O maser sources and includes sources already observed with VERA and sources waiting for VERA observations. We show an example of our monitoring results in Figure 1. This example clearly indicates the intensity variability, the velocity shift of the maser emission, and also the emergence of new velocity components.

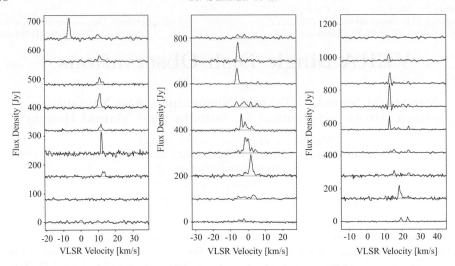

Figure 1. Examples of the results of our monitoring observations for three sources.

2. Survey Observations

Since there were many H_2O maser sources which we could not detect for a long time, survey observations to search for new maser sources are necessary to increase VERA observation candidates.

Various wide-area surveys have been carried out. For example, the far infrared satellite, AKARI, carried out a far infrared survey of all the sky (Murakami *et al.* 2007). The bolometer array camera, Bolocam, carried out the 1.1 mm continuum survey of the galactic plane (Aguirre *et al.* 2011). Such surveys supplied the comprehensive catalogs of young stellar objects (YSOs) and/or dense clumps, such as the AKARI FIS Bright Source Catalogue (AKARI FIS BSC) (Yamamura *et al.* 2010). YSOs and dense clumps in these catalogs are good target sources for searching maser emission.

We selected 901 sources from AKARI FIS BSC. We started the survey observations in August 2009. We are carrying out the survey observations in the intervals between VLBI observations. The specifications of the survey observations were the same of the monitoring observations except for the integration time. Depending on the system noise temperature, we set the integration time of each observation to obtain a 4σ noise level of less than 4 Jy. If the obtained noise level was higher than 4 Jy, we observed the source again. Some of the sources, therefore, were observed more than two times.

We detected the H_2O maser emission from 120 sources. The sources having larger FIS 140 μm flux showed higher detection rates. 61 of the 120 detected sources were new detections. For these new detected sources, we carried out VLBI observations and measured accurate maser positions. These new sources were also added to the monitoring list and we are continuing the monitoring observations.

References

Aguirre, J. E., Ginsburg, A. G., Dunham, M. K., *et al.*, 2011, *ApJS*, 192, 4
Murakami, H., Baba, H., Barthel, P., *et al.*, 2007, *PASJ*, 59S, 369
Yamamura, I., Makiuti, S., Ikeda, N., Fukuda, Y, Oyabu, S, Koga, T., & White, G. J., 2010, AKARI-FIS Bright Source Catalogue Release note Version 1.0

Astrophysical Masers:
Unlocking the Mysteries of the Universe
Proceedings IAU Symposium No. 336, 2017
A. Tarchi, M.J. Reid & P. Castangia, eds.

© International Astronomical Union 2018
doi:10.1017/S1743921317011498

Full polarization analysis of OH masers at 18-cm toward W49 A star forming region

K. Asanok[1,2], B. Hutawarakorn Kramer[1,3], S. Etoka[4], M. Gray[4], A. M. S. Richards[4], N. Gasiprong[5] and N. Naochang[2]

[1] National Astronomical Research Institute of Thailand, 260 Moo 4, Tambol Donkaew, Amphur Maerim, Chiang Mai, 50180, Thailand
email: kitiyanee@narit.or.th

[2] Department of Physics, Faculty of Science, Khon Kaen University, Khon Kaen, 40002, Thailand

[3] Max-Planck-Institut für Radioastronomie, Auf dem Hügel 69, 53121 Bonn, Germany
email: bkramer@mpifr-bonn.mpg.de

[4] Jodrell Bank Center for Astrophysics, School of Physics and Astronomy, The University of Manchester, Alan Turing Building, Oxford Road, Manchester, M13 9PL, United Kingdom
email: sandra.etoka@manchester.ac.uk, malcolm.gray@manchester.ac.uk, a.m.s.richards@manchester.ac.uk

[5] Department of Physics, Faculty of Science, Ubon Ratchathani University, Ubon Ratchathani, 34190, Thailand
email: ngasiprong@yahoo.com

Abstract. W49 A is a star-forming region (SFR) found in the constellation of Aquila. It contains 3 active regions: W49 North (W49 N), W49 South West (W49 SW) and W49 South (W49 S). We present preliminary results from two epochs (e-)MERLIN observations of all ground-state OH masers towards the star-forming region (SFR) complex W49 A. The first epoch of observations was done in full-polarization mode with MERLIN in 2005 while the second epoch was obtained only in dual circular polarization during the test observations of the upgraded e-MERLIN in 2013. The overall maser spatial distributions in both epochs are in good agreement. We found several new high velocity maser features up to $+34$ km s^{-1} and -28 km s^{-1}. The magnetic field strengths are between 1.1 to 10.8 mG. All three sources show evidence of magnetic field reversal.

Keywords. masers, ISM: HII regions, ISM:kinematics and dynamics, ISM:magnetic fields, ISM:molecules, stars: formation

Summary of the results

W49 A is a star-forming region (SFR) complex, containing 3 active regions: W49 North (W49 N), W49 South West (W49 SW) and W49 South (W49 S). The spatial distribution of all four ground-state (1612, 1665, 1667 and 1720 MHz) OH masers in W49 N, W49 S and W49 SW, observed with e-MERLIN in 2013, is shown in Figure 1. The overall spatial distribution of the masers is in good agreement with the previous MERLIN observation in 2005 and the VLA observation in 1991 by Argon *et al.* (2000). The enhanced sensitivity and wider velocity coverage of the e-MERLIN dataset, allowed us to detect new sites of maser emission in all transitions and regions, in particular, masers features up to $+34$ km s^{-1} and -28 km s^{-1} are found in W49 N and W49 S.

Here, we only report the magnetic field and polarization information of the 1612-MHz, 1665-MHz (only in W49 SW) and 1720-MHz transitions observed in full-polarization mode with MERLIN in epoch 2005. The magnetic field strength ranges from 1.1 mG (in W49 S) to 10.8 mG (in W49 N). All three sources show evidence of magnetic field reversal. W49 SW has the highest percentage of linear polarization (up to 40 %) while

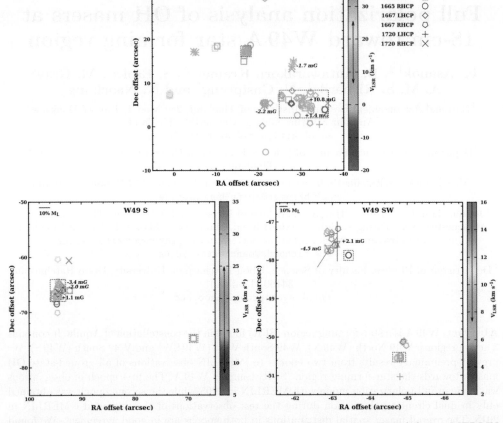

Figure 1. Positions and velocities of all four ground-state OH masers in W49 A from e-MERLIN observations in 2013. The positions are offset from (0,0) at R.A.(J2000)=19h 10m 15.308s, Dec.(J2000)=+09°06′08.″4822. The dash line boxes indicates the location where new maser features were detected. A vertical arrow in the color bar indicates the velocity range covered in Argon *et al.* (2000). The linear polarization vectors and magnetic field strengths obtained from MERLIN (2005). The plus and minus signs indicate whether the direction of the magnetic field is toward (−) or away (+) from us (see the online version for the color figure).

W49 N has the lowest (less than 10 %). Further detailed studies (e.g. comparison of Zeeman measurements in all transitions in both epochs and interpretation of the physical condition of each sources) will be reported in Asanok *et al.* (in prep.).

Reference

Argon, A. L., Reid, M. J., & Menten, K. M. 2000, *ApJ.S.*, 29, 159

Astrophysical Masers:
Unlocking the Mysteries of the Universe
Proceedings IAU Symposium No. 336, 2017
A. Tarchi, M.J. Reid & P. Castangia, eds.

© International Astronomical Union 2018
doi:10.1017/S1743921318000807

VLA Observations of a Sample of Low-Brightness 6.7 GHz Methanol Masers

Luca Olmi[1], Esteban D. Araya[2] and Jason Armstrong[2]

[1] INAF, Osservatorio Astrofisico di Arcetri,
Largo E. Fermi 5, I-50125 Firenze, Italy
email: olmi.luca@gmail.com

[2] Western Illinois University, Physics Department,
1 University Circle, Macomb, IL 61455, USA
email: ed-araya@wiu.edu

Abstract. In 2014 we conducted a survey for 6.7 GHz methanol masers with the Arecibo Telescope toward far infrared sources selected from the Hi-GAL catalog of massive cores. We found a number of sources with weak 6.7 GHz methanol masers, possibly indicating regions in early stages of star formation. Here we describe the results of follow-up observations that were conducted with the Very Large Array in New Mexico to characterize this new population of "weak" 6.7 GHz methanol masers.

Keywords. masers, stars: formation, ISM: molecules, radio lines: ISM

1. Introduction and previous results

Theoretical models and observational studies suggest that Class II methanol masers are exclusively associated with early phases of massive star formation (HMSF). This maser, in particular the $(5_1 - 6_0)$ transition of A^+ methanol at 6668.518-MHz, appears as an ideal tool for detecting a short-lived phase of HMSF, between the end of the large-scale accretion and the formation of massive protostars. With the main goal of determining the physical conditions in the Hi-GAL clumps (Molinari *et al.* 2010) and achieve a better understanding of the evolutionary path of massive starless cores toward massive stars, Olmi *et al.* (2014) carried out an observing program at Arecibo to determine whether a population of low flux-density masers exist and whether they could be used to mark a specific evolutionary phase in HMSF and/or a specific set of physical conditions.

Olmi *et al.* (2014) observed 107 high-mass dust clumps with the Arecibo telescope in search of the 6.7-GHz methanol and 6.0-GHz excited OH masers. The clumps were selected from the Hi-GAL survey to be relatively massive and visible from Arecibo. They detected a total of 37 methanol masers, where 22 sources are new and weak (median peak flux density 0.07 Jy) detections. Most of the methanol masers observed toward the Hi-GAL massive dust clumps by Olmi *et al.* (2014) appeared to be intrinsically weaker than previously observed masers in unbiased surveys (e.g., Pandian *et al.* 2007). However, the physical processes determining these intrinsically lower intensities still remain to be determined.

2. VLA observations

In 2014 we conducted a series of observations with the Karl G. Jansky Very Large Array (VLA) to further investigate several of the regions previously observed by Olmi *et al.* (2014). The main goals of these observations were to get accurate positions of the methanol masers and to determine whether any HII region existed toward the Hi-GAL

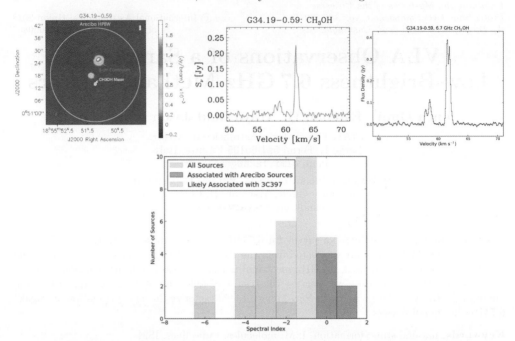

Figure 1. *Top-Left panel.* Continuum (∼ 6 GHz) image of G34.19-0.59. The CH₃OH maser is located near the Arecibo pointing, thus the weak flux density is not caused by pointing offsets. No radio continuum is detected toward the maser region. *Top-Middle panel.* Arecibo original spectrum. *Top-Right panel.* VLA spectrum. Change in relative intensity of components demonstrate variability. *Bottom panel.* Histogram of the number of continuum sources vs. spectral indices detected in a sub-sample of 9 of the pointing positions observed with the VLA. The sample is broken down into the sources associated with Arecibo regions, and sources likely associated with supernova remnant 3C397 (from J. Armstrong's MS thesis).

clumps. We thus observed a total of 22 fields, in continuum and in the 6.7-GHz methanol and 6.0-GHz OH maser lines.

Our preliminary analysis of the VLA maps confirm 8 Arecibo detections (at least two "weak" masers) and we discovered three new 6.7 GHz CH₃OH masers. Our new observations show that the weak flux density of the methanol masers detected with the Arecibo telescope is not always caused by pointing offsets, and also show that several masers (e.g., G43.10+0.04) are variable. The mapped VLA fields also contain other previously known masers and our observations show that some of them are actually composed by multiple spatial components.

We find that not all previously detected masers at Arecibo are associated with continuum emission (at the level of several mJy/beam RMS; e.g., G34.19-0.59, Fig. 1). In those Arecibo-observed sources where continuum emission is detected, we find flat or rising spectral indices (see bottom panel of Fig. 1), suggesting the presence of HII regions. Other continuum sources not associated with previously observed masers show significant negative spectral indices, indicating non-thermal emission, most likely synchrotron emission associated with, e.g., background radio galaxies and supernova remnants.

References

Molinari, S., *et al.* (2010), *PASP*, 122, 314

Olmi, L., *et al.* (2014), *A&A*, 566, A18

Pandian, J. D. *et al.* (2007), *ApJ*, 656, 255

Astrophysical Masers:
Unlocking the Mysteries of the Universe
Proceedings IAU Symposium No. 336, 2017
A. Tarchi, M.J. Reid & P. Castangia, eds.

© International Astronomical Union 2018
doi:10.1017/S1743921317010080

Monitoring and search for periodic methanol masers

M. Olech, M. Szymczak, P. Wolak and A. Bartkiewicz

Centre for Astronomy, Faculty of Physics, Astronomy and Informatics,
Nicolaus Copernicus University, Grudziadzka 5, PL-87-100 Torun, Poland
email: `olech@astro.umk.pl`

Abstract. We summarize a long-term monitoring of 11 periodic 6.7 GHz methanol masers and a search for new periodic sources. Observations were carried out with the Torun 32 m telescope. Periods of observed sources range from 29 to 658 days and the data consist of more than 10 observed cycles for most of the masers. Inspection of archival data resulted in identification of 3 new periodic sources while 2 new periodic objects were found in observations started in 2014.

Keywords. masers, techniques: spectroscopic, radio lines: stars, stars: formation

1. Introduction

Recently a small group of 6.7 GHz methanol maser sources that show periodic changes in emission brightness have been discovered (Araya *et al.* 2010, Fujisawa *et al.* 2014, Goedhart *et al.* 2003, Szymczak *et al.* 2015). About 20 periodic sources of different types of behaviour are known. Most of them show periodic changes in the whole spectrum while for a minority of the sources periodicity is seen for individual features. Sinusoidal, asymmetric and intermittent burst profiles are identified.

There is an ongoing debate on periodicity mechanisms and several models have been proposed. In principle, they can be divided into two groups depending on changes in the flux of seed photons or the maser optical path. Periodic variations of methanol masers can be driven by changes in the background radiation flux in a colliding-wind binary (van der Walt *et al.* 2009). The maser optical depth can be modulated by periodic accretion, pulsation of protostar or accretion in a circumbinary disc (Parfenov & Sobolev 2014). Our recent discovery of anti-correlated bursts of the methanol and 22 GHz water maser lines in G107.298 (Szymczak *et al.* 2016) gives strong evidence that change in the pump rate is a plausible cause. Long-time monitoring and search for new periodic sources are necessary to examine these hypotheses. Here, we present a short summary of long-term monitoring and announce the discovery of new periodic masers.

2. Observations and Results

Monitoring program of known periodic masers included 11 targets visible for the Torun 32 m antenna. Additionally, we inspected our archival data for 139 maser sites observed in 2009-2013 and we found 3 periodic maser sources not reported previously.

In 2014 we started a new program focused on finding new sources with periods shorter than 50 days. A total of 121 objects were monitored. Groups of targets were observed on daily basis for approximately 4 weeks and if significant changes in the flux density were detected the observations were continued. The observations of selected groups were repeated after a longer interval. Two new sources with period significantly longer than 50 days were discovered. Considering these results we conclude that sources with short periods are rare.

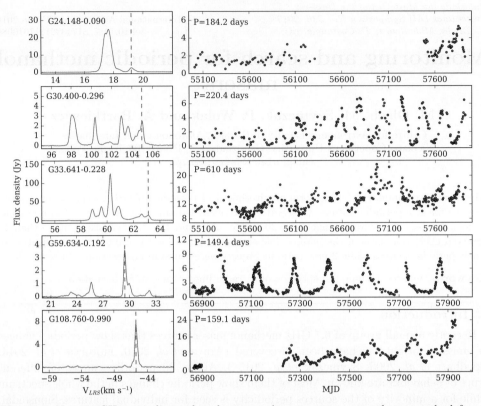

Figure 1. New 6.7 GHz periodic methanol sources. Average spectra are shown on the left and the light-curves of selected features (marked by dashed lines) are shown on the right. For source G30.400−0.296 a periodic behavior is exhibited by spectral features at velocity higher than 104 kms[1]. In the other sources the periodic variations are seen in all the spectral features.

Light curves of new periodic sources are shown in Figure 1. Periods of new sources are longer than 149 days. The intensity of most of the objects was relatively weak and for three objects dropped below a sensitivity limit in quiescent state.

Acknowledgements

The authors acknowledge support from the National Science Centre, Poland through grant 2016/21/B/ST9/01455.

References

Araya, E. D., Hofner, P., Goss, W. M., Kurtz, S., & Richards, A. M. S. 2010, *ApJ*, 717, L133
Fujisawa, K., Takase, G., Kimura, S., Aoki, N., *et al.*, 2014, *PASJ* 68, 78
Goedhart, S., Gaylard, M. J., & van der Walt, D. J. 2003, *MNRAS*, 339, L33
Parfenov, S. Y.u. & Sobolev, A. M. 2014, *MNRAS*, 444, 620
Scargle, J. D., 1982, *ApJ* 263, 835
Szymczak, M., Wolak, P., & Bartkiewicz, A. 2015, *MNRAS*, 448, 2284
Szymczak, M., Olech, M., Wolak, P., & Bartkiewicz, A. 2016, *MNRAS*, 459, L56-L60
van der Walt, D. J., Goedhart, S., & Gaylard, M. J. 2009, *MNRAS*, 398, 961

Astrophysical Masers:
Unlocking the Mysteries of the Universe
Proceedings IAU Symposium No. 336, 2017
A. Tarchi, M.J. Reid & P. Castangia, eds.

© International Astronomical Union 2018
doi:10.1017/S1743921318000108

A Circumstellar Disk in IRAS 23151+5912?

Miguel A. Trinidad, Tatiana Rodríguez-Esnard and Josep M. Masqué

Departamento de Astronomía, Universidad de Guanjuato,
Apdo Postal 144, Guanajuato, México
email: trinidad@astro.ugto.mx, tatiana@astro.ugto.mx, jmasque@astro.ugto.mx

Abstract. We present radio continuum and water maser observations toward the high-mass star-forming region IRAS 23151+5912 from the VLA and VLBA archive, respectively. We detected a continuum source, which seems to be a hypercompact HII region. In addition, a water maser group about 4″ south from the continuum source was detected. We present preliminary results of the analysis of three observations epochs of the water masers, which are tracing an arc-like structure. However, its kinematics is quite complex, since while one section of the structure seems to be moving away from one center, another section seems to be approaching.

Keywords. Stars: formation, HII regions, ISM: individual (IRAS 23151+5912), Masers

1. Introduction

IRAS 23151+5912, a high mass star-forming region, is located at the edge of a molecular cloud in Cepheus. It has a luminosity of $\sim 5 \times 10^5$ L$_\odot$ (at a distance of 5.7 Kpc; Sridharan *et al.* 2002), but only one radio continuum source has been detected, which seems to be consistent with a hypercompact HII region (e.g. Rodríguez-Esnard et al. 2014). In addition, three water masers groups have been detected toward IRAS 23159+5912, which are aligned in the northeast-southwest direction and one of them is spatially associated with the radio continuum source (Rodríguez-Esnard *et al.* 2014). This water maser group seems to be tracing an expanding shell-like structure of about 680 AU.

2. Observations

All observational data were obtained from the VLA archive of the NRAO†. Continuum observations at 1.3, 2 and 6 cm were carried out at B, A and A configuration, respectively (VLA/14A-133 project), while water maser observations were carried out with the VLBA during six epochs (BR145 project) at a rest frequency of 22.2350 GHz.

3. Results

A single radio continuum source was detected in the region at 1.3, 2 and 6 cm, which was reported by Rodríguez-Esnard *et al.* (2014) and is spatially associated with a water maser group. Based on its spectral energy distribution, this continuum source seems to be consistent with a hypercompact HII region (see Figure 1a). On the other hand, high angular resolution observations (~ 0.4 mas) of the water masers, detected about 4″ south from the continuum source, show a rather complex distribution and it is difficult to infer a clear tendency of their kinematics. However, the maser distribution and velocity are compatible with an expanding/rotating shell-like structure of about 680 AU (Figure 1b).

† National Radio Astronomy Observatory is a facility of the National Science Foundation operate under cooperative agreement by Associated Universities

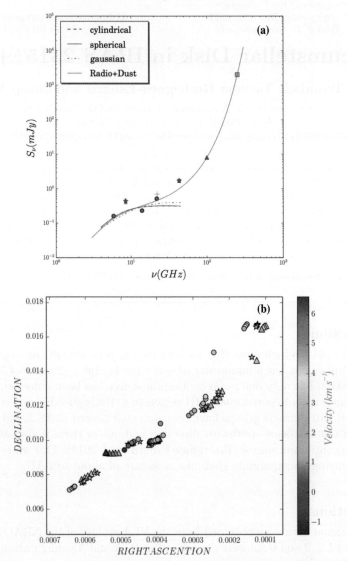

Figure 1. *a)* Spectral energy distribution of the radio continuum source detected in the region IRAS 23151+5912: 1.3, 2 and 6 cm (circles; this work), 0.7 and 3.6 cm (stars; Garay et al. 2007), 3 mm (triangle; Schnee & Carpenter 2009) and 875 μm (square; Beuther *et al.* 2002). The solid line represents the best fit, which considers the contribution of the dust grains. *b)* Spatial distribution of the water masers (only the first three observed epochs are plotted). Circles and triangles represent the first and third epoch, respectively. The systematic velocity of the cloud (-54.4 km s^{-1}) is indicated by zero in the color bar (see the on-line version). Masers do not seem to be tracing a clear structure, however, some subgroups show an expanding structure.

References

Beuther, H., Schilke, P., Menten, K. M., Motte, F., Sridharan, T. K., & Wyrowski, F. 2002, *ApJ*, 566, 945
Garay, G., Rodríguez, L. F., & de Gregorio-Monsalvo, I. 2007, *AJ*, 134, 906
Rodríguez-Esnard, T., Migenes, V. & Trinidad M. A. 2014, *ApJ*, 788, 176
Schnee, S. & Carpenter, J. M. 2009, *ApJ*, 698, 1456
Sridharan, T. K., Beuther, H., Schilke, P., Menten, K. M., & Wyrowski, F. 2002, *ApJ*, 566, 931

Astrophysical Masers:
Unlocking the Mysteries of the Universe
Proceedings IAU Symposium No. 336, 2017
A. Tarchi, M.J. Reid & P. Castangia, eds.

© International Astronomical Union 2018
doi:10.1017/S1743921318000698

Exploring the Nature of MMB sources: A Search for Class I Methanol Masers and their Outflows

Nichol Cunningham[1] , Gary Fuller[2], Adam Avison[2] and Shari Breen[3]

[1]Green Bank Observatory, 155 Observatory Rd, P.O. Box 2, Green Bank, WV, 24944, USA
email: ncunning@nrao.edu

[2]Jodrell Bank Centre for Astrophysics, School of Physics and Astronomy, The University of Manchester, Oxford Road, Manchester, M13 9PL, UK

[3]Sydney Institute for Astronomy (SIfA), School of Physics, University of Sydney, NSW 2006, Australia

Abstract. We present the initial results from a class I 44-GHz methanol maser follow-up survey, observed with the MOPRA telescope, towards 272 sources from the Methanol Multi-beam survey (MMB). Over half (∼60%) of the 6.7 GHz class II MMB maser sources are associated with a class I 44-GHz methanol maser at a greater than 5σ detection level. We find that class II MMB masers sources with an associated class I methanol maser have stronger peak fluxes compared to regions without an associated class I maser. Furthermore, as part of the MOPRA follow-up observations we simultaneously observed SiO emission which is a known tracer of shocks and outflows in massive star forming regions. The presence of SiO emission, and potentially outflows, is found to be strongly associated with the detection of class I maser emission in these regions.

Keywords. masers, stars: formation

1. Introduction

Our lack of understanding of the formation and early evolution of massive stars, in part, reflects that there is not a well established evolutionary categorization in young high mass stars. Establishing such a classification requires large samples of massive young objects, to capture intrinsically rare or short-lived phases of evolution. One possible such probe are the methanol masers associated with young high mass sources. With this in mind, a 44-GHz class I Mopra follow-up survey was performed towards 272 MMB sources with the aim of exploring the environmental and evolutionary nature of regions harboring both class II and class I methanol masers compared with those associated with only class II methanol masers. In addition to the 44-GHz methanol maser, the Mopra survey also included the known shock/outflow tracer SiO (1-0). The observations were selected to sample different spiral arms of the Galaxy, plus all sources identified in the 3kpc arm Green *et al.* (2009).

2. Overview

The overarching goal of this work is to explore the environments and evolutionary nature of class II methanol masers in regions harboring class I methanol masers and if the properties differ to those regions with only class II masers.

Detection Statistics We observed 272 class II MMB maser sources, where approximately 60% have associated class I methanol maser emission at greater than 5σ. For those MMB sources with a known distance in Green & McClure (2011), we find there is no significant difference between the distances to sources with and without a class I methanol maser detection. Thus, for the majority of the non-detections the lack of a class I maser detection does not appear to be a result of distance alone. If we compare the MMB 6.7-GHz class II peak maser fluxes between sources with and without class I detected maser emission, we find that those regions with a class I maser association have higher peak class II fluxes than regions with no detected class I masers. Furthermore, a KS test between the class II peak flux returns a P-value of <0.001 between the class I detected and non-detected samples, suggesting those regions with both class I and class II masers have more luminous class II maser emission.

SiO Outflows. SiO emission is a known tracer of shocks from jets and outflows in massive star forming regions (e.g. Gibb *et al.* 2007, Klaassen *et al.* 2012, Leurini *et al.* 2014, Cunningham *et al.* 2016). With this in mind, we explore the association of SiO (1-0) emission and therefore outflows in these regions. Of the 272 regions observed ~30% have SiO emission detected above 4σ. Furthermore, of those sources with an SiO detection > 80% are associated with regions harboring class I methanol masers.

3. Implications

We have presented the preliminary results from a Mopra 44-GHz class I methanol maser follow-up survey towards 272 MMB 6.7-GHz class II methanol maser sources. We find ~60% of the class II sources have associated class I methanol maser emission at $>5\sigma$, where the distance to non-detected sources does not appear to be a limiting factor in the non-detections. In addition, we observed the shock/jet/outflow tracer SiO (1-0) towards these regions and detect emission towards ~30% of the sample, where > 80% of these detections are associated with the presence of class I maser emission. This highlights that the presence of SiO emission and potentially outflows is strongly associated with class I maser emission in these regions. Previous works by Breen *et al.* (2010) and Jordan *et al.* (2017), have indicated that more luminous 6.7-GHz methanol masers are generally associated with a later evolutionary phase of massive star formation than less luminous 6.7-GHz masers. Towards this sample, we find the class II peak fluxes are stronger in regions with an associated class I maser compared with regions without a class I maser detection, which may indicate that regions with both class I and class II masers are at a later stage of evolution than regions harboring only class II masers.

References

Breen, S. L., Ellingsen, S. P., Caswell, J. L., & Lewis, B. E. 2010, *MNRAS*, 401, 2219
Cunningham, N., Lumsden, S. L., Cyganowski, C. J., *et al.* 2016, *MNRAS*, 458, 1742
Gibb, A. G., Davis, C. J., & Moore, T. J. T. 2007, *MNRAS*, 382, 1213
Green, J. A., Caswell, J. L., Fuller, G. A., Avison, A., *et al.* 2009, *MNRAS*, 392, 783
Green, J. A. & McClure-Griffiths, N. M. 2011, *MNRAS*, 417, 2500
Jordan, C. H., Walsh, A. J., Breen, S. L., Ellingsen, S. P., *et al.* 2017, *MNRAS*, 471, 3915
Klaassen, P. D., Testi, L., & Beuther, H. 2012, *A&A*, 583, A140
Leurini, S., Codella, C., López-Sepulcre, A., Gusdorf, A., *et al.* 2014, *A&A*, 570, A49

Astrophysical Masers:
Unlocking the Mysteries of the Universe
Proceedings IAU Symposium No. 336, 2017
A. Tarchi, M.J. Reid & P. Castangia, eds.

© International Astronomical Union 2018
doi:10.1017/S1743921317009930

Global outburst of methanol maser in G24.33+0.14

P. Wolak, M. Szymczak, M. Olech and A. Bartkiewicz

Centre for Astronomy, Faculty of Physics, Astronomy and Informatics,
Nicolaus Copernicus University, Grudziadzka 5, 87-100 Torun, Poland
email: wolak@astro.umk.pl

Abstract. A strong outburst of 6.7 GHz methanol maser occurred in the high-mass young stellar object (HMYSO) G24.33+0.14 between November 2010 and January 2013. The target was observed with the Torun 32 m radio telescope as a part of a long-term monitoring programme. Almost all twelve spectral features from 108 to $120 \, \text{km s}^{-1}$ varied synchronously with time delays between the flux minima of about two weeks. This may indicate that the variability is driven by global changes in the pump rate. The flare peaks of the two features with the highest relative amplitude of 40-60 are delayed by about 2.5 months while their profiles undergo essential transformation with a velocity drift of $0.23 \, \text{km s}^{-1} \text{yr}^{-1}$. This may suggest that the variability is caused by a rapid increase of the pump rate and excitation of a large portion of the HMYSO environment by an accretion event.

Keywords. masers, stars: formation, ISM: clouds, radio lines: ISM

Target. G24.33+0.14 is a single high-mass hot core associated with the hydroxyl, methanol and water masers (Caswell & Green 2011). The 1667 MHz OH masers extending over more than $20 \, \text{km s}^{-1}$ differs by as much as $50 \, \text{km s}^{-1}$ from the systemic velocity suggesting that it arises in a unique strong blueshifted outflow. The spatial coincidence of methanol and water masers with the OH emission strongly suggests that all three maser species are excited by a single HMYSO.

Maser outburst. The source experienced remarkable synchronised flares lasting 200 -400 days (Fig. 1). For instance, from MJD 55466 to 55780 the emission at $115.36 \, \text{km s}^{-1}$ rose by a factor of 57. This extreme flare was preceded by two weaker bursts peaked at MJD 55314 and 55793 with relative amplitude of 2.7 and 12.7, respectively, and followed by two faint bursts around MJD 56013 and 56122 with relative amplitudes of 3.3 and 0.9, respectively. The feature profile changed during the brightest flare showing a velocity drift of $-0.1 \, \text{km s}^{-1} \text{yr}^{-1}$. The minimum, start, peak and end of the flare were estimated for each maser feature using a synthetic curve which is the best fit of a sum of several Gaussians and one linear function to the data. The minima and peaks are measured directly from the synthetic curve. The start/end of flaring event is the first data point after the minimum/maximum where the flux density is more than 1σ above the quiescent flux density. Figure 1 (Right) and Table 1 give the parameters of flare events for 12 maser features. The mean rise and decline phases of the flare lasted 249 and 312 d, respectively. There is a significant scatter of the times of the events; the times of peaks and ends of flares of individual features differ by 4 and 15 months, respectively. This suggests that the variability is caused by changes in the optical depth of the maser transition that can be related to an accretion event. The observed characteristics of the burst are very similar to those observed for S255 NIRS 3 (Moscadelli *et al.* 2017). If we assume that the time delay between the light curve minima of $2 \times \sigma(\text{minimum mean}) = 10.2 \, \text{d}$ is attributed to the light travel time then we can estimate a linear size of the source of 1900 au. This

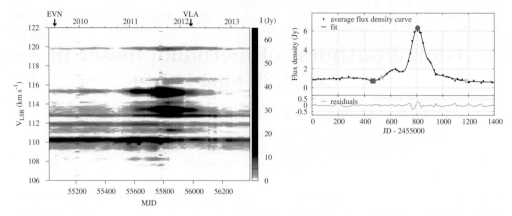

Figure 1. *Left:* Dynamic spectrum of the outburst in G24.33+0.14. The arrows above the upper horizontal axis mark the observations of 6.7 GHz maser emission with the EVN (Bartkiewicz *et al.* 2016) and VLA (Hu *et al.* 2016).*Right:* Flux density curve averaged over the whole velocity range. Times of minimum (square), start (triangle), peak (circle) and end (diamond) are marked.

Table 1. Statistics of flare events. SE and SD denote standard error and standard deviation, respectively.

Event type	mean MJD	SE (d)	SD (d)
Minimum	55470.0	5.1	17.1
Start	55548.7	8.5	28.25
Peak	55797.3	14.9	49.3
End	56109.4	53.4	177.0

value is close to an upper limit of methanol source size (Bartkiewicz *et al.* 2016). High angular resolution observations of 6.7 GHz maser emission (Bartkiewicz *et al.* 2016, Hu *et al.* 2016) have revealed that the angular size of the source is 0.5 arcsec. Thus, for the assumed kinematic distance of 7.7 kpc (Caswell & Green 2011), the linear size of the source would be as large as 3900 au. Our estimate of linear size of 1900 au implies much shorter distance of 3.7 kpc. The emission is highly resolved out when observed with the VLBI beam of 6×5 mas but it is fully recovered with the VLA. It may reflect an intrinsic property of this peculiar source or indicate that the distance adopted in the literature is overestimated.

Conclusions. (i) A global outburst of 6.7 GHz methanol maser of ∼1.5 yr duration occurred in G24.33+0.14. (ii) The relative amplitude of maser intensity was up to 40−60. (iii) It is suggested that maser outburst was induced by a rapid increase of the pump rate.

Acknowledgements

The authors acknowledge support from the National Science Centre, Poland through grant 2016/21/B/ST9/01455.

References

Bartkiewicz, A., Szymczak, M., & van Langevelde, H. J. 2016, *ApJ*, 587, A104
Caswell, J. L. & Green, J. A. 2011, *MNRAS*, 411, 2059
Hu, B., Menten, K. M., Wu, Y., Bartkiewicz, A., Rygl, K., Reid, M. J., Urquhart, J. S., & Zheng, X. 2016, *ApJ*, 833, 18
Moscadelli, L., Sanna, A., Goddi, C., Walmsley, M. C., Cesaroni, R., Caratti o Garatti, A., Stecklum, B., Menten, K. M., & Kraus, A. 2017, *A&A*, 600, L8

Astrophysical Masers:
Unlocking the Mysteries of the Universe
Proceedings IAU Symposium No. 336, 2017
A. Tarchi, M.J. Reid & P. Castangia, eds.

© International Astronomical Union 2018
doi:10.1017/S1743921317009425

Statistical analysis of the physical properties of the 6.7 GHz methanol maser features based on VLBI data

R. Sarniak, M. Szymczak and A. Bartkiewicz

Centre for Astronomy, Faculty of Physics, Astronomy and Informatics,
Nicolaus Copernicus University, Grudziadzka 5, 87-100 Torun, Poland
email: `kain@astro.umk.pl`

Abstract. Methanol masers observed at high angular resolution are useful tool to investigate the processes of high-mass star formation. Here, we present the results of statistical analysis of the 6.7 GHz methanol maser structures in 60 sources observed with the EVN. The parameters of the maser clouds and exciting stars were derived. There is evidence that the emission structures composed of larger number of maser clouds are formed in the vicinity of more luminous exciting stars.

Keywords. stars: formation, ISM: molecules, masers.

1. Context and Methods

High-mass stars play important role in the Galactic evolution and the methanol masers are well known tracer of their formation. Studies of the 6.7 GHz maser emission give us unique insights into regions where processes of accretion and interaction of outflows with the ambient gas are still active. VLBI observations provide the data which allows us to analyze the properties of individual clouds at milliarcsecond scales.

In the present study we used the data from several experiments carried out at 6.7 GHz with the EVN (Bartkiewicz *et al.* 2009, 2014, 2016). Individual maser spots were grouped into the maser cloud structures then their brightness and angular size were measured. We defined cloud size as a distance between the extreme spots projected on a straight line matched to the position of the spots weighted by their brightness. Linear size, luminosity and velocity gradient of maser clouds were determined.

For each source we measured the total spatial size, defined as the diameter of the circle containing all spots. This circle was based on the positions of 2 or 3 extreme spots in the source depending on whether the circle which diameter was equal to the distance of two most extreme spots contains all other spots.

The infrared photometry data of publicly available surveys (2MASS, WISE, GLIMPSE, MSX, MIPSGAL, AKARI, IRAS, Hi-GAL, ATLASGAL, Bolocam) for a total of 33 bands were used to obtain the spectral energy distribution (SED). The package for the SED fitting tool by Robitaille *et al.* (2007) was used. In order to check a quality of IR data, a visual inspection of maps at 23 bands was made.

2. Results and conclusions

Basing on the SED fit, we could verify the distances calculated using Reid *et al.*'s (2009) recipe. Objects with the kinematic distance significantly different from the matched distance were excluded from further analysis or flagged as untrusted. Calculation of typical size of the maser region for objects with well determined distances allowed verification

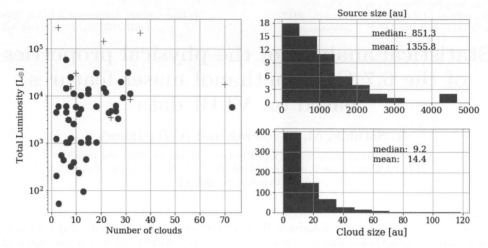

Figure 1. Left: Total luminosity of exciting star vs. the number of maser clouds in the source. Crosses indicate the sources with unreliable SED fit. **Right:** (upper) The histogram of the linear size of methanol maser sources. The extremely large values are not shown. (lower) The histogram of measured sizes of maser clouds.

of possible distance errors for the rest of objects and to identify double or multiple maser sources for which the SED fitting would not be reliable.

Figure 1 shows the total luminosity of exciting star as a function of the number of maser clouds. There is a tendency that the maser sources exciting by less luminous stars have smaller number of the clouds than those powering by more luminous objects. Typical 6.7 GHz methanol maser source size is $\sim$1000 au. The objects of size larger than 2000 au are either double or have an overestimated distance. The mean size of maser cloud is 14.4 au. Weak clouds with a brightness of less than 10 Jy beam^{-1} and typical size of 5-10 au predominate.

We found that the vast majority of maser clouds show linear alignment in spot locations and velocities. Directions of the velocity gradients of maser clouds have been analyzed and for some targets. There is a partial convergence of these velocity gradient vectors with the proper motion measured with the VLBI. The SED analysis revealed that most of the examined objects have accretion discs with a high accretion rate.

Acknowledgements

This material is based upon work supported by the National Science Centre, Poland through grant 2016/21/B/ST9/01455.

References

Bartkiewicz, A., Szymczak, M., van Langevelde, H. J., Richards, A. M. S., & Pihlström, Y. M. 2009, *A&A*, 502, 155
Bartkiewicz, A., Szymczak, M., & van Langevelde, H. J. 2014, *A&A*, 564, A110
Bartkiewicz, A., Szymczak, M., & van Langevelde, H. J. 2016, *A&A*, 587, A104
Reid, M. J., Dame, T. M., Menten, K. M., Zheng, X. W. *et al.* 2009, *ApJ*, 700, 137
Robitaille, T. P., Whitney, B. A., Indebetouw, R., & Wood, K. 2007, *ApJS*, 700, 328

Astrophysical Masers:
Unlocking the Mysteries of the Universe
Proceedings IAU Symposium No. 336, 2017
A. Tarchi, M.J. Reid & P. Castangia, eds.

Probing Early Phases of High Mass Stars with 6.7 GHz Methanol Masers

Sonu Tabitha Paulson and Jagadheep D. Pandian

Indian Institute of Space Science and Technology,
Trivandrum, India
email: sonutabitha.15@res.iist.ac.in,jagadheep@iist.ac.in

Abstract. Methanol masers at 6.7 GHz are the brightest of class II methanol masers and have been found exclusively towards massive star forming regions. These masers can thus be used as a unique tool to probe the early phases of massive star formation. We present here the SED studies of 284 methanol masers chosen from the MMB catalogue, which falls in the Hi-GAL range ($|l| \leqslant 60°$, $|b| \leqslant 1°$). The masers are studied using the ATLASGAL, MIPSGAL and Hi-GAL data at wavelengths ranging from 24−870 micrometers. A single grey body component fit was used to model the cold dust emission whereas the emission from the warm dust is modelled by a black body. The clump properties such as isothermal mass, FIR luminosity and MIR luminosity were obtained using the best fit parameters of the SED fits. We discuss the physical properties of the sources and explore the evolutionary stages of the sources having 6.7 GHz maser emission in the timeline of high mass star formation.

1. Introduction

The 6.7 GHz methanol maser transition is the brightest of class II methanol masers. They have been observed only from high mass star forming regions and have been found to be associated with rapidly accreting massive stars (e.g. Pandian *et al.* 2010). To get a better understanding about the association between high-mass star formation and 6.7 GHz methanol masers, we have determined the SEDs of 284 sources hosting 6.7 GHz methanol masers.

2. Methodology

Our methanol maser sample has been selected from the MMB survey (Green *et al.*2011). We restricted our sample to sources which have been surveyed by Hi-GAL ($|l| \leqslant 60°$ and $|b| \leqslant 1°$) and which have distances available. After eliminating sources that are saturated in Hi-GAL data, extremely crowded fields, and sources with significant uncertainity in their distance (errors $\gtrsim$ 10 kpc), we were left with 284 sources. We have constructed SEDs to these sources from 870 μm to 24 μm using ATLASGAL, Hi-GAL and MIPS-GAL surveys. We extracted the fluxes from ATLASGAL and Hi-GAL images using the Hyper package (Traficante *et al.* 2015). Fluxes were extracted at all wavelengths using the same aperture. A single grey body fit was used to model the cold dust emission of SED whereas the warm dust emission was modelled by a black body.

3. Results

The properties of the clumps hosting methanol masers were determined from the SED parameters. The clump mass ranges from 8330 $M_\odot$ to 6 $M_\odot$ with a mean value of approximately 400 $M_\odot$. About 97% of our sources satisfies the criteria $M(r) \geqslant 580 M_\odot \left(\frac{R_{eff}}{pc} \right)^{1.33}$ (Kauffman *et al.* 2010) and can host atleast one massive star. The surface densities range

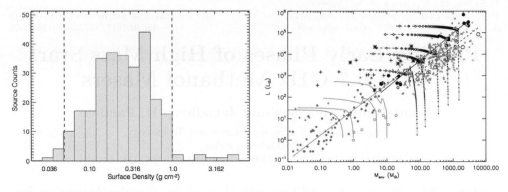

Figure 1. The left panel shows the distribution of surface density of the sources. The two vertical dashed lines are at $\Sigma = 1.0$ and 0.05 g cm^{-2} respectively. The right panel shows the L-M plot with the maser sources being shown as grey triangles.

from 0.03 to 4.8 g cm^{-2} with a mean value of 0.3 g cm^{-2}. Studies by Urquhart *et al.* (2013), show that massive clumps with $\Sigma \gtrsim 0.05$ g cm^{-2} can form high mass stars. The surface density of 98% of our sources exceeds this threshold. The far infrared (FIR) luminosities range from $10^2 - 10^5$ L$_\odot$ with an average value of 10^3 L$_\odot$. Almost 97% of our sources have luminosities greater than 10^3 L$_\odot$, and are hence likely to host at least one massive star (Walsh *et al.* 2003). The FIR luminosity (L$_{FIR}$) is found to have a weak correlation with the methanol maser luminosity (L$_{mmb}$). A partial-Spearman correlation test was done to remove the dependency of parameters on distance and it yielded a value of 0.52 for L$_{mmb}$-L$_{FIR}$.

We plotted the source luminosity as a function of clump mass (right panel of Fig. 1) to investigate the evolutionary stage of the maser hosts. Also shown in the plot are loci of evolutionary tracks of sources based on the turbulent core model (Molinari *et al.* 2008). In the high mass regime (solid black line in the L-M diagram), the clumps in the lower right region are in early stages of evolution with rapid accretion, whereas those in the upper left are in the envelope clearing phase. According to this model all the sources in our sample lie in the high mass regime with 84% being in the accreting phase and the remaining 16% in the envelope clearing phase. This indicates that majority of our methanol maser sources are in early evolutionary stages, in agreement with that found by Pandian *et al.* (2010).

4. Conclusion

We have constructed SEDs from submillimeter to mid-infrared wavelengths for 284 6.7 GHz methanol masers. The SEDs were fit with grey body models with the dust temperature ranging from $13-35$ K. The L-M diagram indicates that the methanol maser sources are at early evolutionary stages with majority of them being in the accretion phase.

References

Green, J. A., & McClure-Griffiths, N. M. 2011, *MNRAS*, 417, 2500
Kauffmann J., Pillai T., Shetty R., Myers P. C., & Goodman A. A. 2010, *ApJ*, 716, 433
Molinari, S., Pezzuto, S., Cesaroni, R., Brand, J., Faustini, F., & Testi, L. 2008, *A&A*, 481, 345
Pandian, J. D., Momjian, E., Xu, Y., Menten, K. M., & Goldsmith, P. F. 2010, *A&A*, 522, A8
Traficante, A., Fuller, G. A., Pineda, J. E., & Pezzuto, S. 2015, *A&A*, 574, A119
Urquhart J. S., Moore T. J. T., & Schuller F. 2013, *MNRAS*, 431, 1752
Walsh A. J., Macdonald G. H., Alvey N. D. S., Burton M. G., & Lee J. K. 2003, *A&A*, 410, 597

Astrophysical Masers:
Unlocking the Mysteries of the Universe
Proceedings IAU Symposium No. 336, 2017
A. Tarchi, M.J. Reid & P. Castangia, eds.

© International Astronomical Union 2018
doi:10.1017/S1743921317010183

Quenching of expanding outflow in massive star-forming region W75N(B)-VLA 2

Soon-Wook Kim[1,2] and Jeong-Sook Kim[1]

[1]Korea Astronomy and Space Science Institute, Daejeon 34055, Republic of Korea
email: skim@kasi.re.kr, evony@kasi.re.kr

[2]Korea University of Science and Technology, Daejeon 34113, Republic of Korea

Abstract. VLBI observation of masers is a powerful mean to understand the early evolutionary phase of massive star formation. A few different scenarios of outflow evolution in the massive protostars have been proposed, and cannot be readily examined because the precise timing of appropriate maser phenomena is difficult. In particular, it has been a matter of debate whether a well-collimated or a less-collimated outflow comes first in the very early phase of the massive protostellar evolution. Long-term, multi-epoch VLBI monitoring is probably the most important method to trace the outflow evolution. Such a monitoring of a massive star-forming region W75N(B) has been very successful. Since the first detection of the expanding water maser shell associated with the star-forming region VLA 2 of W75N(B) in 1999, the observations in 2005 and 2007 displayed that the expanding water maser shell has been evolved to well-collimated from a less collimated morphology. Observations in 2012 also confirmed such a transition. It would be a major breakthrough in our knowledge of the formation and evolution of the first stages of massive protostars. We performed multi-epoch VLBI observations in mid-2014. On the contrary to its expansion for 13 years, the maser shell at VLA 2 observed in 2014 is comparable to the size observed in 2012. The quenching of the maser shell size indicates that the previously expanding outflow has been decelerated plausibly due to the interaction with surrounding interstellar medium.

Keywords. ISM: individual objects (W75N), ISM: jets and outflows, ISM: kinematics and dynamics, stars: formation

1. Introduction: Expanding Outflow in W75N(B)-VLA 2

The active star-forming region (SFR) W75N(B) is located at a distance of 1.32 kpc, and contains three massive young stellar objects within an area of $1.5'' \times 1.5''$ ($\sim$2000 AU$\times$2000 AU): VLA 1, VLA 2 and VLA 3 (after Torrelles *et al.* 2003). VLA 1 and VLA 3 display elongated radio continuum emission, consistent with a thermal radio jet. On the contrary, VLA 2, located between VLA 1 and VLA 3, shows unresolved continuum of unknown nature. The three sources are conjectured to be at different evolutionary stages. VLA 1 and VLA 2 are probably the most and least evolved, respectively. The water maser observations show that VLA 1 traces a collimated thermal radio jet, whereas VLA 2 traces an expanding shell (Torrelles *et al.* 2003 and Surcis *et al.* 2011).

There are two typical types of outflows in massive SFRs: wind-like and bipolar. It has been only conjectured which comes first by comparing two or more SFRs. Based on 1999, 2005 and 2007 VLBI observations of 22 GHz water masers, we first provide a strong evidence of transition from a wind-like to a bipolar-like outflow in a single SFR, VLA 2 (Kim *et al.* 2013; also see Kim & Kim 2014). Only a wind-like feature was detected with an expanding velocity of 40 km s^{-1} during 1999-2005. On the contrary, during 2005-2007, the velocity became 100 km s^{-1}, and the outflow in VLA 2 appeared more collimated. Further observation in 2012 (Surcis *et al.* 2014) show that the water maser distribution

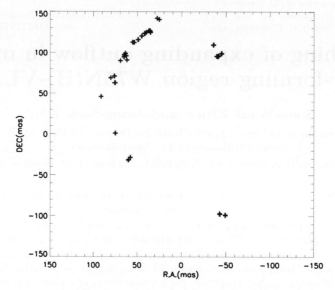

Figure 1. The water maser features around W75N(B)-VLA 2 observed with KaVA in 2014.

around VLA 1 is unchanged from the previous VLBI observations. On the contrary, the shell-like structure in VLA 2 is expanding along the direction parallel to the thermal radio jet of VLA 1. JVLA observations in early 2014 show that the elongation of the continuum emission at 4–48 GHz is in good agreement with that of the water maser distribution in VLA 2 (Carrasco-González et al. 2015).

2. Quenching of expanding outflow during 2012–2014

In April to June 2014, we performed three-epoch KaVA observations of the water masers in VLA 2 (Fig. 1). The elliptical fits show that the sizes of semi-major and semi-minor axes are ~130 and ~70 mas, respectively. The maser shell distribution in 2014 is comparable to the values estimated in the 2012 observations within a few percent (e.g., Kim et al. 2013, Surcis et al. 2014, and Carrasco-Gonzalez et al. 2015). The quenching of the maser shell size in 2012–2014 indicates that the previously expanding outflow has been decelerated plausibly due to interaction with surrounding interstellar medium (Kim et al., in preparation).

References

Carrasco-González, C., Torrelles, J. M., Cantó, J., Curiel, S. Surcis, G., Vlemmings, W. H. T., van Langevelde, H. J., Goddi, C., Anglada, G., Kim, S.-W., Kim, J.-S., & J. F. Gómez, J. F., 2015, *Science*, 348, 114

Kim, J.-S., Kim, S.-W., Kurayama, T., Honma, M., Sasao, T., Surcis, G., Cantó, J., Torrelles, J. M., & Kim, S. J., 2013, *Ap. J.*, 767, 86

Kim, J.-S., & Kim, S.-W., 2014, *Proceedings of the 12th European VLBI Network Symposium and Users Meeting (EVN 2014)*, 42 (https://pos.sissa.it/230/042/pdf)

Surcis, G., Vlemmings, W. H. T., Curiel, S., Hutawarakorn Kramer, B., Torrelles, J. M., & Sanna, A. P., 2011, *A&A*, 527, A48

Surcis, G., Vlemmings, W. H. T., van Langevelde, H. J., Goddi1 , C., Torrelles, J. M., Cantó, J., Curiel, S., Kim, S.-W., & Kim, J.-S., 2014, *A&A*, 565, L8

Torrelles, J. M., Patel, N. A., Anglada, G., Gómez, J. F., Ho, P. T. P., Lara, L., Alberdi, A., Cantó, J., Curiel, S., Garay, G., & Rodríguez, L. F., 2003, *Ap. J.*, 598, L115

Astrophysical Masers:
Unlocking the Mysteries of the Universe
Proceedings IAU Symposium No. 336, 2017
A. Tarchi, M.J. Reid & P. Castangia, eds.

© International Astronomical Union 2018
doi:10.1017/S1743921317009395

Periodic methanol masers and colliding wind binaries

Stefanus P. van den Heever[1], D. J. van der Walt[2], J. M. Pittard[3] and M. G. Hoare[3]

[1]Hartebeeshoek Radio Astronomy Observatory, Krugersdorp, South Africa
email: `fanie@hartrao.ac.za`
[2]Center for Space Research, North-West University, Potchefstroom, South Africa
[3]Dept. Physics and Astronomy, University of Leeds, Leeds, England

Abstract. Since the discovery of periodic variability of Class II methanol masers associated with high-mass star formation, several possible driving mechanisms have been proposed to explain this phenomenon. Here the colliding wind binary (CWB) hypothesis is proposed to describe the periodic variability. It is shown that the recombination of a partially ionized gas describes the flare profiles remarkably well. In addition, the quiescent state flux density is also described remarkably well by the time-dependent change of the electron density. This suggests that the periodicity is caused by the time-dependent change in the radio free-free emission from the background HII regions against which the maser is projected.

Keywords. masers, ISM: HII regions, radio continuum: ISM, stars: mass loss, X-rays: Binaries

1. Introduction

Since the discovery of the widespread class II methanol masers at 12.2 GHz Batrla *et al.* (1987) and 6.7 GHz Menten (1991), a number of these methanol masers have been discovered Goedhart *et al.* (2003) to show periodic/regular variability. Several possible hypotheses have since been proposed to describe the periodic/regular variability. For the masers with flare profiles similar to G9.62+0.20E, we invoke the CWB model proposed by van der Walt *et al.* (2009) and van der Walt (2011), because the decay of the flare profile resemble that of a recombining partially ionized gas. The CWB model describes the flare profile as the time-dependent change in free-free emission from some small volume of partially ionized gas at the ionization front of the background HII region. This is described by the time-dependent change of the electron density from that volume in the optically thin limit (i.e. $I_\nu \propto n_e^2$). The time-dependent electron density at the ionization front is solved for using:

$$\frac{dn_e}{dt} = -\beta n_e^2 + \Gamma n_{H^0} \qquad (1.1)$$

where β is the recombination coefficient, Γ is the ionization rate which is obtained from the equation of ionization balance, and n_e and n_{H^0} have their normal meaning.

2. Results

The top left panel of Figure 1 show the $n_{e,min}^2$'s and $n_{e,max}^2$'s obtained from recombination fits applied to each flare. It also shows a linear regression to both $n_{e,min}^2(t)$ and $n_{e,max}^2(t)$. The top right panel show both the linear regression of $n_{e,min}^2$ to the quiescent state flux density as well as a linear regression to the quiescent state flux density. The

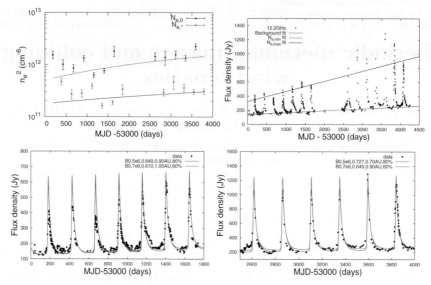

Figure 1. Top left: $n_{e,min}^2$'s and $n_{e,max}^2$'s from recombination fits and linear regression fits. Top right: linear regression fit of $n_{n,min}^2$ fitted to the quiescent state flux density, as well as the linear regression fit of $n_{e,max}^2$ to the observed data at "x". Bottom panels: two CWB models compared with two observed datasets. The legends show several parameters used to obtain the best fit.

gradients are almost identical with values of 0.027 ± 0.010 Jy day^{-1} and 0.026 ± 0.001 Jy day^{-1}, respectively. It shows a remarkable similarity, suggesting that the flare profiles are described by the time-dependent change in the electron density from the background HII region. Additionally, the linear regression of $n_{e,max}^2$ was also applied to the observed flux density at MJD 55000 (indicated by "x") assuming a relative amplitude of $\simeq 2.2$, defined by Goedhart *et al.* (2003). This also describes the increase in the peak flux density remarkably well.

From these results, the best fit peak and quiescent electron densities associated with the flares were used to choose the best fit CWB model. The bottom panels of Figure 1 show the comparisons of two CWB models with the observed flux density. It shows that the time-dependent electron density describes the observed flux density remarkably well. The CWB model also describe the flare profiles of three other sources.

3. Conclusion

This remarkable comparison suggests that the observed flare profiles can be described by the CWB model. The high electron densities (10^{5-6} cm^{-3}) derived suggests a early stage of stellar evolution.

References

Batrla, W., Matthews, H., Menten, K., & Walmsley, C. 1987, *Nature*, 326, 48-51
Menten, K. 1991, *APJL*, 380, L75-L78
Goedhart, S., , P., Zinner, E., & Lewis R. S. 2003, *MNRAS*, 339, L33-L36
van der Walt, D. J., Goedhart, S., & Gaylard, M. J. 2009, *MNRAS*, 398, 961-970
van der Walt D. J. 2011, *AJ*, 141, 152

Astrophysical Masers:
Unlocking the Mysteries of the Universe
Proceedings IAU Symposium No. 336, 2017
A. Tarchi, M.J. Reid & P. Castangia, eds.

© International Astronomical Union 2018
doi:10.1017/S1743921317011425

LBA high resolution observations of ground- and excited-state OH masers towards G351.417+0.645

T. Chanapote[1,2,3], K. Asanok[1,3], R. Dodson[2], M. Rioja[2,4,5],
J. A. Green[6] and B. Hutawarakorn Kramer[3,7]

[1]Department of Physics, Faculty of Science, Khon Kaen University (KKU), Khon Kaen, 40002,
Thailand
email: `t.chanapote@gmail.com`

[2]International Centre for Radio Astronomy Research (ICRAR), The University of Western
Australia (UWA), M468, 35 Stirling Highway, Crawley, Perth, WA 6009, Australia
email: `richard.dodson@icrar.org; maria.rioja@icrar.org`

[3]National Astronomical Research Institute of Thailand (NARIT), 260 Moo 4, Tambol
Donkaew, Amphur Maerim, Chiang Mai, 50180, Thailand
email: `kitiyanee@narit.or.th`

[4]CSIRO Astronomy and Space Science, 26 Dick Perry Avenue, Kensington WA 6151, Australia
[5]Observatorio Astronómico Nacional (IGN), Alfonso XII, 3 y 5, 28014 Madrid, Spain
[6]CSIRO Astronomy and Space Science, Australia Telescope National Facility, PO Box 76,
Epping, NSW 1710, Australia
email: `James.Green@csiro.au`

[7]Max-Planck-Institut für Radioastronomie, Auf dem Hügel 69, 53121 Bonn, Germany
email: `bkramer@mpifr-bonn.mpg.de`

Abstract. We present the results from the Australian Long Baseline Array (LBA) observations
of the ground- and excited-state OH masers at high resolutions towards the massive star-forming
region G351.417+0.645 in 2012. We obtain the most accurate spatial gradient of magnetic fields
at ground state transitions and verify the reliability of magnetic field strengths measured from
previous lower resolution observations. In comparison with previous LBA observations in 2001
at 6.0 GHz, we identified several matched Zeeman pairs. We found that the OH maser features
have no significant change of magnetic field strengths and directions with small internal proper
motions, implying quite stable physical conditions. Additionally, we found that 1665- and 6035-
MHz OH maser features reveal the same trend of reversal of magnetic fields. Moreover, we
also analyzed the physical conditions at different locations from the coincidence of different OH
maser transitions based on current OH maser models.

Keywords. masers, stars: formation, ISM: kinematics and dynamics, ISM: magnetic fields, ISM:
molecules

1. Summary of the results

Our detected 142 OH maser features from both left- and right-handed circular polarization (LHCP and RHCP) from five transitions (1665, 1667, 1720, 6030 and 6035 MHz)
distribute along the Northwest boundary of UC HII region NGC6334F (Fig. 1). We obtained, for the first time, the spatial gradient of magnetic fields and radial velocities
from all five transitions with the most accurate values at $1.6 - 1.7$ GHz. The dominant
magnetic fields are negative (directed towards us) while the reversals of magnetic fields
are found at 1665 and 6035 MHz above Galactic longitude 351.417 deg. In comparison
at 6.0 GHz between epochs 2001 (Caswell *et al.* 2011) and 2012 (our observation), an

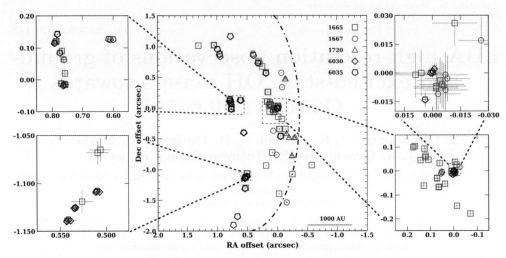

Figure 1. The OH maser distribution from all five transitions along the northwest boundary of the UC HII region NGC6334F shown as dash-dot line adapted from Hunter *et al.* (2006). Square, circle, up-triangle, diamond and pentagonal symbols represent OH maser features from 1665, 1667, 1720, 6030 and 6035 MHz respectively (grey and black colors are used to distinguish between ground– and excited–state transitions). The position offsets are relative to the brightest feature of 6035–MHz transition at R.A.(J2000) = 17h 20m 53.3716s and Dec.(J2000) = $-35°$ 47′ 1.″608.

11-year time span, there was no any significant change, within the uncertainties, in the magnetic fields and radial velocities. Moreover, we have estimated the internal proper motions over the 11-year interval and found the tendency to move downward relative to the brightest feature at the reference center with the maximum velocity of $\sim$11 km s^{-1}. Moreover, we also infer the physical conditions in this region by comparing with Cragg *et al.* (2002)'s OH models based on the coincidences of OH maser features from different transitions.

References

Caswell, J. L., Kramer, B. H., & Reynolds, J. E. 2011, *MNRAS*, 414, 1914

Cragg, D. M., Sobolev, A. M., & Godfrey, P. D. 2002, *MNRAS*, 331, 521

Hunter, T. R., Brogan, C. L., Megeath, S. T., Menten, K. M., Beuther, H., & Thorwirth, S. 2006, *ApJ*, 649, 888

Astrophysical Masers:
Unlocking the Mysteries of the Universe
Proceedings IAU Symposium No. 336, 2017
A. Tarchi, M.J. Reid & P. Castangia, eds.

© International Astronomical Union 2018
doi:10.1017/S1743921318000364

Chemical differentiation in the inner envelope of a young high-mass protostar associated with Class II methanol maser emission

T. Csengeri[1], S. Bontemps[2], F. Wyrowski[1], A. Belloche[1], K. M. Menten[1], S. Leurini[1] and the SPARKS team

[1] Max Planck Institute for Radioastronomy, Auf dem Hügel 69, 53121 Bonn, Germany
email: `csengeri@mpifr.de`

[2] OASU/LAB-UMR5804, CNRS, Université Bordeaux, allée Geoffroy Saint-Hilaire, 33615 Pessac, France

Abstract. We present a case study of a single high-mass protostar associated with an infrared quiet massive clump selected from the ATLASGAL survey. The thermal dust emission reveals a single collapsing object associated with a prominent molecular outflow. We detect bright emission from a torsionally excited state transition of CH_3OH offset from the protostar that is well explained by shocks at the transition from the infalling envelope onto an accretion disk.

Keywords. stars: formation, ISM: kinematics and dynamics, ISM: molecules

1. The SPARKS of high-mass star formation

Whether high-mass star formation proceeds as a scaled-up version of low-mass star formation is an open question in today's astrophysics. The SPARKS project (Search for high-mass protostars up to kpc scales with ALMA) targets at high angular resolution with ALMA 35 of the complete sample of the highest surface density infrared quiet massive clumps selected from the ATLASGAL with $M_{clump} > 650 M_\odot$ up to 5 kpc (Csengeri *et al.* 2017a, Csengeri *et al.* 2017b). The achieved 0.6″ resolution allows us to study the properties of individual collapsing envelopes on ∼2000 au physical scales with a statistical approach. Towards a handful of sources we obtained, however, a considerably higher angular resolution of 0.16″ corresponding to 400 au scales at a distance of 2.5 kpc. We obtained an instantaneous spectral coverage of 7.5 GHz in a frequency range between 333.2 to 337.2 and 345.2 to 349.2 GHz which gives an insight into the molecular composition of the gas in the immediate vicinity of high-mass protostars.

2. Case study of a single high-mass protostar: AGAL 328.25

SPARKS discovered the largest sample of the youngest known precursors of high-mass stars (Csengeri *et al.* in prep). We present here our results towards one object associated with the ATLASGAL clump, AGAL 328.25, located at 2.5 kpc (Fig 1 a). We identify a compact object that stays single down to our resolution of ∼ 400 au corresponding to a massive envelope (Fig 1 b). The highest velocity CO (3–2) emission reveals a prominent bipolar outflow (Fig 1 c), and the bright Class II methanol maser emission associated with the dust continuum suggests the presence of en embedded high-mass protostar (Fig 1 d).

331

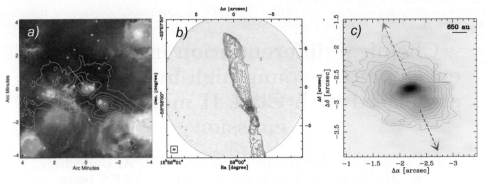

Figure 1. a) Three-color composite image from the Spitzer/GLIMPSE and MIPSGAL surveys (blue: 4.5 μm, green: 8 μm, red: 24 μm, see the on-line version of the figure). Contours show the 870 μm emission from ATLASGAL (Schuller *et al.* 2009, Csengeri *et al.* 2014). The arrow marks the dust continuum peak of the source. b) Line-free continuum emission at 345 GHz measured by the ALMA 12m and 7m array. The high velocity contours from CO (3–2) reveal a prominent bipolar outflow. c) Zoom on the central protostar. The dashed arrows show the direction of the CO outflow. The cross marks the position of the Class II methanol maser from Green *et al.* (2012). Its size shows the estimated accuracy of the maser position.

3. Bright CH$_3$OH emission offset from the protostar

We detect a rotational transition of CH$_3$OH in the $v_t = 1$ state at 334.42 GHz with a high signal-to-noise ratio (Fig. 2). The upper energy level of this transition is at 315 K, which either requires high methanol column density and temperature, or an infrared radiation field at 50 μm to populate this state. Interestingly, the emission drops towards the peak of the continuum, and its spatial morphology shows two prominent emission peaks *offset* from the continuum. These peaks spatially coincide with a spiral structure seen in the dust emission, and their kinematics is consistent with rotational motions. A rotational diagram analysis suggests high methanol column densities at the emission peaks.

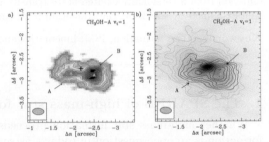

Figure 2. a) Integrated intensity map of the rotational transition of CH$_3$OH in its $v_t = 1$ state at 334.42 GHz. Triangles mark the positions where the spectra have been extracted for the rotational diagram analysis. b) Color scale shows the line-free continuum at 345 GHz (see the on-line version of the figure). The contours are the same as on the left.

4. A shock dominated inner envelope

Based on its spatial distribution and kinematic pattern, we associate the observed CH$_3$OH emission with the inner envelope. The increased CH$_3$OH abundance can be well explained by shocks that are expected to be associated with the transition from the infalling envelope to an accretion disk with a smaller inward motion. This suggests a physical structure that is qualitatively similar to low-mass Class 0 protostars.

References

Csengeri, T., Urquhart, J., Schuller, F. *et al.* 2014, *A&A*, 565, 75

Csengeri, T., Bontemps, S., Wyrowski, F. *et al.* 2017, *A&A*, 601, 60
Csengeri, T., Bontemps, S., Wyrowski, F. *et al.* 2017, *A&A*, 600, L10
Green, J. A., Caswell, J. L., Fuller, G. A. *et al.* 2012, *MNRAS*, 420, 3108
Schuller, F., Menten, K. M., Contreras, Y. *et al.* 2009, *A&A*, 504, 415

Astrophysical Masers:
Unlocking the Mysteries of the Universe
Proceedings IAU Symposium No. 336, 2017
A. Tarchi, M.J. Reid & P. Castangia, eds.

© International Astronomical Union 2018
doi:10.1017/S1743921317010444

Maser Emission in G 339.884−1.259

V. Krishnan[1], L. Moscadelli[1], S. P. Ellingsen[2], H. E. Bignall[3], S. L. Breen[4], R. Dodson[5], L. J. Hyland[2], C. J. Phillips[3], C. Reynolds[3] and J. Stevens[3]

[1]INAF – Osservatorio Astrofisico di Arcetri, [2]University of Tasmania, [3]CSIRO Astronomy and Space Science, [4]University of Sydney, [5]ICRAR University of Western Australia
email: vasaantk@arcetri.astro.it

Abstract. We present multi–epoch VLBI observations of the methanol and water masers in the high–mass star formation region G 339.884−1.259, made using the Australian Long Baseline Array (LBA). Our sub–milliarcsecond precision measurements trace the proper motions of individual maser features in the plane of the sky. When combined with the direct line–of–sight radial velocity (v_{lsr}), these measure the 3 D gas kinematics of the associated high–mass star formation region, allowing us to probe the dynamical processes to within 1000 AU of the core.

Keywords. masers, stars: formation

1. Introduction

G 339.884−1.259 is a prominent source in the study of high–mass star formation. It is relatively nearby at a distance of 2.1 kpc (Krishnan *et al.* 2015), shows intense methanol maser emission at 6.7 GHz of ∼1500 Jy — ranking amongst the strongest at this transition — and strong water maser emission of 39.2 Jy at 22 GHz. Many rare methanol maser transitions (e.g. at 19.9, 37.7, 107.0 and 156.6 GHz) have also been found to be coincident with G 339.884−1.259, suggesting that it has an exceptional environment or might be going through a special or short–lived evolutionary phase. These factors contribute to distinguishing it as an interesting source in characterising high–mass star formation.

2. Observations

We have conducted VLBI observations of the 6.7 GHz methanol and 22 GHz water masers in G 339.884−1.259 using the Australian Long Baseline Array (LBA). Five epochs of observations of 6.7 GHz masers between 2001 July and 2013 November are used for the analysis presented here. Observations of the water maser emission are from 2016 and 2017. Our aim has been to measure the gas kinematics close to the core(s) from the proper motions of individual maser features over time.

3. Proper motions

Figure 1 shows that the distribution of maser emission is perpendicular to the ionised outflow for this source. We identify ten 6.7 GHz methanol maser features which are persistent across three or more continuous epochs for proper motion determination. The proper motions are derived from the procedure in Moscadelli & Goddi (2014) and determined relative to a point which is the mean position of features persistent across all five epochs. The proper motion magnitudes are found to be typically $\lesssim$5 km s^{-1} and primarily in the same direction along the axis of distribution of masers. The exception to this are the features at −39.2, −34.0 and −28.0 km s^{-1}. These three largest proper motions are in agreement with the direction of the inferred axis of the ionised outflow from 8.6 GHz radio continuum measurements by Ellingsen, Norris and McCulloch (1996), made using the Australia Telescope Compact Array (ATCA). Interestingly, polarimetric

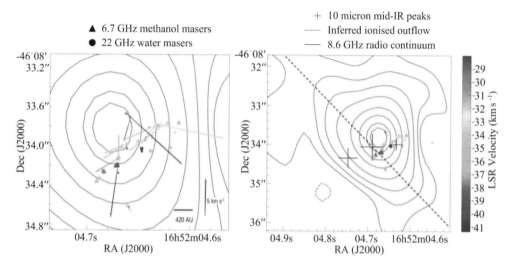

Figure 1. Symbol sizes for the maser features and mid–IR peaks are proportional to their flux density. In the left image is a close-up of the proper motions derived from the 6.7 GHz methanol observations. The 22 GHz water maser emission is shown relative to the methanol masers. The image on the right shows the mid–IR and radio continuum emission with respect to the masers. The positional uncertainty is dominated by the ∼0.2″ astrometry of the continuum and mid-IR measurements compared to ∼5 mas for the masers, which are phase–referenced to a quasar with high positional accuracy. Refer to the electronic copy for a colour version of this figure.

observations by Dodson (2008) show magnetic fields at right angles to the maser distribution, as predicted for a disk model, at the location of the feature at -39.2 km s^{-1}.

4. Fragmentation & evolution in G 339.884−1.259

In Figure 1 (R), we overlay the three 10 μm mid–IR peaks (1.3, 1.0 and 0.6 mJy) from Keck II observations by De Buizer *et al.* (2002), who argue that the peak of the radio continuum (6.16 mJy beam^{-1}) may indicate an additional object. This is therefore a high–mass star formation region with at least four fragmented objects within scales of 2500 AU. Based on the shape of the distribution of the masers and corresponding structure in the mid–IR emission relative to the outflow axis, the sources are in a molecular core elongated along an axis orthogonal to the outflow. There is no detection of 6.7 GHz methanol or 22 GHz water maser emission with the eastern most mid–IR source above a 3σ limit of 90 and 120 mJy. As the mid–IR sources are more evolved and less embedded, we speculate that the obscured source associated with the radio continuum peak — in the midst of the maser emission — is at an earlier stage of evolution.

References

De Buizer, J. M., Walsh, A. J., Piña, R. K., Phillips, C. J., & Telesco, C. M. 2002, *ApJ*, 564, 327

Dodson, R. 2008, *A&A*, 480, 767

Ellingsen, S. P., Norris, R. P. & McCulloch P. M. 1996, *MNRAS*, 279, 101

Krishnan, V., Ellingsen, S. P., Reid, M. J., Brunthaler, A., Sanna, A., McCallum, J., Reynolds, C., Bignall, H. E., Phillips, C. J., Dodson, R., Rioja, M., Caswell, J. L., Chen, X., Dawson, J. R., Fujisawa, K., Goedhart, S., Green, J. A., Hachisuka, K., Honma, M., Menten, K., Shen, Z. Q., Voronkov, M. A., Walsh, A. J., Xu, Y., Zhang, B., & Zheng, X. W. 2015, *ApJ*, 805, 129

Moscadelli, L. & Goddi, C. 2014, *A&A*, 566, A150

Astrophysical Masers:
Unlocking the Mysteries of the Universe
Proceedings IAU Symposium No. 336, 2017
A. Tarchi, M.J. Reid & P. Castangia, eds.

© International Astronomical Union 2018
doi:10.1017/S1743921317011395

The bursting variability of 6.7 GHz methanol maser of G33.641-0.228

Y. Kojima, K. Fujisawa and K. Motogi

Graduated School of Sciences and Technology for Innovation, Yamaguchi University, Yoshida 1677-1, Yamaguchi, 753-8512 Japan

Abstract. From 2014 to 2015, we conducted a total of 469 days observation of the 6.7 GHz methanol maser in a star forming region G33.641-0.228, known to be a bursting maser source. As a result, eleven bursts were detected. On MJD 57364, the flux density grew by more than six times w.r.t the day before. Moreover, during the largest burst, the flux density repeatedly increased and decreased rapidly with time-scale as short as 0.24 day. Since these characteristics of the burst are similar to the solar burst, we speculate that the burst of the 6.7 GHz methanol maser in G33.641-0.228 might occur with a similar mechanism of the solar burst.

Keywords. masers, stars : formation

1. Introduction

G33.641-0.228 is a high-mass star-forming region at the distance of 4 kpc and with a bolometric luminosity of 12000 $L_\odot$. The 6.7 GHz methanol maser of this source was first reported by Szymczak *et al.* (2000). The spectrum of the source consists of four distinct peaks (component I - IV) and two additional separate peaks (component V, VI). Fujisawa *et al.* (2012, 2014) reported that the flux density of the methanol maser spectral component II ($V_{\rm LSR} = 59.6 {\rm kms}^{-1}$) increased seven times within one day, confirming it was a burst event. The time-scale of the burst is the shortest among the variability of the all masers known so far. Moreover, there might be variability with time-scale less than one day, so we conducted continuous observations for several hours by Yamaguchi 32m radio telescopes to investigate the variability during bursts into detail.

2. Results and Discussions

We conducted a total of 469 days observations of G33.641-0.228 from 2014 to 2015. The bursting variabilities of the 6.7GHz methanol maser in G33.641-0.228 were detected 11 times for component II, while the flux density of the other components varied slowly. Fig 2. shows the details of variations of the burst from 57321 to 57329 (MJD) in which short time-scale variations were detected both in rise and fall. The overall time-scale of the burst is about five days, while the time-scale of the fast variability is less than 1 day. The shortest e-folding time was about 0.24 days.

Since these characteristics of the bursting variability are similar to solar radio bursts, the mechanism of the burst could be explained as follows: an energy accumulated by the magnetic fields on the surface of the young stellar object in G33.641-0.228 is released in a short time, and the radiation generated by this energy release is maser-amplified.

3. Summary

We conducted a total of 469 daily observations of the 6.7 GHz methanol maser in G33.641-0.228 from 2014 to 2015, and detected eleven bursts. While the overall time-scale

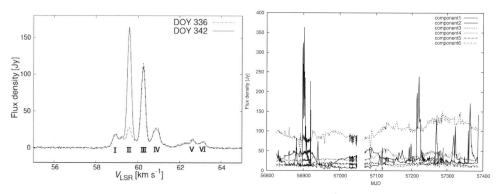

Figure 1. The 6.7 GHz methanol maser in G33.641-0.228. Left) Spectra of G33.641-0.228 at DOY336 (dashed line) and DOY342 (solid line) in 2015. Right) Light curves from 2014 to 2015. Each line shows flux density of each spectral component.

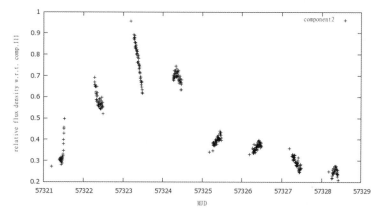

Figure 2. The details of the light curve of a burst from MJD 57321 to 57329. Vertical axis is shown as relative flux density of component II to that of component III in order to remove systematic error.

of the burst was five days, rapid variabilities were detected and the shortest time-scale was 0.24 days. These characteristics suggests that the burst of G33.641-0.228 might occur with a mechanism similar to solar radio burst.

References

Fujisawa, K., Sugiyama, K., & Aoki, N., 2012, *PASJ*, 64, 17
Fujisawa, K., Aoki, N., & Nagadomi, Y., 2014, *PASJ*, 66, 109
Szymczak, M., Hrynek, G., & Kus, A,J., 2000, *A&A*, 143, 269

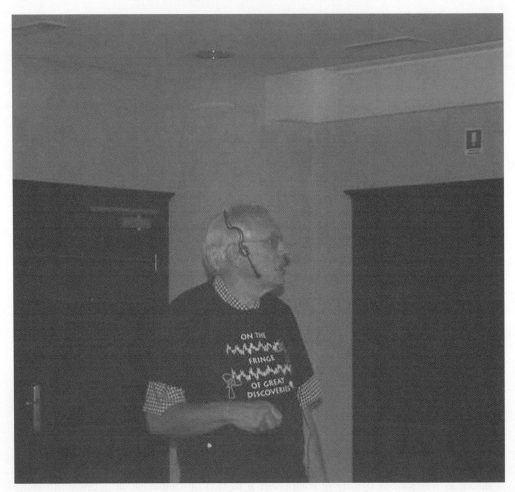

James Moran (photo credit: S. Poppi)

Evolved Stars

Astrophysical Masers:
Unlocking the Mysteries of the Universe
Proceedings IAU Symposium No. 336, 2017
A. Tarchi, M.J. Reid & P. Castangia, eds.

© International Astronomical Union 2018
doi:10.1017/S1743921317009620

Towards continuous viewing of circumstellar maser sources over decades

Hiroshi Imai[1,2]

[1]Science and Engineering Area of the Research and Education Assembly, Kagoshima
University, 1-21-35 Korimoto, Kagoshima 890-0065, Japan
[2]Allround Galactic Astronomy Research Center, Graduate School of Science and Engineering,
Kagoshima University, 1-21-35 Korimoto, Kagoshima 890-0065, Japan
email: hiroimai@sci.kagoshima-u.ac.jp

Abstract. The brightness of maser features are fascinating and give valuable insight for circumstellar physics of oxygen-rich, intermediate-mass stars, in particular the final evolution of circumstellar envelopes (CSEs). The variety of accompanying masers such as SiO, H_2O, and OH in the CSEs may provide unique probes into different stages of rapid CSE evolution. However, with only sparse monitoring of these masers one can sometimes find it difficult to accurately interpret their spatio-kinematics, origins and excitation mechanisms. Examples can be seen in the variety of proposed models for water masers associated with "water fountains" and for silicon-monoxide masers. In order to better understand these issues, one needs to consider continuous monitoring of the individual maser gas clumps over a few stellar cycles or episodic ejection events. Here I present our previous long-term monitoring observations, especially for the water fountain source W43A. Our current efforts involve programs of intensive monitoring observations of circumstellar maser sources over decadal time periods. These programs with the East Asia VLBI Network observe H_2O and SiO maser lines simultaneously mapped at high cadence (2–8 weeks) with VLBI observations.

Keywords. masers, stars: AGB and post-AGB; mass loss, variables)

1. Introduction

Circumstellar envelopes (CSEs) are formed around evolved asymptotic giant branch (AGB) or post-AGB stars that can be isolated or in binary systems (e.g. Richards 2012). They provide opportunities to test maser pumping schemes that may be controlled by outflows in the CSEs and variable stellar radiation. However, these masers are affected by other factors such as inhomogeneity of stellar structure and mass loss (e.g. Zhao-Geisler *et al.* 2011).

The movie of SiO $v = 1$ $J = 1 \rightarrow 0$ masers around TX Cam generated from 78 observations with the VLBA†(Gonidakis *et al.* 2013) demonstrates the uniqueness and importance of long-term (for several years) and intensive (a cadence as short as 2 weeks) VLBI campaigns in order to trace the complicated spatio-kinematical and physical variations of the maser regions. Without a detailed understanding the maser excitation mechanism, it will be difficult to use masers as probes of the mechanism of stellar mass loss, affected by shock wave propagation. In addition, recurrent or episodic major mass ejections are expected in some types of AGB and post-AGB stars, such as red supergiants and stars hosting collimated fast bipolar jets (Sect. 2). The time scales of these events are longer than the stellar pulsation periods, and they may sometimes be related to the periods of binary systems. Because the central stellar systems cannot be spatially resolved except

† The VLBA is operated by the National Radio Astronomy Observatory, a facility of the National Science Foundation operated under cooperative agreement by AUI.

the nearby stars (e.g. Zhao-Geisler *et al.* 2011), circumstellar masers serve as alternative probes of such mass ejections. In these case, monitoring with high angular resolution over years or decades may be crucial for elucidating the origins and effects of ejections.

Although planning decadal long VLBI projects that observe between weekly and monthly intervals are ambitious, the present VLBI networks should encourage such projects. Indeed, the VLBA, the Japanese VLBI Exploration for Radio Astrometry (VERA)‡, and the Korean VLBI Network (KVN)¶ can do this, since are dedicated to regular and/or continuous VLBI operations (Sect. 4).

2. Exploration over decades

Decadal monitoring VLBI observations of circumstellar H_2O masers can lead to improvements or corrections of interpretations of the spatio-kinematics of the masers. One of such examples is the H_2O maser source associated with the water fountain source W43A. The model of a precessing and collimated jet, which was proposed by Imai (2007), based on the maser distributions before 2003, has been superseded by the emergence of new groups of maser features, requiring a new model of cavities of with a wide-angle jet (Fig. 1, Chong *et al.* 2016). The maser distributions found in the recent observations clarify that the maser distributions are point-symmetric and periodic (with a time spacing of 3–4 years). However, they still remain open questions, such as the origin of the intermediate velocity components located at large offsets from the jet major axis, which may be by-products of the collimated jet or a relic envelope formed at an earlier evolutionary stage (i.e. AGB phase) (Imai *et al.* 2013).

New VLBA and VERA observations of H_2O masers in IRAS 18113−2503 (at a trigonometric parallax distance of ∼12 kpc) have yielded striking maser distributions, strongly suggesting periodic generation of "bubbles" (Orosz *et al.* 2018). The evolution of existing bubbles and of newly formed inner bubbles are expected on a decade time scale, taking into account a possible period of bubble developments (∼24 years). Also ballistic motions of clumps of H_2O maser features are traceable over a decade in red supergiants (e.g. S Per, Asaki *et al.* 2018). In such sources, maser motions can exhibit large deviations from radial expansion, as suggested by numerical simulations of shocks in bipolar outflows (e.g., Ostriker *et al.* 2001; Lee *et al.* 2001). Therefore, one expects to directly detect acceleration and/or curving motions of maser with intensive VLBI monitoring observations.

3. Temporal switching of SiO maser pumping mechanisms

The pumping mechanism of circumstellar SiO masers has been a long-standing issue linked to understanding the complicated behavior of the innermost parts of CSEs. The scheme of line-overlapping with mid-infrared radiation from H_2O molecules in the warm dust (Olofsson *et al.* 1981) predicts simultaneous excitation of SiO masers at multiple vibrational levels (v =1, 2, 3, ...). This scheme has been tested in VLBI observations of the multiple maser lines ($v = 1, 2 \; J = 1 \to 0$ and $v = 1 \; J = 2 \to 1$, e.g. Soria-Ruiz *et al.* 2004), but more sensitive tracers of maser actions should be explored. Desnurs *et al.* (2014) show the distributions of SiO $v = 3 \; J = 1 \to 0$ maser emission resembling those of the $v = 1, 2$ lines.

‡ VERA/Mizusawa VLBI observatory and NRO are branches of the National Astronomical Observatory of Japan, an interuniversity research institute operated by MEXT.
¶ The KVN is a facility operated by the Korea Astronomy and Space Science Institute, supervised by the Ministry of Science, ICT and Future Planning.

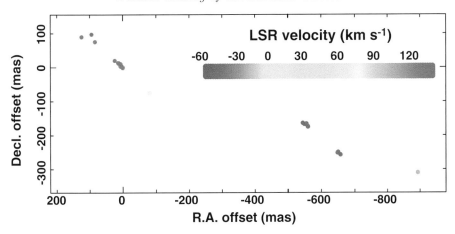

Figure 1. Distribution of H_2O masers in W43A observed on 2014 January 19 with the VLBA. New maser components were found at the south-west edge of the distribution; their locations look roughly consistent with the growth rate of the jet expected from the maser proper motions (Imai 2007). The separation between the brightest blue-shifted and red-shifted clusters of maser features has roughly persisted throughout the monitoring observations. A new intermediate-velocity component was found near the red-shifted lobe. Assuming that it is associated with an intermediate-velocity flow, the flow's expansion velocity is ∼30 km s^{-1}.

Interestingly, recent observations (Imai *et al.* 2010; Imai *et al.* 2012; Oyadomari *et al.* 2016) show the distributions of the $v = 3$ $J = 1 \rightarrow 0$ line significantly deviates from those of the $v = 2$ line (Fig. 2), suggesting that the line-overlapping scheme may be dependent on stellar phase. Although the significance of the spatial differences among the different maser lines should be tested in more sources, continuous tracking of the maser distributions will help to understand the origin of the temporal variations of the maser distributions.

4. Towards intensive VLBI monitoring observations

In the ALMA era, when new discoveries on circumstellar masers are coming, one area of investigation for classical (centimeter to long millimeter) masers will be to fully understand the pumping mechanisms and microscopic behaviors of these masers through conducting long-term and intensive VLBI monitoring observations as described in Sect. 1 for the sources. Towards planning such a legacy program, our ESTEMA (Expanded study on STellar MAsers) project‖ has performed a snap-shot (1–2 hours in integration time per source) imaging survey of circumstellar masers (H_2O in 22.2 GHz, SiO in 43, 86, and 129 GHz) toward 80 stars, using the combined array with the KVN and VERA (KaVA) (e.g. Yun *et al.* 2016). The hybrid observations in ESTEMA were composed of imaging with the full KaVA, dual beam astrometry with VERA, and simultaneous observations in higher frequencies (86 and 129 GHz) with the KVN. The preliminary goal is to select maser sources that are expected to be always visible and exhibit specific maser morphologies (e.g. a ring or arcs shaped by SiO masers). It has focused its main targets on long-period variable stars for further investigation about possible correlation between the maser behaviors and the light curve period and phase. In a preliminary analysis, we found maser detections as summarized in Fig. 3.

‖ https:radio.kasi.re.krkavalarge_programs.php#sh1

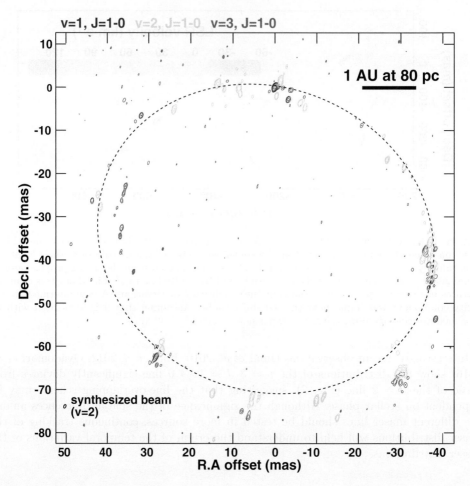

Figure 2. Distribution of SiO $v = 1, 2, 3$ $J = 1 \to 0$ masers around W Hya observed in 2009 February with VERA (Imai *et al.* 2010). A dashed-line ellipse was fitted by eye to the distributions of the $v = 1$ and 2 maser spots, and indicates the existence of a ring structure. Although a large fraction of the extended maser emission would be spatially resolved out, the spatial offsets of the $v = 3$ masers from the $v = 1$ and 2 masers are significant. Such large deviations were confirmed in other sources, while some sources (eg, T Cep) show excellent correlations between $v = 2$ and 3 masers using the same VLBI network (Oyadomari *et al.* 2016).

The results are roughly consistent with our expectations. Roughly half of the observed stars exhibit either H_2O and SiO maser detections, due to the variability correlated with the stellar light curves. Roughly one-third of the stars detected in either H_2O or SiO masers exhibit only one of the two molecular masers, which is attributed to the different evolutionary stages of the AGB stars hosting these masers. The 15 stars that simultaneously exhibited all of H_2O, SiO $J = 1 \to 0$ and $J = 2 \to 1$ masers will be good candidates for the VLBI monitoring observations we have planned. Adding some stars that have been already mapped, we will select the finalists in the monitoring observations from $\simeq 20$ stars for the ESTEMA sample.

We note that the spatial distributions of the H_2O masers are likely to be significantly biased towards one portion of the circumstellar envelope. In fact, all the maps of H_2O masers associated with 20 stars, which were obtained from another VLBI mapping observations with the combined network with VERA, Japanese VLBI Network (JVN)

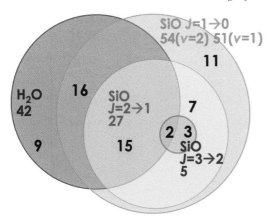

Figure 3. Venn diagram of detections of circumstellar maser lines observed in the KaVA ESTEMA project. The detections were confirmed in the cross-power spectra in scalar (or incoherent) averaging of the VLBI visibility data. The number of the $v = 2$ $J = 1 \rightarrow 0$ SiO maser detections are given in the diagram. The 15 stars hosting H_2O, SiO $J = 1 \rightarrow 0$ and $J = 2 \rightarrow 1$ masers are candidates for the intensive monitoring VLBI observations.

(yielding the shortest baseline of 55 km), and Nobeyama 45 m‡ telescopes (Imai *et al.* in preparation), exhibit such biased maser distributions although these observations were sensitive to more extended maser emission. This implies that registration of H_2O and SiO maser maps is always indispensable to precisely trace the spatio-kinematics of CSE H_2O masers with respect to the central stars surrounded by SiO maser rings.

Acknowledgments

HI deeply acknowledges the opportunities for collaborations in the KaVA ESTEMA, KVN Large Program on circumstellar masers (P.I. S.-H Cho), EAVN Science Working Group on Evolved Stars, HINOTORI (Hybrid Installation project in NObeyama, Triple-band ORIented), and relevant observations and research projects lead by Y. Asaki, R. Burns, G. Orosz, J.-F. Gómez, J. Nakashima, M. Oyadomari, H. Shinnaga, D. Tafoya, and L. Uscanga. HI has been financially supported by the KASI Commissioning Program on KaVA Large Program, JSPS/MEXT KAKENHI (16H02167), the JSPS Foreign Researcher Invitation Program, and Daiwa/Sasagawa Anglo-Japan Foundations (P.I.: J. Th. van Loon).

References

Asaki, Y., *et al.* 2018, in A. Tarchi, M. J. Reid & P. Castangia (eds.), *Astrophysical Masers: Unlocking the Mysteries of the Universe*, Proc. IAU Symposium No. 336 (Cambridge University Press: Cambridge), this volume

Cho, S.-H., Lee, C. W., & Park, Y.-S. 2007, *ApJ* 657, 482

Chong, S.-N., Imai, H., & Diamond, P. J. 2015, *ApJ* 805, 53

Desmurs, J. -F., *et al.* 2014, *A&A* 565, A127

Gonidakis, I., Diamond, P. J., & Kemball, A. J. 2013, *MNRAS* 433, 3151

Imai, H., *et al.* 2013, *ApJ* 773, 182

Imai, H., *et al.* 2012, *PASJ* 64, L6

Imai, H., *et al.* 2010, *PASJ* 62, 431

Imai, H. 2007, in: J. Chapman & W. Baan (eds.), *Astrophysical Masers and their Environments*, Proc. IAU Symposium No. 242 (Cambridge University Press: Cambridge), p. 279

Lee, C.-F. *et al.* 2001, *ApJ* 557, 429

Olofsson, H., *et al.* 1981, *AJ* 247, L81

Orosz, G., *et al.* 2018, in A. Tarchi, M. J. Reid & P. Castangia (eds.), *Astrophysical Masers: Unlocking the Mysteries of the Universe*, Proc. IAU Symposium No. 336 (Cambridge University Press: Cambridge), this volume

Ostriker, E. C., *et al.* 2001, *ApJ* 557, 443

Oyadomari, M., *et al.* 2016, in: *Proc. EVN Symposium 2016*, J. Phys. Conf. Ser. 728, 7

Richards, A. M. S.. 2012, in: Booth, R. S., Humphreys, E. M. L., Vlemmings, W. H. T. (eds.), *Cosmic Masers from OH to H_0* , Proc. IAU Symposium No. 287 (Cambridge University Press: Cambridge), p. 199

Soria-Ruiz, R., *et al.* 2004, *A&A*, 426, 131

Yun, J. Y., *et al.* 2016, *ApJ* 822, 3

Zhao-Geisler, R., *et al.* ,2011, *A&A* 530, A120

Astrophysical Masers:
Unlocking the Mysteries of the Universe
Proceedings IAU Symposium No. 336, 2017
A. Tarchi, M.J. Reid & P. Castangia, eds.

© International Astronomical Union 2018
doi:10.1017/S1743921317010602

Hot and cold running water: understanding evolved star winds

A. M. S. Richards[1], M. D. Gray[1], A. Baudry[2], E. M. L. Humphreys[3], S. Etoka[1], L. Decin[4], I. Marti-Vidal[5], A. M. Sobolev[6] and W. Vlemmings[5]

[1] JBCA, School of Physics & Astronomy, University of Manchester, M13 9PL, UK
contact email: `amsr@jb.man.ac.uk`
[2] Laboratoire d'astrophysique de Bordeaux, Univ. Bordeaux, CNRS, B18N, F-33615 Pessac,
France [4] ESO Karl-Schwarzschild-Str. 2, 85748 Garching, Germany
[4] Instituut voor Sterrenkunde, Katholieke Universiteit Leuven, 3001 Leuven, Belgium
[5] Dept. of Earth, Space and Environment, Chalmers University of Technology, Onsala Space
Observatory, SE 439 92 Onsala, Sweden [6] Ural Federal University, Ekaterinburg, Russia

Abstract. Outstanding problems concerning mass-loss from evolved stars include initial wind acceleration and what determines the clumping scale. Reconstructing physical conditions from maser data has been highly uncertain due to the exponential amplification. ALMA and e-MERLIN now provide image cubes for five H_2O maser transitions around VY CMa, at spatial resolutions comparable to the size of individual clouds or better, covering excitation states from 204 to 2360 K. We use the model of Gray *et al.* 2016, to constrain variations of number density and temperature on scales of a few au, an order of magnitude finer than is possible with thermal lines, comparable to individual cloud sizes or locally almost homogeneous regions. We compare results with the models of Decin *et al.* 2006 and Matsuura *et al.* 2014 for the circumstellar envelope of VY CMa; in later work this will be extended to other maser sources.

Keywords. masers, stars: late-type, stars: winds, outflows

1. Water maser measurements and modelling

The ideal observation of bright circumstellar H_2O masers uses velocity channels $\leqslant 0.1$ km s^{-1} at an angular resolution θ_B of 10–20 mas, sensitive to scales up to ~ 100 mas. This detects, and resolves, all the 22-GHz emission from Asymptotic Giant Branch (AGB) and Red Supergiant (RSG) stars at $\geqslant 0.1$ and 1 kpc, respectively. The channel sampling provides arbitrary slices through the velocity dispersion or gradient in each maser cloud. The position and size s of each maser spot can be found with an accuracy $\sim \theta_B/$(signal-to-noise ratio), for sparse visibility plane coverage. s is the beamed size, but the parent cloud size L can be estimated from the total angular separation of nearby components in adjacent channels, used to find the beaming angle $\Omega = s^2/L^2$ (Richards *et al.* 2011). This distinguishes between $\sim$spherical maser clouds quiescently expanding in the outflow, or shocked slabs and other structures elongated along the line of sight; the brightest masers have smaller spots in the former case whilst in the latter the spot size increases up to the parent cloud size (Elitzur *et al.* 1991). These measurements are possible using e.g. (e-) MERLIN or ALMA† on long baselines; whilst the higher resolution of VLBI is superb for proper motion measurements it usually resolves-out much emission. The ALMA Science

---

† ALMA is a partnership of ESO, NSF (USA) and NINS (Japan), together with NRC (Canada), MOST and ASIAA (Taiwan), and KASI (Republic of Korea), in cooperation with the Republic of Chile. The Joint ALMA Observatory is operated by ESO, AUI/NRAO and NAOJ.

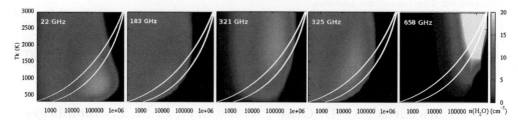

Figure 1. Maser optical depth as a function of o-H_2O number density and T_k. The white lines enclose likely conditions in a CSE, from the star at the top right of each panel.

Table 1. Frequencies and upper energy levels of H_2O maser lines imaged around evolved stars. The spin is given; all are ground vibrational state unless given.

Line (GHz)	22.235079	183.31012	321.22564	325.15292	658.00655
State etc.	o	p	o	p	o, $v2 = 1$
E_U (K)	643	200	1861	454	2360

Verification (SV) results (and MERLIN data for VY CMa, owing to its low declination) presented here have $\theta_B \geqslant 90$ mas, resolving clouds but not individual spots.

22-GHz masers around effectively solitary stars are found in $\approx$spherical shells from ~ 5 to ~ 20 $R_\star$ (about 1 (5) au for AGB (RSG)). The wind shows gradual, radial acceleration, achieving escape velocity during its passage through the 22-GHz shell. Individual cloud are of order 1.5 (15) au around AGB (RSG), consistent with radial expansion from a cloud birth size $\sim 0.1R_\star$ at the stellar surface. The 22-GHz cloud filling factor is $< 1\%$ but the inner radius, determined by the maser quenching density (Cooke & Elitzur 1983), suggests that they are 10–100× overdense, i.e. most of the mass loss is concentrated in a few clouds ejected per stellar pulsation cycle (Richards *et al.* 2012).

Gray *et al.* (2016) comprehensively models of H_2O masers in CSEs (circumstellar envelopes; see that paper for additional references). This predicts 40–50 H_2O maser lines in ALMA bands at energy levels (E_U) $\leqslant 7200$ K, arising from distinct combinations of dust and gas temperatures ($300 < T_K < 3000$ K), gas number density ($10^5 < n < 10^{11}$ cm^{-3}), H_2O fractional abundance (f_{nH2O}), velocity and radiation fields. About 16 of these have been detected spectrally, but only 5 have published images, all (apart from 22 GHz) for VY CMa (Richards et al. 2014; Marti-Vidal *et al.* 2016). These are listed in Table 1. Fig. 1 shows their predicted (negative) maser optical depth as a function of o-H_2O number density and gas kinetic temperature, assuming the ortho:para ratio (o:p) = 3:1 and $f_{nH2O} = 3 \times 10^{-5}$. This shows that these lines sample a variety of physical conditions.

2. VY CMa water maser observation-model comparison

The RSG VY CMa lies at 1.1 kpc (Choi *et al.* 2008), with a 2.2 μm stellar radius $R_\star$ of 5.7 mas (Wittkowski et al. 2012), and an *LSR* velocity $V_\star$=22 km s^{-1}. The 183, 321, 325 and 658 GHz masers were observed for ALMA Science Verification in 2016 and 2013 (Marti-Vidal et al. 2016; Richards *et al.* 2014) The star was found to be at the centre of expansion and used to align the data sets. The 22-GHz masers used here were observed in 1994 (Richards *et al.* 1998) and 2000 using MERLIN and in 2016 as e-MERLIN test observations. The 22-GHz maser Doppler velocities and proper motions (including comparison with Bowers *et al.* 1993) show that, between $\sim$(70–450) mas radius, the wind is accelerating radially away from the star. However, VY CMa has a highly irregular CSE on arcsec scales, with no axis of symmetry, and *HST* line observations suggest clumps are ejected along curved, apparently ballisic trajectories (Humphreys *et al.* 2007).

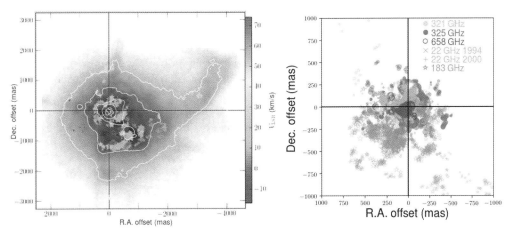

Figure 2. (left) 183 GHz maser components, shaded by velocity, overlaid on an *HST* scattered light image. (right) Masers within 1″ of the star at (0,0); colour and shape represents species.

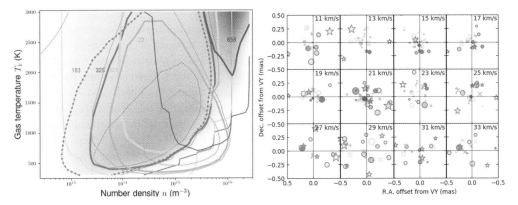

Figure 3. (left) Maser optical depth τ for 5 lines. Heavy contours show 50% maximum inversion. The lowest contour represents the observational sensitivity limit. τ is also shown by a transparent shade for 321, 325 and 658 GHz masers. (right) (sub-)mm maser clump positions within 0.″5 of the star, colour and shape represents species, symbol size proportional to clump size.

On large scales, the 183-GHz masers trace the distribution of small, cool dust grains seen in scattered light with the *HST* (Fig. 2), consistent with the prediction that their excitation extends to cool, low-density conditions. Maser distributions within 1″ of the star are broadly consistent with expectations, in that the 658-GHz emission is found closest to the star, with 321-, 22- and 325-GHz at increasing radii (Fig. 2 (right)). The biggest surprise is that the 658-GHz emission extends > 200 mas from the star, well outside the ~ 70 mas radius of most dust formation. This could be due to shock heating as the wind collides with dust clump **C**, 328 mas SE of the star (O'Gorman *et al.* 2015).

All the masers lie in broad, overlapping, clumpy shells and we compared the $\sim$10-au-scale distribution with model predictions, looking at 2 km s^{-1} velocity bins. We only considered emission within $V_\star \pm 12$ km s^{-1} due to the asphericity of the VY CMa CSE making it difficult to convert $V_{\rm LSR}$ to position along the line of sight, and thus to determine whether apparently cospatial or adjacent emission at a particular velocity was actually at the same depth. We did not use the 22 GHz data as the observations closest in time were taken under poor conditions. We tailored the Gray *et al.* (2016) models to VY CMa by taking the average angular size of the clumps for each line as the propagation depth. Fig 3 (left) shows where different masers are expected to co-propagate or occur alone. Fig 3 (right) shows the observed distribution of 183, 321, 325 and 658 GHz maser clumps in the inner arcsec. The symbol size represents clump size. We matched pairs of clumps of different lines if they overlapped in velocity by half the maximum V_{LSR} span of a clump (i.e. 50% overlap) and by half the sum of their angular sizes (i.e. just touching, to allow for aligment uncertainties of order $R_\star$).

Given the large areas of overlap in Fig. 3, we were surprised to find only 13 groups of 2 or more masers lines appearing to co-propagate (out of $\sim 70 - 170$ features per line). The required ranges of n and T_{k} were determined from Fig. 3 for the combinations detected. Matsuura *et al.* (2014) and Decin *et al.* (2006) derived 1D models for VY CMa. Our results show that the wind acceleration is more gradual than predicted and reaches a higher terminal velocity than their models' gas velocity, possibly due to dust evolution or better momentum coupling in dense clumps. We obtained values of $n > 50\times$ higher close to the star, consistent with maser clumping, with n for 183- and 325-GHz masers at closer to the average at $\geqslant 400$ mas radius, possibly from inter-clump gas. The T_{k} profile roughly supported the Decin model including the effects of a variable mass loss rate.

3. Future work

We have demonstrated the potential of multi-line imaging of H_2O masers to constrain physical conditions on ten-au scales, covering the dust formation and wind acceleration regions of an RSG CSE. The model output will be refined using conditions closer to those in VY CMa. A constant dust temperature of 50 K was used here; a higher value closer to the star would be more realistic. We will also investigate the effect of using f_{nH2O} of 2×10^{-4} (Matsuura *et al.* 2014) or 1.1×10^{-3} (Decin *et al.* 2006). However, two major problems are the uncertainty in estimating the depth of individual maser clumps and their location along the line of sight. At present, we assume that all clumps are spherical so the depth is the same as the angular size in order to estimate the relative maser τ from the flux density and propagation length, but some of the brightest clouds have small angular sizes, leading to unrealistically high τ, probably because they have a much greater depth than assumed. The aspect ratio of clouds is constrained by the relationship between beaming angle Ω and flux density (Section 1), distinguishing between (on average) spherical clouds and shocked slabs. This requires observations using $\sim$8–16 km ALMA baselines. Accurate 3D modelling of the CSE as a whole requires a less aspherical target such as VX Sgr.

References

Bowers, P. F., Claussen, M. J., & Johnston, K. J. 1993, *AJ*, 105, 284
Choi, Y. K., Hirota, T., Honma, M., *et al.* 2008, *PASJ*, 60, 1007
Cooke B. & Elitzur M. 1985, *ApJ*, 295, 175
Decin, L. *et al.* 2006 *A&A* 456, 549
Elitzur, M., Hollenbach, D. J., & McKee, C. F. 1992, *ApJ*, 394, 221
Gray, M. D. *et al.* 2016, *MNRAS*, 456, 374
Humphreys, R. M., Helton, L. A., & Jones, T. J. 2007, *AJ*, 133, 2716
Matsuura et al. 2014, *MNRAS* 437, 532
Marti-Vidal, I., Vlemmings, W. H.T., Carozzi, T. *et al.* 2016,
    https://bulk.cv.nrao.edu/almadata/sciver/VYCMaBand5/VYCMa_Band5_PolCalibrationInformation.pdf
O'Gorman, E. *et al.* 2015, *A&A*, 573, L10
Richards, A. M. S., Yates, J. A., & Cohen, R. J. 1998, *MNRAS* 299, 319
Richards, A. M. S., Elitzur, M., & Yates, J. A. 2011, *A&A*, 525, 56
Richards, A. M. S. *et al.* 2012, *A&A* 546, 16
Richards, A. M. S. *et al.* 2014, *A&A*, 572, L9
Wittkowski, M., Hauschildt, P. H., Arroyo-Torres, B., & Marcaide, J. M. 2012, *A&A*, 540, L12

Astrophysical Masers:
Unlocking the Mysteries of the Universe
Proceedings IAU Symposium No. 336, 2017
A. Tarchi, M.J. Reid & P. Castangia, eds.

© International Astronomical Union 2018
doi:10.1017/S1743921317009589

Bow shocks in water fountain jets

Gabor Orosz[1,6], José F. Gómez[2], Daniel Tafoya[3], Hiroshi Imai[1], José M. Torrelles[4], Ann Njeri Ngendo[5] and Ross A. Burns[6,1]

[1]Kagoshima University, Kagoshima (Japan)
email: gabor.orosz@gmail.com

[2]IAA (CSIC), Granada (Spain)

[3]Chalmers, OSO, Onsala (Sweden)

[4]ICE (CSIC-IEEC), Barcelona (Spain)

[5]University of Nairobi, Nairobi (Kenya)

[6]JIVE, Dwingeloo (Netherlands)

Abstract. We briefly introduce the VLBI maser astrometric analysis of IRAS 18043–2116 and IRAS 18113–2503, two remarkable and unusual water fountains with spectacular bipolar bow shocks in their high-speed collimated jet-driven outflows. The 22 GHz H_2O maser structures and velocities clearly show that the jets are formed in very short-lived, episodic outbursts, which may indicate episodic accretion in an underlying binary system.

Keywords. masers, techniques: interferometric, astrometry, stars: AGB and post-AGB

1. Introduction

Post-AGB stars are important transitory objects as they are the link between AGB stars and PNe, two late stellar stages of evolution with strikingly different characteristics, yet barely separated in time. Explaining the formation and shaping of PNe, whose morphologies depart significantly from spherical symmetry, is one of the puzzling questions in late stellar evolution (see review in Balick & Frank 2002). It involves a sudden change in the mass-loss mode from spherical to bipolar/multipolar, which occurs on timescales of only a few hundreds of years.

Among the most popular mechanisms that have been proposed to explain the presence of post-AGB axisymmetric lobes involve magnetic fields, torus-like equatorial density enhancements, and collimated bipolar jets. There is a growing consensus that these mechanisms cannot operate within single-star scenarios. Sensitive high-angular resolution observations have shown that many post-AGB objects exhibit high-speed collimated outflows, which seem to be carving bipolar cavities within slowly-expanding circumstellar envelopes (e.g., Sahai *et al.* 2005). It is thought that the bipolar structures seen in PNe are cavities created in the post-AGB phase and blown out by ionization and the fast wind from the hot central star.

In order to understand the origin of stellar high-speed collimated outflows, it is necessary to study in detail their kinematics, energetics, and physical conditions. Fortunately, there exists a group of post-AGB objects whose collimated outflows are traced by 22 GHz H_2O maser emission which can be well characterized via interferometric and VLBI observations: the so-called Water Fountain stars (WF, see review in Imai 2007).

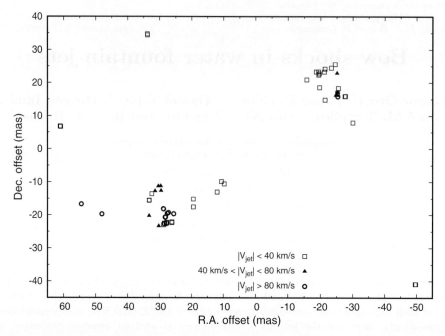

Figure 1. Spatial distribution of 22 GHz H_2O masers towards the WF IRAS 18043–2116, derived from VLBA observations. The symbols are related to maser line-of-sight velocities relative to the systemic velocity of the star, i.e. $V_{jet} = V_{lsr,obs} - V_{star}$, with circles representing the fastest and squares the slowest masers in the outflows. The map origin is at the geometric center of the bipolar masers and the approximate location of the central object.

2. The water fountain IRAS 18043–2116

The 22 GHz H_2O masers of IRAS 18043–2116 were first mapped by Walsh *et al.* (2009) using the ATCA and covering a velocity range of $\sim$400 km s^{-1}. They found a bipolar maser distribution with an outflow axis aligned close to the line of sight and confirming the source as a WF. Pérez-Sánchez *et al.* (2017) detected radio continuum emission that they interpreted as the result of an ionizing shock front of a collimated high-velocity jet. In addition, Tafoya *et al.* (2014) found that the source also hosts H_2O masers at 321 GHz. The authors propose that these sub-mm H_2O masers arise in the same regions as their 22 GHz counterparts (with a peak flux ratio of $\sim$1) and might be indicators of multiple jet launching events (see contribution by Tafoya *et al.*, this volume).

To study the proposed jet scenarios, we mapped the region using archival 22 GHz H_2O maser data taken with the VLBA (project BP150, Day *et al.* 2011). Figure 1 shows the detected maser features in the first two epochs. The masers trace bipolar arc-shaped structures along P.A.$\sim$128° that are blue- and redshifted relative to an LSR velocity of 87 km s^{-1} (Deacon *et al.* 2004). There are also three compact features that are spatially well removed from the bipolar structure, perhaps related to a relic AGB wind. In the eastern blueshifted lobe, the masers span line-of-sight velocities from -111 km s^{-1} to 78 km s^{-1}, with the slower masers tracing an arc and the faster masers located mainly inside this structure. The redshifted lobe ranges from 94 km s^{-1} to 176 km s^{-1}. Unfortunately at the time of the VLBA observations the true velocity extent of the source was not yet known, so the most redshifted masers above 245 km s^{-1} were not observed, and we cannot tell whether or not they also arise inside a slower arc.

We also measured proper motions, showing that the masers are expanding with an average outflow speed of $\sim$1 mas yr^{-1}. The kinematic age of the outflow is estimated to

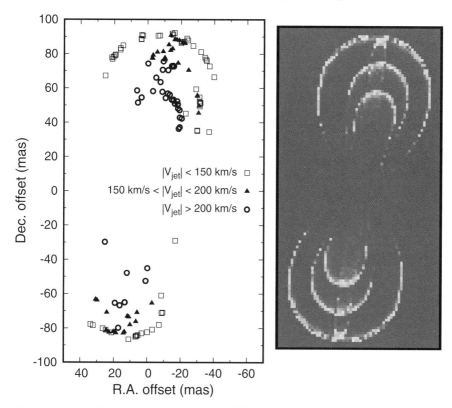

Figure 2. *Left:* Spatial distribution of 22 GHz H_2O masers towards the WF IRAS 18113–2503, derived from VLBA observations. The symbols are related to maser line-of-sight velocities relative to the systemic velocity of the star (same as Fig. 1), with circles representing the fastest and squares the slowest masers in the outflows. The map origin is at the geometric center of the masers and the approximate location of the central object. *Right:* A simple rendered 3D morpho-kinematic "Shape" model made up of three bipolar lobes and a central conical jet. The pixel values have an arbitrary scale and are proportional to the calculated column density.

be $\lesssim 30$ years. The maser proper motions can be well explained with a ballistic bow-shock model of a jet-driven outflow (see Ostriker *et al.* 2001, Lee *et al.* 2001). H_2O masers can trace both the regions around the working surface of a collimated jet and the shocked shell at the interaction with the ambient material. It is therefore possible that the fastest masers (see circles on Fig. 1) inside the eastern arc trace the tip of the high-speed jet, while the slower masers are related to a bow shock and a shell of ambient shocked gas. It is also possible that the faster masers are due to projection effects (as the jet axis is close to the line of sight), and the masers can be explained by the cavity model scenario introduced for W43A (Chong *et al.* 2015).

3. The water fountain IRAS 18113–2503

IRAS 18113–2503 was confirmed by Gómez *et al.* (2011) to be a post-AGB WF with 22 GHz H_2O masers spanning a very large 500 km s^{-1} line-of-sight velocity range. The masers were found to be in two spatially separated clusters, with each having a high velocity dispersion of approximately 170 km s^{-1}. In order to characterize the physical size and 3D velocity field of IRAS 18113–2503, we started VLBI astrometric campaigns with the VLBA (project BG231) and VERA (project VERA13-85) arrays, and measured the proper motions and annual parallax of the H_2O masers. The left side of Fig. 2 shows the spatial distribution of the detected masers in the VLBA experiments. We found that the source is located at $\sim$12 kpc towards the Galactic Centre in the Galactic thick disk, and that the masers trace a nested axisymmetric bow shock structure along a position angle of $\sim$168°, with a clear velocity gradient across the different arcs. Analyzing the 3D maser motions, we find that the jets slow down and get less collimated further from the star, from $\sim$270 km s^{-1} for the innermost jet to $\sim$140 km s^{-1} in the outermost part.

A possibility is that IRAS 18113–2503 hosts an episodic jet and the H_2O masers trace very short-lived, episodic outbursts. The right side of Fig. 2 illustrates the rendering of a simple 3D morpho-kinematic model using the "Shape" program (Steffen *et al.* 2011), with parameters derived from the maser maps. Here we assume a high-speed conical jet and three axisymmetric lobes with slightly different orientations, caused by a possible preccesing motion in the jet. Shape uses LTE radiative transfer codes for rendering, so the amplitudes cannot be directly compared to the masers. However, the model highlights the areas with the highest column density and as such the preferential places of the shock-excited H_2O masers. The goal is to visualize the 3D patterns that might be responsible for the complicated maser distribution on the sky.

In an episodic jet scenario, the more recent ejections could be faster and more colli-mated due to previous ejections sweeping up material and thus allowing the new ejections to expand more freely. It is also possible that the slowing shock fronts are not tracing three ejections but only the progression of a single ejection through an inhomogeneous medium, that could be associated with a layered relic AGB circumstellar envelope. How-ever, it is interesting to note that while the three bow shocks are not spaced equally, the kinematic age difference between them are about 10 years and equal to within a year. Such episodic behavior might indicate that the underlying driving source of the bipolar mass ejections is a binary system with a period on the same timescale.

Acknowledgements

GO acknowledges the support of the Monbukagakusho:MEXT scholarship, Kagoshima University, the Joint Institute for VLBI ERIC and the Konkoly Observatory.

References

Balick, B. & Frank, A. 2002, *ARA&A*, 439, 40
Chong, Sz.-N., Imai, H., & Diamond, P. J. 2015, *ApJ*, 805, 53
Day, F. M., Pihlstroem, Y. M., Sahai, R., & Claussen, M. J. 2011, *APN5 Conf. Proc.*, Ebrary, 9D
Deacon, R. M., Chapman, J. M., & Green, A. J. 2004, *ApJS*, 155, 595
Gómez, J. F., Rizzo, J. R., Suárez, O., *et al.* 2011, *ApJ*, 739, L14
Imai, H. 2007, *IAUS*, 242, 279
Lee C.-F., Stone J. M., Ostriker E. C., & Mundy L. G. 2001, *ApJ*, 557, 429
Ostriker E. C., Lee C.-F., Stone J. M., & Mundy L. G. 2001, *ApJ*, 557, 443
Pérez-Sánchez, A. F., Tafoya, D., Garc'ia López, R., *et al.* 2017, *A&A*, 601, A68
Sahai, R., Le Mignant, D., Sánchez Contreras, C., & Campbell, R. D. 2005, *ApJ*, 622, L53
Steffen, W. 2011, *IEEE Trans. Vis. Comput. Graphics*, 17, 454
Tafoya, D., Franco-Hernández, R., Vlemmings, W. H. T., *et al.* 2014, *A&A*, 562, L9
Walsh A. J., Breen, S. L., Bains, I., & Vlemmings, W. H. T. 2009, *MNRAS*, 394, L70

Astrophysical Masers:
Unlocking the Mysteries of the Universe
Proceedings IAU Symposium No. 336, 2017
A. Tarchi, M.J. Reid & P. Castangia, eds.

© International Astronomical Union 2018
doi:10.1017/S1743921317009309

A detailed study toward the Water fountain IRAS 15445-5449

Andrés F. Pérez-Sánchez[1], Rebeca García López[2], Wouter Vlemmings[3] and Daniel Tafoya[3]

[1]European Souther Observatory, Alonso de Córdova 3107, Vitacura, Casilla 19001, Santiago de Chile email: aperezsa@eso.org
[2]Dublin Institute for Advance Studies, 31 Fitzwilliam Place, Dublin 2, Ireland
[3] Department of Space, Earth and Environment, Chalmers University of technology, Onsala Space Obsevatory, 439 92 Onsala, Sweden.

Abstract. Post-Asymptotic giant branch (post-ABG) sources with high-velocity spectral features of H_2O maser emission detected toward their circumstellar envelopes (CSEs) are known as Water Fountain (WF) nebulae. These are low- or intermediate-mass Galactic stellar sources that are undergoing the late stages of an intense mass-loss process. The velocity and the spatial distribution of the H_2O maser spectral features can provide information about the kinematics of the molecular gas component of their CSEs. Hence, observational studies toward WF nebulae could help to better understand the formation of the asymmetric structures (hundred to thousand AUs) commonly seen toward Planetary nebulae (PNe). Here we present preliminary results of observations done toward the WF IRAS 15445-5449 using the Australia Telescope Compact Array (ATCA) and the Very Large Telescope (SINFONI/VLT). Assuming that the pumping of the H_2O maser transitions is a consequence of shocks between different velocity winds, the spatial distribution of the emission shed light on the scales of the regions affected by the propagation of the shock-fronts.

Keywords. stars: AGB and Post-AGB, Masers, stars: late-type.

1. Introduction

The WF nebulae IRAS 15445-5449 was classified as a post-AGB given its position in the MSX color-color diagram (Deacon *et al.* (2007)). Detection of 1612 MHz line was reported by Sevenster *et al.* (1997). Main line observations (1665 MHz and 1667 MHz) were carried out in 1998 (Deacon *et al.* (2007); and references therein). The profile of the spectra detected in all the lines observed (1612 MHz, 1665 MHz, 1667MHz) were classified as irregular, given their broad line profiles ($\Delta v \geqslant 80$km s^{-1}). Although none of the OH line profiles are double-peak, the central velocity of the OH lines was assumed as the systemic velocity of the source ($v_{sys} = -150$ km s^{-1}). Moreover, Deacon *et al.* (2007) reported the detection of high-velocity H_2O maser emission at 22 GHz. The H_2O spectral features were found to be red-shifted with respect to the OH maser lines. Therefore the emission was thought to arise from a red-shifted jet-like structure. Bains *et al.* (2009) reported the detection of radio continuum emission at centimeter wavelengths with negative spectral index ($\alpha < -0.1$) in the frequency range from 4.8 GHz to 8.6 GHz. This result confirmed the negative spectral index retrieved from radio continuum observations carried out between 1998/1999 at 2 GHz, 5 GHz, and 10 GHz (Deacon *et al.* (2007)). The synchrotron jet was confirmed by observations of a spatially resolved radio continuum emission at 22 GHz (Pérez-Sánchez *et al.* (2013)). The detection of a strong non-thermal component in the radio continuum implies a strong magnetic field interacting with relativistic particles. The distribution of the emission, elongated in the north-south direction,

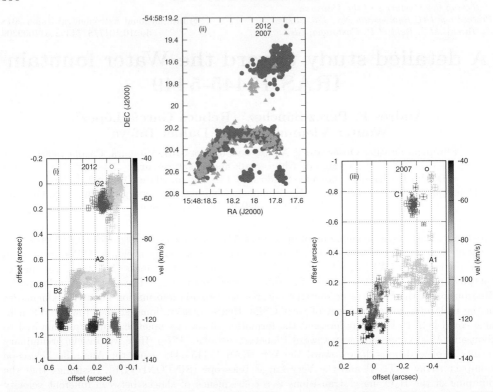

Figure 1. Spatial distribution of the 22 GHz H$_2$O maser spectral features detected in 2012 (panel i), 2007 (panel iii), and both data sets overlied using the emission in regions B1 and B2 for alignment (panel ii). In panels (i) and (iii) the offsets are with respect to the position of the brightest spectral feature detected in each data set, and the color bar indicates the velocity of the spectral features with respect to $v_{sys} = -96.0$ km s^{-1}. In panel (i) the strongest spectral feature is detected toward region C2, while in the panel (iii) the strongest feature arises from region B1. Panel (ii) shows that the spatial distribution is consistent between the 2012 and 2007 data sets.

is consistent with the hourglass morphology seen in the mid-IR (Lagadec *et al.* 2011). Radio continuum emission, OH and H$_2$O maser emission; atomic and molecular Hydrogen lines, and CO overtone emission are direct evidence of the energetic activity taking place in large spatial scales within the CSE of the WF IRAS 15445−5449 (Pérez-Sánchez *et al.* in prep).

2. Observations and results

The observing programs were carried out with ATCA in 2007 and 2012, using, respectively, the 6B and 6A array configurations. In both cases the amplitud (1934-638) and phase (1613-586) calibrators observed were the same. The calibration and the imaging of the data was done using MIRIAD (Sault *et al.* 1995). The synthesized beam are $\theta_{2007} = 0.46"\times0.34"$ and $\theta_{2012} = 0.5"\times0.35"$. The rms in line free channels are rms$_{2007} \approx 50$ mJy beam^{-1} and rms$_{2012} \approx 30$ mJy beam^{-1}. The AIPS task SAD was used to fit 2-D gaussian components to the emission above 3-σ within each spectral channel (component fitting). The spectral channel with the strongest peak flux in each spectrum was self-calibrated. With this, the position offset retrieved from each spectral channel is relative to the position of the brightest spectral feature. Pérez-Sánchez *et al.* (2011)

reported 11.8 Jy beam^{-1} as the brightest spectral feature in the 2007 spectrum. From the new component fitting results, to retrieve the position of the different spectral features allowed us to measure the flux of the blended spectral features within the strongest feature reported in 2011. Hence, the brightest of the blended features has peak flux of 6.8±0.2 Jy beam^{-1} at -119.8 km s^{-1}, and arises toward region B1 (Fig. 1). In the 2012 data set, the brightest spectral feature has a peak flux of 6.9±0.07 Jy beam^{-1} and was detected toward region C2 at -98 km s^{-1}. SINFONI/VLT observations yielded spatially resolved emission of shock-excited ro-vibrational transitions of molecular hydrogen (H$_2$) (Pérez-Sánchez *et al.* in prep). The brigthest spectral feature in the SINFONI data cube is the H$_2$ 1-0 S(1) line (Fig. 2). Its peak wavelength (λ_{peak}) was determined in each pixel along the λ axis of those pixels with emission over 3-σ. The mean peak wavelength ($\bar{\lambda}_{peak}$ = 2.1211 μm) was estimated in order to calculate the Doppler shifting with respect to the rest wavelength of the transition (λ_{rest}=2.1218μm). The resolution unit of the SINFONI cube is 2.45×10^{-4} μm. With this we calculated the systemic velocity v$_{sys}$ = -96.4±34 km s^{-1}. Despite the large error, we estimated the kinematic distances to the source using the velocity retrieved from the H$_2$ line. Hence, using the kinematic distance calculator (Reid *et al.* 2014) near and far distances are 5.4 ± 0.4 kpc and 8.5 ± 0.4 kpc, respectively.

3. Discussion and conclusions

With the systemic velocity calculated, the interpretation of the velocity and spatial distribution of the H$_2$O spectral features needs to be revisited. The spatial distribution of the spectral features detected in 2007 displays three different regions in emission (A1, B1, and C1, panel (iii) in Fig. 1). The high-velocity spectral features are found over the same velocity range toward regions B1 and C1. The spectral features in the velocity range from -100 km s^{-1} to -96 km s^{-1} are detected toward the elongated region A1. With respect to v$_{sys}$ = -96 km s^{-1}, the 2007 data set displays only blue-shifted spectral features †. The velocity and spatial distribution of the emission detected in 2012 is consistent with the 2007 distribution. The regions B2 and C2 contains the high-velocity spectral features, again, with features over the same velocity range. All the red-shifted features are detected towards region D2, which has the largest positional offset with respect to B2, A2 and C2. The spectral features within the velocity range from -100 km s^{-1} to -87 km s^{-1} arise from region A2, an elongated structure with similar shape to that of A1. In Fig 1 (panel ii), both data sets were overlied for ilustration. The emission in regions B1 and B2 have similar angular extension and velocity distribution. Most likely the foreground pumped gas moves along the line of sight toward the observer. In this case, the masers in regions B trace a molecular wind with a larger velocity component along the radial direction of a cylindrical reference frame.

The offset between the centroids of regions C1 and C2 is larger than 0.1 arcsec in Dec and ≈ 0.15 arcsec in RA. Such offset could suggest a precessing collimated wind propagating throughout the ionized cavity and colliding the dense internal wall of the hourglass morphology at different heights. Another possible scenario could involve episodic ejections propagating as high-velocity shock-fronts throughout the ionize cavity, which eventually collide with the internal walls of the hourglass structure. In either case, the maser spectral features indicates the large activity and interaction between the ionized cavity and the molecular layers that surrounds it. Hollenbach *et al.* (2013) have shown

† Because the spectral setup, the red-shifted spectral features were not detected in 2007 observations.

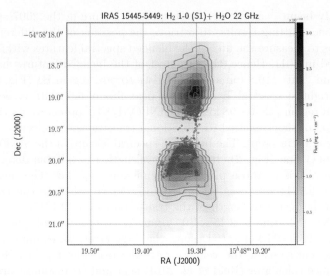

Figure 2. Spatial distribution of the H_2O maser emission (2012, dots) overlaid on the spatial distribution of the H_2 1-0 S(1) ro-vibrational line detected with SINFONI/VLT (gray scale and contours, Pérez-Sánchez *et al.* in prep). The position of the maser features were shifted 0.08 arcsec in RA and 0.4 arcsec in Dec. Assuming that the error in the position of the SINFONI data is $\approx$1 arcsec, the spatial distribution suggest a strong correlation between the H_2 peaks and the position of the regions with maser emission, therefore supporting a scenario where the maser emission trace postshocks regions of J-type shocks (Hollenbach *et al.* (2013)).

that 22 GHz H_2O maser emission can be generated in the postshock region of J-type shocks ($v_s > 30$ km s^{-1}). The precense of dust grains in the postshock region is key. Molecular hydrogen (H_2) re-forms onto grains surface, and then are ejected to the gas phase. In high postshock densities (n$>10^6$cm^{-3}) the H_2 ro-vibrational levels are collisionally de-excited. This process mantain the postshock gas temperature in the range 300 K - 400 K, which favor the formation of H_2O. A number of H_2O IR transitions involved in the pumping scheme of the $J_{k_a K_c} = 6_{16}$-5_{23} H_2O upper level are involved in the cooling of the postshock region, leading to an effective inversion of the level population of the 22 GHz H_2O transition. In the case of the WF IRAS 15445-5449, we might be looking at the emission arising from the postshock regions at the foreground molecular layers. Within the error of SINFONI/VLT, the offset between the peak flux of the H_2 ro-vibrational line, and the offsets between regions C, A and B in each data set can be correlated (Fig. 2). This suggest that the emission is being generated in the postshock region following the reformation of H_2 as described by Hollenbach *et al.* (2013).

References

Bains, I., Cohen, M. *et al.* 2009, *MNRAS*, 397, 1386
Deacon, R. M., Chapman, J. M., Green, A. J., & Sevenster, M. N. 2007, *ApJ*, 658, 1096
Hollenbach, D., Elitzur, M., & McKee, C. F. 2013, *ApJ*, 773, 70
Lagadec, E., Verhoelst, T., Mékarnia, D., *et al.* 2011, *MNRAS*, 417, 32
Pérez-Sánchez, A. F., *et al.* 2013, *MNRAS*, 436, L79
Pérez-Sánchez, A. F., Vlemmings, W. H. T., & Chapman, J. M. 2011, *MNRAS*, 418, 1402
Reid, M. J., Menten, K. M., Brunthaler, A., *et al.* 2014, *ApJ*, 783, 130
Sault, R. J. *et al.* 1995, *Astronomical Data Analysis Software and Systems IV*, 77, 433
Sevenster, M. N., Chapman, J. M., Lindqvist, M. *et al.* 1997, *A&AS*, 124, 509

Astrophysical Masers:
Unlocking the Mysteries of the Universe
Proceedings IAU Symposium No. 336, 2017
A. Tarchi, M.J. Reid & P. Castangia, eds.

© International Astronomical Union 2018
doi:10.1017/S1743921317010274

A study on evolved stars by simultaneous observations
of H_2O and SiO masers using KVN

Se-Hyung Cho[1] , Youngjoo Yun[1], Jaeheon Kim[2], Dong-Hwan Yoon[1],
Dong-Jin Kim[1,3], Yoon Kyung Choi[1], Richard Dodson[4], María Rioja[4]
and Hiroshi Imai[5]

[1]Korea Astronomy and Space Science Institute,
776, Daedukdae-ro, Yuseong-gu, Daejeon, 34055, Republic of Korea
email: cho@kasi.re.kr

[2]Shanghai Astronomical Observatory, Chinese Academy of Sciences, China
[3]Department of Astronomy, Yonsei University, Republic of Korea
[4]International Center for Radio Astronomy Research, The University of Western Australia,
Australia
[5]Department of Physics and Astronomy, Kagoshima University, Japan

Abstract. The Korean VLBI Network (KVN) is a unique millimeter VLBI system which is
consisted of three 21 m telescopes with relatively short baselines. We present the preliminary
results of simultaneous monitoring observations of the 22.2 GHz H_2O and 43.1/42.8/86.2/129.3
GHz SiO masers based on the KVN Key Science Project (KSP). We obtained the astrometrically
registered maps of the H_2O and SiO masers toward nine evolved stars using the source frequency
phase referencing method (SFPR). The SFPR maps of the H_2O and SiO masers enabled us to
investigate the spatial structure and kinematics from the SiO to H_2O maser regions including
the development of an outward motion from the ring-like or elliptical structures of SiO masers
to the asymmetric structures of the 22.2 GHz H_2O maser features. In particular, the 86.2/129.3
GHz SiO (v=1, J=2–1 and J=3–2) masers were clearly imaged toward several objects for the
first time. The SiO v=1, J=3–2 maser shows different distributions compared to those of the
SiO v=1, 2, J=1–0 and v=1, J=2–1 masers implying a different physical condition.

Keywords. masers, radiative transfer, atmospheric effects, techniques: interferometric, stars:
AGB and post-AGB, stars: atmospheres, stars: circumstellar matter, stars: fundamental parameters, stars: individual, stars: mass loss, supergiants

1. Introduction

Many oxygen-rich evolved stars exhibit a SiO maser emission together with a H_2O
maser. The SiO masers arise from the infall and outflow regions inside the dust formation
layer, while 22.2 GHz H_2O maser arises from radially accelerated regions outside the
dust layer (Chen *et al.* 2007). The intensity variations of both the SiO and H_2O masers
show a correlation with the optical light curve of a stellar pulsation. Therefore, it is
important to perform simultaneous observations of both the SiO and H_2O masers to
obtain homogeneous data-sets for both masers. However, previous maser observations
were always performed separately due to the lack of those system. In addition, the VLBI
observations of the 86.2 GHz SiO maser were very limited and those of the 129.3 GHz
SiO maser have not been performed yet. The KVN operates at the H_2O 22 GHz and
SiO 43/86/129 GHz bands simultaneously (Han *et al.* 2008) which enabled us to perform
combined studies of the H_2O and SiO masers including the SiO J=2–1 and J=3–2 masers.
Herein, the KVN results of evolved stars are introduced mainly based on the Key Science

Project (KSP) "Simultaneous monitoring observations of KVN 4 bands toward evolved stars" (2015, P. I. Se-Hyung Cho). The two main scientific goals of the KSP are the following. One is to investigate the spatial structure and dynamical effect from SiO to H_2O maser regions according to stellar pulsation. We also investigate the pulsation and shock wave effect from the SiO to H_2O maser region via the dust layer together by tracing the development of the asymmetric outflow motion. Another goal is to investigate the correlation and difference of the maser properties among the SiO J=1–0, J=2–1, and J=3–2 masers to provide the constraints for SiO maser pumping models.

2. Source selection and observations

At the first stage of the KSP (2015-2017), 16 target sources (`https://radio.kasi.re.kr/kvn/ksp.php`) were selected based on the KVN single dish survey (Kim *et al.* 2010, Cho & Kim 2012). Toward 16 target sources, KVN single dish monitoring was performed every two-months. Simultaneous KVN VLBI monitoring observations of H_2O 6_{16}–5_{23} (22.235080 GHz), SiO v=1, 2, J =1–0, and SiO v=1, J=2–1, 3–2 (43.122080 GHz, 42.820587 GHz, 86.243442.8 GHz and 129.363359 GHz) maser lines were performed toward the KSP sources. The VLBI observations were carried out every 1-3 months from November 2014 to May 2017. The angular resolution is 5.9 mas at 22 GHz, 3.0 mas at 43 GHz, 1.5 mas at 86 GHz, and 1.0 mas at 129 GHz (`https://radio.kasi.re.kr/kvn/status_report_2017/angular_resolution.html`). The DiFX correlator was used for the correlation with 1 s averaging and velocity resolutions of 0.2 km s^{-1} and 0.1 km s^{-1} at 22 and 43 GHz. Data reduction was performed using the NRAO AIPS software package. In addition, the SFPR method (Dodson *et al.* (2014)) and SFPR pipeline (Y. J. Yun *et al.* in preparation) were also adopted for registrations of the H_2O and SiO masers.

3. Preliminary results of several individual sources

Astrometrically registered simultaneous maps of the H_2O and SiO masers were obtained from 9 sources (VY CMa, VX Sgr, IK Tau, W Hya, WX psc, R Crt, V1111 Oph, V5102 Sgr, and V627 Cas) among the 16 KSP sources. In particular, the registered maps of the H_2O and SiO masers including the 86.2/129.3 GHz SiO (v=1, J=2–1 and J=3–2) masers were clearly imaged toward 5 sources (VY CMa, VX Sgr, IK Tau, W Hya, and WX psc) for the first time.

3.1. *Supergiant VX Sgr*

VX Sgr is a red supergiant with an optical pulsation period of 732 days (Kukarkin at al. 1970). The 4 band observational results based on five epoch data with the detection of the 129.3 GHz SiO maser (D. H. Yoon's oral presentation, in preparation) are as follows. Fig. 1 shows the astrometrically registered integrated intensity maps of the 22.2 GHz H_2O and 43.1/42.8/86.2/129.3 GHz SiO masers observed on March 27, 2016 (ϕ=0.67). The SiO maser displays a typical ring-like structure, while the H_2O maser has an asymmetric structure spread slightly in the NW and SE direction in 350 mas. The ring fitting results based on the astrometrically registered maser spot maps of SiO masers give an accurate position of the central star. The ring fitting results also show that the 42.8 GHz SiO maser is located inside the 43.1 GHz maser and the 86.2 GHz maser located at the outer region of the 43.1 GHz maser as previously reported. However, the 129.3 GHz SiO maser was located at the outermost region compared to the 43.1/42.8/86.2 GHz masers. Fig. 1 shows the variations in the ring radius of the 129.3 GHz SiO maser with respect to the

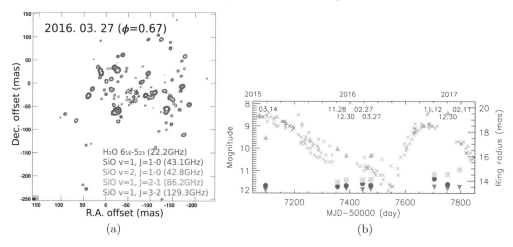

Figure 1. (a) Astrometrically registered integrated intensity maps (moment zero map) of the 22.2 GHz H_2O and 43.1/42.8/86.2/129.3 GHz SiO masers toward VX Sgr obtained on March 27, 2016 (ϕ=0.67). (b) Variation of the SiO ring radius with the optical light curve. Green triangle, yellow square, blue circle, and red inverted triangle indicate the 129.3, 86.2, 43.1, and 42.8 GHz SiO masers.

optical light curve. The ring size of the 129.3 GHz maser increases around the optical maximum suggesting that radiative pumping is dominant in this maser.

3.2. *Supergiant VY CMa*

VY CMa is one of the very important objects to study the evolution of high-mass stars showing localized mass ejections (Decin *et al.* 2016). The 4 band monitoring observations were performed in 29 epochs from September 2014 to June 2017. The preliminary results based on 5 epoch data (Cho *et al.* in preparation) are presented here together with their movie. The movie for the SiO masers shows an elliptical pulsation motion in the NE-SW direction different from the ring-like pulsation motion. Fig. 2 shows representative 4 band SFPR maps of the H_2O and SiO masers observed on Dec. 1, 2016. The SiO 43/42 GHz SiO masers are active in the NE direction accompanying the isotopic ^{29}SiO v=0, J=1–0 maser. The 129.3 GHz maser arises from the inner region of the 43.1/42.8/86.2 maser features in the NE direction, while this maser arises from the outer region with extended features in the SW direction. These results may be associated with the high and low density bipolar axis oriented in the NE-SW direction (Smith *et al.* 2009, Decin *et al.* 2016). The registered maps of the SiO J=1–0, J=2–1, and J=3–2 maser lines show the different locations in the SiO maser distributions (Fig. 2). This result suggests that there are different excitation conditions and pumping mechanisms according to the different transitions. In addition, the elliptical pulsation motion in the NE-SW direction in the movie of the SiO maser seems to be related with the different pulsation mode of VY CMa compared to the fundamental mode of Mira variables.

3.3. *Mira variable IK Tau*

IK Tau is a well-studied oxygen-rich AGB star in a large number of thermal and maser lines. Mid-infrared interferometric observations show a substantially asymmetric dust shell (Weiner *et al.* 2006). We performed 4 band monitoring in 23 epochs from August 2014 to June 2017. Five epoch data (January, March, and April, 2016, and January and February, 2017) were used for the maser movie. The 43.1/42.8 GHz SiO masers were active in the SW direction while the 86.2 GHz SiO maser was active in the South. Fig. 3

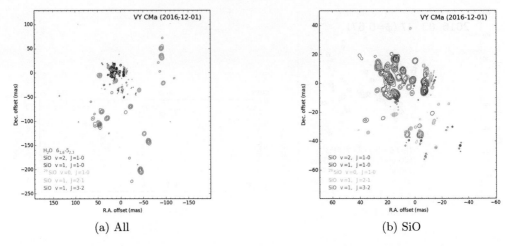

Figure 2. Astrometrically registered integrated intensity maps (moment zero map) of the 22.2 GHz H$_2$O and 43.1/42.8/86.2/129.3 GHz SiO masers toward VY CMa obtained on December 1, 2016.

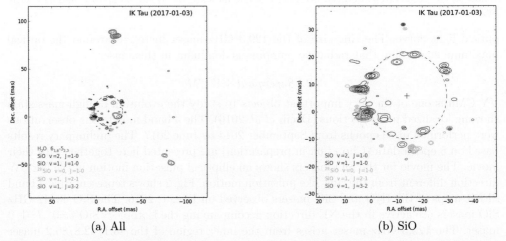

Figure 3. Astrometrically registered integrated intensity maps (moment zero map) of 22.2 GHz H$_2$O and 43.1/42.8/86.2/129.3 GHz SiO masers toward IK Tau obtained on January 3, 2017.

shows the astrometrically registered integrated intensity maps of the 22.2 GHz H$_2$O and 43.1/42.8/86.2/129.3 GHz SiO masers obtained on January 3, 2017. The isotopic ^{29}SiO $v=0$, $J=1$–0 maser was located at the active region of the 43.1/42.8 GHz SiO masers. The 129.3 GHz SiO maser was located at the outer most region compared to the 43.1/42.8/86.2 GHz SiO masers. We need to compare these results of the 129.3 GHz SiO maser with those of WX Psc (Y. J. Yun *et al.* in preparation) according to the optical phases.

3.4. *Semi-regular variable R Crt*

R Crt is classified as SRb type semi-regular variable star. The SRb type variables show uncertain or superimposed periodicity which indicates a complex pulsation mode such as one or more overtones. Eleven epoch monitoring observations were performed from October 2014 to Febuary 2016. Here we present the results of three epoch data which include the successful registered maps of the 22.2 GHz H$_2$O and 43.1/42.8/86.2 GHz SiO

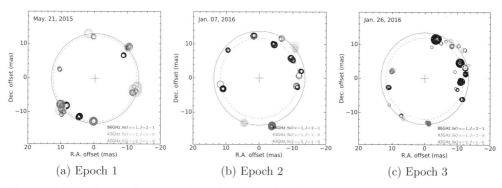

(a) Epoch 1   (b) Epoch 2   (c) Epoch 3

Figure 4. Ring fitting of the 43.1 (green) and 86.2 (grey) GHz SiO masers based on the maser spot-velocity maps.

masers (D. J. Kim *et al.* in preparation). The 129.3 GHz SiO maser was not detected due to its week intensity. The H_2O maser features are distributed in the southern part of the ring-like structure of the SiO masers showing a high asymmetric feature. We determined the dynamical center of the H_2O maser with the rind fitting center of the SiO masers. Fig. 4 shows the spot-velocity maps and ring fitting of the 43.1 and 86.2 GHz SiO masers. As shown in Fig. 4, the 86.2 GHz SiO maser spots are located more in the inner regions compared to those of the 43.1 GHz SiO maser. The previous observations (Soria-Ruiz *et al.* 2004; Soria-Ruiz *et al.* 2007) and our VX Sgr results for the 86.2 GHz SiO maser showed that the 86.2 GHz maser was distributed in the outer regions compared to those of the 43.1 GHz SiO maser. Therefore, D. J. Kim *et al.* discussed that these peculiar features of the 86 GHz SiO maser from R Crt seem to be originated from the complex dynamics caused by the overtone pulsation mode of the SRb type R Crt.

3.5. *Symbiotic star V627 Cas*

V627 Cas was included on the Belczynski *et al.* (2000) list of stars suspected to be symbiotic. The pulsation period is not regular, and the brightness varies much indicating the possible presence of flickering (Kolotilov *et al.* 1996). We obtained successful registered maps of both the H_2O and SiO masers by the SFPR method in five epochs among 14 epoch monitoring observations (H. Yang *et al.* in preparation). The registered maps including the 86.2 GHz SiO maser were obtained in three epochs. The spot distributions of the H_2O maser showed very rapid variations according to observational epochs. Therefore, we are investigating whether these variations of the H_2O maser are associated with the orbital motion of a hot companion. In addition, the spot distributions of the SiO masers did not show the ring-like structure differently from those of Mira variable stars.

4. Summary of and issues with the first stage KSP results

Based on the astrometrically registered simultaneous maps of the H_2O and SiO masers from 9 KSP sources, we are investigating the spatial structure and kinematics from the SiO to the H_2O maser regions including the development of outward motion from the ring-like or elliptical structures of the SiO masers to the asymmetric structures of the 22.2 GHz H_2O maser features. We also need to investigate the evolution of the asymmetric clumpy structure and the role of the dust layer from the SiO to the H_2O maser regions. To interpret the registered maps of the SiO and H_2O masers, both SiO and H_2O maser models coupled to the hydrodynamic atmosphere are required as well as dust shell models.

We need collaborations with the maser theory (Gray *et al.* 2009) and VLTI (Wittkowski *et al.* 2012) teams.

The movie of the SiO masers around the supergiant VY CMa shows an elliptical pulsation motion in the NE-SW direction different from the ring-like pulsation motion in the movie of the SiO masers around the Mira variable IK Tau.

There are different features in the registered maps between the H_2O and SiO masers and also different features among the SiO maser spot distributions according to the KSP sources. We are investigating the relation with the characteristics of the sources and the evolutionary stage including the relation with the optical light curve and pulsation mode (especially, in the case of a different shape for the SiO maser features).

Based on the registered maps of the SiO J=1–0, J=2–1 and J=3–2 masers, we find that the SiO maser spot distributions are different among the SiO J=1–0, J=2–1, J=3–2 maser lines. These results imply that there are different dominant pumping mechanisms according to the different type of sources and SiO transitions.

Acknowledgements

We are grateful to all the staff members at KVN who helped operate the array and the single dish telescope, and correlate the data. The KVN is a facility operated by the Korea Astronomy and Space Science Institute, which is under the protection of the National Research Council of Science and Technology.

References

Belczynski, K., Mikolajewska, J., Munari, U. *et al.* 2000, *A&AS*, 146, 407
Chen *et al.* 2007, *ChJAA*, 7, 531
Cho, S.-H & Kim, J. 2012, *AJ*, 144, 129
Decin, L., Richards, A. M. S., Millar, T. J. *et al.* 2016, *A&A*, 592, A76
Dodson R. *et al.* 2014, *AJ*, 148, 97
Gray, M. D., Wittkowski, M., Scholz, M. *et al.* 2009, *MNRAS*, 394, 51
Han S.-T., Lee, J.-W. Kand, J. *et al.* 2008, *IJIMW*, 29, 69
Kim, J., Cho, S.-H,. Oh, C.-S., & Byun, D.-Y. 2010, *ApJS*, 188, 209
Kukarkin, B. V., Kholopov, P. N., & Efremov, Yu. N. *et al.* 1970, *General Catalogue of Variable Stars (Moscow: Acad. Sci. USSR)*
Kholopov, P. N., Samus', N. N., Kazarovets, E. V., & Kireeva, N. N. 1987, *Information Bulletin on Variable Stars*, 3058, 1
Kolotilov, E. A., Munari, U., Yudin, B. F., & Tatarnikov, A. M. 1996, *Astron. Rep.*, 40, 812
Smith, N., Hinkle, K. H., & Ryde, N. 2009, *AJ*, 137, 3558
Soria-Ruiz, R., Alcolea, J., Colomer, F. *et al.* 2004, *A&A*, 426, 131
Soria-Ruiz, R., Alcolea, J., Colomer, F. *et al.* 2007, *A&A*, 468, L1
Weiner, J., Tatebe, K., & Hale, D. D. S. *et al.* 2006, *ApJ*, 636, 1067
Wittkowski, M., Boboltz, D. A., Gray, M. D. *et al.* 2012, *Proceedings IAU Symposium*, 287, 209

Astrophysical Masers:
Unlocking the Mysteries of the Universe
Proceedings IAU Symposium No. 336, 2017
A. Tarchi, M.J. Reid & P. Castangia, eds.

© International Astronomical Union 2018
doi:10.1017/S1743921317009449

Astrometric VLBI Observations of the Galactic LPVs, Miras, and OH/IR stars

Akiharu Nakagawa[1], Tomoharu Kurayama[2], Gabor Orosz[1], Ross A. Burns[3], Tomoaki Oyama[4], Takumi Nagayama[4], Takashi Miyata[5], Mamoru Sekido[6], Junichi Baba[7] and Kiichi Wada[1]

[1]Kagoshima University, 1-21-35, Korimoto, Kagoshima-shi, Kagoshima, Japan
email: nakagawa@sci.kagoshima-u.ac.jp

[2]Teikyo University of Science, 2-2-1 Sakuragi, Senjyu, Adachi-ku, Tokyo, Japan

[3]Joint Institute for VLBI ERIC (JIVE), Postbus 2, 7990 AA Dwingeloo, the Netherlands

[4]National Astronomical Observatory of Japan, Mizusawa VLBI Observatory, 2-12 Hoshigaoka-cho, Mizusawa-ku, Oshu, Iwate, Japan

[5]Institute of Astronomy, The University of Tokyo, 2-21-1Osawa, Mitaka, Tokyo, Japan

[6]National Institute of Information and Communications Technology, Kashima Space Technology Center, 893-1 Hirai, Kashima, Ibaraki, Japan

[7]National Astronomical Observatory of Japan, 2-21-1 Osawa, Mitaka, Tokyo, Japan

Abstract. Studies of Galactic LPVs based on astrometric VLBI are presented. We use a VLBI array, "VERA", to measure parallaxes and calibrate the K-band period luminosity relation (PLR) of the Galactic Miras. Since the PLR offers a distance indicator, its calibration is crucial to reveal their spatial distribution. Parallaxes of dozens of LPVs are presented. For the longer period stars, the mass-loss is high and the stars are obscured and recognized as OH/IR stars. We estimated mid-infrared absolute magnitudes of dozens of OH/IR stars and found that they show a loose concentration around -14 mag at λ of 11.6 μm, indicating an existence of PLR for OH/IR stars. Astrometry of OH/IR stars will also help us to study non-steady spiral arms as proposed from the latest simulation study of the galactic dynamics. We will start astrometric VLBI observation of two OH/IR stars NSV25875 and OH127.8+0.0 at 43 GHz with VERA.

Keywords. VLBI, Astrometry, Maser, Mira, OH/IR, AGB.

1. Introduction

Long Period Variables (LPVs) are low- to intermediate-mass $(1 - 8 M_\odot)$ asymptotic giant branch (AGB) stars that pulsate with a period range of $100 - 1000$ days. They are surrounded by large and extended dust and molecular shells and, in sources with mass-loss rate higher than $\sim 10^{-7} M_\odot \mathrm{yr}^{-1}$, we find maser emission of H_2O, SiO, or OH. The well-known relation between K-magnitude and period of Mira variables, derived from studies of Miras in the LMC (Feast *et al.* 1989), is used as a distance indicator to obtain source distances from their periods and magnitudes. Since there is a metallicity difference between LMC and our Galaxy, it is also important to establish this relation using sources in our Galaxy. Since LPVs are very bright in infrared, we can use them to probe a region where interstellar extinction is strong, such as the direction of the Galactic Center and Galactic plane. Making use of the high performance of the VERA array (Kobayashi *et al.* 2003), which is a Japanese VLBI project dedicated to the Galactic astrometry, we aim to construct the $M_K - \log P$ relation for Galactic LPVs.

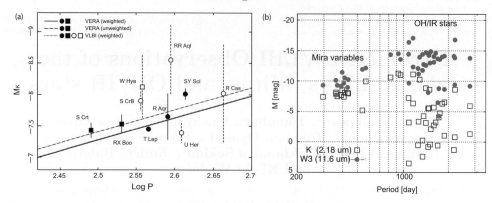

Figure 1. (a) K-band period luminosity relation of the Galactic Mira variables (Nakagawa *et al.* 2014). (b) Same relation for both Miras and OH/IR stars in wider period range. Absolute magnitudes at near(●)- and mid(□)-infrared bands are presented.

2. Period luminosity relation of Galactic LPVs

Mira and semi-regular variables.
In table 1, we show parallaxes of 26 Galactic LPVs whose distances were measured with VLBI observations using VERA and VLBA. Some red giants and semiregular variables are also included. Absolute magnitudes are estimated from their apparent magnitudes and parallax distances. Figure 1(a) shows a PLR derived from some results in table 1. Filled and open symbols indicate results from H_2O and OH maser, respectively. Circles and squares correspond to Mira and semiregular variables. The solid line is an unweighted-least squares fit for an $M_K - \log P$ relation using these results from VERA, $M_K = -3.52 \log P + (1.09 \pm 0.14)$ (Nakagawa *et al.* 2016). New results will be added soon and more accurate calibration can be expected.

OH/IR stars with very long periods ($P \gtrsim 1000$ days).
Compared to Mira variables, OH/IR stars tend to show longer pulsation periods, and sometimes it extends 1000 days. We compiled OH/IR stars with such very long periods from literatures (e.g. Engels & Bunzel 2015). Distances of some sources were determined by the "phase-lag method" (Engels *et al.* 2015). For sources with no estimated distances, we derived kinematic distances using their radial velocities. Then, apparent magnitudes were converted to absolute ones and presented in figure. 1(b). The K-band absolute magnitudes of LPVs are presented with open squares. Since OH/IR stars are surrounded by thick circumstellar dust shells, K-band absolute magnitudes become lower. And also we can see scattering of the magnitudes. We think this is due to an anisotropic distribution of the circumstellar dust. At longer wavelengths, re-radiation from the dust shell becomes dominant. To minimize circumstellar extinction, we estimated absolute magnitudes in the mid-infrared band ($\lambda = 11.6\mu m$, W3-band) using data from the Wide-field Infrared Survey Explorer (WISE; http://wise.ssl.berkeley.edu/index.html). Absolute magnitudes in the W3-band of Mira, semiregular, and OH/IR stars are presented with filled circles in figure 1(b). Although it is difficult to find clear relation for OH/IR stars with $P \gtrsim 1000$ days in K-band, it becomes narrower in the W3-band, and some relation can be implied. It is well known that there are deep silicate absorption features at $\lambda \simeq 8 - 18\mu m$, and so a more appropriate waveband should be considered or calibration of the absorption is needed. If the mid-infrared PLR is confirmed, this can be used as a new distance indicator for sources along the Galactic plane or deeply obscured by dust.

Table 1. Parallax of the Galactic LPVs determined with VLBI astrometry.

Source	Type	Parallax [mas]	P [day]	LogP	m_K [mag]	M_K [mag]	Maser	Reference (Parallax[1], m_K[2])
RW Lep	SRa	1.62±0.16	150	2.176	0.639	-8.31 ± 0.22	H_2O	kam14, a
S Crt	SRb	2.33±0.13	155	2.190	0.786	-7.38 ± 0.12	H_2O	nak08, a
RX Boo	SRb	7.31±0.5	162	2.210	−1.96	-7.64 ± 0.15	H_2O	kam12, b
T UMa	Mira	0.96±0.15	257	2.410	2.60	-7.49 ± 0.44	H_2O	**in prep.**, a
Y Lib	Mira	1.24±0.13	276	2.441	3.16	-6.37 ± 0.23	H_2O	**in prep.**, a
R UMa	Mira	1.92±0.05	302	2.480	1.19	-7.39 ± 0.06	H_2O	nak16, d
SY Aql	Mira	1.10±0.07	356	2.551	2.36	-7.43 ± 0.14	H_2O	**in prep.**, a
R Cnc	Mira	3.84±0.29	357	2.553	−0.97	-8.05 ± 0.16	H_2O	**in prep.**, a
W Hya	SRa	10.18±2.36	361	2.558	−3.16	-8.12 ± 0.51	OH	vle03, c
S CrB	Mira	2.39±0.17	360	2.556	0.21	-7.90 ± 0.15	OH	vle07, c
T Lep	Mira	3.06±0.04	368	2.566	0.12	-7.45 ± 0.03	H_2O	nak14, c
R Peg	Mira	3.98±0.21	378	2.577	0.45	-6.55 ± 0.11	H_2O	**in prep.**, a
R Hya	Mira	8.96±0.51	380	2.580	-2.51	-7.75 ± 0.12	H_2O	**in prep.**, a
R Aqr	Mira	4.7±0.8	390	2.591	−1.01	-7.65 ± 0.37	SiO	kam10, c
R Aqr	Mira	4.59±0.24	390	2.591	−1.01	-7.70 ± 0.11	SiO	min14, c
RR Aql	Mira	1.58±0.40	396	2.598	0.46	-8.55 ± 0.56	OH	vle07, c
U Her	Mira	3.76±0.27	406	2.609	−0.27	-7.39 ± 0.16	OH	vle07, c
SY Scl	Mira	0.75±0.03	411	2.614	2.55	-8.07 ± 0.09	H_2O	nyu11, b
R Cas	Mira	5.67±1.95	430	2.633	−1.80	-8.03 ± 0.78	OH	vle03, c
U Lyn	Mira	1.27±0.06	434	2.637	1.533	-7.95 ± 0.10	H_2O	kam15, a
OH231.8+4.2	OH/IR	0.55±0.05	551	2.741	$\cdots$	$\cdots$	H_2O	**in prep.**
UX Cyg	Mira	0.54±0.06	565	2.752	1.40	-9.94 ± 0.24	H_2O	kur05, a
OZ Gem	Mira	1.00±0.18	598	2.777	3.00	-7.00 ± 0.40	H_2O	**in prep.**, a
S Per	SRc	0.413±0.017	822	2.915	1.33	-10.59 ± 0.09	H_2O	asa10, b
PZ Cas	SRc	0.356±0.026	925	2.966	1.00	-11.24 ± 0.16	H_2O	kus13, b
VY CMa	SRc	0.88±0.08	956	2.980	−0.72	-11.00 ± 0.20	H_2O	cho08, b
NML Cyg	$\cdots$	0.62±0.047	1280	3.107	0.791	-10.25 ± 0.16	H_2O	zha12, a

Notes: [1]References of the parallax are as follows : (kam14) Kamezaki *et al.* 2014, (nak08) Nakagawa *et al.* 2008, (kam12) Kamezaki *et al.* 2012, (nak16) Nakagawa *et al.* 2016, (vle03) Vlemmings *et al.* 2003, (vle07) Vlemmings & van Langevelde 2007, (nak14) Nakagawa *et al.* 2014, (kam10) Kamohara *et al.* 2010, (min14) Min *et al.* 2014, (nyu11) Nyu *et al.* 2011, (kam15) Kamezaki *et al.* 2015, (kur05) Kurayama *et al.* 2005, (asa10) Asaki *et al.* 2010, (kus13) Kusuno *et al.* 2013, (cho08) Choi *et al.* 2008, and (zha12) Zhang *et al.* 2012. [2]References of the apparent magnitudes (m_K) are as follows : (a) The IRSA 2MASS All-Sky Point Source Catalog (Cutri *et al.* 2003), (b) Catalogue of Stellar Photometry in Johnson's 11-color system (Ducati(2002)), (c) Photometry by (Whitelock & Feast 2000), and (d) Photometry using Kagoshima 1m telescope.

Table 2. Correspondence table of model and observation in galactic dynamics.

Age	Physics	Target	Model		Observation
$\sim 10^6$ yr	Spiral arm	SFR, Giants	✓	↔	✓
$\sim 10^8$ yr	Bifurcating/merging arm	Heavy OH/IR star ?	✓	↔	No
$\sim 10^9$ yr	Relaxed system	Mira	✓	↔	✓

3. Astrometry for heavy OH/IR stars for galactic dynamics

The OH/IR stars showing longer periods are thought to have larger mass, i.e. variable stars with $P \simeq 1000$ days have initial masses of $\simeq 4.0 M_\odot$ (Feast *et al.* 1989). Their ages are thought to be on the order of 10^8 yr. The latest study of galactic spiral arms based on three-dimensional *N*-body simulations supports a picture of non-steady spiral arms (e.g. Baba *et al.* 2013). As a characteristic behavior, it is predicted that spiral arms bifurcate or merge on a time scale of 10^8 yr. Now, we can find that this time scale is similar to the age of OH/IR stars with very long periods ($P \gtrsim 1000$ days). From recent VLBI

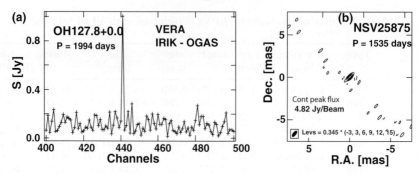

Figure 2. (a) SiO maser cross correlated spectrum of OH127.8+0.0 detected at Iriki-Ogasawara baseline of VERA. (b) Self-calibrated map of SiO maser in NSV25875.

observations, astrometric results of many SFRs and Mira variables are obtained. But the ages of the sources are in the order of 10^6 yr and 10^9 yr, respectively. So, observations of sources with various ages are needed to fully understand the mechanism of spiral arm formation, and OH/IR stars can become a good probe to compensate the time scale gap in these studies (Table 2). Astrometric VLBI is a promising method to determine three-dimensional positions and motions of the OH/IR stars. We selected several OH/IR stars with $P \gtrsim 1000$ days and detected two sources of 43 GHz SiO maser emission using VERA. Phase-referencing VLBI monitoring for two pairs, "OH127.8+0.0 & J0128+6306" and "NSV25875 & J2231+5922" have just started in Oct. 2017. Figure 2(a) and (b) show results of a test VLBI observation prior to the monitoring VLBI. Parallax measurements of the two sources will also be used for the PLR study proposed in the previous section.

References

Asaki, Y., Deguchi, S., Imai, H., *et al.* 2010, *APJ*, 721, 267

Baba, J., Saitoh, T. R., & Wada, K. 2013, *APJ*, 763, 46

Choi, Y. K., Hirota, T., Honma, M., *et al.* 2008, *PASJ*, 60, 1007

Cutri, R. M., Skrutskie, M. F., van Dyk, S., *et al.* 2003, VizieR Online Data Catalog, 2246,

Engels, D. & Bunzel, F. 2015, *A&A*, 582, A68

Engels, D., Etoka, S., Gérard, E., & Richards, A. 2015, Why Galaxies Care about AGB Stars III: A Closer Look in Space and Time, 497, 473

Feast, M. W., Glass, I. S., Whitelock, P. A., & Catchpole, R. M. 1989, *mnras*, 241, 375

Kamezaki, T., Nakagawa, A., Omodaka, T., *et al.* 2012, *PASJ*, 64, 7

Kamezaki, T., Kurayama, T., Nakagawa, A., *et al.* 2014, *PASJ*, 118

Kamezaki, T., Nakagawa, A., Omodaka, T., *et al.* 2015, *PASJ*, 208

Kamohara, R., Bujarrabal, V., Honma, M., *et al.* 2010, *A&A*, 510, A69

Kobayashi, H., *et al.* 2003, ASP Conference Series, 306, 48P

Kurayama, T., Sasao, T., & Kobayashi, H. 2005, *APJ*, 627, L49

Kusuno, K., Asaki, Y., Imai, H., & Oyama, T. 2013, *APJ*, 774, 107

Min, C., Matsumoto, N., Kim, M. K., *et al.* 2014, *PASJ*, 66, 38

Nakagawa, A., Tsushima, M., Ando, K., *et al.* 2008, *PASJ*, 60, 1013

Nakagawa, A., Omodaka, T., Handa, T., *et al.* 2014, *PASJ*, 66, 101

Nakagawa, A., Kurayama, T., Matsui, M., *et al.* 2016,*PASJ*, 68, 78

Nyu, D., Nakagawa, A., Matsui, M., *et al.* 2011, *PASJ*, 63, 63

Vlemmings, W. H. T., van Langevelde, H. J., Diamond, P. J., Habing, H. J., & Schilizzi, R. T. 2003, *A&A*, 407, 213

Vlemmings, W. H. T. & van Langevelde, H. J. 2007, *A&A*, 472, 547

Whitelock, P. & Feast, M. 2000, *MNRAS*, 319, 759

Zhang, B., Reid, M. J., Menten, K. M., Zheng, X. W., & Brunthaler, A. 2012, *A&A*, 544, AA42

Astrophysical Masers:
Unlocking the Mysteries of the Universe
Proceedings IAU Symposium No. 336, 2017
A. Tarchi, M.J. Reid & P. Castangia, eds.

© International Astronomical Union 2018
doi:10.1017/S1743921317010572

Submillimeter H$_2$O maser emission from water fountain nebulae

Daniel Tafoya[1], Wouter H. T. Vlemmings[1] and Andres F. Pérez-Sánchez[2]

[1]Chalmers University of Technology, Onsala Space Observatory, 439 92 Onsala, Sweden
email: `daniel.tafoya@chalmers.se`
[2]European Southern Observatory, Alonso de Córdova 3107, Vitacura, Casilla 19001, Santiago, Chile

Abstract. We present the results of the first detection of submillimeter water maser emission toward water-fountain nebulae. Using APEX we found emission at 321.226 GHz toward two sources: IRAS 18043−2116, and IRAS 18286−0959. The submillimeter H$_2$O masers exhibit expansion velocities larger than those of the OH masers, suggesting that these masers, similarly to the 22 GHz masers, originate in fast bipolar outflows. The 321 GHz masers in IRAS 18043−2116 and IRAS 18286−0959, which figure among the sources with the fastest H$_2$O masers, span a velocity range similar to that of the 22 GHz masers, indicating that they probably coexist. The intensity of the submillimeter masers is comparable to the 22 GHz masers, implying that the kinetic temperature of the region where the masers originate is T$_k$>1000 K. We propose a simple model invoking the passage of two shocks through the same gas that creates the conditions for explaining the strong high-velocity 321 GHz masers coexisting with the 22 GHz masers in the same region.

Keywords. submillimeter, ISM: jets and outflows, stars: AGB and post-AGB, masers

1. Introduction

Water maser emission from the transition $6_{16} \rightarrow 5_{23}$ at 22 GHz has proven to be a valuable tool to study the kinematics of the gas in the circumstellar envelope (CSE) of evolved stars. In the envelopes of AGB stars, the 22 GHz H$_2$O masers exhibit typical expansion velocities of ∼10 km s^{-1} and trace clumpy spherical structures located at a distance of ∼100 AU from the star, where the stellar wind is accelerated (e.g. Richards *et al.* 2012). In the water-fountain nebulae (wf-nebulae), a subgroup of post-asymptotic giant branch (post-AGB) objects, the water masers trace collimated structures at larger distances from the star ($\gtrsim$500 AU) and they expand with larger velocities ($\gtrsim$100 km s^{-1}; Likkel & Morris 1988; Imai *et al.* 2002). From the expansion velocity and proper motion of the maser spots in wf-nebulae, it has been calculated a kinematical time scale for the jet-like outflows in these sources of ∼100 years, assuming that the expansion velocity has been constant. In some cases, the masers seem to be tracing a precessing jet. This phenomenon has been attributed to a binary companion (e.g. Imai *et al.* 2002). The magnetic field that is thought to be responsible for the collimation of the jet-like outflows has been measured via the Zeeman effect on the H$_2$O masers by Vlemmings *et al.* (2006). Despite of the valuable information that 22 GHz H$_2$O masers have revealed, due to the particular excitation conditions, they do not probe the entire physical conditions of the CSE.

Apart from the H$_2$O ($6_{16} \rightarrow 5_{23}$) line, other water maser lines, most of them at submillimeter wavelengths, have been detected toward star forming regions and late-type stars (Menten *et al.* 1990a,b; Melnick *et al.* 1993; Patel *et al.* 2007). Some transitions of

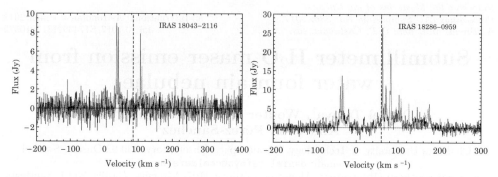

Figure 1. Spectra of the H_2O ($10_{29} \rightarrow 9_{36}$) maser emission in two water-fountain nebulae. The gray line indicates the spectrum obtained in the epoch May 15, 2013, and the black line indicates the spectrum obtained in the epoch July 6, 2013. The systemic velocity reported in the literature is indicated with a vertical dashed line.

the submillimeter water masers have upper levels with energies above the ground state higher than for the 22 GHz masers, for which E/k=643 K. In particular, the upper level of the 321 GHz water maser transition has an energy E/k=1861 K above the ground state. Consequently, these masers can be used as tools to trace dense gas with relatively high temperatures.

2. Pilot search for submillimeter water masers

In May (epoch 1) and July (epoch 2) 2013 we carried out observations with APEX toward seven wf-nebulae that exhibit relatively strong 22 GHz maser to search for sub-millimeter H_2O maser emission. 321 GHz water maser emission was detected for the first time in two wf-nebulae. The details of the observations and the results are reported by Tafoya *et al.* (2014). The peak fluxes and rms noise levels are listed in Table 1 and the spectra of the masers are shown in Fig. 1. Although the emission showed variability by a factor of up to $\sim$10 in some spectral features, the masers were clearly detected toward the two sources in both observation epochs. The sources with no detection are listed in Table 2 of Tafoya *et al.* (2014). The 321 GHz water emission from IRAS 18043−2116 and IRAS 18286−0959 consists of clusters of spectral lines spread over a velocity range of $\gtrsim$200 km s^{-1}. The narrow width of the lines suggests that the emission is indeed amplified by the maser effect. The emission from IRAS 15445−5449 exhibits a broader line width (see Fig. 1a of Tafoya *et al.* 2014). Tafoya *et al.* (2014) proposed that the emission from the latter source is submillimeter water masers that originate in a region different from where the 22 GHz are located, possibly at a distance closer to the star.†

3. Discussion and conclusions

The most striking result from our observations is that both masers, 321 GHz and 22 GHz, span similar velocity ranges and that several spectral features appear at the same velocity. Thus, it is likely that these masers originate in the same gas. The models that explain the water maser emission in evolved stars assume that the emission origi-nates in the expanding CSE created by the massive wind at the end of the AGB phase

† From recent ALMA observations it has been confirmed that the emission in from IRAS 18043−2116 and IRAS 18286−0959 is maser, but the emission from IRAS 15445−5449 is thermal SO_2 ($18_{0,18} \rightarrow 17_{1,17}, v = 0$), ν_0=321.3301645 GHz, (Pérez-Sánchez in prep., private communication).

Table 1. Sources with detected 321 GHz water maser emission

Source	RA(J2000)	Dec(J2000)	line peak (epoch 1)	rms (epoch 1)	line peak (epoch 2)	rms (epoch 2)
IRAS name	h m s	° ′ ″	Jy	Jy	Jy	Jy
18043−2116	18 07 21.10	−21 16 14.2	8.5	0.9	4.2	0.7
18286−0959	18 31 22.93	−09 57 19.8	25.2	1.0	11.1	0.7

Notes:
Line peak and rms noise values for a spectral resolution of 0.6 km s^{-1}.

(Cooke & Elitzur 1985; Neufeld & Melnick 1990). However, the origin of the H_2O maser emission in wf-nebulae is associated to fast collimated outflows that interact with the slowly expanding CSE. Therefore, it is more appropriate to interpret the water maser emission in a similar way to that of the outflows in star-forming regions. According to Elitzur *et al.* (1989), the water maser emission in star forming regions arise after the passage of a dissociative shock ($v_s \gtrsim 50$ km s^{-1}) through the interstellar medium. Behind the shock, a layer of high-density gas with a temperature of ~ 400 K forms, where the conditions for 22 GHz maser emission are optimal. Neufeld & Melnick (1990) showed that under the physical conditions of the post-shock region described by Elitzur *et al.* (1989) 321 GHz emission can be produced with a luminosity ratio $L_p(22\ \text{GHz})/L_p(321\ \text{GHz}) \gtrsim 5$. They also suggested that values of the ratio <5 could be attained with slower non-dissociative shocks that would heat the molecules to temperatures up to $T_k = 1000$ K (Kaufman & Neufeld 1996).

For the case of the water-fountain nebulae presented in this work, the expansion velocity of the 22 GHz and 321 GHz H_2O masers is ~ 100 km s^{-1}, implying a dissociative shock. According to the model proposed by Elitzur *et al.* (1989), when the shocked material cools down, H_2 and H_2O molecules form in gas that is maintained at $T_k = 400$ K. As mentioned above, for this temperature the luminosity ratio $L_p(22\ \text{GHz})/L_p(321\ \text{GHz})$ is expected to be $\gtrsim 5$. Comparing the intensities of the 22 GHz masers (see Deacon *et al.* 2007; Walsh *et al.* 2009; Yung *et al.* 2011; Pérez-Sánchez *et al.* 2017) and the 321 GHz masers of IRAS 18043−2116 and IRAS 18286−0959, we find a luminosity ratio ≈ 1. This implies a kinetic temperature $T_k > 1000$ K for the gas (Neufeld & Melnick 1990; Yates *et al.* 1997). But this temperature indicates the presence of a relatively slow non-dissociative shock, in contradiction to the relatively high velocity of the masers. Therefore, to explain the coexistence of strong 321 GHz masers with 22 GHz masers, there should be a mechanism that accelerates the gas to the observed high velocities ($v_{exp} \sim 100$ km s^{-1}), favoring the creation of water molecules, while maintaining the gas at high temperatures ($T_k > 1000$ K).

We propose scenario that includes the passage of two shocks with different speeds through the same material, as it is shown schematically in Fig. 2. The first shock is due to the collision between a fast collimated wind and the slowly expanding CSE, which produces a J-type dissociative shock. If the post-shock density is much higher than for the pre-shock gas, then the velocity of the shocked gas would be very low in the frame of reference of the shock. Thus, in the frame of reference of the star, the shocked gas, where the 22 GHz water masers originate, moves almost at the same speed as the shock ($v_{exp} \sim 100$ km s^{-1}). Subsequently, we consider a collision between the fast collimated wind and the shocked material. Since the shocked material is already moving at $v_{exp} \sim 100$ km s^{-1}, the collision occurs at a slower relative velocity. This produces a slower C-type non-dissociative shock, which raises the temperature of the shocked gas

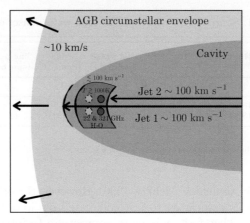

Figure 2. Schematic representation of the two jets scenario proposed to explain the coexistence of the 22 GHz and 321 GHz water masers. The circles with dashed lines represent the 22 GHz masers and the circles with solid lines represent the 321 GHz masers.

to a higher value, where the 321 GHz transition is inverted more efficiently (Kaufman & Neufeld 1996). The 321 GHz water masers would move at the same velocity as the 22 GHz masers, which is the velocity of the shocked gas.

If the C-type shock does not occur or if there is an efficient cooling mechanism in the shocked material then the 321 GHz maser emission will be negligible, which would explain the non-detections toward the other wf-nebulae of our sample. In this regard, the sources with strong, high-velocity 321 GHz water masers represent a subclass of post-AGB stars that could be referred to as *hot-water fountain nebulae*. It is clear that our observations pose a challenge for the current water maser excitation models. Higher angular resolution observations with ALMA and more theoretical models are required to fully understand the presence of high-velocity 321 GHz masers in these sources.

References

Cooke, B. & Elitzur, M. 1985, *ApJ*, 295, 175

Deacon, R. M., Chapman, J. M., Green, A. J., & Sevenster, M. N. 2007, *ApJ*, 658, 1096

Elitzur, M., Hollenbach, D. J., & McKee, C. F. 1989, *ApJ*, 346, 983

Imai, H., Obara, K., Diamond, P. J., Omodaka, T., & Sasao, T. 2002, *Nature*, 417, 829

Kaufman, M. J. & Neufeld, D. A. 1996, *ApJ*, 456, 250

Likkel, L. & Morris, M. 1988, *ApJ*, 329, 914

Melnick, G. J., Menten, K. M., Phillips, T. G., & Hunter, T. 1993, *ApJ* (Letters), 416, L37

Menten, K. M., Melnick, G. J., & Phillips, T. G. 1990a, *ApJ* (Letters), 350, L41

Menten, K. M., Melnick, G. J., Phillips, T. G., & Neufeld, D. A. 1990b, *ApJ* (Letters), 363, L27

Neufeld, D. A. & Melnick, G. J. 1990, *ApJ* (Letters), 352, L9

Patel, N. A., Curiel, S., Zhang, Q., *et al.* 2007, *ApJ* (Letters), 658, L55

Pérez-Sánchez, A. F., Tafoya, D., García López, R., Vlemmings, W., & Rodríguez, L. F. 2017, *A&A*, 601, A68

Richards, A. M. S., Etoka, S., Gray, M. D., *et al.* 2012, *A&A*, 546, A16

Tafoya, D., Franco-Hernández, R., Vlemmings, W. H. T., Pérez-Sánchez, A. F., & Garay, G. 2014, *A&A*, 562, L9

Vlemmings, W. H. T., Diamond, P. J., & Imai, H. 2006, *Nature*, 440, 58

Walsh, A. J., Breen, S. L., Bains, I., & Vlemmings, W. H. T. 2009, *MNRAS*, 394, L70

Yates, J. A., Field, D., & Gray, M. D. 1997, *MNRAS*, 285, 303

Yung, B. H. K., Nakashima, J.-i., Imai, H., *et al.* 2011, *ApJ*, 741, 94

Astrophysical Masers:
Unlocking the Mysteries of the Universe
Proceedings IAU Symposium No. 336, 2017
A. Tarchi, M.J. Reid & P. Castangia, eds.

© International Astronomical Union 2018
doi:10.1017/S1743921317010547

Registration of H₂O and SiO masers in the Calabash Nebula, to confirm the Planetary Nebula paradigm

R., Dodson[1], M. Rioja[1,2,3], V. Bujarrabal[3], J. Kim[4], S. H. Cho[5], Y. K. Choi[5] and Y. Youngjoo[5]

[1]International Centre for Radio Astronomy Research, UWA, 35 Stirling Hwy, Western Australia
[2]CSIRO Astronomy and Space Science, 26 Dick Perry Avenue, Kensington WA 6151, Australia
[3]Observatorio Astronómico Nacional (IGN), Alfonso XII, 3 y 5, 28014 Madrid, Spain
[4]Shanghai Astronomical Observatory, Chinese Academy of Sciences, Shanghai 200030, China
[5]Korea Astronomy and Space Science Institute 776, Daedeokdae-ro, Yuseong-gu, Daejeon, 34055, Republic of Korea

Abstract. We report on the astrometric registration of VLBI images of the SiO and H₂O masers in OH 231.8+4.2, the iconic Proto-Planetary Nebula also known as the Calabash nebula, using the KVN and Source/Frequency Phase Referencing. This, for the first time, robustly confirms the alignment of the SiO masers, close to the AGB star, which drives the bi-lobe structure with the water masers in the out-flow.

Keywords. stars: individual: QX Pup – stars: AGB and post-AGB – masers – stars: evolution

1. Introduction

OH 231.8+4.2 (OH 231) is perhaps the best studied proto-Planetary Nebulae (pPN), and in many ways provides the prototype for the class. pPN develop from Asymptotic Giant Branch (AGB) stars, into the environment seeded by the previous copious AGB mass-loss, which forms a thick spherical circumstellar envelope (CSEs). These pPN occur when the AGB completes throwing off the outer stellar layers, when the stellar core becomes exposed becoming the new central star. This new star is very compact, rapidly evolving to the blue and white dwarf phase. In at least some cases there are also fast moving jets carrying significant angular momentum. At present, the only way to explain the origin of such energetic flows is to assume that a fraction of the ejected CSE is re-accreted by the central star or a companion through a rapidly rotating disk. The CSE is shocked by very fast bipolar outflows from the dwarf, with axial velocities as high as several hundred km s⁻¹. These generate a series of axial shocks that cross the massive CSE, inducing high axial velocities in it. The blue dwarf at the centre is able to significantly ionize the nebula, revealing its wide bipolar shape and producing some of the most beautiful objects in the sky.

SiO masers are known to form close to AGB stars, and water masers form in outflows so together they probe the crucial jet/shock regions. However as soon as the star+nebula system leaves the AGB phase the observation of the masers associated with them becomes more and more difficult. The reason is that the inner circumstellar masers are becoming more and more diffuse at the same time as the mass-loss rate decreases, therefore SiO masers are very rarely observed. However there is one, paradigmatic, example showing both SiO and H₂O masers: the strongly bipolar nebula OH 231, the Calabash Nebula, which shows the classic bi-polar structure. OH 231 has a binary central source, which has been identified through optical spectroscopy; a M9-10 III Mira variable (i.e. an AGB star) (Cohen 1981) and a A0 main sequence companion (Sánchez Contreras *et al.* 2004). This remarkable bipolar nebula shows all the signs of post-AGB evolution: fast bipolar outflows with velocities $\sim 200 - 400$km s^{-1}, shock-excited gas and shock-induced chemistry.

VLBI observations of SiO and H$_2$O masers with the VLBA have yielded a number of important results (for example, see Sánchez Contreras *et al.* 2002; Desmurs *et al.* 2007; Choi *et al.* 2012; Leal-Ferreira *et al.* 2012). Water vapour emission comes from two regions in opposite directions along the nebula axis and presumably the H$_2$O clumps represent the inner nebula, at the base of the bipolar flow. SiO masers occupy smaller regions and lie almost exactly perpendicular to the axis. The movements depicted by the observations of SiO are compatible with a disk orbiting the central star(s) (Sánchez Contreras *et al.* 2002). In principle, we are seeing in this object the whole central structure of disk plus outflow that would confirm our ideas on the post-AGB nebular dynamics. However the astrometric information in these observations is either missing or poor; the SiO observations of Sánchez Contreras *et al.* (2002) and H$_2$O observations of Leal-Ferreira *et al.* (2012) were self-calibrated, so have no absolute positions. The joint observations of H$_2$O and SiO by Desmurs *et al.* (2007) are phase referenced, but the SiO v=2 detection is only 'tentative', and no image nor spectrum is provided.

The SiO transition, because of its high excitation energy, is believed to always mark the location of the central AGB-star (Elitzur 1992). Therefore, whilst it is logical that the SiO/AGB-star is at the center of nebulae, between the H$_2$O clusters, definitive astrometry is required to allow solid conclusions on the disk/outflow association. Efforts over the last decade to provide the registration have been unsuccessful. The main reason for this is that conventional phase referencing at mm-wavelengths is extremely challenging and that the masers weaken as they expand away from the central star. In this paper we present bona-fide astrometric registration of the SiO emission to the positions of the H$_2$O masers using Source Frequency Phase Referencing (SFPR), with the H$_2$O registered to an absolute frame using conventional phase referencing.

2. Observational Details

OH 231 was observed by the Korean VLBI Network (KVN) on 25 Jan, 2017 (N17RD01A) with simultaneous 12mm and 7mm frequencies, each recording 4, dual polarisation, Intermediate Frequency (IF) bands of 16-MHz width. These were spread to cover the H$_2$O and v=2,1 J→0 transitions at 22235.044, 42820.57 and 43122.09-MHz respectively and provide the maximum frequency span compatible with the backend (64-MHz at 22-GHz and 382-MHz at 43-GHz), to provide accurate delay measurements.

The KVN has a truly unique capability: simultaneous frequency phase referencing between bands, which registers the high frequency, mm-wave image against the low frequency image, allowing the measurement of the change of source structure across the frequency bands. Examples demonstrated include core-shifts for AGNs (Rioja *et al.* 2014, 2015, 2017) and spatial relationships between maser transitions (Dodson *et al.* 2014; Yoon *et al.* 2017, for H$_2$O and SiO). We used the latter to derive the relative astrometric separation of the H$_2$O and SiO masers in OH 231.

3. Results and Discussion

We measured the residual delays from J0746-1555 at 22GHz, and the phases from a point source model-fit to the strongest channel of the H$_2$O maser. These were applied to the whole H$_2$O maser dataset. These results were also scaled up and applied to the

SiO maser datasets, following the method presented in Dodson *et al.* (2014), placing the H_2O and SiO masers on a common reference frame. Figure 1 shows the relative phase referenced positions of the detected emission at epoch 2017.07 in J2000. The absolute positions of the strongest features are in Table 1.

Absolute position errors are those for the calibrator, which is 0.3mas. The relative position error for referencing between 22/43-GHz would be dominated by the absolute position error of the H_2O maser position and the fractional bandwidth $\Delta\nu/\nu$, that is 0.15 mas. However the dominant relative position error between SiO features will be that from the beam size over the SNR for the individual spots. For the strongest feature this is $\sim$0.2mas, but for the median spot flux this is $\sim$0.5mas.

Maser Transition	Velocity (km s^{-1})	RA 07:42:s	Dec -14:42:s	Error (μas)
H_2O	27.5–29.5	16.91525	50.02167	40
SiO v=1	35–38	16.915379	50.06995	10
SiO v=2	33–38	16.915371	50.06963	10

Table 1: Absolute positions for strongest integrated maser feature as observed on 2017/01/25, with the velocity range of the feature and fitting errors.

We find, as expected and suggested tentatively in Desmurs *et al.* (2007), that the SiO masers are placed in the center of the H_2O distribution, but slightly shifted to the North with respect to the H_2O centroid. This result is compatible with the general trend of this nebula to show more extended southern lobes, notably in the wide optical and CO images. The SiO+H_2O images (Fig. 1) is amazingly similar to a reduced version of the optical image, scaled down by a factor $\sim$ 500. Our data also confirm the general structure of the SiO-emitting region, Fig. 2: elongated and perpendicular to the nebular axis found at larger scales. We can so confirm the detection of an equatorial torus-like structure placed in the very center of this strongly bipolar nebula. If, as usually assumed, the SiO spots are placed in a region tightly surrounding the late-type star, that star is shown to be accurately placed in the center of the nebula.

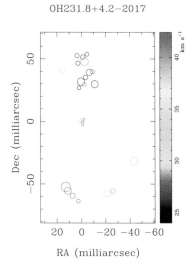

OH231.8+4.2−2017

Figure 1. *Phase Referenced spot map of H_2O, SiO v=1 and v=2 J=1→0 masers in OH 231. The SiO emission is at the centre with the H_2O emission in clusters to the North and South. See Dodson* et al. *(2017) for details.*

We are able to detect emission at $\sim$28 and $\sim$36 km s^{-1} for both the J=1→0, v=1 and 2 masers. The lines themselves are very broad ($\sim$0.4km s^{-1}), and may well be blended because of the KVN resolution, which is 4×3 mas at 43 GHz. However, we did not detect the emission at 40–43 km s^{-1} found by Sánchez Contreras *et al.* (2002) and other authors. The KVN resolution, even at 43 GHz, does not allows a detailed investigation of the structure of the SiO-emitting region. Moreover, the whole emission in our data just occupies about 3 mas (Fig. 2), much smaller than the total region detected by Sánchez Contreras *et al.* (2002),

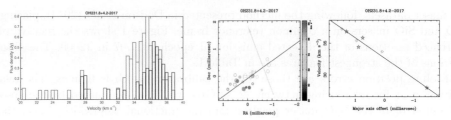

Figure 2. *a) The spectrum of the detected spots (formed from the clean components) in SiO v=1 (blue) and v=2 (red) masers, clipped at 0.05Jy, showing that both transitions have emission around 28 and 36 km s^{-1}. b) The positions of both v=1,2 SiO masers coloured to match the velocity scale on the side bar. All the detected channels that contribute to the spot features are plotted separately, with v=1 in open circles and v=2 in filled circles. The circle size is proportional to the flux. The light blue line marks the expected axis of symmetry, based on the large scale structure. c) the velocity-position plots of the SiO maser emission at J=1→0, v=1 (green cross) and 2 (red star). The enclosed mass, assuming that all emission is detected, would be 0.05 M$_\odot$.*

~ 8 mas. It is obvious that only a fraction of the torus found by those authors is detected in our data. We only can conclude that our observations are not incompatible with their model torus.

These issues, and measurements of the temporal development of this iconic pPN will be addressed with future deeper, higher resolution, observations with the VLBA, also using SFPR.

References

Choi, Y. K., Brunthaler, A., Menten, K. M., & Reid, M. J. 2012 (July). *Pages 407–410 of:* Booth, R. S., Vlemmings, W. H. T., & Humphreys, E. M. L. (eds), *Cosmic Masers - from OH to H0*. IAU Symposium, vol. 287.

Cohen, M. 1981. *PASP*, **93**(June), 288–290.

Desmurs, J.-F., Alcolea, J., Bujarrabal, V., Sánchez Contreras, C., & Colomer, F. 2007. *A&A*, **468**(June), 189–192.

Dodson, R., *et al.*, 2014. *AJ*, **148**(Nov.), 97.

Dodson, R., Rioja, M. J., Bujarrabal, V., Kim, J., Cho, S. H., Choi, Y. K. & Youngjoo, Y., 2017. *MNRAS, Submitted*

Elitzur, M. (ed). 1992. *Astronomical masers*. Astrophysics and Space Science Library, vol. 170.

Leal-Ferreira, M. L., Vlemmings, W. H. T., Diamond, P. J., Kemball, A., Amiri, N., & Desmurs, J.-F. 2012. *A&A*, **540**(Apr.), A42.

Rioja, M., Dodson, R., Gómez, J., Molina, S., Jung, T., & Sohn, B. 2017. *Galaxies*, **5**(Jan.), 9.

Rioja, M. J., 2014. *AJ*, **148**(Nov.), 84.

Rioja, M. J., Dodson, R., Jung, T., & Sohn, B. W. 2015. *AJ*, **150**(Dec.), 202.

Sánchez Contreras, C., Desmurs, J. F., Bujarrabal, V., Alcolea, J., & Colomer, F. 2002. Submilliarcsecond-resolution mapping of the 43 GHz SiO maser emission in the bipolar post-AGB nebula OH231.8+4.2. *A&A*, **385**(Apr.), L1–L4.

Sánchez Contreras, C., Gil de Paz, A., & Sahai, R. 2004. *ApJ*, **616**(Nov.), 519–524.

Yoon, D. H., Cho, S. H., Yun, Y. J., Choi, Y.H, Dodson, R., Rioja, M., Kim, J., Kim, D., Yang, H., H., Imai, & Byun, D. Y. 2017. *Nature, Submitted*.

Astrophysical Masers:
Unlocking the Mysteries of the Universe
Proceedings IAU Symposium No. 336, 2017
A. Tarchi, M.J. Reid & P. Castangia, eds.

© International Astronomical Union 2018
doi:10.1017/S1743921317009577

Water masers as signposts of extremely young planetary nebulae

José F. Gómez[1], Luis F. Miranda[1], Lucero Uscanga[2] and Olga Suárez[3]

[1]Instituto de Astrofísica de Andalucía, CSIC
Glorieta de la Astronomía s/n, 18008 Granada, Spain
email: jfg@iaa.es

[2]Departamento de Astronomía, Universidad de Guanajuato
A.P. 144, 36000 Guanajuato, Gto., Mexico

[3]Université Côte d'Azur, OCA, CNRS, Laboratoire Lagrange, F-06304 Nice, France

Abstract. Only five planetary nebulae (PNe) have been confirmed to emit water masers. They seem to be very young PNe. The water emission in these objects preferentially traces circumstellar toroids, although in K 3-35 and IRAS 15103-5754, it may also trace collimated jets. We present water maser observations of these two sources at different epochs. The water maser distribution changes on timescales of months to a few years. We speculate that these changes may be due to the variation of the underlying radio continuum emission, which is amplified by the maser process in the foreground material.

Keywords. masers, stars: AGB and post-AGB, planetary nebulae: general

1. Introduction

Planetary nebulae (PNe) represent one of the last stages of evolution of low and intermediate mass stars ($\simeq 0.8 - 8$ M$_\odot$). The PN phase takes place after the Asymptotic Giant Branch (AGB) and a short ($10^2 - 10^4$ yr) transitional post-AGB phase. During the post-AGB phase the central start contracts and increases its temperature until it is hot enough to photoionize the circumstellar envelope expelled in previous evolutionary stages. This photoionization marks the entrance of a source to the PN phase.

Maser emission of different molecules (e.g., OH, H$_2$O, SiO) is widespread in oxygen-rich AGB stars (Lewis 1989), mainly tracing the (roughly spherical) expansion of the circumstellar envelope. Masers tracing collimated jets are observed in post-AGB stars, as in the case of "water fountain" stars (Imai 2007). However, maser emission is rare in PNe. So far, only 7 and 5 sources have been confirmed to harbor OH emission (Uscanga et al. 2012, Qiao et al. 2016) and H$_2$O masers (Miranda et al. 2001, de Gregorio-Monsalvo et al. 2004, Gómez et al. 2008, Uscanga et al. 2014, Gómez et al. 2015), respectively. No SiO maser has ever been detected in a PN.

In the particular case of water masers, those seen in AGB stars are expected to fade out in 100 yr after the end that phase (Lewis 1989). This timescale is shorter than the duration of the post-AGB phase. Therefore, the detected water masers in PNe are not the remnant of those pumped during the AGB, but are related to later outflow events. The scarce number of water-maser-emitting PNe (H$_2$O-PNe) and their tendency to be optically obscured (Suárez et al. 2009), strongly suggest that this emission is produced only during the short time period in the early stages of PN evolution.

Photoionization in PNe is a key differential factor with respect to sources in previous evolutionary phases that, in turn, can have a fundamental effect on maser emission. The presence of ionized material implies the emission of free-free radiation at radio wavelenghts. This emission can be strong, and favors the presence of maser emission since it acts as a background that can be amplified by foreground parcels of gas with inverted populations. Thus, maser emission can be present in PNe under physical conditions that would not produce it in the AGB or post-AGB phases. This complicates the interpretation of the morphology and kinematics of maser emission in PNe, since maser spots would tend to be distributed with the morphology of the background ionized region, but with the kinematics of the foreground molecular gas. Moreover, radio continuum emission in PNe is expected to significantly vary, as the photoionization proceeds along the envelope. This implies variation in flux density and distribution of the observed maser emission.

All H_2O-PNe have clear bipolar morphologies in radio, optical, and or/infrared images (Miranda *et al.* 2001, Gómez *et al.* 2008, Lagadec *et al.* 2011, Uscanga *et al.* 2014). Moreover, water masers in these sources tend to trace equatorial (toroidal) structures (with the noticeable exception of the two sources discussed below). This suggest that these particular sources are the result of the evolution of binary systems.

2. Monitoring water maser emission in PNe

We have observed water maser emission at 22 GHz in several H_2O-PNe over different epochs, using radio interferometers. Here we present some results for two of them.

2.1. *IRAS 15103-5754*

This is probably the youngest PN known. It is the only one whose water maser emission is spread over a large velocity range ($\simeq$ 75 km s^{-1}, Suárez *et al.* 2009, Gómez *et al.* 2015), significantly larger than expected from the expansion of a circumstellar envelope. Therefore, it is the only water fountain that has already entered the PN phase. Moreover, it is the first PN in which non-thermal radio continuum emission has been confirmed (Suárez *et al.* 2015). This radio continuum emission shows a significant variability both in flux density and spectral index, and it has been interpreted as synchrotron emission being suppressed by a growing ionized region.

The maser distribution observed in 2010-2011 (Gómez *et al.* 2015) was dominated by a blueshifted jet to the northeast, nearly aligned with the infrared nebula (Lagadec *et al.* 2011). It showed a linear velocity gradient with highest velocities farther away from the star, which suggest an explosive collimated mass-loss event.

We have monitored the water maser emission in IRAS 15103-5754 with the Australia Telescope Compact Array (ATCA). The spatial and kinematical distribution changes significantly in timescales < 1 year (Fig. 1). While it shows an elongated distribution, suggestive of a jet, its orientation changes gradually with time. It is still roughly aligned with the infrared nebula in August 2014, but is almost perpendicular to it in May 2016. This could indicate a large precession of a jet, but the well-defined orientation of the innermost region of the infrared nebula does not seem consistent with a such a precessing jet. We speculate that the changes in the emission are due to changes in the background radio continuum emission. In this scenario, the radio continuum would trace shocks in a collimated jet in the first epochs. A growing ionization of a circumstellar torus would induce an increasing contribution of free-free emission in the equatorial direction. This change in background continuum would produce a change in the masing areas of the foreground surrounding medium. A continuing monitoring of the maser emission would help to ascertain whether we are witnessing the transition of maser distribution from tracing a jet (as in post-AGB water fountains) to a toroid (as in other water-maser-emitting PNe).

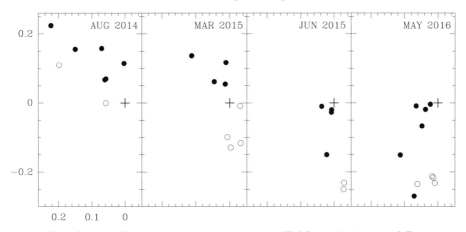

Figure 1. Distribution of water maser components in IRAS 15103-5754 in different epochs. Filled and open circles represent blue- and redshifted components, with respect to the central LSR velocity of the system (-33 km s^{-1}, Gómez *et al.* 2017). The cross marks the centre of the radio continuum emission at 22 GHz. Axis coordinates are offsets (in arcseconds) with respect to the radio continuum position.

2.2. *K 3-35*

K 3-35 was the first PN in which the presence of water maser emission was confirmed (Miranda *et al.* 2001). This emission traces an equatorial torus in the central region, similar to other H$_2$O-PNe. However, some maser components are located at the tip of the ionized nebula traced by the radio continuum emission.

The water masers also show an interesting short-time evolution, which we have studied with observations with the Very Large Array (VLA). Regarding the masers at the central core of the object (Fig. 2, left), those observed in 1999 can be fitted to an expanding and rotating toroid, with velocities 1.4 and 3.1 km s^{-1}, respectively (Uscanga *et al.* 2008). However, later observations show a distribution that is not consistent with the masers tracing the same spatial and kinematical structure (de Gregorio-Monsalvo *et al.* 2004). Our newest VLA observations (carried out in 2015) still show an equatorial structure, but extending over a larger extent ($0.07'' = 290$ AU, at a distance of 3.9 kpc, Tafoya *et al.*2011) than expected from the expansion velocity determined by Uscanga *et al.* (2008). A possible explanation is that the changes of the maser distribution are not due to motions of the masers themselves, but to changes in the excitation conditions in the gas and/or in the background continuum. In this scenario, the masers could be tracing shocked regions progressively farther away from the central star, as the ionization front progresses along a circumstellar toroid. Alternatively, it could be foreground gas that produces detectable maser emission as it amplifies a growing ionized region in a toroid.

In addition to these central water masers, the observations in 1999 revealed some components on both sides of the bipolar nebula, at the tips of the radio continuum emission, which coincide with bright optical knots in optical images. This bipolar maser distribution did not appear in the observations in 2002, nor in the VLBI observations of Tafoya *et al.* (2011). However, it reappeared in our latest 2015 observations (Fig. 2, right), at positions close to, but not exactly coincident with those seen in 1999. The most straightforward explanation would be that these masers are tracing shocks at the tips of a bipolar jet. However, a proper interpretation depends on an accurate determination of the velocity of the central star. As discussed in Qiao *et al.* (2016), it is unclear whether the central LSR velocity is 23 or 10 km s^{-1}, although these authors favored the later, which is

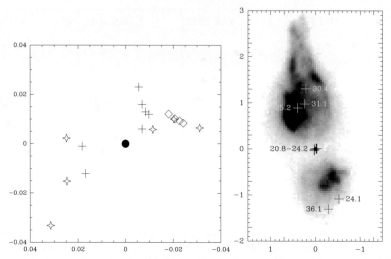

Figure 2. Water maser distribution in K 3-35. Left: maser components in 1999 (crosses, Miranda *et al.* 2001), 2002 (squares, de Gregorio-Monsalvo *et al.* 2004) and 2015 (starred polygons) at the central part of the nebula. The filled circle represents the center of the radio continuum emission. Right: Water maser components in K 3-35 (crosses) observed in 2015, overlayed on a Hubble Space Telescope Image in the F658N filter (which mainly covers the [NII] 6583 Å emission line). The labels represent the LSR velocity of the maser components. Axis coordinates in both panels are offsets (in arcseconds) with respect to the radio continuum position.

based on optical spectroscopy (Miranda *et al.* 2000). If this is the case, all water masers in K 3-35 are redshifted, which challenges any easy explanation. Assuming a velocity of 23 km s^{-1} seems easier to reconcile with the central structure being a circumstellar toroid, but the masers at the lobes are still redshifted on both sides of the nebula. A possible explanation could be that the masers at the lobes are tracing shocks of a wind with a large opening angle, and that only the rear side of the wind is exciting masers, due to an asymmetric density distribution in the circumstellar gas.

These results suggest that short-time variations of the distribution of water masers may be a specific differential characteristic of H$_2$O-PNe.

References

de Gregorio-Monsalvo I., Gómez Y., Anglada G., *et al.* 2004, *ApJ*, 601, 921
Gómez, J. F., Suárez, O., Gómez, Y., *et al.* 2008, *AJ*, 135, 2074
Gómez, J. F., Suárez, O., Bendjoya, Ph., *et al.* 2015, *ApJ*, 799, 186
Gómez, J. F., *et al.* 2017, in preparation.
Imai, H., 2007, IAU Symp 242, 279
Lagadec, E., Verhoelst, T., Mékania, D., *et al.* 2011, *MNRAS*, 417, 32
Lewis, B. M. 1989, *ApJ*, 338, 234
Miranda, L. F., Fernández, M., & Alcalá, J. M.. 2000, *MNRAS*, 311, 748
Miranda, L. F., Gómez, Y., Anglada, G., & Torrelles, J. M. 2001, *Nature*, 414, 284
Qiao, H. H., Walsh, A. J., Gómez, J. F., *et al.* 2016, *ApJ*, 817, 37
Suárez, O., Gómez, J. F., Miranda, L. F., *et al.* 2009, *A&A*, 505, 217
Suárez, O., Gómez, J. F., Bendjoya, P., *et al.* 2015, *ApJ*, 806, 105
Tafoya, D., Imai, H., & Gómez, Y. 2011, *PASJ*, 63, 71
Uscanga, L., Gómez, Y., Raga, A. C., *et al.* 2008, *MNRAS*, 390, 1127
Uscanga, L., Gómez, J. F., Suárez, O., & Miranda, L. F. 2012, *A&A*, 547, A40
Uscanga, L., Gómez, J. F., Miranda, L. F., *et al.* 2014, *MNRAS*, 444, 217

Astrophysical Masers:
Unlocking the Mysteries of the Universe
Proceedings IAU Symposium No. 336, 2017
A. Tarchi, M.J. Reid & P. Castangia, eds.

© International Astronomical Union 2018
doi:10.1017/S1743921317009826

Distances of Stars by mean of the Phase-lag Method

Sandra Etoka[1,2], Dieter Engels[2], Eric Gérard[3] and Anita M. S. Richards[1]

[1] Jodrell Bank Centre for Astrophysics, University of Manchester, UK
email: Sandra.Etoka@googlemail.com

[2] Hamburger Sternwarte, Universität Hamburg, Germany

[3] GEPI, Observatoire de Paris-Meudon, France

Abstract. Variable OH/IR stars are Asymptotic Giant Branch (AGB) stars with an optically thick circumstellar envelope that emit strong OH 1612 MHz emission. They are commonly observed throughout the Galaxy but also in the LMC and SMC. Hence, the precise inference of the distances of these stars will ultimately result in better constraints on their mass range in different metallicity environments. Through a multi-year long-term monitoring program at the Nancay Radio telescope (NRT) and a complementary high-sensitivity mapping campaign at the eMERLIN and JVLA to measure precisely the angular diameter of the envelopes, we have been re-exploring distance determination through the phase-lag method for a sample of stars, in order to refine the poorly-constrained distances of some and infer the currently unknown distances of others. We present here an update of this project.

Keywords. masers, stars: late-type, stars: variables: OH/IR, stars: distances

1. Introduction

Evolved stars at the tip of the AGB for low- and intermediary-mass stars experience heavy mass loss surrounding the star with a circumstellar envelope (CSE), which ultimately become opaque to visible light. These enshrouded OH/IR stars commonly exhibit strong periodic (ranging typically from 1 to 6 yr) ground-state OH maser emission in the 1612-MHz transition. Over 2000 OH masers of stellar origin are currently known in the Milky Way (Engels & Bunzel, 2015) and it is anticipated that the SKA will detect thousands of OH maser sources of stellar origin in the anti-solar Galactic hemisphere and Local Group of galaxies (Etoka *et al.* 2015). This makes OH/IR stars potentially valuable objects for a wide range of studies in our Galaxy but also for stellar-evolution metallicity-related studies.

Because OH/IR stars are optically thick, their distances cannot be inferred using optical parallaxes. The period-luminosity relation found towards Miras (Whitelock, Feast & Catchpole 1991) breaks down for P> 450 days. Kinematic distances can be very imprecise due to peculiar motions (Reid *et al.* 2009). As it has been extensively demonstrated in this symposium, maser emission at high(er) frequency from e.g. water and methanol species is successfully used to infer distances via parallax measurements towards distant Galactic star forming regions. The use of ground-state OH maser to infer distances of AGB stars via parallax measurements has also been successfully done but only for objects relatively nearby (i.e., ⩽2 kpc; Vlemmings & van Langevelde 2007; Orosz *et al.* 2017). Another alternative to distance determination for more distant evolved stars is the use of the "phase-lag" method.

2. Method and Observations

The determination of the distance of an OH/IR star via the phase-lag method relies on the measurement of the linear and angular diameter of its OH-maser CSE which are both obtained independently. OH/IR stars typically exhibit a double-peaked spectral profile where the blueshifted peak ("blue" peak here after) emanates from the front cap of the CSE and the redshifted peak ("red" peak here after) emanates from the back cap of the CSE while the faint interpeak emission emanates from the outer part of the CSE. We measure the phase lag (τ_0) of a source with no external fitting function, using simply the shape of the light curve, by scaling and shifting the integrated-flux light curves of the blue peak F_b with respect to the red one F_r, minimizing the function $\Delta F = F_r(t) - a \cdot F_b(t - \tau_0) + c$ (where a and c are constants for the amplitude and mean flux) leading to the measurement of the linear diameter of the OH shell of the star. The angular diameter is obtained from interferometric mapping.

Schultz, Sherwood & Winnberg (1978) performed the first phase-lag measurements, and in the 1980's, Herman & Habing (1985) and van Langevelde, van der Heiden & van Schooneveld (1990) explored this method to retrieve distances from OH/IR stars, but there are discrepancies in the phase-lag measurement of a good fraction of the sources in common in these 2 works. In an attempt to constrain the distance uncertainties achievable with this method we are re-exploring it. Our sample consists of 20 OH/IR stars that we have been monitoring with the Nançay Radio Telescope (NRT) in order to measure their phase-lags. About half of the sample is composed of sources for which phase-lags were determined in the 1980's, the ones for which both works are in agreement serving as benchmark objects while for the objects with clear discrepancy the aim being re-determination of their phase-lag. The rest of the sample consists of objects with no recorded phase-lag measurements. About half of the sources of the sample have been previously imaged but the interferometric observations were taken at a random time and/or with poor sensitivity. We are currently in a process of imaging all the sources in the sample with either eMERLIN or JVLA around the OH maxima of each source, as predicted from the NRT light curves, in order to improve the angular diameter determination by detecting the faint interpeak signal and better constrain the shell (a)symmetries.

Past reports of the method and status of the project were presented in Engels *et al.* (2012, 2015) while a detailed description of the applicability of the method in measuring distances for objects beyond the solar vicinity can be found in Etoka *et al.* (2014). In the next section an update and discussion based on the results obtained so far is presented.

3. Discussion

Table 1 presents the summary of the results obtained so far. The first half of the table gives the range of periods, phase-lags and linear diameters inferred from the NRT monitoring from all the sources of the sample. The second half of the table gives the range of angular diameters and inferred phase-lag distances. The phase-lags measured account for linear shell diameter of ∼1700 to ∼19000 AU. Generally, the diameter is larger for longer-period objects. Distances ranges from 0.5 to 10.6 kpc. But, it has to be noted, that at the time of writing, although 70% of the sources of the sample have a measured angular diameters, 60% are from the literature including these 2 extremas.

Figure 1 presents the NRT monitoring and interferometric mapping for OH 83.4-0.9 and OH 16.1-0.3, two objects of the sample for which there was no previous imaging. We mapped both objects around their OH maximum with eMERLIN and the JVLA

Table 1. Summary of the periods, phase-lags, linear & angular diameters and distances inferred for the entire sample

		min.	max.
P^a	[yrs]:	1.16	6.05
$\tau_0{}^a$	[days]:	< 10	110
$2\,R_{OH}{}^a$	[10^3 AU]:	< 1.7	19
ϕ	["]:	0.8^b	8.0^b
D	[kpc]:	0.5^b	10.6^b

Table 2. Results for OH 83.4-0.9 & OH 16.1-0.3

Object	P [yrs]	τ_0 [days]	$2\,R_{OH}$ [10^3 AU]	ϕ ["]	D [kpc]
OH 83.4-0.9	4.11	30	5.2	~1.8	~3.0
OH 16.1-0.3	6.03	110	19.0	~3.5	~5.5

a: All the periods, phase lags and corresponding linear diameters are inferred from our NRT monitoring. The status of which can be followed here: http://www.hs.uni-hamburg.de/nrt-monitoring
b: from the literature

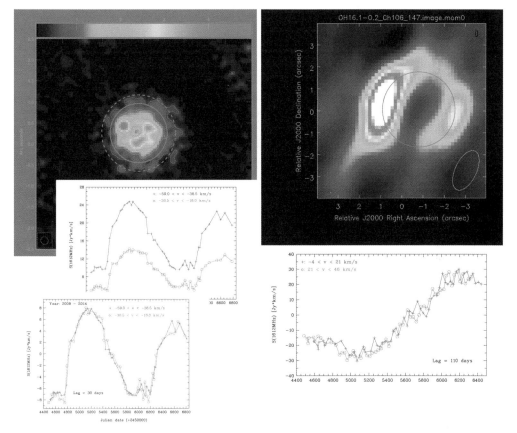

Figure 1. *Left main-panel:* eMERLIN map of the integrated emission over the inner part of the spectrum of OH 83.4-0.9 covering the velocity range $[-55; -23]$ km s^{-1}. The magenta full and dotted circles delineate the outer part of the shell. *main-panel insert:* spectrum. *middle:* "raw" blue-peak and red-peak light curves of OH 83.4-0.9 obtained from the NRT monitoring over a period of 7 years. *bottom:* scaled and shifted light curves for phase-lag determination. *Right main-panel:* JVLA map of the integrated emission over the red inner part of the spectrum of OH 16.1-0.3 covering the velocity range $[+27; +37]$ km s^{-1}. The magenta circle presents the best fit of the projected diameter of the shell. *main-panel insert:* Spectrum. *bottom:* scaled and shifted light curves for phase-lag determination.

respectively, which allowed us to retrieve a substantial amount of the faint interpeak emission. For both objects, the channel maps obtained are in agreement with the shells being spherically-thin in uniform radial expansion. As an illustration of the phase-lag determination method explained in Section 2, the left middle- and bottom-panels show the "raw" blue-peak and red-peak light curves of OH 83.4-0.9 and the scaled and shifted light curves leading to the phase-lag measurement. The period, phase lags and corresponding linear & angular diameters measured and subsequent inferred distances for these 2 objects are summarized in Table 2. For these 2 objects, we estimated the uncertainty of the linear diameter to be within 10%, while that of the angular diameter to be within 15%, leading to a distance determination uncertainty of less than 20%. On the other hand, the distance determinations given in Table 1 are still questionable as strongly dependent on the degree of exploration for the shell extent determination, i.e., faint tangential emission, which not only allow to better constrain the actual extent of the shell, but also its actual geometry, as in particular, a strong deviation from a spherically thin shell in uniform radial expansion can lead to distance uncertainty greater than 20% (Etoka & Diamond, 2010).

4. Closing Notes

The main contribution to the early phase-lag inconsistencies could be due to: incomplete coverage of lightcurves; inhomogeneous sampling; use of analytical functions to fit the lightcurves. Phase-lag distances can be determined with an uncertainty of less than 20%, provided that a good constraint on both the linear and angular diameter determinations can be achieved. The main factors for doing so are the following:

- the shape of the light curves must be well defined. This can be obtained with high cadence monitoring observations (i.e., typically with 0.03 P);
- the light curves cover more than one period;
- the faint tangential emission tracing the actual full extent of the shell can be imaged via high-sensitivity interferometric observations better retrieved around the maximum of the OH cycle;
- significant shell asymmetries can be excluded or modelled.

References

Engels, D. & Bunzel, F. 2015 *A&A*, 582A, 68
Engels, D., Etoka, S., Gérard, E., & Richards, A. M. S. 2015, *ASPC*, 497, 473
Engels, D., Gérard, E., & Hallet, N., 2012, *IAUS* 287, 254
Etoka, S., Engels, D., Imai, H. *et al.* 2015, *Proc. Science*, (AASKA14), 125
Etoka, S., Engels, D., Gérard, E., & Richards A. M. S. 2014, *evn conf*, 59
Etoka, S. & Diamond, P. D. 2010, *MNRAS*, 406, 2218
Herman, J. & Habing, H. J. 1985, *A&AS*, 59, 523
Orosz, G., Imai, H., Dodson, R. *et al.* 2017, *AJ*, 153, 119
Reid, M. J., Menten, K. M., Zheng, X. W. *et al.* 2009, *ApJ*, 700, 137
Schultz, G. V., Sherwood, W. A., & Winnberg, A. 1978, *A&A*, 63L, 5
van Langevelde, H. J., van der Heiden, R., & van Schooneveld, C. 1990, *A&A*, 239, 193
Vlemmings, W. H. T.. & van Langevelde, H. J. 2007, *A&A*, 472, 547
Whitelock, P., Feast, M., & Catchpole, R. 1991, *MNRAS*, 248, 276

Astrophysical Masers:
Unlocking the Mysteries of the Universe
Proceedings IAU Symposium No. 336, 2017
A. Tarchi, M.J. Reid & P. Castangia, eds.

© International Astronomical Union 2018
doi:10.1017/S1743921317010262

Excited OH Masers in Late-Type Stellar Objects

A. Strack[1], E. D. Araya[1], M. E. Lebrón[2], R. F. Minchin[3], H. G. Arce[4], T. Ghosh[3], P. Hofner[5,6], S. Kurtz[7], L. Olmi[8], Y. Pihlström[9,6] and C. J. Salter[3]

[1] Physics Department, Western Illinois University, 1 University Circle, Macomb, IL 61455, USA.
[2] University of Puerto Rico at Rio Piedras, San Juan, PR 00931, USA.
[3] Arecibo Observatory, NAIC, HC03 Box 53995, Arecibo, PR 00612, USA.
[4] Department of Astronomy, Yale University, New Haven, CT 06511, USA.
[5] Physics Department, New Mexico Institute of Mining and Technology, 801 Leroy Place, Socorro, NM 87801, USA.
[6] National Radio Astronomy Observatory, 1003 Lopezville Road, Socorro, NM 87801, USA.
[7] Instituto de Radioastronomía y Astrofísica, Universidad Nacional Autónoma de México, Morelia 58090, Mexico.
[8] INAF, Osservatorio Astrofisico di Arcetri, Largo E. Fermi 5, I-50125 Firenze, Italy.
[9] The Department of Physics and Astronomy, The University of New Mexico, Albuquerque, NM 87131, USA.

Abstract. The final stages of low-mass stellar evolution are characterized by significant mass loss due to stellar pulsations during the AGB phase, which lead to the development of planetary nebulae. Molecular masers of H_2O, SiO, and ground state OH transitions are commonly detected in oxygen-rich late-type stars (OH/IR objects). In contrast, *excited* OH maser transitions are rare. We discuss our study of the carbon-rich pre-planetary nebula CRL 618 (a prototypical post-AGB star). Observations conducted in May 2008 with the 305m Arecibo Telescope resulted in the first detection of a 4765 MHz OH maser line in a late-type stellar object; the detection was confirmed a few months later also with Arecibo. Subsequent observations in 2015 and 2017 resulted in non-detection of the 4765 MHz OH line. Our observations indicate that the 4765 MHz OH maser in CRL 618 is highly variable, possibly tracing a short-lived phenomenon during the development of a pre-planetary nebula.

Keywords. masers, stars: AGB and post-AGB, circumstellar matter, stars: individual (CRL 618)

1. Introduction and Observations

Late-type solar-like stars evolve from asymptotic giant branch (AGB) to the planetary nebula phase. The transition between AGB stars and planetary nebulae is important in the development of asymmetries observed in many planetary nebulae. This phase of evolution is known as pre-planetary nebula (PPN, also known as post-AGB stars).

We observed the PPN CRL 618 to investigate the presence of excited OH masers. The observations were conducted with the 305m Arecibo Telescope in 2008, 2015 and 2017. In addition to the 4765 MHz OH line, we searched for emission/absorption of all other OH transitions between 1 and 9 GHz in October 2008 and 2015.

2. Results and Discussion

An excited 4765 MHz OH emission line was detected in May and October 2008. The 4765 MHz OH emission line was not detected in 2015 or 2017 (Figure 1). No other OH

385

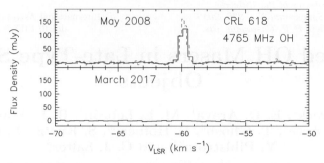

Figure 1. Example spectra from two epochs, first detection (upper panel; see observation details in Araya *et al.* 2015) and 2017 (lower panel) non-detection. The upper panel shows a high spectral resolution spectrum (blue-dashed) and the smoothed spectrum (black-solid).

transition was detected at rms levels of $\lesssim 5$ mJy. Excited OH transitions in late-type stellar objects are extremely rare. Before this work, the only confirmed excited OH masers were toward the young planetary nebulae Vy2-2 and K3-35 (6035 MHz OH, Desmurs *et al.* 2010). Unconfirmed detections include the 4750 MHz OH line in the Mira star AU Gem, and the 6030 and 6035 MHz OH lines in the hypergiant NML Cygni (Zuckerman *et al.* 1972, Claussen & Fix 1981, Jewell *et al.* 1985, Sjouwerman *et al.* 2007). In CRL 618, the velocity difference between the 4765 MHz line (-60 km s^{-1}) and the systemic velocity (-21.5 km s^{-1}, Sánchez Contreras *et al.* 2004) suggests that the OH maser is associated with the bipolar outflow. The production of the OH could be from photo-dissociation of H_2O (e.g., Netzer & Knapp 1987). In K3-35, the OH maser likely occupies the same region as the H_2O masers and the masers have similar velocities (Miranda *et al.* 2011).

Acknowledgements

This work has made use of the computational facilities donated by Frank Rodeffer to the WIU Astrophysics Research Laboratory. A.S. acknowledges support from the WIU College of Arts and Sciences, a M. & C. Wong RISE Travel Grant, and support from this conference. The Arecibo Observatory is operated by SRI International under a cooperative agreement with the National Science Foundation (AST-1100968), and in alliance with Ana G. Méndez-Universidad Metropolitana, and the Universities Space Research Association.

References

Araya, E. D., Olmi, L., Morales Ortiz, J., *et al.* (2015), *ApJS*, 221, 10
Claussen, M. J., & Fix, J. D., (1981), *ApJ*, 250, L77
Desmurs, J.-F., Baudry, A., Sivagnanam, P., Henkel, C., Richards, A. M. S., & Bains, I., (2010), *A&A*, 520, A45
Jewell, P. R., Schenewerk, M. S. & Snyder, L. E., (1985), *ApJ*, 295, 183
Miranda, L. F., Suárez, O, & Gómez, J. F., (2011), *arXiv*:1101.2837
Netzer, N., & Knapp, G. R., (1987), *ApJ*, 323, 734
Sánchez Contreras, C., Bujarrabal, V., Castro-Carrizo, A., Alcolea, J., & Sargent, A., (2004), *ApJ*, 617, 1142
Sjouwerman, L. O., Fish, V. L., Claussen, M. J., Pihlström, Y. M. & Zschaechner, L. K., (2007), *ApJ*, 666, 101
Zuckerman, B., Yen, J. L., Gottlieb, C. A., & Palmer, P., (1972), *ApJ*, 177, 59

Astrophysical Masers:
Unlocking the Mysteries of the Universe
Proceedings IAU Symposium No. 336, 2017
A. Tarchi, M.J. Reid & P. Castangia, eds.

© International Astronomical Union 2018
doi:10.1017/S1743921317009553

Missing flux in VLBI observations of SiO maser at 7 mm in IRC+10011

J.-F. Desmurs[1], J. Alcolea[1], V. Bujarrabal[1], F. Colomer[1,2] and R. Soria-Ruiz[1]

[1] Observatorio Astronómico Nacional (OAN/IGN), Spain

[2] JIVE, The Netherlands

Abstract. VLBI observations of SiO masers recover at most 40-50% of the total flux obtained by single dish observations at any spectral channel. Some previous studies seems to indicate that, at least, part of the lost flux is divided up into many weak components rather than in a large resolved emission area. Taking benefit of the high sensitivity and resolution of the HSA, we investigate the problem of the missing flux in VLBI observations of SiO maser emission at 7 mm in the AGB stars and obtain a high dynamic range map of IRC+10011. We conclude that the missing flux is mostly contained in many very weak maser components.

Keywords. stars: AGB and post-AGB, instrumentation: interferometers, masers.

1. Introduction

VLBI observations of SiO masers are providing extremely valuable information on the inner circumstellar shells around AGB stars, the regions where dust grains are not yet formed and mass ejection originates, after a complex pulsational dynamics. These data are also very useful to understand the pumping mechanisms responsible for this widespread emission in AGB envelopes. The J=1–0 maser lines (in the v=1 and v=2 vibrationally excited states), at 7 mm wavelength, systematically yield ring-like flux distributions, with diameters of about 10^{14} cm (equivalent to a few stellar radii, see Diamond *et al.* 1994, Desmurs *et al.* 2000).

One of the main problems that persists in the study of the circumstellar SiO masers is the significant amount of flux lost when long baseline interferometry observations are performed. For 7 mm lines, up to about one half of the line emission is usually lost, as it is also the case at 3 mm (see Colomer *et al.* 2017), a problem that is not present in VLA observations. This missing flux could be due to over-resolution, i.e. when the emission is produced on larger scales than those corresponding to the shortest projected baselines of the array. However, another possible explanation could be that this missing flux, or at least part of it, consists of a multitude of compact but weak undetected maser components (at the noise level of the resulting map).

2. Observations

To check if part of the missing flux is contained in many very weak maser components (see Soria-Ruiz *et al.* 2004) or not, we took advantage of HSA capabilities at 7 mm that give a better UV-coverage, higher sensitivity and higher resolution. We observed in dual circular polarization with a velocity resolution (i.e. channel width) of 0.2 km/s and a total velocity coverage of about 55 km/s. Using all VLBA antennas, the VLA, the GBT and Effelsberg, we obtained maps of IRC+10011 of the two ^{28}SiO transitions v=1 and v=2, J=1–0 with a high spatial resolution and a high dynamic range (see Fig. 1).

388 J.-F. Desmurs *et al.*

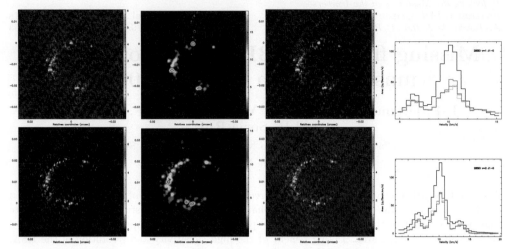

Figure 1. *Top* ^{28}SiO v=1, J=1–0 (at 43.122 GHz) transition and *Bottom* ^{28}SiO v=2, J=1–0 (at 42.820 GHz) transition. From left to right, maps of the two maser transitions obtained respectively with, case A, the full HSA array with full sensitivity and spatial resolution (baselines up to ∼10500 km, restoring beam 0.2 mas), case B, with a subset of antennas forming a very sensitive short array (with baselines <2500 km) and, case C, with all HSA antennas (and full sensitivity) but with a degraded restoring beam of 1 mas (low resolution). At right, flux density comparison between the autocorrelation flux intensity of the reference antenna used for the flux calibration (black line) and the integrated flux recovered in the maps in case A (red line), in case B (green line), and case C (blue line).

3. Preliminary results

Our preliminary results tend to show very similar results for the proportion of missing flux measured in these observations and in previous works. About half of the flux is still missing! The high sensitivity, we reach an rms of about 5 mJy/beam per channel for ^{28}SiO v=2, J=1–0, and high resolution of HSA (∼0.2 mas) do not allow us to significantly recover a higher percentage of flux. Moreover, either using the full spatial resolution of HSA with baseline of up to 10500 km or a compact array with baselines shorter than 2500 km (including short baseline highly sensitive VLBA-PT/VLA), do not significantly change this result. Even degrading the resolution (using a restoring beam 5 times larger), the small flux increase measured in a couple of channels (@ 10-11 km/s) for the v=1 map and corresponding to the arc like structure seen on the east side of the map is not significant, there is no difference in the recovering flux. Our main idea to explain these results is that the missing flux must be spread in a multitude of weak components undetected in our observations.

References

Colomer, F., Desmurs, J.-F., Bujarrabal, V. *et al.* 2017, *HSA IX, proc. SEA held on jul 18-22, 2016 in Bilbao*, p361-366
Desmurs, J.-F., Bujarrabal V., Lindqvist M., *et al.* 2000, *A&A*, 565, 127
Diamond, P. J., Kemball, A. J., Junor, W., *et al.* 1994, *ApJ*, 430, L61
Soria-Ruiz, R., Alcolea, J., Colomer, F. *et al.* 2004, *A&A*, 426, 131

Astrophysical Masers:
Unlocking the Mysteries of the Universe
Proceedings IAU Symposium No. 336, 2017
A. Tarchi, M.J. Reid & P. Castangia, eds.

© International Astronomical Union 2018
doi:10.1017/S1743921318000339

OH masers as probes: How does the variability fade away during the AGB - post-AGB transition?

D. Engels[1], S. Etoka[2], M. West[3] and E. Gérard[4]

[1]Hamburger Sternwarte, Universität Hamburg, Germany,
email: dengels@hs.uni-hamburg.de
[2]Jodrell Bank Centre for Astrophysics, University of Manchester, UK,
email: sandra.etoka@googlemail.com
[3]Hartebeesthoek Radio Astronomy Observatory, South Africa,
email: marion@hartrao.ac.za
[4]GEPI, Observatoire de Paris, Meudon, France,
email: eric.gerard@obspm.fr

Abstract. We are currently performing a monitoring program of the 1612 MHz OH maser emission of several dozen Galactic disk OH/IR stars with the Nancay Radio Telescope (NRT). They are complemented by several OH/IR stars toward the Galactic center, which were monitored with the Hartebeesthoek radio telescope. We use the maser variations to probe the underlying stellar variability. As early monitoring programs already have shown, some stars are large amplitude variables with periods up to 7 years, others show small or even no amplitude variations. This dichotomy in the variability behaviour is assumed to mark the border between the AGB and the post-AGB stages. With the current program, we wish to find objects in transition and to describe their variability properties. We consider the fading out of pulsations with steadily declining amplitudes as a viable process. Promising candidates in the disk are the small-amplitude variables OH 138.0+7.2 and OH 51.8−0.2. 'Non-variable' OH/IR stars in the Galactic center region may be as frequent as in the disk.

Keywords. stars: AGB and post-AGB, masers, stars: evolution

During the AGB – post-AGB evolutionary transition stars stop pulsating. While they are observed as large-amplitude variables on the AGB (L-AGB) they are almost non-variable (S-pAGB or 'non-variable' OH/IR stars) at the onset of the post-AGB phase. In both phases the stars are still deeply embedded in their dusty circumstellar shell. H_2O and OH maser emission are present in both phases. In particular the "water fountains", post-AGB stars with jets traced by high-velocity H_2O masers, have attracted much attention in recent years (e.g., Orosz *et al.*, Perez-Sanchez *et al.*, Tafoya *et al.*, in this volume). However, high-velocity H_2O masers are present only in few post-AGB stars, indicating that they document a brief phase at the beginning of post-AGB evolution and may possibly not be representative for the AGB – post-AGB transition.

To obtain a less biased view on the transition process, we are studying a sample of bright OH/IR stars with a flux limit of ~ 4 Jy compiled by Baud *et al.* (1979). In this sample the L-AGB and S-pAGB stars are almost of equal number (Herman & Habing 1985). Assuming similar bolometric luminosities, this implies that the "pulsating" phase connected to relatively high mass-loss rates ($\dot{M} > 10^{-5}$ $M_\odot$/yr) is of similar duration to the early post-AGB phase (Engels 2002). The absence of stars known to be currently in transition, indicate a rather fast transition time ($\leqslant 2000$ yr). To find transition objects we started a monitoring program of OH 1612 MHz masers, in 2013 with the NRT, to probe

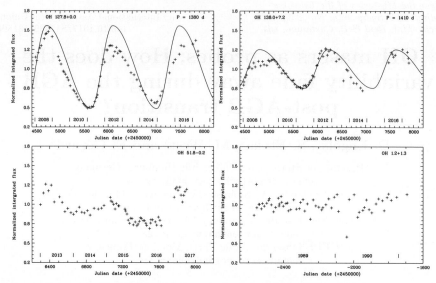

Figure 1. Lightcurves of the 1612 OH maser emission of OH/IR stars monitored at the NRT since 2013 and Hartebeesthoek (1985–1996). NRT lightcurves extending back to 2008 are from a legacy program (Etoka *et al.*, this volume). OH 127.8+0.0 (upper left) is a L-AGB star, and OH 1.2+1.3 (lower right) is a 'non-variable' post-AGB star. The other two OH/IR stars might be objects in transition. The lightcurves give the summed integrated flux of the blue and red emission peaks divided by the mean integrated flux. The stars with periodic variability were fitted by an asymmetric sine-curve.

the underlying stellar variability. A similar monitoring program of OH/IR stars toward the Galactic bulge was made at Hartebeesthoek 1985–1996 (Gaylard & West 1995).

We present here a few examples of OH maser lightcurves obtained so far (Fig. 1). A classical large-amplitude variable AGB star is OH 127.8+0.0 (P$\sim$ 3.8 years). OH 138.0+7.2 still shows periodic variability (P$\sim$ 3.9 years) albeit with significantly smaller amplitude. The type of variability (i.e. periodic or not) of OH 51.8−0.2 is uncertain. OH 1.2+1.3 is a S-pAGB star in the Galactic center region. OH 138.0+7.2 and OH 51.8−0.2 are possible examples for objects in transition between large-amplitude variability and absence of variations. OH 1.2+1.3 shows, that 'non-variable' OH/IR stars are present in the Galactic center region. Summing up all infrared non-variable and non-detected OH/IR stars among the sample monitored in the infrared by Wood *et al.* (1998), the fraction of 'non-variable' OH/IR stars toward the Galactic center is 27%. This is of the same order as found in the disk. Up to now no evidence of stars with short-period, small amplitude pulsations have been found, as assumed to exist as transition objects by Blöcker (1995). While an instantaneous cessation of the pulsation (Vassiliadis & Wood, 1993) cannot be ruled out yet, we consider the fading out of pulsations with steadily declining amplitudes (damped oscillator) as a viable process.

References

Baud, B., Habing, H. J., Matthews, H. E., & Winnberg, A., 1979, *A&AS*, 36, 193
Blöcker, T., 1995, *A&A*, 297, 727
Engels, D. 2002, *A&A*, 388, 252
Gaylard, M. J., & West, M. E., 1995, *ASPC*, 83, 411
Herman, J., & Habing, H. J., 1985, *A&AS*, 59, 523
Vassiliadis, E., & Wood, P. R., 1993, *ApJ*, 413, 641
Wood, P. R., Habing, H. J., & McGregor, P. J., 1998, *A&A*, 336, 925

Astrophysical Masers:
Unlocking the Mysteries of the Universe
Proceedings IAU Symposium No. 336, 2017
A. Tarchi, M.J. Reid & P. Castangia, eds.

© International Astronomical Union 2018
doi:10.1017/S1743921317010948

Strong magnetic field of the peculiar red supergiant VY Canis Majoris

Hiroko Shinnaga[1], Mark J. Claussen[2], Satoshi Yamamoto[3] and Shimojo Masumi[4]

[1] Department of Physics and Astronomy, Faculty of Science, Kagoshima University
1-21-35 Korimoto Kagoshima, Kagoshima 890-0065 Japan
email: `shinnaga@sci.kagoshima-u.ac.jp`

[2] National Radio Astronomy Observatory (NRAO) Box 1003 Lopezville Rd. Socorro NM 87801
U.S.A. email: `mclausse@nrao.edu`

[3] Department of Physics, University of Tokyo 7-3-1 Hongo, Bunkyo-ku, Tokyo, 113-0033, Japan
email: `yamamoto@taurus.phys.s.u-tokyo.ac.jp`

[4] National Astronomical Observatory of Japan (NAOJ) Osawa, Mitaka, Tokyo 181-8588, Japan
email: `masumi.shimojo@nao.ac.jp`

Abstract. We report on magnetic field measurements associated with the well-known extreme red supergiant (RSG), VY Canis Majoris (VY CMa). We measured both linear and circular polarization of the SiO $v = 0$, $J = 1 - 0$ transition using a sensitive radio interferometer. The measured magnetic field strengths are surprisingly high. A lower limit for the field strength is expected to be at least ~ 10 Gauss based on the high degree of linear polarization. Since the field strengths are very high, the magnetic field must be a key element in understanding the stellar evolution of VY CMa as well as the dynamical and chemical evolution of the complex circumstellar envelope of the star.

Keywords. magnetic fields, polarization, masers, stars: individual (VY CMa), stars: mass loss

1. Introduction

VY CMa is one of the most luminous ($L_* \sim 2.7 \times 10^5$ $L_\odot$; Wittkowski *et al.* 2012) evolved stars known in the Galaxy. Despite its high luminosity, it has a low effective temperature ($\sim 3{,}500$ K), which makes the star's spectral class M5 Ib with a mass of 25 $M_\odot$. Since the high-mass RSG has evolved quickly, it is still within the natal HII region Sharpless 310 (Lada & Reid 1978). The high mass-loss rate of $\sim 6 \times 10^{-4}$ $M_\odot$/yr (Shenoy *et al.* 2016) contributes to the rapid evolution of the star into the next evolutionary phase − a core-collapse supernova.

One of the unusual characteristics of VY CMa is that many SiO transitions associated with the star show a high degree of polarization, even in the ground vibrational state ($v = 0$; Shinnaga *et al.* 1999, 2003). Some velocity components of the SiO $v = 0$ low J transitions of this star show a particularly high degree of linear polarization, up to $\sim 70\%$ (Shinnaga *et al.* 1999, 2003), indicating that the SiO line even in the ground vibrational state is partly of maser origin. Highly linearly polarized emission (up to several $\times$ 10%) can originate only by maser action (e.g., Western & Watson 1984).

2. Observations

We have investigated the physical mechanism of very highly polarized SiO $v = 0$, $J = 1 - 0$ emission at 43.424 GHz (i.e., λ 6.9039 mm) in the B configuration of Very Large Array (VLA) in Socorro, New Mexico (U.S.A.), which is operated by NRAO. The observations had been done on March 16 and 23 in 2001. The angular resolution was $0.''29 \times 0.''12$. The first setting offered a bandwidth of 12.5 MHz and 32 channels,

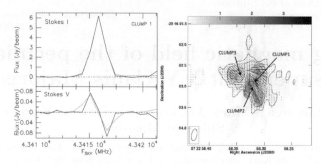

Figure 1. Left: Stokes I and V spectra of SiO $v = 0$, $J = 1 - 0$ transition towards Clump 1 (Shinnaga *et al.* 2017). Black lines represent measured spectrum. **Right:** Polarimetric image of VY CMa in SiO v=0, J=1-0 line taken with the VLA (Shinnaga *et al.* 2017). The polarization vectors of the line (black/white bars) are plotted over the systemic velocity channel map. The lowest contour is at 9 times of 6.8×10^{-3} Jy/beam (which corresponds to 1 σ noise level). The circle marks the location of the star measured with ALMA at 321GHz (Shinnaga *et al.* 2017).

yielding a velocity resolution and coverage of 2.7 and 87 km/s, respectively. The second setting had the same bandwidth but 16 channels with full polarization, yielding a velocity resolution of 5.4 km/s.

3. Results and discussion

Unlike the large scale complicated structures seen in the optical, the inner CSE traced with SiO $v = 0$ $J = 1 - 0$ emission is found to be concentrated near the star (Fig 1) and seems to show a bipolar outflow emanating from the star (Shinnaga *et al.* 2003, 2004).

In carefully examining the data taken with the VLA in 2001, we discovered that the Stokes V spectra towards three clumps in the circumstellar envelope (CSE) have the characteristic S-shaped profile of a single line split into the Zeeman pattern due to a very strong magnetic field, up to 150 − 650 Gauss (an example is shown in Fig 1; Shinnaga *et al.* 2017). The measured linear polarization pattern indicates that large scale well-ordered magnetic fields exist in the CSE of the star. A theoretical study (Western & Watson 1984) predicts that linearly polarized emission of 50 % requires a magnetic field strength greater than 10 G (Western & Watson 1984), which may correspond to a lower limit of the field strength of the CSE of VY CMa.

The detected strong magnetic field associated with VY CMa is a challenge to our current knowledge of the stellar evolution of RSGs and the mass-loss processes in the presence of strong magnetic fields. One possibility for an explanation of the intense field is that, since a supergiant has a convective shell around the helium core, the convective shell may play a critical role to generate the strong magnetic field (Maeder & Meynet 2014), with an extraordinarily high mass-loss rate (Shenoy *et al.* 2016).

References

Lada, C. J. & Reid, M. J. 1978, *ApJ*, 219, 95
Maeder, A. & Meynet, G. 2014, *ApJ*, 793, 123
Shenoy, D., Humphreys, R. M., Jones, T. J., *et al.* 2016, *AJ*, 151, 51
Shinnaga, H., & Tsuboi, M. Kasuga, T. 1999, *PASJ*, 51, 175
Shinnaga, H., Claussen, M. J., Lim, J., *et al.* 2003, *Mass-Losing Pulsating Stars and their Circumstellar Matter*, 283, 393
Shinnaga, H., Moran, J. M., Young, K. H., & Ho, P. T. P. 2004, *ApJ*, 616, L47
Shinnaga, H., Claussen, M. J., Yamamoto, S., & Shimojo, M. 2017, *PASJ*, 69, L10
Western, L. R., & Watson, W. D. 1984, *ApJ*, 285, 158

Astrophysical Masers:
Unlocking the Mysteries of the Universe
Proceedings IAU Symposium No. 336, 2017
A. Tarchi, M.J. Reid & P. Castangia, eds.

© International Astronomical Union 2018
doi:10.1017/S1743921317009450

Variability of water masers in evolved stars on timescales of decades

Jan Brand[1], Dieter Engels[2] and Anders Winnberg[3]

[1]INAF-Istituto di Radioastronomia & Italian ALMA Regional Centre, Bologna, Italy
email: brand@ira.inaf.it
[2]Hamburger Sternwarte, Hamburg, Germany; [3]Onsala Rymdobservatorium, Onsala, Sweden

Abstract. For several decades (1987-2015) we have been carrying out observations of water masers in the circumstellar envelopes (CSE's) of Mira variables, Red Supergiants (RSG's) and Semi-Regular Variables (SRV's) with the Medicina 32-m and Effelsberg 100-m antennas. The single-dish monitoring observations provide evidence for strong H_2O maser profile variations, which likely are connected to structural changes in the maser shells. Such variations include strong flares in intensity lasting several (tens of) months and systemic velocity gradients of maser components developing over years, as well as other secular variations which are superimposed on periodic variations following the stellar light variations.

When complemented with interferometric observations, it is possible to derive the 3-D distribution of the maser spots, and their lifetime, as we have done for RX Boo (Winnberg *et al.* 2008) and U Her (Winnberg *et al.* 2011; Brand *et al. in prep.*).

Keywords. stars: AGB and post-AGB; stars: variables: other; masers; flare

1. Observations

Observations were carried out with the Medicina 32-m antenna between 1987 and 2015, of a representative sample of late-type stars (see Fig.1b), at 1-4 month intervals. The resolution was typically 0.13 km/s (9.8 kHz); the sensitivity improved over the years, and was around $0.5 - 1.5$ Jy for most of the time. With the Effelsberg 100-m dish occasional observations were made between 1987 and 1999; typical resolution and sensitivity were 0.08 km/s and $0.2 - 0.4$ Jy, respectively.

2. Variability; periodic and not

Plots such as shown in Fig.1a are an efficient way to capture the results of the monitoring observations. Virtually all Mira's show periodicity in their maser emission, with the same periods as in the optical. The maser emission lags behind by about 0.2 ± 0.1 in phase. Also the maser emission from some RSG's shows periodicity, but irregularly and not always with a constant period. Individual velocity components may exhibit secular variations in flux density on top of periodic variations.

3. Bursts of maser emission: flares

Long-term monitoring also enabled the detection of infrequent, irregular outbursts or flares (see Fig.1c). Flares occur in all the types of stars we monitored, with an average of once every 5.6 years, and average duration of 18 ± 7 months and an increase in integrated flux by factors of $2 - 20$. Often outbursts occur in individual components, rather than in all components together, and if they do, there may be a delay between the flares in different components. Mira's have the most intense flares in terms of increase in flux.

393

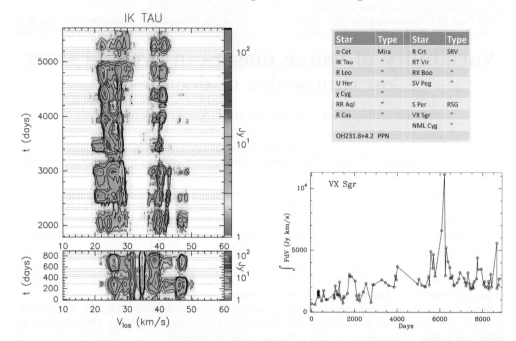

Figure 1. a (left). Flux density versus velocity, as a function of time, for IK Tau. Each horizontal dotted line indicates an observation (spectra within 4 days from each other were averaged). Data are resampled to 0.3 km/s and only emission at levels $\geqslant 3\sigma$ is shown. Day $0 = 12/12/95$. **b (top right).** Observed targets. **c (bottom right).** A flare in VX Sgr; the integrated flux increased by a factor of $4-5$, and declined, in a period of about 2.4 yrs (2003-05). Note the general increase of integrated emission over time (by a factor of $\geqslant 2$). Day $0 = 28/3/87$.

4. Velocity drifts

Not only the intensity of the maser emission, also the velocities of the emission components change with time. See for example Fig.1a. In this, and similar cases we measure gradients of $0.05 - 0.25$ km/s/yr, about an order of magnitude smaller than gradients observed in star-forming regions (see Brand *et al.* 2007). This can be interpreted as a maser region moving outwards in the shell, and would imply lifetimes of several decades.

5. Conclusions

Long-term single-dish monitoring of water masers in the CSE's of late-type stars unveils both periodic and erratic behaviour of the emission. Whereas single-epoch observations are snapshots of the maser activity and are not necessarily representative for the general behaviour, long-term monitoring can reveal (persistent) profile changes, velocity gradients and maser outbursts. The observed phenomena point to reconfiguration of emission regions; regular interferometric observations are required to study these structural changes. The use of multiple transitions to constrain the physical conditions in the CSE's requires the near-simultaneous (within months) observations of all transitions involved.

References

Brand, J., Felli, M., Cesaroni, R., *et al.* 2007, *IAU Symp. 242 (eds. Chapman & Baan)*, p. 223
Winnberg, A., Brand, J., & Engels, D. 2011, *ASP Conf. Ser. 445 (eds. Kerschbaum, Lebzelter, & Wing)*, p. 375
Winnberg, A., Engels, D., Brand, J., Baldacci, L., & Walmsley, C. M. 2008, *A&A*, 482, 831

Astrophysical Masers:
Unlocking the Mysteries of the Universe
Proceedings IAU Symposium No. 336, 2017
A. Tarchi, M.J. Reid & P. Castangia, eds.

© International Astronomical Union 2018
doi:10.1017/S1743921317008948

The Extensive Database of Astrophysical Maser Sources (eDAMS): the First Release on Circumstellar Maser Sources

Nakashima, J.[1], Engels, D.[2], Hsia, C.-H.[3], Imai, H.[4],
Ladeyschikov, D. A.[5], Sobolev, A. M.[5], Yung, B. H. K.[6] and
Zhang, Y.[7]

[1]Department of Astronomy and Geodesy, Ural Federal University,
Lenin Avenue 51, 620000, Ekaterinburg, Russia
email: nakashima.junichi@gmail.com

[2]Hamburger Sternwarte, Gojenbergsweg 112, D-21029 Hamburg, Germany

[3]Space Science Institute, Macau University of Science and Technology,
Avenida Wai Long, Taipa, Macau, China

[4]Graduate School of Science and Engineering, Kagoshima University,
1-21-35 Korimoto, Kagoshima 890-0065, Japan

[5]Astronomical Observatory, Ural Federal University,
Lenin Avenue 51, 620000, Ekaterinburg, Russia

[6]N. Copernicus Astronomical Center, Rabiańska 8, 87-100 Toruń, Poland

[7]School of Physics and Astronomy, Sun Yat-sen University, Zhuhai 519082, China

Abstract. We introduce the newly developed database of circumstellar maser sources. Until now, the compilations comprehensively including the three major maser species in evolved stars (i.e., SiO, H_2O, OH) has been practically limited only to the Benson's catalog (Benson *et al.* 1990), which was published more than a quarter of a century ago. For OH masers alone, there exists the University of Hamburg (UH) database, but there is no updated compilation work for H_2O and SiO masers. In order to utilize the information of masers in actual studies, it is highly desirable to have a database containing all the three masers. We are currently constructing a database covering SiO, H_2O and OH masers. This database consists of a web-service, which accesses compiled maser observations in available archives and combines them with the data we newly collected and IR databases. The archives currently used are the OH maser archive from Engels & Bunzel (2015), and H_2O and SiO archives, which are currently under construction. So far, the information of about 27,000 observations (about 10,000 objects) has been implemented. We also have a plan to extend the database by including higher transitions and other types of objects, such as young stellar objects, in future. In this paper, we briefly summarize, (1) outline of the data collected, and (2) future development plans of the eDAMS system. The URL of the database is as follows: http://maserdb.ins.urfu.ru/

Keywords. masers, astronomical data bases: miscellaneous, catalogs, stars: AGB and post-AGB, stars: late-type

1. Summary of the Collected Data

The initial release of eDAMS† is dedicated to the circumstellar maser sources of evolved stars mainly in the following maser lines: SiO $J = 1-0$, $v = 1$ & 2 (43 GHz), H_2O 22 GHz, OH 1612, 1665, 1667 MHz. The data are taken mainly from 5 published/unpublished

† The research was supported by the Ministry of Education and Science of the Russian Federation Agreement no. 02.A03.21.0006.

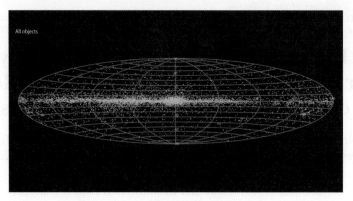

Figure 1. Distribution in Galactic coordinates of > 10000 objects included in the eDAMS database. These objects are observed, at least, in one of the OH, H_2O and SiO maser lines, and the both detections and non-detections are included.

compilation catalogs (see, the eDAMS web for the details of the used catalogs). The OH data are based on the OH maser archive from Engels & Bunzel (2015). The H_2O data are based on an on-going compilation work (PI: Engels, D.). A significant amount of additional data of other maser transitions (for example, SiO $J = 1 - 0$ $v = 0$ & 3, SiO $J = 2 - 1$, $v = 1$ & 2, ^{29}SiO $J = 1 - 0$ $v = 0$, etc.) are also included in the database, but the data survey for these lines are still not completed (the data will keep updating). We note that a non-negligible number of unpublished data of the Nobeyama SiO maser survey project are released to the public for the first time (the number of unpublished Nobeyama observations is about 400). In addition to the basic line parameters (such as intensity, velocity, line-profile, etc.), for a part of the observations, spectral data in ascii format are available, so that users could process the spectral data for their own purposes. In total, at this moment, 10466 objects, which have been observed, at least, in one of the OH, H_2O or SiO maser lines or in the multiple maser lines, are included in the database (the distribution of the objects in the Galactic coordinates is given in Figure 1). Among the 10466 sources, the number of objects observed in the SiO, H_2O and OH maser lines are 3745, 3863, and 6372 respectively (overlaps exist between different maser species).

2. Future Development Plans

The eDAMS project has following future development plans: (1) Add the data of the higher-J transition lines of circumstellar maser sources, so that the system would be useful for potential users of latest sub-mm telescopes, such as ALMA and SOFIA. (2) Increase the number of ascii spectral data, so that the users could process the data for their purposes. (3) Add the data of other kinds of astrophysical objects. For the moment, we have a plan to add the data of methanol masers of young stellar objects (YSO), of which the data collection has been basically already finished. (4) Add the reduced FITS images of the KaVA ESTEMA project, which is a VLBI imaging survey of circumstellar masers of mira-type variables. (5) Additionally, we will keep adding new data whenever the new data are published/released. We would very much appreciate if you could inform us when you publish new papers, which include maser observations.

References

Benson, P. J., *et al.* 1990, *ApJS*, 74, 911
Engels, D. & Bunzel, F. 2015, *A&A*, 582, A68

Astrophysical Masers:
Unlocking the Mysteries of the Universe
Proceedings IAU Symposium No. 336, 2017
A. Tarchi, M.J. Reid & P. Castangia, eds.

© International Astronomical Union 2018
doi:10.1017/S1743921317010092

Magnetic fields and radio emission processes in maser-emitting planetary nebulae

L. Uscanga [1], J. F. Gómez[2], J. A. Green[3], O. Suárez[4], H.-H. Qiao[5],
A. J. Walsh[6], L. F. Miranda[2], M. A. Trinidad[1], G. Anglada[2] and
P. Boumis[7]

[1]University of Guanajuato, Mexico; email: lucero@astro.ugto.mx;
[2]IAA−CSIC, Spain; [3]CASS, Australia; [4]OCA, France; [5]NTSC, CAS, China;
[6]ICRAR, CU, Australia; [7]IAASARS, NOA, Greece

Abstract. We present polarimetric observations of the 4 ground-state transitions of OH, toward a sample of maser-emitting planetary nebulae (PNe) using the Australia Telescope Compact Array. This sample includes confirmed OH-emitting PNe, confirmed and candidate H_2O-maser-emitting PNe. Polarimetric observations provide information related to the magnetic field of these sources. Maser-emitting PNe are very young PNe and magnetic fields are a key ingredient in the early evolution and shaping process of PNe. Our preliminary results suggest that magnetic field strengths may change very rapidly in young PNe.

Keywords. magnetic fields – masers – polarization – stars: AGB and post-AGB – planetary nebulae: general.

1. Introduction

Previous polarimetric studies of a few OHPNe with single-dish telescopes shown that circularly polarized OH features are common in OHPNe, but the detection of Zeeman pairs is far more elusive (Szymczak & Gérard 2004, Wolak *et al.* 2012, Gonidakis *et al.* 2014). However, new interferometric observations have shown more promising candidates. The only OHPN in which Zeeman splitting has been clearly detected is IRAS 16333-4807 at 1720 MHz (Qiao *et al.* 2016), with a derived magnetic strength of $\simeq 11$ mG. Another possible Zeeman pair has been reported in K3-35 at 1665 MHz (Gómez *et al.* 2009), giving $B \simeq 0.9$ mG. Both objects are probably among the youngest PNe, since both are OH and H_2O maser emitters. Gómez *et al.* (2016) presented full-polarizations observations of the 4 OH transitions, toward 5 confirmed and 1 candidate OHPNe. We detected significant circular and linear polarization, in 4 and 2 objects, respectively. Possible Zeeman pairs were seen in JaSt 23 and IRAS 17393-2727, with magnetic field strengths 0.8−24 mG.

2. Observations and Results

New observations were carried out with the ATCA, in its 6A configuration, on 2017 February 13−14. We obtained data in full polarization (two linear polarizations and their corresponding cross-polarizations). Broadband continuum data cover a bandwidth of 2 GHz, centered at 2.1, 5.5 and 9 GHz. We also observed the OH ground-level transitions of rest frequency 1612, 1665, 1667, and 1720 MHz with a spectral resolution of 0.5 kHz ($\simeq 0.09\ \mathrm{km\,s^{-1}}$). For these observations, our main targets were IRAS 16333-4807 (I16333), IRAS 17393-2727 (I17393), and IRAS 18061-2505 (I18061). Radio continuum emission is detected in all three sources. We detected OH maser emission in I16333 and I17393, while no OH maser emission was detected (< 0.06 Jy) in I18061.

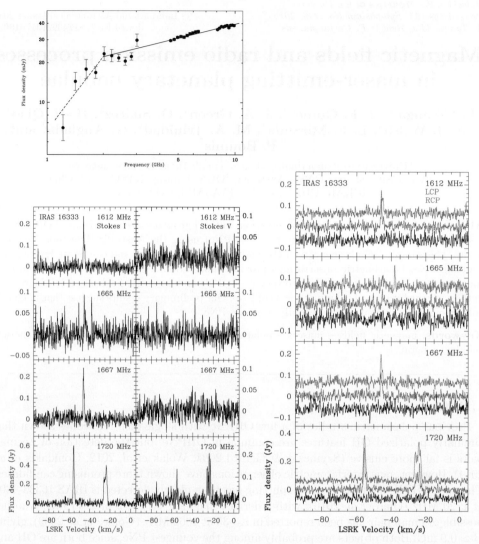

Figure 1. *Top left:* Flux density of I16333 as a function of frequency. The emission at < 2 GHz is optically thick (spectral index $\simeq 1.87$) and partially thick ($\simeq 0.36$) at higher frequencies. *Bottom left:* OH maser spectra in Stokes I and V. *Bottom right:* circular (red and blue) and linear (black) polarizations.

Here we present preliminary results of I16333 (Fig. 1). The flux densities are consistent with the values measured previously by Qiao *et al.* (2016). These spectra do not show a clear evidence of Zeeman pairs as in previous observations (Qiao *et al.* 2016). Our results may indicate a fast change in the magnetic field strength within a few years.

References

Gómez, J. F., Uscanga, L., Green, J. A., *et al.* 2016, *MNRAS*, 461, 3259

Gómez, Y., Tafoya, D., Anglada, G., *et al.* 2009, *ApJ*, 695, 930

Gonidakis, I., Chapman, J. M., Deacon, R. M., & Green, A. J. 2014, *MNRAS*, 443, 3819

Qiao, H.-H., Walsh, A. J., Gómez, J. F., *et al.* 2016, *ApJ*, 817, 37

Szymczak, M. & Gérard, E. 2004, *A&A*, 423, 209

Wolak, P., Szymczak, M., & Gérard, E. 2012, *A&A*, 537, A5

Astrophysical Masers:
Unlocking the Mysteries of the Universe
Proceedings IAU Symposium No. 336, 2017
A. Tarchi, M.J. Reid & P. Castangia, eds.

© International Astronomical Union 2018
doi:10.1017/S1743921317010109

Simultaneity and Flux Bias between 43 and 86 GHz SiO Masers

Michael C. Stroh[1], Ylva M. Pihlström[1,2] and Lorant O. Sjouwerman[2]

[1]Department of Physics & Astronomy,
The University of New Mexico, Albuquerque, NM 87131
email: mstroh@unm.edu

[2]National Radio Astronomy Observatory

Abstract. Using quasi-simultaneous observations of 86 stars with known SiO maser emission, we searched for systematic differences between the strengths of the 43 and 86 GHz v=1 maser lines. Although for individual stars there is wide scatter between the line strengths spanning nearly an order of magnitude, there is no evidence of a systematic difference between these line strengths for the entire sample.

Keywords. Masers, stars: late-type, infrared: stars, radio lines: stars

1. Introduction

The Bulge Asymmetries and Dynamical Evolution (BAaDE) project aims to explore the complex structure of the inner Galaxy and Galactic Bulge, by observing SiO maser lines in red giant stars using the 43 GHz receivers at the Very Large Array (VLA) and the 86 GHz receivers at the Atacama Large Millimeter/submillimeter Array (ALMA). A fundamental assumption for the BAaDE project is that stars emitting 43 GHz SiO maser emission also harbor 86 GHz masers, and vice versa. This appears to be a commonly accepted fact, supported by, for example, Sjouwerman *et al.* (2004) who noted that out of 39 sources displaying 86 GHz SiO maser emission, 38 also produced 43 GHz masers. What is less clear, however, is whether there is a statistically significant difference in the flux density of 43 GHz versus 86 GHz masers. Such a difference could have an impact on the analysis of the BAaDE sample, as the VLA and ALMA samples are observed to the same noise levels and the ALMA sample covers the far side of the bar ($-110° < l < 0°$) while the VLA sample covers the near side. Here we present near-simultaneous observations of the 43 GHz and 86 GHz SiO maser lines in a set of 86 stars to test whether one line is consistently brighter than the other.

2. Observations

The Australia Telescope Compact Array (ATCA) allows the 43 GHz and 86 GHz SiO transitions to be observed quasi-simultaneously. Given that the 86 GHz system at ATCA requires longer integration times compared to the 43 GHz to reach similar rms noise, the targets selected were the brightest masers in our sample. Whereas this selects for mostly nearby sources, it also selects for an average metallicity.

For the detected maser lines, self calibration was applied and the data was smoothed into 1 km/s bins for final line detections. A line is considered a detection if the velocity integrated flux density in at least one 1 km/s bin is at least five times the rms noise. For each detected line, a final velocity integrated flux density is calculated by integrating over a 3 km/s bin centered on the peak.

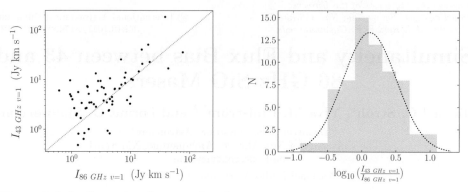

Figure 1. Left: Comparison between the velocity integrated flux densities for the two $v=1$ lines. Right: Histogram of ratios with a best-fit Gaussian over-plotted.

3. Results

We detected the SiO 43 and 86 GHz $v=1$ lines in 81 and 66 sources respectively and 86 GHz $v=1$ non-detections are likely due to poor observing conditions resulting in high rms noise. On average, the relative strengths of the 43 GHz and 86 GHz v=1 lines are equal based on the scatter plot of the line intensities (Fig. 1a) and the histogram of the line ratio distribution (Fig. 1b). While the data shows a very broad range of line ratio values, the scatter is centered around a line ratio of one and the center of the Gaussian fit is consistent with a ratio of one. The center and standard deviations of the Gaussian from the least squares fit to the logarithm of the line ratios are 0.12 ± 0.05 and 0.39 ± 0.05 respectively.

As we observed thinner shells, our results follow the trend outlined by Nyman *et al.* (1993), where the 43 GHz line gets successively weaker in thinner envelopes. However, our results do not demonstrate a turnover into a consistently brighter 86 GHz line for Mira-type envelopes, as has been previously suggested by modeling of SiO maser emission in Miras (Humphreys *et al.* 2002) and in observations of smaller Mira samples (e.g. Pardo *et al.* 1998). We note that our sample is much larger than other samples studied, and the large spread of line ratios likely can explain some earlier observational results indicating the 86 GHz line would be brighter.

Acknowledgements

The Australia Telescope Compact Array is part of the Australia Telescope National Facility which is funded by the Australian Government for operation as a National Facility managed by CSIRO.

This material is based upon work supported by the National Science Foundation under Grant Number 1517970.

References

Humphreys, E. M. L., Gray, M. D., Yates, J. A., Field, D., Bowen, G. H., & Diamond, P. J. 2002, *A&A*, 386, 256
Nyman, L. A., Hall, P. J., & Le Bertre, T. 1993, *A&A*, 280, 551
Pardo, J. R., Cernicharo, J., Gonzalez-Alfonso, E., & Bujarrabal, V. 1998, *A&A*, 329, 219
Sjouwerman, L. O., Messineo, M., & Habing, H. J. 2004, *PASJ*, 56, 45

Jürgen Ott and Katharina Immer (photo credits: S. Poppi)

New facilities

Astrophysical Masers:
Unlocking the Mysteries of the Universe
Proceedings IAU Symposium No. 336, 2017
A. Tarchi, M.J. Reid & P. Castangia, eds.

© International Astronomical Union 2018
doi:10.1017/S1743921317010778

Masers and ALMA

Alison B. Peck[1] and C. M. Violette Impellizzeri[2]

[1] Gemini North Operations Center, 670 N. A'ohoku Pl.
Hilo, Hawaii, USA,
email: `apeck@gemini.edu`

[2] NRAO/JAO,
Alonso de Cordova 3107, Santiago, Chile
email: `violette.impellizzeri@alma.cl`

Abstract. Masers have been well-known phenomena for decades, but water masers at 183, 321, 325 and 658 GHz have only been detected since the 1990s. Early detections came from single-dish telescopes with follow-up observations from the PdBI and the Submillimeter Array. Detecting them at these short wavelengths has been very difficult due to water in our atmosphere, meaning that even in very good weather, one can only detect very bright masers, such as those in stellar atmospheres. In the last 7 years, a new window on submillimeter water masers, both Galactic and now extragalactic, has opened. Located at high altitude, above a large fraction of the Earth's atmosphere, ALMA sits on the edge of the driest desert on the planet, meaning that the air that does remain above the telescope is frequently extremely low in water vapor content. Combine this with sensitive, stable receivers covering a number of masing transitions from 183-658 GHz and you have an excellent machine for detecting and characterizing submillimeter water masers. In addition, other molecules also exhibit maser emission in the ALMA observing bands, such as SiO and HCN.

Keywords. instrumentation: interferometers, masers, submillimeter

1. The Attraction and Tribulations of Submm Maser Emission

Early theoretical predictions showed that many maser molecules would have most of their rotational transitions in the submillimeter range (e.g. Neufeld & Melnick 1991, Yates, Field & Gray 1997). More recently, these models were improved by Gray *et al.* (2016) (see his contribution in these proceedings) who performed a thorough exploration of the relevant parameter space (i.e., gas density, kinetic temperature, and dust temperature) for the water molecule specifically and found that many of the submillimeter transitions have emission matching that at 22 GHz. Thus, observing a number of transitions at the same time offers the opportunity to put tight constraints on the physical conditions in the masing regions.

Submillimeter masers hold the promise to constrain and refine the radiative transfer models in regions so far unexplored in cm-wave observations. The amplified emission of several molecules provides a means to probe source temperature and density distributions in a variety of astrophysical sources. Cosmic masers allow us to study a large range of environments, from newly formed stars to the envelopes of evolved stars to nearby or high redshift active galactic nuclei. Maser emission provides unique information, particularly at very high angular resolution owing to its extremely high brightness temperature ($> 10^{10}$ K). Unfortunately, technical and sensitivity limitations, as well as the lack of angular resolution, have rendered observations of these lines extremely difficult until quite recently. Most of the existing maser studies have been carried out in the cm-wave regime where very strong masers like OH, H_2O, and CH_3OH can be studied both by single dish telescopes and with powerful interferometric techniques. However,

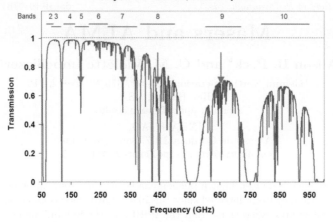

Figure 1. The transmissivity at the ALMA site in reasonably good weather. The purple bars indicate the frequency ranges of the ALMA receivers. The red arrows show the locations of common water maser lines.

until now extending the study of masers to the submillimeter has been challenging in a number of ways. Two big limitations have hindered observations of these transitions. The first one is sensitivity : low sensitivity has been the result of the lack of sensitive receivers that work at higher frequencies and the fact that the radio transmission at those frequencies is severely affected by the atmosphere (Fig. 1). The latter is especially true for the water maser transitions, for example, at 183 GHz and 325 GHz because of their low energies above ground state. The second limitation for the study of sub-millimeter masers has been the lack of high angular resolution observations. Relating cm-wave maser emission observed on milliarcsecond scales with that of the submillimeter maser emission on many tens of arcseconds, has made it difficult to constrain and test the radiative transfer models. The arrival of the SMA, the first submillimeter imaging array capable of sub-arcsecond resolution, and APEX, with more modern and sensitive receivers, have by and large enabled progress in this area in the last decade, especially for the stronger masers.

Now, the Atacama Large Millimeter/submillimeter Array (ALMA), located at an elevation of 5000 m in northern Chile, is in the process of revolutionising the study of astrophysical masers by providing sensitive and stable receivers covering the frequency range from 100-900 GHz, long baselines for high angular resolution and an unparalleled site for observing water and other molecules. Currently carrying out the 5th cycle of science observations, ALMA also offers Band 5 (which includes the promising 183 GHz water maser transition) and baselines of up to ~16 km. The ALMA archive and science verification observations are readily available and any data that is not proprietary may be downloaded.

2. The Atacama Large Millimeter/submillimeter Array

The Atacama Large Millimeter/submillimeter Array (ALMA) consists of two arrays of high-precision antennas (see Fig. 2). The first one, made up of twelve 7-meter diameter antennas operating in closely-packed configurations of about 50 meters in diameter, is known as the ALMA Compact Array (ACA), also called the Morita Array. The second array is made up of fifty 12-meter antennas arranged in configurations with diameters

Figure 2. The ALMA telescope. Photo by Pablo Carrillo.

Figure 3. ALMA cryostat showing several receiver inserts. Photo credit: ESO/JAO

ranging from about 150 meter to ~16 km. Four further 12-meter antennas, for use in conjunction with the ACA, provide "zero-spacing" information, critical for making accurate images of extended objects. The antennas are all equipped with sensitive millimeter-wave receivers covering most of the frequency range 84 to 950 GHz (see Fig. 3). State of the art microwave, digital, photonic and software systems capture the signals, transfer them to the central building and correlate them, while maintaining accurate synchronization.

The ALMA array is situated at 5000 m above sea level, on the Chajnantor plateau in the desert of Atacama, in the northern Chile. The high altitude and dry air are essential for mitigating the phase fluctuations introduced by water vapor and turbulence in the atmosphere. In order to further reduce phase fluctuations in the incoming signals, we use radiometers on each antenna to make measurements of the water vapor content along each line of site through the atmosphere. These radiometers operate at 183 GHz where there is a sufficiently strong emission line of water to give an accurate reading even on short timescales (less than a second) so that the fast fluctuations can be modeled and removed in post-processing.

Antenna integration, commissioning and science verification of ALMA started in 2007, and the telescope was officially inaugurated in September 2013. The first cycle of proposals, Cycle 0 - or Early Science - was offered during the commissioning stage in 2011, with only 16 12-m antennas. Even during Early Science, ALMA represented the most powerful submillimeter interferometer available. Full science operations are now underway, and in August 2017 - just one month before this meeting - the ALMA residencia was opened to host astronomers and engineers working around the clock at the site (see Fig. 4).

Figure 4. The new ALMA residencia. This building houses dormitories, cafeteria and other facilities. Photo by Pablo Carrillo.

ALMA Specifications

- 54 12-m antennas and 12 7-m antennas at 5000m site
- Surface accuracy $<25\mu$m, 0.6″ reference pointing in 9m/s wind, 2″ absolute pointing
- Array configurations between 150m and ~15-18km.
- Angular resolutions ~40mas at 100 GHz (5mas at 900 GHz)
- 10 bands covering 31-950 GHz (not yet all in production) + 183 GHz WVR.
- 8 GHz BW, dual polarization.
- Interferometry, mosaicing & total-power observing.
- 4096 channels/IF (multiple spectral windows, mixed configurations), full Stokes.
- Data rate: 6MB/s average; peak 64 MB/s.
- All data archived (raw + images), pipeline processing.

We are now currently in Cycle 5, with 43 12 m antennas being offered and with bands (3-10) extending continuously from (84-950 GHz), baselines of up to 16 km (yielding up to ~20 mas resolution; see Figure 5), and full polarization products also in Bands 4 and 5. Additionally, as part of a broader collaboration with the Global mm-VLBI Array (GMVA, at Band 3) and Event Horizon Telescope (EHT, at Band 6), ALMA can be used as a single phased array in VLBI mode, achieving resolutions of μarcsecs. The ALMA-VLBI observations will be run for the second time during this cycle (April 2018) but thus far it can only be used for continuum observations. Though still in the commissioning stage, it is foreseeable that spectral line ALMA-VLBI observations will truly revolutionise our maser observations.

3. How Can I Get ALMA Data?

All of the information needed to propose for ALMA can be found at: https://almascience.nrao.edu/proposing/learn-more.

It is important to keep in mind that ALMA has a very high oversubscription rate, and if a proposal is not successful on the first try, it is worth proposing again. At present, masers observations are still perceived as difficult, but we can see from the impressive results that have been published so far that ALMA will be an important tool in maser studies moving forward. The best way to ensure that maser observations become standard is to keep proposing, observing and publishing outstanding results.

In the meantime, it is also possible to download ALMA data directly from the Archive if they were taken in the course of Science Verification, or during an Early Science cycle that

Config	Lmax / Lmin	Band / Freq	Band 3 100 GHz	Band 4 150 GHz	Band 5 183 GHz	Band 6 230 GHz	Band 7 345 GHz	Band 8 460 GHz	Band 9 650 GHz	Band 10 870 GHz
7-m Array	45 m	AR	12.5"	8.4"	6.8"	5.4"	3.6"	2.7"	1.9"	1.4"
	9 m	MRS	66.7"	44.5"	36.1"	29.0"	19.3"	14.5"	10.3"	7.7"
C43-1	161 m	AR	3.4"	2.3"	1.8"	1.5"	1.0"	0.74"	0.52"	0.39"
	15 m	MRS	29.0"	19.0"	15.4"	12.4"	8.3"	6.2"	4.4"	3.3"
C43-2	314 m	AR	2.3"	1.5"	1.2"	1.0"	0.67"	0.50"	0.35"	0.26"
	15 m	MRS	22.6"	15.0"	12.2"	9.8"	6.5"	4.9"	3.5"	2.6"
C43-3	500 m	AR	1.4"	0.94"	0.77"	0.62"	0.41"	0.31"	0.22"	0.16"
	15 m	MRS	16.2"	10.8"	8.7"	7.0"	4.7"	3.5"	2.5"	1.9"
C43-4	784 m	AR	0.92"	0.61"	0.50"	0.40"	0.27"	0.20"	0.14"	0.11"
	15 m	MRS	11.2"	7.5"	6.1"	4.9"	3.3"	2.4"	1.7"	1.3"
C43-5	1.4 km	AR	0.54"	0.36"	0.30"	0.24"	0.16"	0.12"	0.084"	0.063"
	15 m	MRS	6.7"	4.5"	3.6"	2.9"	1.9"	1.5"	1.0"	0.77"
C43-6	2.5 km	AR	0.31"	0.20"	N/A	0.13"	0.089"	0.067"	0.047"	0.035"
	15 m	MRS	4.1"	2.7"		1.8"	1.2"	0.89"	0.63"	0.47"
C43-7	3.6 km	AR	0.21"	0.14"	N/A	0.092"	0.061"	0.046"	0.033"	0.024"
	64 m	MRS	2.6"	1.7"		1.1"	0.75"	0.56"	0.40"	0.30"
C43-8	8.5 km	AR	0.096"	0.064"	N/A	0.042"	0.028"	N/A	N/A	N/A
	110 m	MRS	1.4"	0.95"		0.62"	0.41"			
C43-9	13.9 km	AR	0.057"	0.038"	N/A	0.025"		N/A	N/A	N/A
	368 m	MRS	0.81"	0.54"		0.35"				
C43-10	16.2 km	AR	0.042"	0.028"	N/A	0.018"		N/A	N/A	N/A
	244 m	MRS	0.50"	0.33"		0.22"				

Figure 5. Receivers and associated configurations available in 2017. For more information, see http://www.almaobservatory.org/en/audience/science/

has passed its proprietary period. The ALMA Archive is searchable by science target or several other parameters, and the interface is located at: http://almascience.nrao.edu/aq/

3.1. *ALMA Science Verification using Masers*

During the commissioning stage of ALMA, a number of science verification observations were taken to test newly integrated components by verifying previously published science results. These observations were usually carried out as a new receiver was installed or a brand new mode was offered to the community. The list of all science verification data acquired to date can be found at: https://almascience.nrao.edu/alma-data/science-verification.

We would like to highlight here two of the science verification results that easily show the ALMA potential to study submillimeter masers.

The first were observations in Bands 7 and 9 of the water masers in VY Canis Majoris (VY CMa; Richards *et al.* 2014). VY CMa was observed on August 2013 using 16-20 12 m antennas on baselines from 0.014-2.7 km. Three water maser transitions were

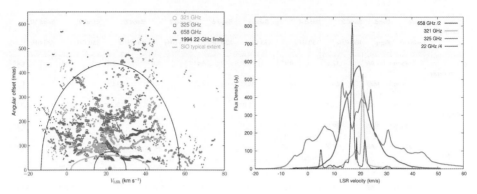

Figure 6. Science Verification results showing the sub-millimeter water masers in VY CMa in Bands 7 and 9. (Richards *et al.* 2014)

observed, namely the 321, 325, and 658 GHz transitions. The velocity resolution after Hanning smoothing was $0.45\,\mathrm{km\,s^{-1}}$ at 321 and 658 GHz, and $0.9\,\mathrm{km\,s^{-1}}$ at 325 GHz. All three sets of maser lines were found at increasing distance from the star, with the 658 GHz masers being closest and 325 GHz further away (see Fig. 6). The masers reached unexpectedly large separations from the central source and the different transitions form clumps but did not overlap even when found at similar separations. This was the first map of submillimeter water masers made at such high precision. The 658 GHz masers displayed half the flux density as the 22 GHz ones (see Fig. 6, right), however observations were not taken at the same time.

The second set of science verification observations we would like to mention were taken during the exciting integration of ALMA Band 5 (163-211 GHz) which was commissioned during Cycle 4 and is finally available this year for the first time.

Observations of the H_2O $3_{13} - 2_{20}$ line at 183 GHz rest frequency were made toward Arp 220 on July 2016 (König *et al.* 2017). The array was composed of 12 antennas equipped with Band 5 receivers in a configuration with baselines ranging from 30 m to 480 m. The emission at 183 GHz was detected in both nuclei of Arp 220. The brightest flux was found in Arp 220 West, with Arp 220 East about three times less luminous.

A comparison of spectra of the 183 and 325 GHz water lines observed at different dates shows that the emission was not variable given the velocity resolution and sensitivity of the data. The 22 GHz observations suggest that the lack of emission in the western nucleus at this frequency is most likely not intrinsic to the physics of the water line, but a result of the strong ammonia absorption. The observed line intensity ratios are not compatible with a pure thermal origin of the water emission, hence may be due to maser emission.

References

Gray, M. D., Baudry, A., Richards, A. M. S. *et al.* 2016, *MNRAS*, 374, 456
König, S., Martín, S., Muller, S. *et al.* 2017, *A&A*, 602, 42
Neufeld, D. A. & Melnick, G. J. 1991, *ApJ*, 368, 215
Richards, A. M. S. and Impellizzeri, C. M. V. and Humphreys, E. M. *et al.* 2014, *A&A*, 572, 9R

Astrophysical Masers:
Unlocking the Mysteries of the Universe
Proceedings IAU Symposium No. 336, 2017
A. Tarchi, M.J. Reid & P. Castangia, eds.

© International Astronomical Union 2018
doi:10.1017/S1743921317009887

Masers! What can VLBI do for you?

Francisco Colomer[1,2] and Huib van Langevelde[1,3]

[1] Joint Institute for VLBI ERIC (JIVE)
PostBus 2, 7990 AA Dwingeloo, the Netherlands
emails: `colomer@jive.eu`, `langevelde@jive.eu`

[2] Observatorio Astronómico Nacional (OAN-IGN)
Calle Alfonso XII 3-5, E-28014 Madrid, Spain

[3] Sterrewacht Leiden, Leiden University
Postbus 9513, 2300 RA Leiden, the Netherlands

Abstract. Very Long Baseline Interferometry (VLBI) is providing key information to the study of maser processes in the Universe, from star formation regions or circumstellar envelopes around evolved stars, to Galactic structure and cosmology, through precise astrometry. VLBI networks offer various capabilities and, most importantly, support to users, to ensure that these infrastructures are fully accesible and that the best science can emerge. In this paper we describe the advances in VLBI that enable exciting maser studies.

Keywords. masers, radio lines: ISM, radio lines: stars, radio lines: galaxies, stars: evolution, techniques: high angular resolution, techniques: interferometric

1. Introduction

Special conditions in the interstellar medium make possible microwave amplification by stimulated emission of radiation (maser) processes in the Universe. These can be found in star formation regions, in circumstellar envelopes around evolved stars (mainly AGB and post-AGB stars), and in our and other galaxies. Since maser emission can become very bright and compact, high resolution VLBI observations are a tool to study their distribution and other characteristics, therefore providing clues to derive the conditions in the regions where they occur.

The advent of recent technological developments and instruments make it now possible to study cosmic masers in a more efficient way. For example, the capability to observe several maser lines simultaneously provides unique information to constrain models of the maser emission.

2. What can VLBI do for masers?

Very Long Baseline Interferometry (VLBI) provides the sharpest view of cosmic objects in the Universe. For maser research, there are many applications where the VLBI technique produces superbly high spatial resolution data, complementary to that of other astronomical techniques. A few examples are described in Sec. 3. We now present which VLBI networks are available, in some detail.

2.1. Available VLBI networks

Nowadays there are several VLBI networks operating around the world, each with their own specific characteristics (number and size of radio telescopes, frequency coverage, location and governance). Also they are technically compatible to operate as a global VLBI array, when required. The largest ones are the European VLBI Network (EVN,

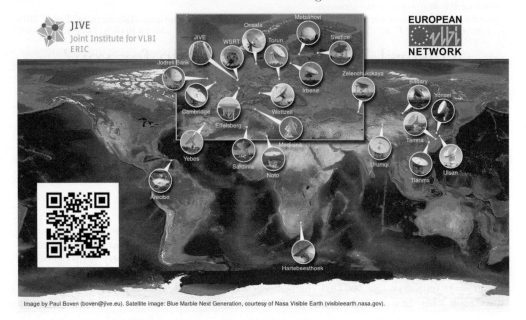

Image by Paul Boven (boven@jive.eu). Satellite image: Blue Marble Next Generation, courtesy of Nasa Visible Earth (visibleearth.nasa.gov).

Figure 1. The European VLBI Network (EVN/JIVE).

with its correlator at the Joint Institute for VLBI ERIC - JIVE), and the Very Long Baseline Array (VLBA, operated by the US Long Baseline Observatory, LBO).

The EVN is a network of radio telescopes located primarily in Europe and Asia, with additional antennas in South Africa and Puerto Rico. It is the most sensitive VLBI array in the world, thanks to the collection of extremely large telescopes that contribute to the network (Effelsberg 100-m, Lovell 76-m, SRT 64-m, Yebes 40-m, etc.; see Fig. 1). It recently introduced real-time capability (eEVN). The EVN operates in frequency bands from 1.4 to 45 GHz (wavelenghts from 21 cm to 7 mm).

The EVN calls for observing proposals on Feb 1, June 1 and Oct 1, scheduling observations in three sessions each year. Selection of frequency bands for each session is based on proposal pressure. "Target of Opportunity" (ToO) and short observations can be submitted at any time. The EVN facility is open to all astronomers, and selection of proposals is based only on scientific merit and technical feasibility. Use of the network by astronomers not specialised in VLBI techniques is encouraged (see Sec. 2.2). The current Call for Proposals is available at: http://www.evlbi.org/proposals/

JIVE is the central organisation in the EVN. Its primary missions are to operate and further develop the EVN VLBI Data Processor and provide user support. The Institute also carries out a broad range of research and development activities in VLBI-related fields, such as radio astronomy data processing, and innovative applications of VLBI and radio astronomy technologies. JIVE staff carry out a range of cutting-edge research in various fields of Galactic and extragalactic radio astronomy and planetary and space sciences. The Institute is actively involved in a number of large international projects, such as the SKA. JIVE acts as the coordinator of several projects funded by the European Commission. JIVE is located in Dwingeloo, the Netherlands, and is hosted by ASTRON – the Netherlands Institute for Radio Astronomy. For more information, visit the websites of the EVN (http://www.evlbi.org/) and JIVE (http://www.jive.eu/).

The Very Long Baseline Array (VLBA) is a network of 10 identical antennas spread across the USA. Combining the VLBA with the phased Jansky VLA, the GBT, Arecibo, and Effelsberg, defines the High Sensitivity Array (HSA), offered for proposals with the same open-skies policy as the EVN. Combination of the VLBA with the EVN defines Global VLBI, while with some other European antennas capable of observations at 3mm wavelength collaborate in the Global Millimeter VLBI Array (GMVA). For more information, visit the websites of the GMVA (`http://www3.mpifr-bonn.mpg.de/div/vlbi/globalmm/`) and the LBO/VLBA (`https://www.lbo.us/`).

A new capability for multiline studies has been introduced through technological development in broadband receivers and the capability to perform simultaneous observations in different frequency bands. One very successful case is the Korean VLBI Network (KVN, Han *et al.* 2013), where receivers centered at 22, 43, 86 and 129 GHz allow the study of H_2O and several lines of SiO at the same time. This instrument is producing a great advance in the observational field of masers in AGBs (see e.g. Yun *et al.* 2016). However it lacks very long baselines; in this context, it should be noted that a system has been installed at the IGN Yebes 40-m radio telescope (in Guadalajara, Spain) that allows the simultaneous observation of the frequency bands centered at 22 and 43 GHz. The receiver for 86 GHz is expected to be installed in 2018, which will enhance the technical capabilities for these kinds of studies. The KVN multifrequency receivers are being installed also at the VERA array in Japan, which together with antennas in China, constitute the KaVa and East-Asia VLBI Network (EAVN). For more information, visit the websites of the KVN (`https://www.kasi.re.kr/eng/pageView/89`), VERA (`http://veraserver.mtk.nao.ac.jp/outline/index-e.html`), KaVa (KVN+VERA, `https://radio.kasi.re.kr/kava/about_kava.php`).

All these VLBI networks may cooperate in the near future to construct the *Earth VLBI Array*, a coordinated facility where astronomers can submit proposals for optimum assignment of resources and user support.

2.2. *Support to (new) VLBI users*

JIVE's mission includes supporting EVN users and operations of the EVN as a facility. These activities are conducted by a team of JIVE Support Scientists, and include proposal preparation (to be submitted using the tool NorthStar), scheduling, quality assurance for correlator data products, and/or data analysis. In this way, usage of the EVN becomes easier for astronomers not specialised in the VLBI technique. For more information, see: `http://www.jive.eu/european-vlbi-network-user-support-jive` and `http://proposal.jive.nl`

2.3. *Developments*

Under the umbrella of the EC H2020 Radionet project, a new prototype receiver for the $1.5 - 15.5$ GHz band is being developed. This receiver, known as *BRAND EVN*, will be capable of registering many maser lines simultaneously, in particular those of OH (1.6, 1.7, 4.9 and 6.0 GHz) and Methanol (6.7 and 12 GHz). The prototype will be available in 2020, and then to be considered for installation on telescopes of the EVN.

A compact version of the KVN multifrequency system can be developed for those antennas of the EVN (or other VLBI networks) which are capable of observing at higher frequencies. As mentioned above, the IGN Yebes 40-m radio telescope has installed this capability, but for other antennas, a smaller version is needed. Combined with *BRAND*, these two receivers would be all that is needed to cover the frequencies of interest between 1.5 and 110 GHz, including bands in which masers of water (22 GHz), SiO (43 and 86 GHz), HCN (89 GHz) and also methanol (44 and 95 GHz, maybe also 36 GHz) can be simmultaneously studied.

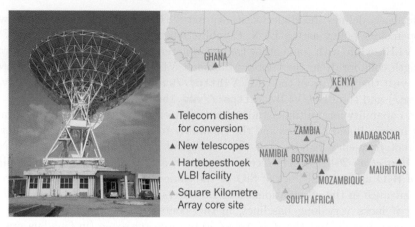

Figure 2. The Kutunse antenna in Ghana, first of the African VLBI Network (AVN) under construction.

JUMPING JIVE (for "Joining up Users for Maximizing the Profile, the Innovation and Necessary Globalization of JIVE") is an EC H2020 project that aims to enhance VLBI, and advocate JIVE and the EVN as globally recognized centres of excellence in radio astronomy. The project, coordinated by JIVE, brings together scientists and engineers to define the future of VLBI for scientific applications, and identify the necessary technological innovations. The project work packages cover a number of topics. Some of them are strategic, like encouraging existing telescopes to join the EVN, and finding new JIVE partners to expand the membership base. Others are aimed at improving the user experience, for example by developing new global interfaces (SCHED, remote control of systems), geodetic/astrometric capabilities, connecting with future instruments (like VLBI with the SKA, and training to the staff of the African VLBI Network). In addition, there are resources for dedicated outreach efforts. For more information, see http://www.jive.eu/jumping-jive

The African VLBI Network (AVN) aims to establish a 30-m class radio telescope in each of the partner countries (Botswana, Ghana, Kenya, Madagascar, Mauritius, Mozambique, Namibia, and Zambia) and link these together and with South Africa in a VLBI network, which will operate in tandem with the EVN (Fig. 2). This will be achieved through a combination of converting ex-telecommunications dishes and newly built antennas. The AVN is a vital part of the effort towards building SKA on the African continent over the next decade. The AVN dishes will provide a focus for the development of radio astronomy in each partner country so that a skilled local team is ready to install, maintain and operate the SKA outstations when they are deployed. Moreover, the aim is to establish astrophysics education and research communities in these counties as a springboard for wider technical and economic developments. The efforts are supported by the NEWTON, DARA (Developments in Africa with Radio Astronomy) networks, and JUMPING JIVE (see above) programs. The conversion at the Kutunse site in Ghana reached an important milestone with the first detection of VLBI fringes, announced in July 2017.

In order to get VLBI ready for future users, data processing also needs to be addressed. ParselTongue is developed as a scripting language based on Python that allows one to process VLBI data with classic AIPS using a modern programming language, thus making complex automated data reduction possible. The excellent support for today's web standards in Python facilitates the development of pipelines that interact easily with other programs. There is also support for accessing data in FITS files, and full access to the visibilities in AIPS UV data is also available. For more information, see: http://www.jive.nl/jivewiki/doku.php?id=parseltongue:parseltongue The Common Astronomy Software Applications package (CASA) is being developed by NRAO and collaborators. The package can process both interferometric and single dish data. JIVE is involved in tasks specific to VLBI, such as calibration and fringe fitting. For more information, see: https://casa.nrao.edu/

3. (Some) VLBI success cases on maser research

The present conference proceedings include some very good examples of the application of VLBI to the study of astrophysical masers.

Late–type stars on the Asymphotic Giant Branch (AGB) have circumstellar envelopes (CSEs) rich in molecules, in different layers, whose masers are being studied by VLBI (mostly SiO, H_2O and OH in the case of O–rich envelopes). Observations of SiO masers performed in various vibrational and rotational transitions by VLBI techniques have provided extremely valuable information on the spatial structure and dynamics of the inner circumstellar shells around AGB stars (see e.g. Desmurs *et al.* 2000, Diamond & Kemball 2003, Soria-Ruiz *et al.* 2004, Desmurs *et al.* 2014, Yun *et al.* 2016, etc.). VLBI mapping systematically shows emission clumps distributed in ring-like structures, consistent with tangential ray amplification at about 10^{14} cm from the star (~ 2 stellar radii). Comparing the observed brightness distributions in different vibrational/rotational states can be indicative of which excitation mechanisms dominate. The v=1,2 J=1→0 maser lines often occupy similar regions, but their clumps are rarely spatially coincident and the v=2 emission ring tends to be closer to the star. Recent observations by KaVA confirm this scenario by precise astrometric analysis (Yun *et al.* 2016). The standard theoretical models predict that the v=1 J=1→0 and the v=1 J=2→1 lines, with nearby energy levels and thus requiring a similar pumping mechanism, must come from the same clumps; however, observations show that the J=2→1 maser clumps occupy a clearly larger shell in the circumstellar envelope (Soria-Ruiz *et al.* 2004). However, all VLBI observations cann be affected by missing maser flux (Desmurs, these proceedings).

The observed total intensities and spatial distributions of all lines are being accurately measured only recently thanks to very good relative astrometry (see e.g. Imai, Yoon, these proceedings). Line overlap seems to be a basic phenomenon that can explain observed properties and models seem to work, at least qualitatively. In particular, they allow one to reconcile the 43 and 86 GHz observed maser distributions. These models, however, do not include the important clumpiness actually observed in VLBI maps. SiO masers can be used as well to map the magnetic field in the near stellar environment, since highly polarized ($> 20\%$) masers are probably probing the magnetic field within a few stellar radii (Vlemmings *et al.* 2005; Tobin, these proceedings).

Water masers at 22 GHz have been extensively studied by VLBI, in particular by the EVN and in combination with MERLIN. Recent developments in instrumentation explained in Sec. 2.3 now make possible simultaneous observations of multiple SiO masers (Cho, Dodson, these proceedings).

Many interesting results on star formation are being provided by instruments like VERA, in combination or cooperation with the EVN and VLBA. A multiepoch VLBI study of 22 GHz water masers in the Orion KL region detected the annual parallax to be 2.29 ± 0.10 mas, corresponding to a distance of 437 ± 19 pc from the Sun, a much more accurate value than previously obtained, with an uncertainty of only 4% (Hirota *et al.* 2007). VERA operates a dual-beam receiving system, which provides simultaneous phase-referencing VLBI astrometry. In addition, absolute proper motions of the maser features suggest an outflow motion powered by the radio Source I. Magnetic fields can also be measured using water masers, for example for the synchrotron protostellar jet in W3(H_2O) (Goddi, these proceedings).

Methanol class II masers at 6.7 GHz are well known tracers of high-mass star-forming regions. However, their origin is still not clearly understood. Studies with the EVN have provided high sensitivity images with milliarcsecond angular resolution (Bartkiewicz *et al.* 2016; Moscadelli, these proceedings). VLBI imaging of a 44 GHz class I methanol maser was performed by KaVa (Matsumoto, 2014; Kim, these proceedings), and polarization has been studied for G10.34-0.14 (Kang, these proceedigns). Attempts to detect Methanol Class-I masers at 95 GHz with VLBI have not been successful so far.

Hundreds of trigonometric parallaxes and proper motions for masers associated with young, high-mass stars have been measured with the VLBA, EVN and VERA, some with accuracies of ± 10 microarseconds (Reid *et al.* 2014; Honma in these proceedings). These measurements provide strong evidence for the existence of spiral arms in the Milky Way, accurately locating many arm segments, with the widths of spiral arms increasing with distance from the Galactic center.

The combination of gas and stellar astrometry (as being provided by Gaia) will be a powerful tool to distinguish several dynamical models of galaxy rotation. Long Period Variables (LPVs), showing a well defined Period Luminosity Relation, are important distance indicators. VLBI observations of OH, H_2O and SiO masers can provide accurate distances to significant numbers of LPVs and a critical check on Gaia parallaxes of LPVs (Zhang, these proceedings).

Water megamasers can be used to test the unified model for AGNs, the need for a torus, and the physics of the central engine; actually they currently provide the only way to map the structure of circumnuclear accretion disks within a parsec of AGN supermassive black holes. Maser distance estimations can also be used to measure H_0 accurately and constrain cosmological parameters. This is the aim of the MCP (Braatz, these proceedings). OH megamasers can probe magnetic fields in starbust galaxies (Robishaw, these proceedings). SiO, H_2O and OH masers in AGB stars in the Magellanic Clouds have been used to study dust formation and mass-loss under low–metallicity conditions (van Loon *et al.* 2001). No significant differences between Galactic and MC CSEs have been found.

References

Bartkiewicz, A., Szymczak, M. & van Langevelde, H. J., 2016 *A&A* 587, 104

Colomer, F., Desmurs, J. F., Bujarrabal, V. *et al.*, 2017, In: *Highlights on Spanish Astrophysics IX*, p. 361-366

Desmurs, J.-F., Bujarrabal, V., Colomer, F., & Alcolea, J., 2000, *A&A* 360, 189 366

Desmurs, J.-F., Bujarrabal, V., Lindqvist, M. *et al.*, 2014, *A&A* 565, 127

Diamond, P. J. & Kemball, A. J., 2003, *ApJ* 599, 1372

Han, S.-T., Lee, J.-W., Kang, J., *et al.*, 2013, *PASP* 125 (927), 539

Hirota, T., Bushimata, T., Choi, Y.-K. *et al.*, 2007. *PASJ* 59, 897

Matsumoto, N., Hirota, T., Sugiyama, K. *et al.*, 2014. *ApJL* 789, L1

Reid, M. J., Menten, K. M., & Brunthaler, A., 2014. *ApJ* 783, 130

Soria-Ruiz, R., Alcolea, J., Colomer, F., *et al.*, 2004, *A&A* 426, 131

van Loon, J. Th., Zijlstra, A. A., Bujarrabal, V., & Nyman, L.-A., 2001 *A&A* 368, 950

Vlemmings, W. H. T., van Langevelde, H. J. & Diamond, P. J., 2005, *A&A* 434, 1029

Yun, Y., Cho, S.-H., Imai, H. *et al.* 2016, *ApJ* 822, 3

Astrophysical Masers:
Unlocking the Mysteries of the Universe
Proceedings IAU Symposium No. 336, 2017
A. Tarchi, M.J. Reid & P. Castangia, eds.

© International Astronomical Union 2018
doi:10.1017/S1743921317011401

RadioAstron space-VLBI project: studies of masers in star forming regions of our Galaxy and megamasers in external galaxies

A. M. Sobolev[1], **N. N. Shakhvorostova**[2], **A. V. Alakoz**[2] and **W. A. Baan**[3] on behalf of the RadioAstron maser team

[1] Astronomical Observatory, Ural Federal University,
Lenin Ave. 51, Ekaterinburg 620083, Russia
email: `Andrej.Sobolev@urfu.ru`

[2] Astro-Space Center of LPI RAS, Moscow, Russia

[3] ASTRON, Netherlands

Abstract. Observations of the masers in the course of RadioAstron mission yielded detections of fringes for a number of sources in both water and hydroxyl maser transitions. Several sources display numerous ultra-compact details. This proves that implementation of the space VLBI technique for maser studies is possible technically and is not always prevented by the interstellar scattering, maser beaming and other effects related to formation, transfer, and detection of the cosmic maser emission. For the first time, cosmic water maser emission was detected with projected baselines exceeding Earth Diameter. It was detected in a number of star-forming regions in the Galaxy and two megamaser galaxies NGC 4258 and NGC 3079. RadioAstron observations provided the absolute record of the angular resolution in astronomy. Fringes from the NGC 4258 megamaser were detected on baseline exceeding 25 Earth Diameters. This means that the angular resolution sufficient to measure the parallax of the water maser source in the nearby galaxy LMC was directly achieved in the cosmic maser observations. Very compact features with angular sizes about 20 μas have been detected in star-forming regions of our Galaxy. Corresponding linear sizes are about 5-10 million kilometers. So, the major step from milli- to micro-arcsecond resolution in maser studies is achieved by the RadioAstron mission. The existence of the features with extremely small angular sizes is established. Further implementations of the space–VLBI maser instrument for studies of the nature of cosmic objects, studies of the interaction of extremely high radiation field with molecular material and studies of the matter on the line of sight are planned.

Keywords. masers, stars: formation, galaxies: fundamental parameters, techniques: high angular resolution

1. Introduction

The space-ground interferometer RadioAstron allows observations with the longest-ever baselines exceeding the size of the Earth by more than an order of magnitude. Maser sources represent one of the main targets of the RadioAstron (RA) science program along with active galactic nuclei and pulsars. The RadioAstron project allows us to observe maser emission in one quantum transition of water at 22.235 GHz and two transitions of hydroxyl at 1.665 and 1.667 GHz. Water and hydroxyl masers are found in star-forming regions of our and nearby galaxies, around mass-loosing evolved stars, and in accretion discs around super-massive black holes in external galaxies.

Masers have small angular sizes (a few milli-arcsec and smaller), very high flux densities (up to hundreds of thousand Jy), and small line widths (normally about 0.5 km/s and smaller). Because of that masers proved to be precise instruments for studies of kinematics and physical parameters of the objects in our and other galaxies.

The space radio interferometer RadioAstron provides a record of high angular resolution. This provides tight limits on the sizes of the most compact maser spots and estimates of their brightness temperatures, which are necessary input for the studies of the pumping mechanisms.

Typical values of the minimal flux density detectable with RA for the water masers at 22 GHz and hydroxyl masers at 1.665/1.667 GHz are 15 Jy and 3.5 Jy, respectively. These values were calculated for a typical line width of 0.1 km/sec and coherent accumulation time of 100 sec and 600 sec for 22 GHz and 1.665/1.667 GHz, respectively. However, when we use the large ground-based antenna, for example 100-m GBT, and the line is broad, RA proved ability to detect 3-4 Jy water maser source.

2. Statistics of maser observations for the first 6 years of operation

Maser observation program. During the period from November 2011 to May 2012 interferometric mode of RA operation was tested. For that purpose, a number of bright quasars and the brightest and most compact sources of maser emission were selected (Kardashev *et al.* 2015). Basic conditions for choosing these sources were the existence of details that remain compact (i.e. unresolved) on the longest baseline projections and the highest brightness temperatures measured during VLBI and VSOP surveys. The first positive detections of maser sources by space interferometer were achieved for W51 (water) and W75N (hydroxyl) in two sessions in May and July 2012. Baseline projections were $1.0-1.5$ and $0.1-0.8$ Earth diameters (ED), respectively. Later, more sophisticated data analysis led to even more positive results in these test sessions: compact water maser features were detected in W3 IRS5 and W3(OH) in two sessions in February 2012. Baseline projections were $3.7-3.9$ ED.

After the first successful tests, the early science program started. The main purpose of these observations was to obtain first astrophysical results and measurements of the main parameters of the operating interferometer. The list of observed sources was significantly expanded, objects of other types were included in addition to the star-forming regions. Stellar masers in S Per, VY CMa, NML Cyg, U Her and extragalactic masers in Circinus and N113 were observed. It was proved that RadioAstron can observe cosmic masers with very high spectral resolution. This was not obvious at the beginning and indicates the presence of the ultra-fine structure in the maser images, and that interstellar scattering does not prevent observations of masers in the galactic plane (Kardashev *et al.* 2015). Positive detections for stellar and extragalactic masers were not obtained during the early science program.

The early science program was followed by the key and general research programs which were conducted (and the general program continues at the moment) on the basis of the open call for proposals received from research teams around the world. Details of the preparation and the conditions of the call for proposals are published at the RadioAstron project site (*http://www.asc.rssi.ru/radioastron*). The main objectives of this phase of the maser program are studying the kinematics and dynamics of the compact sources of maser emission in star-forming regions, as well as the study of extragalactic masers. As a result, the signals from extragalactic masers NGC3079 and NGC 4258 were detected along with star-forming regions. The maser of NGC 4258 is associated with the accretion disk around super-massive black hole at the center of this galaxy. Projected baselines

Table 1. Maser sources detected on the space–ground baselines

Source	Projected baseline length, ED	Best angular resolution, μas
Galactic H_2O masers	**0.4–10.0**	**22**
W3 Irs5	2.5−2.8; 3.5; 3.9; 5.4; 6.0; 6.0−10.0	22
W49 N	2.2−3.0; 4.5; 7.9; 9.4	23
W3(OH)	3.9	56
Cepheus A	0.9−1.7; 1.1; 3.1−3.5	62
Orion KL	1.9; 3.4	64
W51 E8	0.4−2.3; 1.3; 1.4−1.8; 1.7	95
G43.8−00.13	1.2; 1.2	182
Extragalactic H_2O masers	**1.3 − >25**	**<9**
NGC 4258	1.3; 1.7; 6.5; 9.5; 11.6; 11.8; 12.2; 19.2; 19.5, >25	<9
NGC 3079	1.9; 1.9	115
Galactic OH masers	**0.1–1.9**	**1540**
Onsala 1	0.2−0.7; 1.0−1.9	1540
W75 N	0.1−0.3; 0.1−0.8	3660

in these observing sessions were up to 2.0 ED (for NGC3079) and exceeding 25 ED (for NGC 4258) corresponding to angular resolutions of 115 μas and 8 μas, respectively. The latter resolution represents a record of angular resolution in astronomy, the previous record of 11 μas was reported by us in (Sobolev *et al.* 2017). The new record resolution is formally sufficient to measure the parallax of the water maser source in the nearby galaxy LMC.

General statistics of maser source detections. This section provides statistics from the beginning of RA maser observations (Nov 2011) up to present time (Oct 2017). During this period a large amount of data was accumulated. 154 maser observation sessions were conducted, and 32 sources were observed. The majority of masers observed in RA program is related to star-forming regions – 20 sources in total. 8 maser sources in the envelopes of late-type stars of the Galaxy were observed, and 4 extragalactic masers in star-forming regions and circum-nuclear disks of external galaxies were observed.

The scientific data have been corrupted or lost in 10 sessions out of total 154 due to technical problems. 141 observations of the remaining 144 sessions at the moment (Oct 2017) are processed on the ASC software correlator (Likhachev *et al.* 2017), positive detections are obtained in 38 sessions. Thus, the current detection rate of fringes at space-ground baselines is about 27 %.

All of the successful fringe detections for galactic masers at space-ground baselines were obtained for the sources associated with star-forming regions, – 26 of all 38 positive

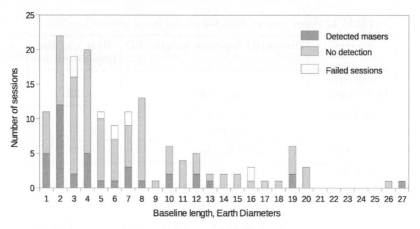

Figure 1. Statistics of maser observation results over projected baseline length of the space interferometer RadioAstron.

detections. 12 detections were obtained for extragalactic masers. No fringes for the stellar masers were obtained at space-ground baselines yet.

Table 1 gives information on the observational sessions which provided positive fringe detection with the space-ground interferometer. The columns show source names, projected lengths of the space-ground baselines at which the interferometric detection was obtained and the last column shows the best angular resolution achieved for each source. Each baseline (or baseline interval) corresponds to one observational session with positive detection.

The distribution of the number of detections depending on the length of the baseline projection is shown in the Figure 1, which presents statistics of observational sessions for the whole set of database lengths. It is seen that most of the positive detections fall in the range from 1 to 4 ED.

3. Summary

The main conclusions of the work are the following:

1. Space-VLBI observations of the water and hydroxyl masers show that the bright details of the masers in galactic star-forming regions often remain unresolved at projected baseline which exceeds Earth diameter many times.

2. Record angular resolution better than 9 μas was archieved in the observation of the water maser in NGC 4258. Scintillation does not prevent fringe detection.

3. Very compact water maser features with the angular sizes of about $20-60$ μas are registered in galactic star-forming regions. Their brightness temperatures range from 10^{14} up to 10^{16} K (Shakhvorostova *et al.* 2017). The best linear resolution better than 4 million km (a few solar diameters) was achieved for the maser in Orion. For Galactic masers, the highest angular resolution (23 micro-arcsec) was achieved for W49 N. This source is located in the galactic plane about 11 kpc away. So, detection provides important input for the theory of interstellar scattering in the Galaxy.

So, the major step from milli- to micro-arcsecond resolution in maser studies is done in the RadioAstron mission. The existence of the features with extremely small angular sizes is established. Further implementations of the space–VLBI maser instrument for studies of the nature of cosmic objects, studies of the interaction of extremely high radiation field with molecular material and studies of the matter on the line of sight are planned.

Acknowledgements

The RadioAstron project is led by the Astro Space Center of the Lebedev Physical Institute of the Russian Academy of Sciences and the Lavochkin Association of the Russian Federal Space Agency, and is a collaboration with partner institutions in Russia and other countries. This research is partly based on observations with the 100 m telescope of the MPIfR at Effelsberg; radio telescopes of IAA RAS (Institute of Applied Astronomy of Russian Academy of Sciences); Medicina & Noto telescopes operated by INAF; Hartebeesthoek, Torun, WSRT, Yebes, and Robledo radio observatories. The National Radio Astronomy Observatory is a facility of the National Science Foundation operated under cooperative agreement by Associated Universities, Inc. Results of optical positioning measurements of the Spektr-R spacecraft by the global MASTER Robotic Net, ISON collaboration, and Kourovka observatory were used for spacecraft orbit determination in addition to mission facilities. AMS was financially supported by the Russian Science Foundation (project no. 15-12-10017).

References

Kardashev, N. S., Alakoz, A. V., Kovalev, Y. Y. *et al.* 2015, *Solar System Research*, 49, 573
Sobolev, A. M., Shakhvorostova, N. N., Alakoz, A. V., & Baan, W. 2017, *ASP-CS*, 510, 27
Likhachev, S. F., Kostenko, V. I., Girin, I. *et al.* 2017, *J. Astron. Instrum.*, 6, 1750004-131
Shakhvorostova, N. N., Alakoz, A. V., & Sobolev, A. M. 2017, *Proc. IAU*, this volume

Astrophysical Masers:
Unlocking the Mysteries of the Universe
Proceedings IAU Symposium No. 336, 2017
A. Tarchi, M.J. Reid & P. Castangia, eds.

© International Astronomical Union 2018
doi:10.1017/S1743921317010869

H$_2$O MegaMasers: RadioAstron success story

Willem Baan[1,2], Alexey Alakoz[3],
Tao An[4], Simon Ellingsen[5], Christian Henkel[6,7], Hiroshi Imai[8],
Vladimir Kostenko[3], Ivan Litovchenko[3], James Moran[9],
Andrej Sobolev[10] and Alexander Tolmachev[3]

[1]Netherlands Institute for Radio Astronomy, Dwingeloo, The Netherlands,
email: baan@astron.nl
[2]XinJiang Astronomical Observatory, Chinese Academy of Sciences, Urumqi, PR China
[3]AstroSpace Center, Lebedev Physical Institute, Moscow, Russia,
email: alexey.alakoz@gmail.com
[4]Shanghai Astrophysical Observatory, Chinese Academy of Sciences, Shanghai, PR China
[5]University of Tasmania, Hobart, Australia
[6]Max Planck Institut für Radioastronomie, Bonn, Germany
[7]Astron. Dept., King Abdulaziz Univ., Jeddah, Saudi Arabia
[8]Kagoshima University, Kagoshima, Japan
[9]Center for Astrophysics, Cambridge MA, USA
[10]Ural Federal University, Ekaterinburg, Russia

Abstract. The RadioAstron space-VLBI mission has successfully detected extragalactic H$_2$O MegaMaser emission regions at very long Earth to space baselines ranging between 1.4 and 26.7 Earth Diameters (ED). The preliminary results for two galaxies, NGC 3079 and NGC 4258, at baselines longer than one ED indicate masering environments and excitation conditions in these galaxies that are distinctly different. Further observations of NGC 4258 at even longer baselines are expected to reveal more of the physics of individual emission regions.

Keywords. galaxies: nuclei, galaxies: ISM, masers, radio lines: ISM

1. Introduction

The Space Radio Telescope or the RadioAstron Observatory (RAO) is an international space-VLBI project led by the Astro Space Center of PN Lebedev Physical Institute. The RadioAstron payload on board of the Spectr-R mission has been equipped with a 10 meter antenna, two hydrogen masers, and receivers in P, L, C, and K-band (Kardashev *et al.* 2013). This paper presents some recent results of RAO observations of H$_2$O MegaMasers (MM) using the highest window of the Multi-Frequency Synthesis (MFS) system at 22 GHz, which covers a redshift range z = 0.0 - 0.0053 for H$_2$O MM emission studies. A total of 24 known H$_2$O MM have a redshift falling within this window but only 7 of them are deemed strong enough for detection with RadioAstron. The strong nearby sources that may be searched and their observing status are:

NGC3079 - 3.5 Jy - masering material shocked ISM in nucleus - detected at 1.6 - 2.3 ED
NGC4258 - 9.8 Jy - maser regions in compact nuclear disk - detected at 1.4 - 26.7 ED
NGC4945 - 8.5 Jy - masers in nuclear disk - not yet searched
N133 and 30Dor - 70 & 3 Jy - star formation regions in LMC - not yet detected
Circinus - 4.2 Jy - masers in Keplerian disk & bi-conical outflow - not yet detected
IC133 - 1.5 Jy - star formation region in M33 - not yet searched
NGC1068 - 0.65 Jy - Keplerian nuclear disk - potential candidate

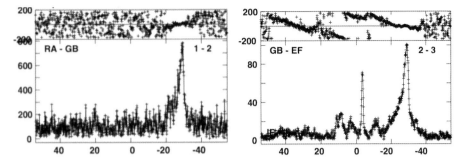

Figure 1. RadioAstron detection of NGC 3079 with Green Bank Telescope (GBT) at 2.3 ED. (left) The uncalibrated cross-correlation and fringe phase spectrum on the RAO-GBT baseline. (right) The uncalibrated cross spectrum and the fringe phase on the GBT - Effelsberg baseline. The axes are in arbitrary units of flux density and the velocity scale is centered at 984 km s^{-1}.

The initial detections of both NGC 3079 and NGC 4258 were obtained from observations in late 2014. After the update of the orbital elements for the ASC correlator (Likhachev *et al.* 2017), ten more new detections have been obtained for NGC 4258 up to a baseline of 26.7 Earth Diameters (ED).

2. NGC 3079 results

The high-brightness maser components in the H$_2$O MM NGC 3079 form an arc that is offset from the triple components of the Compact Symmetric Object (CSO) at the nuclear center (Kondratko, Greenhill and Moran 2005). The systemic velocity of NGC 3079 is 1116 km s^{-1} at a distance of 15.2 Mpc, while the main maser components are blueshifted between 950 - 990 km s^{-1}. Although initially the string of maser components has been interpreted as part of a rotating disk, the component velocities and the offset arc-structure do not support that picture. Instead, it appears that the maser components are a shocked part of the nuclear ISM that is also seen in blueshifted OH and HI components (Hagiwara, Klöckner and Baan 2004). These components are possibly connected to the two super-starburst regions found East of the core (Middelberg *et al.* 2007), which appear to be associated with the nuclear blowout seen in the optical. Shocks passing through the nuclear ISM provide for the H$_2$O population inversions resulting in the amplification of diffuse radio background across the nuclear region (see Baan and Irwin 1995), which in turn will result in concentrated regions of diffuse and compact emission.

The cross-correlation spectrum of NGC 3079 from the RAO-GBT observation at 2.3 ED has been presented in Figure 1a and shows two features peaking at 963 and 955 km s^{-1}. The strength of the features on the space-Earth baseline is significantly lower than obtained on the terrestrial baseline between GBT and Effelsberg in Figure 1b. The other features in the terrestrial spectrum were not detected on the space-Earth baseline and no detections have been made for NGC 3079 at any longer baselines.

The decrease in strength of the detected features and the fact that no further detections were made at longer baselines would indicate that the maser emission is mostly extended at a 2.3 ED baseline, and appears completely resolved at longer baselines. The beam size at 2.3 ED suggests that the strongest masering components in the nuclear medium are larger than 1400 AU at the distance of NGC 3079, which is consistent with amplification by diffuse medium. Although the association with shocks resulting from the super-starburst regions is not confirmed, any change in the velocity and spatial structure of the maser components found in past VLBI observations may help to confirm the nature of the excitation of these masering regions.

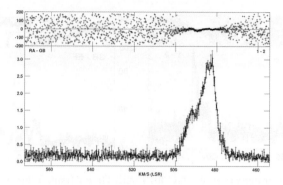

Figure 2. RadioAstron detection of NGC 4258. The uncalibrated cross-correlation spectrum of the RAO-GBT baseline of 1.9 ED. Flux density (arbitrary units) is plotted versus radial velocity.

3. NGC 4258 results

The H_2O MM emission regions in NGC 4258 are confined to a nearly edge-on disk of 0.5 pc surrounding the nuclear AGN (Herrnstein *et al.* 1998), also qualified as a CSO. The orbiting molecular regions within the disk drift in front of the southern part of the CSO radio continuum and amplify this continuum. Because of the orbital motion in the disk, the masering components drift across the spectrum from low velocity to high, at approximately 8.1 km s^{-1} yr^{-1} across the velocity range 440 - 550 km s^{-1} (Haschick, Baan and Peng 1994, Humphreys *et al.* 2008). The systemic velocity of NGC 4258 is 472 km s^{-1} at a distance of (approximately) 7 Mpc and at half the distance to NGC 3079.

At the time of this writing, the H_2O MM emission in NGC 4258 has been detected with 11 RadioAstron experiments, the first dating back to 2014. While fringes were initially found in observational data at a baseline of 1.9 ED, the updated orbital model of RAO at the ASC correlator resulted in subsequent detection of fringes up to baselines of 26.7 ED (corresponding to 340,000 km). The detection of fringes of the H_2O MM emission on this long RAO-GBT baseline constitutes an absolute record of 8 μas in angular resolution.

The RAO-GBT cross correlation spectrum for NGC 4258 at a baseline of 1.9 ED is presented in Figure 2. This (uncalibrated) spectrum shows a two-component profile that resembles the one obtained with terrestrial baselines, except for the lower amplitude on the space-Earth baseline. The resolution obtained for this baseline is 110 μas, which corresponds to 790 AU at the distance of NGC 4258. The profile also shows that a large fraction of the emission regions has not yet been resolved at this resolution, which is different from the results obtained for NGC 3079 at lower spatial resolution.

At higher resolution an increasing part of the diffuse maser components in NGC 4258 will be resolved, and only more compact components will remain unresolved. This is evident in the fringe amplitude plot of the detection with the 26.7 ED RAO-Medicina baseline displayed in Figure 3. Three individual components may be identified in this plot with a spatial resolution of 56 AU at the distance of NGC 4258. The mere detection of such compact masering components in NGC 4258 provides stringent limits on the degree of saturation and the excitation process. In addition, these more compact masering regions are likely to have less tangled magnetic fields and may allow detection of the magnetic field strength by its polarization properties.

4. Overview

The RadioAstron space-VLBI mission has successfully detected extragalactic H_2O MegaMaser emission regions, at space-Earth baselines ranging from 1.4 to 26.7 ED. The

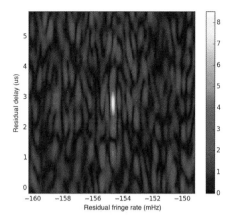

Figure 3. The fringe amplitude plot of the RAO-Medicina detection of NGC 4258 at 26.7 ED. The ratio of the interferometer fringe amplitude to the average noise amplitude is plotted against residual delay and fringe rate.

preliminary results for NGC 3079 and NGC 4258 at shorter baselines already indicate masering environments and excitation conditions that are distinctly different for the two galaxies. Although NGC 3079 has not been detected at baselines longer than 2.3 ED, early results for NGC 4258 suggest that individual masering regions can be detected at longer baselines up to 340,000 km.

5. Acknowledgements

The RadioAstron project is led by the Astro Space Center of the Lebedev Physical Institute of the Russian Academy of Sciences and the Lavochkin Scientific and Production Association under a contract with the Russian Federal Space Agency, in collaboration with partner organizations in Russia and other countries. These results are based partly on observations with the 100-m telescope of the MPIfR (Max-Planck-Institute for Radio Astronomy) at Effelsberg, on observations with the Medicina telescope operated by INAF - Istituto di Radioastronomia, and on observations with the 110-m Green Bank Observatory (GBT), which is a facility of the National Science Foundation operated by Associated Universities, Inc., under a cooperative agreement. Results from the optical positioning measurements of the Spektr-R spacecraft by the global MASTER Robotic Net (Lipunov *et al.* 2010), the ISON collaboration, and the Kourovka observatory were used for spacecraft orbit determination.

References

Baan, W. A. & Irwin, J. A. 1995, *ApJ*, 446, 602
Haschick, A. D., Baan, W. A., & Peng, E., 1994, *ApJ*, 437, L35
Hagiwara, Y., Klöckner, H.-R., & Baan, W. A., 2004, *MNRAS*, 353, 1055
Herrnstein, J. R., Greenhill, L. J., Moran, J. M., *et al.* 1998, *ApJ*, 497, L69
Humphreys, E. M. L., Reid, M. J., Greenhill, L. J., *et al.*, 2008, *ApJ*, 672, 800
Kardashev, N. S., Khartov, V. V., Abramov, V. V., *et al.*, 2013, *Astronomy Reports*, 57, 153
Kondratko, P. T., Greenhill, L. J., & Moran J. M. 2005, *ApJ*, 618, 618
Likhachev, S. F., Kostenko, V. I., Girin, I. A., *et al.* 2017, *J. Astron. Instrum.*, 6, 3, 1750004
Lipunov, V., Kornilov, V., Gorbovskoy, E., *et al.*, 2010, *Advances in Astronomy*, 2010, 30L
Middelberg, E., Agudo, I., Roy, A. L., & Krichbaum, T. P., 2007, *MNRAS*, 377, 731

Astrophysical Masers:
Unlocking the Mysteries of the Universe
Proceedings IAU Symposium No. 336, 2017
A. Tarchi, M.J. Reid & P. Castangia, eds.

© International Astronomical Union 2018
doi:10.1017/S1743921317009838

A next-generation Very Large Array

Eric J. Murphy
(on behalf of the ngVLA community)

National Radio Astronomy Observatory, 520 Edgemont Road, Charlottesville, VA 22903, USA
email: `emurphy@nrao.edu`

Abstract. In this proceeding, we summarize the key science goals and reference design for a next-generation Very Large Array (ngVLA) that is envisaged to operate in the 2030s. The ngVLA is an interferometric array with more than 10 times the sensitivity and spatial resolution of the current VLA and ALMA, that will operate at frequencies spanning $\sim$1.2 – 116 GHz, thus lending itself to be highly complementary to ALMA and the SKA1. As such, the ngVLA will tackle a broad range of outstanding questions in modern astronomy by simultaneously delivering the capability to: unveil the formation of Solar System analogues; probe the initial conditions for planetary systems and life with astrochemistry; characterize the assembly, structure, and evolution of galaxies from the first billion years to the present; use pulsars in the Galactic center as fundamental tests of gravity; and understand the formation and evolution of stellar and supermassive blackholes in the era of multi-messenger astronomy.

Keywords. instrumentation: high angular resolution, instrumentation: interferometers

1. Introduction

Inspired by dramatic discoveries from the Jansky VLA and ALMA, a plan to pursue a large collecting area radio interferometer that will open new discovery space from protoplanetary disks to distant galaxies is being developed by the astronomical community and NRAO. Building on the superb cm observing conditions and existing infrastructure of the VLA site, the current vision of a next-generation Very Large Array (ngVLA) will be an interferometric array with more than 10 times the sensitivity area and spatial resolution of the current VLA and ALMA, that will operate at frequencies spanning $\sim$1.2 – 116 GHz. The ngVLA will be optimized for observations at wavelengths between the exquisite performance of ALMA in the submm, and the future SKA1 at decimeter to meter wavelengths, thus lending itself to be highly complementary with these facilities. This is illustrated in Figure 1 where the effective collecting area for a number of radio/mm facilities expecting to be operational in the 2030s is shown. The ngVLA clearly sits in the frequency range outside of the core science missions for both the SKA1 and ALMA, creating an overwhelming set of new synergistic opportunities with these other facilities.

As such, the ngVLA will open a new window on the universe through ultra-sensitive imaging of thermal line and continuum emission down to milliarcecond resolution, as well as deliver unprecedented broad band continuum polarimetric imaging of non-thermal processes yielding a broad range of scientific discoveries (e.g., planet formation, signatures of pre-biotic molecules, cosmic cycling of cool gas in galaxies, massive star formation in the Galaxy etc.) Additionally included in current ngVLA planning is an option to greatly expand current U.S. VLBI capabilities by both replacing existing VLBA antennas/infrastructure with ngVLA technology and adding additional stations on $\sim$1000 km baselines to bridge the gap between ngVLA and existing VLBA baselines. A second science option to provide access to the low frequency sky (i.e., 5 – 800 MHz) in a commensal fashion is also being explored.

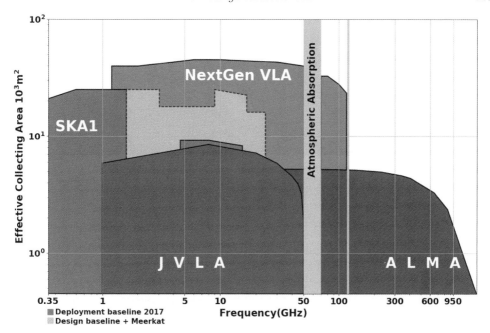

Figure 1. A comparison of effective collecting area plotted against frequency for various radio/mm facilities that will be operating in the 2030s, illustrating how complementary the ngVLA is to both ALMA and the SKA1.

2. ngVLA Key Science Goals

The ngVLA Science Advisory Council (SAC), a group of experts appointed by NRAO, in collaboration with the broader international astronomical community, recently developed over 80 compelling science cases requiring observations between 1.2 – 116 GHz with sensitivity, angular resolution, and mapping capabilities far beyond those provided by the Jansky VLA, ALMA, and the SKA1. These science cases span a broad range of topics in the fields of planetary science, Galactic and extragalactic astronomy, as well as fundamental physics. Consequently, the primary science requirement for the ngVLA has overwhelmingly been determined to be flexible enough to support the wide breadth of scientific investigations that will be proposed by its highly creative user base over the full lifetime of the instrument. This in turn makes the ngVLA a different style of instrument than many other facilities on the horizon (e.g., SKA1, LSST, etc.), which are heavily focused on carrying out large surveys. However, each of the individual science cases were objectively reviewed and thoroughly discussed by the different Science Working Groups within the ngVLA-SAC, ultimately distilling a finite list of key scientific goals for a future radio/mm telescope. The initial set of key science goals, along with the results from the entire list of science use cases, were then presented and discussed with the broader community at the ngVLA Science and Technology Workshop June 26 – 29, 2017 in Socorro, NM as a means to build consensus around a single vision for the key science mission of the ngVLA†. These key science goals are described in detail in ngVLA SAC (2017) and include:

- Unveiling the Formation of Solar System Analogs
- Probing the Initial Conditions for Planetary Systems and Life with Astrochemistry

† https://science.nrao.edu/science/meetings/2017/ngvla-science-program/index

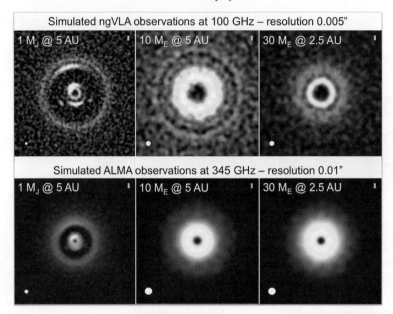

Figure 2. Simulated observations (ngVLA on top row, ALMA on bottom row) of the continuum emission of a protoplanetary disk perturbed by a Jupiter mass planet orbiting at 5 au (left), a 10 Earth mass planet orbiting at 5 au (center), and a 30 Earth-mass planet orbiting at 2.5 au (right). The ngVLA observations at 100 GHz were simulated assuming an angular resolution of 5 mas and a rms noise level of 0.5 μJy beam^{-1}. ALMA observations at 345 GHz where simulated assuming the most extended array configuration comprising baselines up to 16 km and a rms noise level of 8 μJy beam^{-1}. From Ricci *et al.* (2018).

- Charting the Assembly, Structure, and Evolution of Galaxies from the First Billion Years to the Present
- Using Pulsars in the Galactic Center to Make a Fundamental Test of Gravity
- Understanding the Formation and Evolution of Stellar and Supermassive Black Holes in the Era of Multi-Messenger Astronomy

In Figure 2 we highlight the power of the ngVLA to execute one of these science goals. The figure shows a comparison of ngVLA- (top row) and ALMA- (bottom row) simulated observations of continuum emission of a protoplanetary disk perturbed by a Jupiter mass planet orbiting at 5 au (left column), a 10 Earth mass planet orbiting at 5 au (center column), and a 30 Earth mass planet orbiting at 2.5 au (right column). ALMA is only able to achieve au-scale resolution at $\gtrsim$345 GHz, where the associated emission from such disks is optically-thick. The ngVLA will deliver the requisite combination of frequency coverage, resolution, and sensitivity to pierce into these highly enshrouded regions and directly image planet formation in the terrestrial zone. Even more spectacular, by imaging such sources with a monthly cadence over many years, the ngVLA effectively turns planet formation into a time-domain science, by being able to track the planetary orbits and associated tidal debris and see how they evolve. Such observations are highly synergistic with the major goals of both ground-based 30 m glass optical, and currently discussed next-generation UV/optical/NIR space missions, which include direct imaging of planets in the terrestrial zone.

The ngVLA will also be able to uniquely access and characterize a large range of predicted, yet undetected, pre-biotic molecules to try to discern the necessary requirements for habitability in proto-stellar regions. This is a second key science goal of the ngVLA.

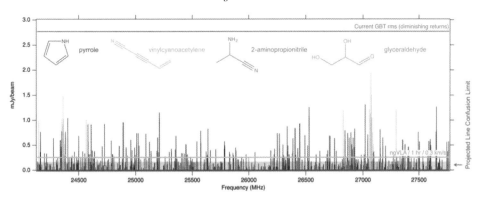

Figure 3. A conservative simulation of 30 complex organic molecules not currently detected in the ISM (in black), but which are good candidates for detection by the ngVLA. If present, such species are expected to be detectable by the ngVLA above the confusion limit of a typical ngVLA line-survey. A few key molecules are highlighted in color, which are important for understanding chemical evolution (e.g., increasingly complex carbon-nitrogen chains), probing chirality (e.g., glyceraldehyde), are part of largely unknown cyclic/aromatic chemistry (e.g., pyrrole), or are excellent candidates for being direct precursors to amino acids and other biogenic species (e.g., aminopropionitrile). (Credit: B. McGuire)

A simulated spectrum of 30 such complex organic molecules is shown in Figure 3 (Credit: B. McGuire). Each of these molecules are currently undetected in the ISM and clearly out of reach using current facilities. However, if present, such species are expected to be detectable by the ngVLA at frequencies where line blending is not an issue. A few key molecules are highlighted in color (see Figure 3 caption). The simulation assumes that the emission is coming from a compact, $1''$ hot core in Sgr B2(N), from molecules with column densities of $10^{12} - 10^{14}$ cm^{-2}, at $T = 200$ K, and with a linewidth of 3 km s^{-1}.

While we have highlighted two of the key ngVLA science goals here, as stated above, the primary mission of the ngVLA is to be flexible enough to deliver capabilities, enabling a broad range of science. A brief list includes: gravitational wave EM follow-up, extrasolar space weather, studying the bursting universe (FRB, GRB, TDE, etc.), low surface brightness H$\textsc{i}$ and CO, obscured black hole growth and associated AGN physics, quasar-mode feedback and the SZ effect, black hole masses and H$_0$ with mega-masers, μas Astrometry: ICRF, Galactic structure, etc., solar system remote sensing: passive and active radar, spacecraft telemetry, tracking: movies from Mars, and much more.

3. Reference Design Summary

The ngVLA concept began to rapidly converge post the June 2017 workshop in both its requirements and a matching concept that was able to support more than 80% of all of the science use cases submitted by the community (see Selina, Murphy, & Erickson 2017). A more detailed description of the design and the arrays performance can be found in Selina & Murphy (2017). Here, we simply highlight a number of the key elements.

The array configuration (Figure 4) shows a notional picture of the extent of the dish locations having a longest baselines of nearly 1000 km. The array collecting area is distributed to provide high surface brightness sensitivity on a range of angular scales spanning from approximately 1000 to 5 mas. In practice, this means a core with a large fraction ($\approx$50%) of the collecting area in a randomized distribution to provide high snapshot imaging fidelity, and arms/rings extending asymmetrically out to $\approx$1000 km baselines, filling out the (u, v)-plane with Earth rotation and frequency synthesis. The exact configuration

430   E. J. Murphy

Figure 4. Stations in the present ngVLA configuration (Greisen & Owen). The compact core
is located at the apex of the present VLA, and antennas are populated along the VLA arms.
Long baseline stations radiate primarily south and east from the VLA. The locations of the long
baseline stations are approximate, but account for land ownership and available infrastructure
including roads, electrical distribution lines, and fiber optics networks.

remains a work in progress, as we try to identify a version that yields the highest quality
synthesized beam for a large range of angular resolutions.

The present concept is a homogeneous array of 214 18m apertures, which is supported
by an internal parametric cost estimation. The antenna locations are fixed, and each is
outfitted with front ends that provide access to the atmospheric windows spanning 1.2
– 50.5 GHz and 70 – 116 GHz. The current front-end concept uses 6 single-pixel feeds in
two dewar packages, but a single dewar solution is actively being pursued. The antenna
optical configuration favors unblocked apertures, with an Offset Gregorian feed-low design offering synergy with the front-end concepts under consideration, while additionally
accounting for maintenance and operational concerns. The antenna surface error will be
limited to of order 160 μm rms, ensuring that the antenna Ruze efficiency is better than
50% at the 116 GHz upper operating limit. The correlator is an FX design, possibly with
a distributed F-engine. The correlator will support up to 64 k channels at coarse time
resolution, or up to 1 msec time resolution for time-domain science. The design will incorporate other necessary observing modes such as phased arrays for VLBI and pulsar
timing or searches.

While there has been significant convergence for many of the design parameters, there
still remains a number of open questions in the reference design that are actively being
investigated. These include determining the best strategy for phase calibration, an optimized array configuration that delivers a high-quality synthesized beam for a large range
of angular scales, and the need and associated requirements for a short-spacing mas array
(e.g., 16 6 m antenna array combined with 18 m total power antennas) to recover missing
flux.

4. Additional Science Options

4.1. *Expansion of VLBI Capabilities*

We are additionally looking at a possible science option that will upgrade/replace existing VLBA antennas/infrastructure with ngVLA technology. Such an option would be outside of the ngVLA project (construction and operations), and may largely depend on resources external from the astronomy community.

A potential configuration for such an option is to have $10 - 20\%$ of the ngVLA collecting area be placed in ≈ 8 stations of 3 antennas each. This will introduce new $\sim 1000\,\mathrm{km}$ baseline stations that will bridge the gap between the ngVLA core's "short" ($50 - 500\,\mathrm{km}$) spacings and the current continental-scale VLBA baselines. New stations, and any VLBA antenna replacements would leverage the ngVLA antenna design and receiver package, with all data being electronically transferred using the full ngVLA bandwidth.

Such a facility would significantly improve upon the current VLBA capabilities (sensitivity and astrometric accuracy), yielding a number of exciting new scientific possibilities. These include, sub- 1% constraints on H_0 measurements using megamasers, an order of magnitude improvement in General Relativity parameters from light bending, and routine distance measurements across the Milky Way. Such capabilities may even open the door to imaging the surfaces of 10's to 100's of stars. which is highly synergistic with current optical interferometry goals.

4.2. *Commensal Low Frequency Science*

There is also an investigation looking into the possibility for commensal low frequency capabilities, which is currently being called the next-generation LOw Band Observatory (ngLOBO; see Taylor *et al.* 2017). ngLOBO would provide access to the low frequency sky (i.e., $5 - 800\,\mathrm{MHz}$) in a commensal fashion, operating independently from the ngVLA, but leveraging common infrastructure (e.g., land, fiber, power, etc.) . This approach provides continuous coverage through an aperture array (called ngLOBO-Low) below $150\,\mathrm{MHz}$ and by accessing the primary focus of the ngVLA antennas (called ngLOBO-High) above $150\,\mathrm{MHz}$.

ngLOBO has three primary scientific missions: (1) Radio Large Synoptic Survey Telescope (Radio-LSST): a commensal, continuous synoptic survey of large swaths of the sky for both slow and fast transients; (2) Complementary low frequency images of all ngVLA targets and their environments to enhance their value; (3) Independent beams from the ngLOBO-Low aperture array for research in astrophysics, Earth science and space weather applications, engaging new communities and attracting independent resources. The ngVLA will be a superb, high frequency instrument and ngLOBO will additionally allow it to participate in the worldwide renaissance of low frequency science.

5. Summary

The ngVLA is being designed to tap into the astronomical community's intellectual curiosity and enable a broad range of scientific discoveries (e.g., planet formation, signatures of pre-biotic molecules, cosmic cycling of cool gas in galaxies, massive star formation in the Galaxy etc.). Based on community input to date, the ngVLA is the obvious next step to build on the VLA's legacy and continue the U.S.'s place as a world leader in radio astronomy. Presently, there have been no major technological risks identified. However, the project is continually looking to take advantage of major engineering advancements seeking performance and operations optimizations. As the project moves forward in preparation for the U.S. 2020 Astronomy Decadal Review, we will continue to refine the ngVLA science mission and instrument specifications/performance through a detailed science book and reference design study. The ultimate goal of the ngVLA is to give the U.S. and international communities a highly capable and flexible instrument to pursue their science in critical, yet complementary ways, with the large range of multi-wavelength facilities that are on a similar horizon.

References

ngVLA Science Advisory Council. 2017, *ngVLA Memo Series*, #19
Ricci L. *et al.* 2018, *ApJ*, 853, 110
Selina, R. & Murphy, E. 2017, *ngVLA Memo Series*, #17
Selina, R., Murphy, E., & Erickson, A 2017, *ngVLA Memo Series*, #18
Taylor, G., *et al.* 2017, *ngVLA Memo Series*, #20

Astrophysical Masers:
Unlocking the Mysteries of the Universe
Proceedings IAU Symposium No. 336, 2017
A. Tarchi, M.J. Reid & P. Castangia, eds.

© International Astronomical Union 2018
doi:10.1017/S174392131800008X

Maser science with the Square Kilometre Array

Anna Bonaldi, on behalf of the SKA science team

SKA Organization, Jodrell Bank, Lower Withington, Macclesfield,
Cheshire, SK11 9DL, United Kingdom
email: a.bonaldi@skatelescope.org

Abstract. The Square Kilometre Array (SKA), reaching a collecting area of one square kilometre, will be the world's largest radio telescope. Even in its first stage of deployment (SKA1, whose construction will be completed in 2026) it will enable transformational science on a very broad range of scientific objectives. Amongst them, there is the investigation of several Galactic and extra-galactic Masers. In this paper I will present the status of the SKA project and I will describe the capabilities of the SKA, with a focus on those that are more relevant for Maser science.

Keywords. intrumentation: interferometers, techniques: spectroscopic, physical and data processes: masers, Galaxy: kinematics and dynamics, radio lines: galaxies

1. The SKA concept

The SKA concept has been developed in order to answer to a set of fundamental questions:

- *The Cradle of life and Astrobiology:* How do planets form? Are we alone?
- *Strong-field Tests of Gravity with Pulsars and Black Holes:* Was Einstein right with General Relativity?
- *The Origin and Evolution of Cosmic Magnetism:* What is the role of magnetism in galaxy evolution and the structure of the cosmic web?
- *Galaxy Evolution probed by Neutral Hydrogen:* How do normal galaxies form and grow?
- *The Transient radio Sky:* What are Fast Radio Bursts? What haven't we discovered?
- *Galaxy Evolution probed in the radio Continuum:* What is the star-formation history of normal galaxies?
- *Cosmology and Dark Energy:* What are dark matter and dark energy? What is the large-scale structure of the Universe?
- *Cosmic Dawn and the Epoch of Reionization:* How and when did the first stars and galaxies form?

Such a broad range of science can be explored only by a very broad frequency range, which is probed by two separate instruments: the Low-frequency Array (SKA Low), from 50 to 350 MHz, as a single band, and the Mid-frequency array (SKA Mid), from 350 MHz to 24 GHz, as 5 bands, as detailed in Table 1. The former will consist of stations of log-periodic antennas and will be built in Australia; the latter will be an array of dishes built in South Africa. Both those sites have been selected as having very low Radio Frequency Interference (RFI), a feature that will be preserved in the decades to come, thanks to suitable agreements with the local governments. They already host radio astronomical facilities, including the SKA precursors MWA (Murchison Widefield Array, http://www.mwatelescope.org), ASKAP (Australian SKA Precursor,

Table 1. SKA frequency bands

Telescope name	Band name	Frequency range	Bandwidth	Notes
SKA Low	SKA Low	50–350 MHz	300 MHz	(1)
SKA Mid	Band 1	0.35–1.05 GHz	1 GHz	(1)
	Band 2	0.95–1.76 GHz	1 GHz	(1)
	Band 3	1.65–3.05 GHz	1 GHz	(2)
	Band 4	2.80–5.18 GHz	2.5 GHz	(2)
	Band 5a	4.60–8.50 GHz	2×2.5 GHz	(1)
	Band 5b	8.30–15.3 GHz	2×2.5 GHz	(1)
	Band 5c	15–24 GHz	2×2.5 GHz	(2)
	Band A	1.6–5.2 GHz	2.5 GHz	(2)
	Band B	4.6–24 GHz	2×2.5 GHz	(2)

Notes:
[1] Part of the baseline design and deployed as a top priority
[2] Deployed as an upgrade path (to be confirmed)

https://www.atnf.csiro.au/projects/askap/index.html) and HERA (Hydrogen Epoch of Reionization Array, http://reionization.org/) in Australia, and MeerKat (http://www.ska.ac.za/gallery/meerkat/) in South Africa.

2. SKA project schedule

The deployment of the telescope has been staged in two phases. In the first phase (SKA phase 1, or SKA1) around 131,000 of the total 1 million elements of the Low Frequency Aperture Array, and around 200 (64 of which will be integrated from Meerkat) of the total 2500 dishes of the Mid Frequency Array will be deployed. On phase 2, the remaining elements of the Low and Mid arrays will be deployed, as well as new technology (Phased Array Feeds and Mid-Frequency Aperture Array concepts are currently being developed). The deployment of receivers for the 5 bands of SKA Mid has been prioritised according to the science objectives they enable, with Bands 1, 2 and 5a/5b being the top priorities (see notes in Table 1).

The design process of SKA1 is expected to be complete in mid/late 2018, with construction beginning in late 2019 and ending in 2026. Early operations will begin in 2026/27 and routine operations in 2028.

3. SKA performance

The SKA will have a very good coverage in the visibility plane already in its first phase of deployment. This will guarantee excellent image quality, as illustrated in Figure 1. Figure 2 nicely illustrates the progress in sensitivity to be achieved in the next decades from the current state-of-the-art (LOFAR, uGMRT, JVLA and ALMA) with the deployment of SKA1 first, and, finally, SKA2. The analysis of the first prototypes for the Mid dishes (shown in Figure 3) has shown a total surface error RMS $< 350\,\mu$m and a relative pointing error RMS < 1.3 arcsec, which would guarantee a good high-frequency performance. This opens up the opportunity of extending the frequency coverage of the SKA beyond 24 GHz, subject to the interest of the community and the availability of funds.

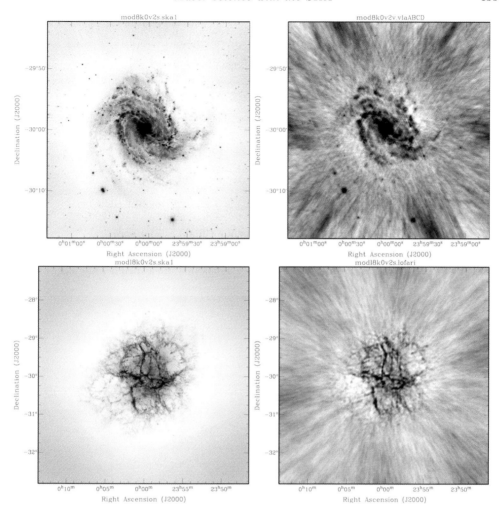

Figure 1. Image quality comparison between SKA phase 1 and the current state-of-the-art. *Top*: a single snapshot of SKA1-Mid (left) vs combination of snapshots in each of VLA A+B+C+D configurations (right). *Bottom*: a single snapshot of SKA1-Low (left) vs LOFAR-INTL (right).

4. SKA specifications for Maser observations

The relevant instrument for Maser science is SKA Mid, which will therefore be the focus of the next sections. In Table 2 we show some of the Maser lines that would be observable with the SKA. The design for the correlator of SKA1-Mid will be able to process frequency "slices" of 200 MHz with 65,536 channels, which corresponds to a channel width of 12.5 KHz. However, there will be also the possibility of "zoom windows", where a smaller frequency range will be processed with the same number of channels, thus allowing "zooming-in" any relevant spectral feature, as detailed in Table 3. The frequency centre of zoom windows within an SKA band will be completely flexible.

The SKA will have VLBI capabilities for both Low and Mid. It will be possible to use either the whole array or a subset (subarray) of the SKA for VLBI purposes. The frequency centre and resolution of the VLBI observation will be set independently. High data rate recording and/or eVLBI interface will be used to make the data available for the VLBI analysis, outside of the SKA correlator.

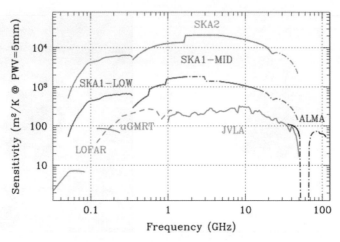

Figure 2. Sensitivity comparison between the SKA and the current state-of-the-art. The dot–dashed line on the SKA curves indicates frequency bands that are available as an upgrade path.

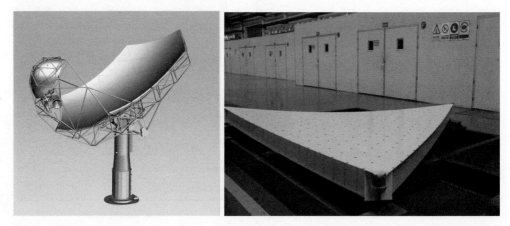

Figure 3. SKA-Mid dish prototyping. *Left*: model of the dish. *Right*: one of the panels, having an average surface accuracy of $\sim 250\,\mu m$ RMS.

5. SKA proposals

Two kinds of proposals will be invited for the SKA: "standard" PI-led proposals and Key Science Project (KSP) proposals. KSP proposals will typically require significantly more observing time and/or resources than PI-led proposals. As a general guideline, PI-led proposals would typically be observed within a single observing cycle while KSPs will typically run over multiple observing cycles. Time for both proposals will be allocated through normal submission and review procedures. The current working assumption is that between 50 and 75% of the time in the first five years of routine operations will be dedicated to KSPs.

One interesting feature that the SKA observations will have is the possibility of "commensality", which consists in multiple proposals potentially sharing the same data, with usage rights restricted to their science case. For example, a wide survey in continuum to investigate the star-formation history of the Universe could be used at the same time as a cosmological survey to investigate the large-scale-structures or to measure gravitational lensing. This will maximise the scientific output of the SKA telescope time and enable

Table 2. Some of the Maser lines observable with the SKA

Molecule	Frequency [GHz]	SKA Band
OH	1.612	Band 2
	1.665	Band 2
	1.667	Band 2
	1.720	Band 2
OH*	4.750	Band 4
	6.030	Band 5
	6.035	Band 5
	13.44	Band 5
H_2CO	4.829	Band 4
CH_3OH	6.668	Band 5
	12.18	Band 5
H_2O	22.23	Band 5
NH_3	23.87	Band 5

Table 3. Spectral resolution with SKA-Mid *zoom windows*

Window [MHz]	200	100	50	25	12.5	6.25	3.125
Channel width [KHz]	12.5	6.25	3.12	1.56	0.78	0.38	0.19

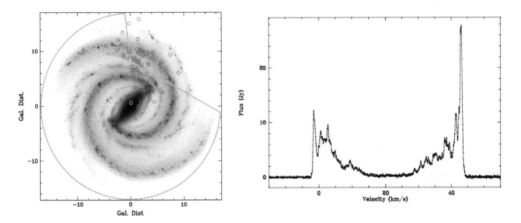

Figure 4. *Left*: Masers with astrometric parallaxes from the BeSSel survey and other observations (Reid *et al.* 2014) overlaid on the artist impression of the Milky Way, from Green *et al.* (2015). *Right*: 1612-MHz OH maser of the long-period variable star OH 16.1-0.2, from Etoka *et al.* (2015).

progressing with several KSPs simultaneously. The data will be also make completely public after a proprietary period, therefore constituting a legacy archive.

6. The SKA scientific community and Maser science

There is a large scientific community behind the SKA, who recently collaborated to produce an updated version of the science cases, published in the books *Advancing Astrophysics with the Square Kilometre Array*, available on http://skatelescope.org/books/. The community is organised in several science working groups, whose list is available on the SKA science website, astronomers.skatelescope.org. Participation to the working groups is open, and expressions of interest should be sent to the working group chairs, whose contact details are also provided in the SKA science website.

There is a huge potential in the SKA for doing Maser science, thanks to the the excellent sensitivity and image quality, good spectral resolution with flexible setup, VLBI capabilities and full polarization information. Some SKA Maser science cases are described in Green *et al.* (2015) and Etoka *et al.* (2015) (see also Figure 4). Moreover, the concept of commensality of SKA observations will allow using wide SKA surveys for blind Maser detection.

References

Braun R., Bourke T., Green J. A., Keane E., & Wagg J., 2015, *aska.conf*, 174

Green, J., Van Langevelde, H. J., Brunthaler, A., Ellingsen, S., Imai, H., Vlemmings, W. H. T., Reid, M. J., & Richards, A. M. S., 2015, *aska.conf*, 119

Etoka, S., Engels, D., Imai, H., Dawson, J., Ellingsen, S., Sjouwerman, L., & van Langevelde, H., 2015, *aska.conf*, 125

Reid, M. J., Menten, K. M., Brunthaler, A., *et al.* , 2014, *ApJ*, 783, 130

Astrophysical Masers:
Unlocking the Mysteries of the Universe
Proceedings IAU Symposium No. 336, 2017
A. Tarchi, M.J. Reid & P. Castangia, eds.

© International Astronomical Union 2018
doi:10.1017/S1743921317010560

MultiView High Precision VLBI Astrometry at Low Frequencies

M. Rioja[1,2,3], R. Dodson[1], G. Orosz [4] and H. Imai [4,5]

[1] International Centre for Radio Astronomy Research, UWA, 7 Fairway, Western Australia
[2] CSIRO Astronomy and Space Science, 26 Dick Perry Avenue, Kensington WA 6151, Australia
[3] Observatorio Astronómico Nacional (IGN), Alfonso XII, 3 y 5, 28014 Madrid, Spain
[4] Graduate School of Science and Engineering, Kagoshima University, 1-21-35 Korimoto, Kagoshima 890-0065, Japan
[5] Science and Engineering Area of Research and Education Assembly, Kagoshima University, 1-21-35 Korimoto, Kagoshima 890-0065, Japan

Abstract. Observations at low frequencies ($< 8GHz$) are dominated by distinct direction dependent ionospheric propagation errors, which place a very tight limit on the angular separation of a suitable phase referencing calibrator and astrometry. To increase the capability for high precision astrometric measurements an effective calibration strategy of the systematic ionospheric propagation effects that is widely applicable is required. The MultiView technique holds the key to the compensation of atmospheric spatial-structure errors, by using observations of multiple calibrators and two dimensional interpolation. In this paper we present the first demonstration of the power of MultiView using three calibrators, several degrees from the target, along with a comparative study of the astrometric accuracy between MultiView and phase-referencing techniques. MultiView calibration provides an order of magnitude improvement in astrometry with respect to conventional phase referencing, achieving $\sim$ 100micro-arcseconds astrometry errors in a single epoch of observations, effectively reaching the thermal noise limit.

Keywords. astrometry - techniques: high angular resolution - techniques: interferometric

1. Introduction

Very Long Baseline Interferometry observations hold the potential to achieve the highest astrometric accuracy in astronomy. The development of advanced phase referencing (PR) techniques to compensate for the tropospheric propagation errors have led to routinely achieving micro-arcsecond (μas) astrometry at frequencies between $\sim$ 10 and a few tens of GHz using alternating observations of the target and a nearby calibrator (Reid & Brunthaler 2004; Honma *et al.* 2008). More recently, the development of phase calibration techniques using (nearly) simultaneous observations at multiple mm-wavelengths, that is Source Frequency Phase Referencing (SFPR) (Rioja & Dodson 2011) and Multi Frequency Phase Referencing (MFPR) (Dodson *et al.* 2017), have extended the capability to measure μas astrometry up to mm-wavelengths. This capability has resulted in a wide scientific applicability (Reid & Honma 2014, and references therein). Nevertheless the application of these advanced PR techniques to relatively low frequencies $\leqslant$ 8 GHz are hindered by the ionospheric propagation effects, increasingly dominant at lower frequencies. Therefore, a new strategy is required to overcome the limitations imposed by the ionospheric propagation medium. In this paper we present results from the MultiView (MV) technique which, by deriving 2-D phase screens from observations of three or more calibrators, achieves a superior mitigation of atmospheric errors that results in increased precision astrometry, along with wide applicability. For a complete review of MV and application to maser astrometry see Rioja *et al.* (2017) and Orosz *et al.* (2017).

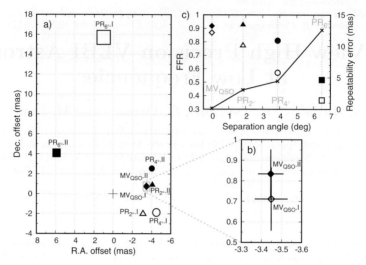

Figure 1. Sky distribution of the sources observed with the VLBA at 1.6 GHz. Dashed lines and arrows mark the source switching order during the observations with 5-min duty cycles. Star and solid symbols mark the simultaneously observed OH–C4 pair, with the VLBA antennas pointed halfway between the two. The two concentric circles represent the half-power beamwidth and full beamwidth of the antennas. Both OH and C4 are targets in the astrometric analyses, and C2, C1 and C3 are the calibrators $\sim 2^{\circ}, 4^{\circ}$ and 6° away, respectively.

2. Observations

We conducted two epochs of observations with the Very Long Baseline Array (VLBA) separated by one month, on 2015 June 8 and July 7 at 1.6 GHz (obs. ID: BO047A7, BO047A4). Both epochs used identical setups with a duration of ~ 4 hours.

The observations consisted of alternated scans switching between the sources shown in Fig. 1 with a duty cycle of ~ 5 minutes. The two sources in the centre of the distribution, the OH maser and the quasar C4, were observed simultaneously since they lie within the primary beam of the antennas. They are the targets of the analyses presented in this paper, allowing the MultiView calibration to be tested for both a maser line and quasar continuum observations simultaneously. The other three sources act as calibrators.

3. Basis of Astrometric Technique: MultiView

The MV calibration strategy corrects for the direction dependent nature of the ionospheric phase errors by using simultaneous or near-simultaneous observations of multiple calibrators around the target. Then we use a two dimensional (2-D) interpolation of the antenna phases, estimated along the directions of all calibrators, to provide corrections along the line of sight of the target observations. This is realized by a weighted average of the complex antenna gains, representing the relative source distribution in the sky, as shown in Fig. 1 for the case of interest to this paper. This is equivalent to the treatment of the propagation medium as a wedge-like spatial structure, up to several degrees in size, above each antenna (Fomalont & Kopeikin 2002; Rioja *et al.* 2002). The temporal structure of the propagation medium effects is best calibrated using simultaneous observations of the calibrators and the target sources in MV observations. However when this observing configuration is not possible one can use alternating observations of the sources, as long as the duty cycle is less than the atmospheric coherence time. Our implementation of the MV direction dependent calibration strategy is more complicated

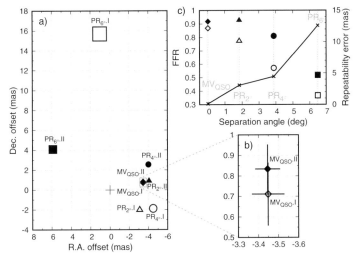

Figure 2. *a)* Astrometric offsets in the angular separations measured with MV (with three calibrators), and with PR_{2^o}, PR_{4^o} and PR_{6^o} analyses (with C2, C1, and C3 as calibrator, respectively), from the observations of quasars at the two epochs (I and II). The size of the plotted symbol corresponds to the estimated thermal noise error in each case. The labels describe the analysis id. and epoch of observations. *b)* Zoom for MV astrometric solutions. The error bars are the thermal noise errors. Both epochs agree within the error bars. *c)* Solid line shows the corresponding repeatability astrometric errors versus the angular separation between target and calibrator for PR analysis, and for an effective 0° separation for MV. Filled and empty symbols show the Flux Fractional Recovery (FFR) quantity versus angular separation for MV (diamond), PR_{2^o} (triangle), PR_{4^o} (circle), and PR_{6^o} (square) analyses, for epochs I (empty) and II (filled).

than a basic bilinear interpolation. It includes a correction for untracked 2π-ambiguities inherent in the measured calibrator phases, which, if present, would lead to errors in the interpolated phases. We have carried out a comparative study between MV and PR astrometric analysis, the latter using one calibrator at a time. We have used the repeatability of the measured positions between the two epochs of observations, which provide independent measurements of the relative source position, as an empirical estimate of the precision of the calibration method. Note that while there is a limited sample of two epochs the different analyses are carried out on the same observations, enabling a direct comparison of the compensation efficiency of the systematic errors under the same weather conditions.

4. Results and Discussion

Fig. 2a shows the astrometric measurements of the target quasar C4 using PR (analysis id. PR_{2^o}, PR_{4^o}, PR_{6^o}) and MV (analysis id.: MV_{QSO}) at the two epochs of observations. Fig. 2b shows an expansion of the area around the MV measurements at the two epochs. The astrometric uncertainties in the plot are 1-σ thermal noise error bars. For stationary sources, as it is the case for quasars, no or negligible position changes are expected between the two epochs. Therefore one can estimate the astrometric error (σ_{rep}) using the repeatability between the two epochs. It is immediately obvious that σ_{rep} are much larger for PR, compared to those for MV. Also, that σ_{rep} are larger than the thermal noise errors for PR; instead they are within the 1σ thermal noise error bars for MV. Fig. 2c displays σ_{rep} as a function of source pair angular separations, for PR analysis. This linear trend is as expected from PR analysis, as closer angular separations provide a better atmospheric compensation. The MV σ_{rep} values are the smallest, more than one order of magnitude smaller than those for the closest pair with PR, and are equivalent to those from a very close pair of sources (i.e. close to zero angular separation) in PR analysis. Also in this figure are shown the coherence losses (i.e. FFR) in the images undergone in each analysis, for both epochs. It is worth highlighting that MV and PR$_{2^\circ}$ images are of similar quality and have similar FFR values. This underlines the insensitivity of the PRed images to large systematic errors. Also, underlines the superior quality of the calibration of atmospheric errors using multiple calibrators, compared to that achieved with a single calibrator with the same range of angular separations, and that MV analysis leads to higher precision astrometry. This is in agreement with the findings from our previous simulation studies, where we concluded that using multiple calibrator sources with MV resulted in one order of magnitude improvement compared to PR with a single calibrator (Jimenez-Monferrer *et al.* 2010; Dodson *et al.* 2013).

5. Conclusion

We have presented a demonstration of the capability of the MultiView technique to achieve a superior mitigation of atmospheric errors that results in increased precision astrometry, along with wide applicability by relaxing the constraints on the angular separation up to few degrees, and does not require alignment of sources. The scope of application is for the low frequency regime where the perfomance of PR is degraded due to the spatial structure of the ionospheric dominant errors. We believe that the implementation of MultiView techniques will enhance the performance of VLBI observations, by providing higher precision astrometric measurements of many targets at low frequencies, in particular for methanol and OH maser astrometry.

MultiView will achieve its full potential with the enhanced sensitivity and multibeam capabilities of SKA and the pathfinders, which will enable simultaneous observations of the target and calibrators. Our demonstration indicates that the 10 micro-arcseconds goal of astrometry at ~ 1.6GHz using VLBI with SKA is feasible using the MultiView technique.

References

Dodson, R., Rioja, M., Asaki, Y., Imai, H., Hong, X.-Y., & Shen, Z. 2013. *AJ*, **145**, 147.
Dodson, R., Rioja, M., Molina, S., & Gómez, J. L. 2017. *ApJ*, 834, 177.
Fomalont, E. B., & Kopeikin, S. 2002. *Proc. 6th EVN Symp.*
Honma, M., Tamura, Y., & Reid, M. J. 2008. *PASJ*, **60**, 951.
Jimenez-Monferrer, S., Rioja, M. J., Dodson, R., Smirnov, O., & Guirado, J. C. 2010. *Proc. 10th EVN Symp.*
Orosz, G., Imai, H., Dodson, R., Rioja,M. J., Frey, S., Burns, R. A., Etoka, S. Nakagawa, A., Nakanishi, H., Asaki, Y., Goldman, S. R. & Tafoya, D. 2017 *AJ*,**153**, 119
Reid, M. J., & Brunthaler, A. 2004. *ApJ*, **616**, 872.
Reid, M. J., & Honma, M. 2014. *ARAA*, **52**, 339.
Rioja, M., & Dodson, R. 2011. *AJ*, **141**, 114.
Rioja, M. J., Porcas, R. W., Desmurs, J.-F., Alef, W., Gurvits, L. I., & Schilizzi, R. T. 2002. *Proc. 6th EVN Symp.*
Rioja, M. J., Dodson, R., Orosz, G., Imai, H., & Frey, S. 2017. *AJ*,**153**, 105

Astrophysical Masers:
Unlocking the Mysteries of the Universe
Proceedings IAU Symposium No. 336, 2017
A. Tarchi, M.J. Reid & P. Castangia, eds.

© International Astronomical Union 2018
doi:10.1017/S1743921317011383

Peculiarities of Maser Data Correlation / Postcorrelation in Radioastron Mission

I. D. Litovchenko, S. F. Likhachev, V. I. Kostenko, I. A. Girin, V. A. Ladygin, M. A. Shurov, V. Yu. Avdeev and A. V. Alakoz

Astro Space Center, Lebedev Physical Institute
119991, 53, Leninskiy Prospekt, Moscow, Russia
email: grosh@asc.rssi.ru

Abstract. We discuss specific aspects of space-ground VLBI (SVLBI) data processing of spectral line experiments (H_2O & OH masers) in Radioastron project. In order to meet all technical requirements of the Radioastron mission a new software FX correlator (ASCFX) and the unique data archive which stores raw data from all VLBI stations for all experiments of the project were developed in Astro Space Center. Currently all maser observations conducted in Radioastron project were correlated using the ASCFX correlator. Positive detections on the space-ground baselines were found in 38 sessions out of 144 (detection rate of about 27%). Finally, we presented upper limits on the angular size of the most compact spots observed in two galactic H_2O masers, W3OH(H_2O) and OH043.8-0.1.

Keywords. Space vehicles, radio lines: ISM, instrumentation: high angular resolution.

1. Correlator and Data Archive Overview

ASCFX correlator (first Fourier transform and then cross-spectrum multiplication) developed in Astro Space Center has the following properties (Likhachev *et al.* 2017): (1) Full support of all modern VLBI data formats: Radioastron data format (RDF), Mark5A, Mark5B, VDIF, VLBA, K5; (2) Computing cluster with 1 Tflop/s performance (about 100 processor cores); (3) Online raw data storage of 90 TB for correlation; (4) Online data storage for correlated observations – 250 TB; (5) Raw data archive of 3000+3000 TB: 1st copy on hard drives and 2nd copy on tapes; (6) Correlator can operate in "Continuum", "Spectral Line" and "Pulsar" modes. (7) The unique data archive stores all raw data (including data from ground stations) for all observations. This allows to re-correlate the data of any experiment with the updated information, like improved reconstructed orbit, different coordinates of phase center, different coordinates for ground stations, clock offsets etc.

2. Specific aspects of calibration and data processing

In comparison to the ground VLBI, SVLBI observations require correlator to take into account: relativistic effects of highly elongated elliptical orbit, significantly wider range of possible values for geometric delay, delay rate and acceleration term (see Likhachev *et al.* 2017). Galactic masers often characterized by the presence of a large number of features scattered over a relatively large area, so that at longer baselines it become important to perform initial fringe search in a very wide window and to correlate the data with multiple phase centers corresponding to different parts of a source.

Amplitude calibration & Bandpass calibration. Typically in VLBI observations amplitude calibration of each band is mainly based on T_{sys} measurements done by inserting noise calibration signal during the gaps between observing scans. In case of strong maser lines better precision might be achieved using a template calibration method based on total power spectra obtained from a reliable antenna. Amplitude bandpass calibration can be derived from spectral observations itself by flagging channels with maser emission and from observations of continuum calibrator or off-source position. Performing complex bandpass is possible only for short baselines where a very high SNR could be obtained in observations of strong continuum sources.

Fringe fitting & delay calibration. Since the baselines for SVLBI are significantly longer than the Earth diameter (ED) at 22 GHz we have a limited sample of unresolved continuum sources which can be used as delay calibrators for spectral line soures. In significant part of observations we should rely on the reconstructed orbit, that currently provides an accuracy of 200 m in the position (0.7 μs in delay) (Likhachev *et al.* 2017), and other information is known a priori. Fringe rate fitting is possible using spectral line data itself.

3. Preliminary results: estimation of angular size of H_2O maser spots

To estimate the size of the most compact features we used the data from two 1 hour long observations of two galactic masers (W3OH(H2O) and OH 043.8-0.1) at 22 GHz water line. Experiments were conducted within the maser survey during the Early Science Program and Announcement of Opportunity 2 period and then processed using the ASCFX correlator. Even in case of short observations with a limited sampling of the UV-plane it is still possible to estimate some parameters of a source using simplifed assumption like a circular Gaussian brightness distribution. After performing the calibration procedure, we constructed the "visibility amplitude versus baseline projection" dependence for the brightest detected features and then calculated their angular size using the expression: $\Theta_r = \frac{2\sqrt{ln2}}{\pi}\frac{\lambda}{B}\sqrt{ln(\frac{V_0}{V_q})}$, where B - projected baseline length, V_0 and V_q - visibilities at zero-space and current baseline respectively.

W3OH(H2O), experiment RAES02A, ground stations: Effelsberg, Yebes. A two-component model can be assumed, which is composed of an extended "halo" and a bright compact detail. Unfortunately, because of the big gap between the space and ground baselines, it's impossible to perform reliable fit for these two components together. We performed separate fitting for each potential component and obtained their sizes: extended "halo" $- \approx 740$ μas and the upper limit for the size of compact detail $- \approx 42$ μas.

OH043.8–0.1, experiment RAGS11AZ, ground stations - Medicina, Yebes, HartRAO 26m. In this experiment the length of space-ground baselines was smaller than 1 ED and the visibility data well fitted with a single component providing the angular size of 73 μas.

Acknowledgements

The RadioAstron project (Kardashev *et al.* 2013) is led by the Astro Space Center of the Lebedev Physical Institute of the Russian Academy of Sciences and the Lavochkin Scientific and Production Association under a contract with the Russian Federal Space Agency, in collaboration with partner organizations in Russia and other countries. This research is partly based on observations with the 100 m telescope of the MPIfR at Effelsberg; Medicina telescope operated by INAF; Yebes 40m (operated by IGN), and HartRAO 26m radio telescopes.

References

Kardashev, N. S., Khartov, V. V., Abramov, V. V. *et al.* 2013, *Astron. Rep.*, 57, 153
Likhachev, S., Kostenko, V., Girin, I. *et al.* 2017, *J. Astron. Instrum.*, 6, 3

Astrophysical Masers:
Unlocking the Mysteries of the Universe
Proceedings IAU Symposium No. 336, 2017
A. Tarchi, M.J. Reid & P. Castangia, eds.

© International Astronomical Union 2018
doi:10.1017/S1743921317011346

First Galactic Maser Observations on Ventspils Radio Telescopes – Instrumentation and Data Reduction

Ivar Shmeld, Artis Aberfelds, Kārlis Bērziņš, Vladislavs Bezrukovs, Mārcis Bleiders and Artūrs Orbidans

Engineering Research Institute Ventspils International Radio Astronomy Centre (VIRAC),
Ventspils University College,
Inzenieru 101,LV-3601, Ventspils, Latvia
email: ivarss@venta.lv

Abstract. Ventspils International Radio Astronomy Centre (VIRAC) has two fully steerable Cassegrean System 32 and 16 m radio telescopes. After renovation and modernization program the Galactic masers, particularly CH_3OH research and monitoring program became one of the most important realized on these telescopes. Both telescopes are equipped with broadband cryogenic receivers covering 4.5-8.8 GHz frequency band. Digital backend consisting from DBBC-2 (*Digital Base Band Convertor* developed by HAT-LAB, Italy) and FLEXBUFF (data storage system based on commercially available server system) is used for data digitalization and registration. A special program complex for spectral line data reduction and correction was developed and implemented.

Keywords. instrumentation: miscellaneous, methods: data analysis, ISM: molecules – lines and bands

1. Instrumentation and data reduction

Instrumentation. Ventspils International Radio Astronomy Center (VIRAC) operates with two radio telescopes RT-16 and RT-32 accordingly with 16 and 32 m fully steerable Cassegrain type antennas. Recently both antennas have been upgraded, new tracking and reception systems and even new dish for RT-16 were installed.

The main receiving system of both telescopes are cryogenic receivers with 4.5 – 8.8 GHz frequency range. Antennas are equipped with modern data registration units: *Digital Base Band Converters* (DBBC2) as analog to digital converter and two data recorders *Mark5C*, and *Flexbuff* – data storage system based on commercially available server system. Available configuration allows parallel recording of two circular polarizations up to 1 GHz bandwidth with data rate up to 4 Gbps. The broad-band network connection allows to stream recorded data flow to the data processing centres in real time.

Antena calibration. The one of common methods for amplitude calibration is total power measurements of cold sky used with calibrated noise diode integrated in the receiver system. However on the moment our calibration of the noise diode seems to be too uncertain for our maser lines' calibration and its improvement to the required precision is the task for near future. Therefore, for estimation of absolute values of registered maser lines, as a calibrator was taken source G32.745-0.076 simultaneously observed by Torun 32 m telesscope, and its spectral features with velocities 30.4 and 39.2 km/s. Average amplitude between this two features are known as 15 Jy (M. Olech, priv. comm.).

Data processing. The digital output of DBBC2 in our case is data stream with 2 MHz bandwidth in two channels for left and right circular polarizations. As autocorrelation

446   I. Shmeld *et al.*

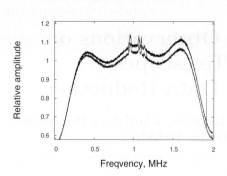

Figure 1. Input spectra. Two lines are LCP and RCP

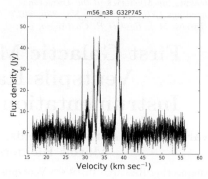

Figure 2. Gain dependence from frequency corrected, the region with maser lines excreted

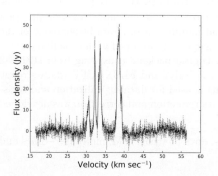

Figure 3. Averaging between LCP and RCP and primary noise filtering done

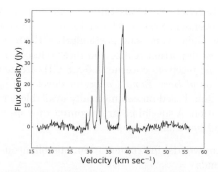

Figure 4. Final spectra: noise filtering by using Gaussian kernel done

spectrometer, with resolution power between data points 488 Hz, the spectrometric program from program package *MARK5Access* was used. The further data processing was done by program complex developed in VIRAC *sprli.py* with following steps shown in figures below: (i)converting data in Jy; (ii)cutting (under manual control) the spectral region with maser lines; (iii)the zero level of signal is found and gain dependence of system gain is compensated; (iv)averaging between left and right polarization channels; (v)noise filtering by using Gaussian 1DK kernel from *Astropy* program package.

The overview of the CH_3OH observation program and its first results is given at Aberfelds *et al.* (2018).

2. Acknowledgements

This work is financed by ERDF project No. 1.1.1.1/16/A/213, being implemented in Ventspils University College.

Reference

Aberfelds, A., Shmeld, I., & Berzins, K. 2018, these proceedings

Astrophysical Masers:
Unlocking the Mysteries of the Universe
Proceedings IAU Symposium No. 336, 2017
A. Tarchi, M.J. Reid & P. Castangia, eds.

© International Astronomical Union 2018
doi:10.1017/S174392131701047X

Brightness temperatures of galactic masers observed in the RadioAstron project

Nadezhda N. Shakhvorostova[1], Alexey V. Alakoz[1] and Andrej M. Sobolev[2]

[1] Astro-Space Center of Lebedev Physical Institute,
84/32 Profsoyuznaya st., Moscow, GSP-7, 117997, Russia,
email: nadya@asc.rssi.ru, alexey.alakoz@gmail.com

[2] Ural Federal University, 19 Mira street, Ekaterinburg, 620002, Russia
email: Andrej.Sobolev@urfu.ru

Abstract. We present estimates of brightness temperature for 5 galactic masers in star-forming regions detected at space baselines. Very compact features with angular sizes of $\sim$23-60 μas were detected in these regions with corresponding linear sizes of $\sim$4-10$\times 10^6$ km. Brightness temperatures range from 10^{14} up to 10^{16} K.

Keywords. masers, techniques: high angular resolution

1. Maser observations in the RadioAstron project

Galactic masers have been observed in the RadioAstron (RA) project during 6 year of operation since the launch in July 2011 (Kardashev *et al.* 2015). The satellite is equipped with receivers allowing observations of strong maser lines at 22235, 1665 and 1667 MHz. The space interferometer provides a record angular resolution up to 7 μas at 22 GHz. So, we can put tight limits on the sizes of very compact maser spots, estimate their brightness temperatures and, thus, obtain important parameters for maser models.

The sensitivity of the RA together with the 100-m Effelsberg radio telescope at 22 GHz is $\sim$10 Jy (at 6σ) with a coherent integration time $\sim$600 sec and typical line width $\sim$0.4 km/s. Observations carried out for H_2O masers on the RA indicate in the most cases only a small contribution from ultra-compact components to the total flux of the separate spatial-kinematic features. Thus, the W3 IRS 5, observed on a baseline of 2.5 Earth Diameter (ED), showed a visibility function amplitude of only 1% of the total flux density (Sobolev *et al.* 2017).

Such super-compact H_2O features were successfully detected in 7 galactic star-forming regions regions and in 2 extragalactic masers in NGC4258 and NGC3079. Current statistics of RA observations can be found in (Sobolev *et al.* 2017). In the present work we consider 5 galactic H_2O masers and obtain upper limits on the angular size of the most compact components and lower limits on the brightness temperature. We used the data processed on the ASC software correlator for the RA mission (Likhachev *et al.* 2017).

2. Brightness temperatures from the interferometric visibilities

Normally, brightness temperature T_b can be obtained from imaging with a long period of observations and many telescopes involved. But there are a lot of short observations ($\sim$1 hour) in the early RA maser survey with a few baseline sets, about 3 to 6. In this case it is possible to estimate brightness temperature of a source using some assumptions.

447

Table 1. Brightness temperature for compact H_2O masers observed in the RA project.

Source	RA (J2000) hh mm ss.ss	DEC (J2000) ∘ ′ ″	Baseline, ED	Resolution, μas	$T_{b,min}$, K	T_b, K
Orion KL	05 35 14.13	−05 22 36.48	3.3	66	1.2×10^{15}	6×10^{15}
Cepheus A	22 56 17.97	62 01 48.75	3.4	64	1.2×10^{14}	3×10^{14}
W3 OH	02 27 04.84	61 52 24.61	3.8	58	2.1×10^{14}	7×10^{14}
W3 IRS5	02 25 40.71	62 05 52.52	5.4	40	1.5×10^{15}	8×10^{15}
W49 N	19 10 13.41	09 06 12.80	9.6	23	4.5×10^{14}	3×10^{15}

Thus, without a priory information about brightness distribution, we may use a circular Gaussian and estimate T_b and size of a source as proposed in (Lobanov 2015):

$$T_b = \frac{\pi}{2k}\frac{B^2 V_0}{\ln(V_0/V_q)}[K],\qquad(2.1)$$

where V_q is the visibility amplitude, V_0 is the space-zero visibility, B is the baseline length, $q = B/\lambda$. It was shown in (Lobanov 2015) that T_b is at its lowest when $V_0/V_q = e$. This provides the minimal brightness temperature given the baseline length and correlated flux obtained from data processing using PIMA package (Petrov *et al.* 2011):

$$T_{b,min} \approx 3.09\,(B[\mathrm{km}])^2\,(V_q[\mathrm{mJy}])\,[K].\qquad(2.2)$$

3. Results and conclusions

Results of our calculations of T_b and $T_{b,min}$ according to Eqs. 2.1 and 2.2 are given in Table 1. Columns contain from left to right: (1) source name, (2) RA and DEC coordinates (J2000), (3) baseline in units of Earth diameters (ED), (4) corresponding resolution in μas (this value can be considered as an upper limit of angular size of the compact feature), (5) $T_{b,min}$ (the lower limit of T_b) and (6) T_b.

Main conclusions. In star-forming regions very compact maser features with angular sizes of 23-60 μas were observed, which correspond to ∼4-10×10^6 km. The best linear resolution was obtained for the H_2O maser in Orion – 4 million km. The best angular resolution for Galactic masers is 23 μas for W49 N (the distance is ∼11 kpc). Brightness temperatures for the most compact maser features range from 10^{14} to a few of 10^{15}.

Acknowledgements

The RadioAstron project is led by the Astro Space Center of the Lebedev Physical Institute of the RAS and the Lavochkin Association of the Russian Federal Space Agency, and is a collaboration with partner institutions in Russia and other countries: IAA RAS, MPIfR, INAF, NRAO, MASTER Robotic Net, ISON and others.

References

Kardashev, N., Alakoz, A., Kovalev, Y. *et al.* 2015, *Solar System Research*, 49, 573
Sobolev, A., Shakhvorostova, N., Alakoz, A., & Baan, W. 2017, *ASP-CS*, 510, 27
Likhachev, S., Kostenko, V., Girin, I. *et al.* 2017, *J. Astron. Instrum.*, 6, 1750004-131
Lobanov, A. 2015, *A&A*, 574, A84
Petrov, L., Kovalev, Y., Fomalont, E., & Gordon D. 2011, *AJ*, 142, 35

Astrophysical Masers:
Unlocking the Mysteries of the Universe
Proceedings IAU Symposium No. 336, 2017
A. Tarchi, M.J. Reid & P. Castangia, eds.

© International Astronomical Union 2018
doi:10.1017/S1743921317010468

H$_2$O maser observation using the 26-meter Nanshan Radio Telescope of the XAO

Yu-Xin He[1,2], **Jarken Esimbek**[2,1], **Jian-Jun Zhou**[2,1], **Gang Wu**[2,1], **Xin-Di Tang**[3,2,1], **Wei-Guang Ji**[2,1], **Ye Yuan**[2,1] **and Da-Lei Li**[2,1]

[1]Xinjiang Astronomical Observatory, Chinese Academy of Sciences,
Urumqi 830011, P. R. China
email: `heyuxin@xao.ac.cn`

[2]Key Laboratory of Radio Astronomy, Chinese Academy of Sciences,
Urumqi 830011, P. R. China
emails: `jarken@xao.ac.cn`, `zhoujj@xao.ac.cn`

[3]Max-Planck-Institut für Radioastronomie, Auf dem Hügel 69, 53121 Bonn, Germany

Abstract. In the past few years, we have performed a 22 GHz H$_2$O maser survey towards hundreds of BGPS sources using the 25-meter Nanshan Radio Telescope (NSRT) of the Xinjiang Astronomical Observatory, and detected more than one hundred masers. Our aim is to study star formation activities associated with these sources, as well as search for any correlations that may exist between 22 GHz H$_2$O masers and the evolutionary stage of high-mass star formation regions. The NSRT has been upgraded and have now an effective diameter of 26 meter. Besides, cryogenically cooled dual-beam receiver systems covering seven millimeter-wave observing bands have been installed on the NSRT. For the next step of maser observation, we will continue to do H$_2$O and SiO masers survey of massive dust clumps and monitor some maser sources.

Keywords. masers, stars: formation, radio lines: ISM

1. Introduction

Many unbiased pilot surveys for Galactic water masers suggest that interstellar H$_2$O masers are very abundant, are the most popular maser species, and one of the most sensitive and reliable tracers of newly forming massive young stellar objects in the Milky Way (MW) (e.g., Caswell & Breen 2010, Walsh *et al.* 2011). At the early stage of star formation, H$_2$O maser act as a good probe for discovering invisible stars which are shielded by the dust in the surrounding molecular cloud, and also for measuring the physical parameters of the environment. Very long baseline interferometry (VLBI) observations of H$_2$O masers are used to measure accurate distances via their trigonometric parallax (Hachisuka *et al.* 2006). Moreover, they have also been successfully used to infer properties of the magnetic field in the massive star-forming region W75N (Surcis *et al.* 2011). Therefore, finding more interstellar H$_2$O masers is important to understand high-mass star formation and construct the MW's structure.

In this paper, we focus on 22 GHz H$_2$O masers that were observed using the previous 25-meter Nanshan Radio Telescope (NSRT) of the Xinjiang Astronomical Observatory. In the past few years, we have performed a maser survey towards hundreds of Bolocam Galactic Plane Survey (BGPS) sources (Xi *et al.* 2015, Xi *et al.* 2016). After the completion of the upgrade, the NSRT has become a 26 meter effective diameter radio telescope with high-precision tracking and pointing. We have also installed a cryogenically cooled dual-beam receiver systems covering seven millimeter-wave observing bands on it. In the next months, we are planning to do an H$_2$O maser survey in single-pointing mode towards 481 molecular clouds identified by Du *et al.* (2016), of which 457 probably belong

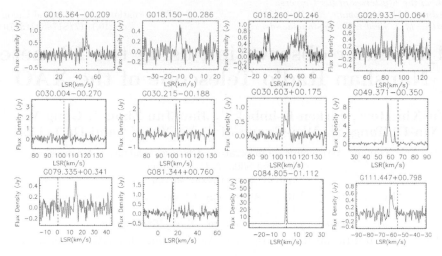

Figure 1. Examples of newly detected H_2O masers.

to the Outer arm. The molecular clouds with H_2O masers will then be used to delineate the Outer arm through VLBI observation using the trigonometric parallax method.

2. Results

We have performed an H_2O maser survey towards BGPS sources twice, once for 221 BGPS sources in the v1.0 catalogue that have 1.1mm fluxes exceeding 5 Jy and once for 274 BGPS sources located at $85° < 1 < 193°$. Finally, we detected 132 H_2O masers in total, of which 17 are new (Fig. 1). The detection rates of H_2O masers are 48.4 per cent and 9 per cent, respectively for the two samples (see Xi *et al.* 2015 and Xi *et al.* 2016 for details). In addition, we obtained the position of the new H_2O masers using OTF observations. Moreover, we found that the detection rate of H_2O masers is correlated with the continuum emission fluxes of BGPS sources at FIR, sub-millimetre and millimetre wavelengths.

Acknowledgements

This work was funded by the Program of the Light in China's Western Region under grant 2016-QNXZ-B-22, the National Natural Science foundation of China under grant 11703073, 11433008 and 11373062. This work is based on observations made with the NSRT, which is operated by the Key Laboratory of Radio Astronomy, Chinese Academy of Sciences.

References

Caswell & Breen 2010, *MNRAS*, 407, 4
Du, Xinyu, Xu, Ye, Yang, Ji, Sun, Yan, Li, Facheng, Zhang, Shaobo, & Zhou, Xin 2016, *ApJS*, 224, 1
Hachisuka, K., Brunthaler, A., Menten, K. M., *et al.* 2006, *ApJ*, 645, 1
Surcis, G., Vlemmings, W. H. T., Curiel, S., Hutawarakorn Kramer, B., Torrelles, J. M., & Sarma, A. P. 2011, *A&A*, 527, A48
Walsh, A. J., Breen, S. L., Britton, T., *et al.* 2011, *MNRAS*, 416, 3
Xi, Hongwei, Zhou, Jianjun, Esimbek, Jarken, Wu, Gang, He, Yuxin, Ji, Weiguang, & Tang, Xiaoke 2015, *MNRAS*, 453, 4
Xi, Hong-Wei, Zhou, Jian-Jun, Esimbek, Jarken, Wu, Gang, He, Yu-Xin, Ji, Wei-Guang, Tang, Xiao-Ke, & Yuan, Ye 2016, *RAA*, 16, 6

Astrophysical Masers:
Unlocking the Mysteries of the Universe
Proceedings IAU Symposium No. 336, 2017
A. Tarchi, M.J. Reid & P. Castangia, eds.

© International Astronomical Union 2018
doi:10.1017/S1743921318000145

Conference Summary

Philip J. Diamond

SKA Organisation, Jodrell Bank
Macclesfield, Cheshire, SK11 9DL, United Kingdom
email: `p.diamond@skatelescope.org`

Abstract. IAU Symposium 336, Astrophysical Masers: Unlocking the Mysteries of the Universe, took place between 4 - 8 September, 2017 in Cagliari, on the beautiful island of Sardinia. The Symposium, the fifth focusing on masers as a tool for astrophysics, was dedicated to our friend and colleague Malcolm Walmsley, who sadly passed away shortly before the meeting. To quote Karl Menten: "Malcolm made numerous fundamental contributions to our understanding of the physics and chemistry of star formation and the interstellar medium. He was an exceptional scientist, a highly esteemed colleague and a true gentleman". Vale Malcolm.

The topics discussed at the symposium covered a huge range, from star-formation, evolved stars, galaxies and their constituents, super-massive black-holes to cosmology.

1. Introduction

The previous symposium in the series (IAUS 287; Booth, Vlemmings & Humphreys 2012) took place in Stellenbosch, South Africa, five years ago; looking back on the topics covered then, they were broadly similar to those discussed in Cagliari but with important and major differences. First, we are seeing the culmination of major, long-term monitoring programmes; secondly, we are seeing the massive impact of ALMA and the JVLA; thirdly, it is clear that GAIA is a game-changer for galactic science; and, finally, the panchromatic information that is now available is enabling a much deeper view of the physical conditions and overall environments in which masers exist than was previously available. And, it is only going to get better.

This summary will not describe every paper presented but will be a personal view of the key advances since the Stellenbosch meeting. In the text below, when referring to the work of colleagues I will, in general, reference the papers from this meeting, except where this is inappropriate.

The Scientific Organising Committee had arranged for several broad reviews, which were uniformly excellent and set the scene and the context for the conference papers. Karl Menten kicked off the meeting with a review of *50+ Years of Maser Research*, this described the state of the art in maser astrophysics prior to this Symposium. Christian Henkel summarised the current situation with surveys of extragalactic masers (Henkel, Greene & Kamali 2018), while Till Sawala provided an excellent review on the Local Group from a perspective outside of radio astronomy. As the meeting focused on the structure of the Milky Way as revealed by maser observations, Ortwin Gerhard set the scene with a review of galactic structure, again from a broader perspective (Gerhard 2018). The two final reviews covered both the birth of stars (Beltran 2018) and the end of their lives, the latter was presented by Perrin.

2. Key advances

Extragalactic Megamasers. The Megamaser Cosmology Project (MCP) is a multi-year VLBI project with the aim of measuring the Hubble Constant through the determination of the geometric distances to circumnuclear 22 GHz H_2O megamasers in galaxies within the Hubble flow. Braatz, on behalf of the MCP team, presented the latest results of this key project (Braatz *et al.* 2018).

178 galaxies are currently known to host H_2O megamasers; their detection rate peaked in the years 2006-2009, with the majority of recent discoveries being made with the Green Bank telescope. Of the 37 megamasers known to be in disk galaxies, 10 have been targeted for distance measurements with a global, highly-sensitive VLBI array. Four galaxies have had their distances measured to date, the current estimate of H_O from those galaxies is $69.3 \pm 4.2 \mathrm{km\ s}^{-1}\ \mathrm{Mpc}^{-1}$. The MCP team expect to determine the distances of five more galaxies in the next year and should achieve a total uncertainty of $\sim 4\%$.

The detection rates of H_2O megamasers from carefully designed surveys has been quite low, typically $\sim 15\%$. However, in two separate papers, Zhang *et al.* (2018) and Panessa *et al.* (2018) have experimented with different selection criteria, namely Seyfert 2 galaxies, and galaxies selected through their hard X-ray emission, respectively. The detection rate from a complete sample approximately doubled, which is encouraging for the future.

There has been surprisingly little progress over recent years in our understanding of OH Megamasers (OHM). The principle result reported at this meeting was the study of OHM magnetic fields reported by Robishaw. 77 OHM have been searched for evidence of Zeeman splitting, 14 have been detected, with a median magnetic field strength of 12mG. To date, little polarization-sensitive VLBI imaging has been conducted so the structure of the magnetic fields remain obscure.

The Structure of the Milky Way. The study of galactic structure through astrometric measurements of the parallax of masers has been one of the highlights of the field over most of the last decade. Two projects, the VLBA key science project, BeSSeL, reported on by Reid (2018) and its equivalent being undertaken with VERA in Japan, reported on by Honma *et al.* (2018), have both been magnificent success stories and have significantly improved our understanding of the size and structure of the Milky Way.

These projects are delivering results at the same time as the ESA satellite GAIA is flying. As reported by Mignard on behalf of the GAIA team, GAIA will determine parallaxes and proper motions for over a billion stars, with associated accurate photometry. With such an enormous database, GAIA will provide a superb view of the structure of the Milky Way in the optical domain, which will provide a highly complimentary picture to that determined through the maser astrometry.

Surveys. Recent improvements in technology and in the capabilities of observing facilities have resulted in a renaissance in surveys of maser sources. I'd like to call out a few specific examples, a list which does not pretend to be complete. Breen, in an invited talk, described the current status of recent southern hemisphere maser surveys (Breen 2018). Two major surveys have been the Methanol Multi-Beam (MMB), which also has a northern hemisphere component, and MALT45, a survey for mm-wave masers. The MMB has detected almost 1000 6.7GHz methanol masers within the galactic plane; like most similar surveys, there have been numerous spin-off projects, many of which have themselves resulted in interesting science and new discoveries. MALT45, an ATCA legacy survey, is observing 90 square degrees of the Southern galactic plane at 7mm, searching for Class 1 methanol masers, SiO masers and recombination lines.

A fascinating new survey is BAADE, which was reported on by Sjouwerman *et al.* (2018). BAADE is a survey of $\sim 34,000$ AGB/RGB stars in the galactic plane and bulge for SiO masers with the JVLA and ALMA. The principal aim of the survey is to significantly improve models of the dynamics and structure of the inner Galaxy, through the use of the stellar masers as radio-detected point-sources. The current status is that all of the JVLA data has been taken, the ALMA data acquisition is underway and pilot follow-up VLBI observations with the EVN, VLBA and VERA have been undertaken.

Star Formation Masers. As is by now normal in any maser symposium, a large fraction of the talks and posters are focused on increasing our understanding of the physics of star-formation as revealed through maser observations. The basic evolutionary sequence for high-mass star-formation, which has been obscure for many years, now seems to be yielding to the weight of observation. In his review, Menten discussed that time sequence in broad terms, which sees H_2O masers associated with the earliest stages, then Class II radiatively pumped CH_3OH masers, followed by OH maser emission.

CH_3OH featured in a major way at this meeting, with 21 talks and numerous posters discussing both Class I and Class II masers. As mentioned above Class II sources are strongly associated with regions of high-mass star-formation, whereas Class I masers are well-known to be associated with outflow sources.

Evolved Stars. Another old favourite for attendees at maser symposia is the now traditional session on evolved stars, and area in which I declare an interest lasting almost 30 years. Imai provided an excellent invited talk describing the rationale behind multi-year, indeed multi-decadal, single dish and VLBI monitoring of circumstellar maser sources. He pointed out that long-term VLBI monitoring of such sources is essential for our understanding of the mass ejections processes and the nature of the environments around AGB stars (Imai 2018).

One particular source highlighted by Imai in his talk was the water fountain in W43A. He and his colleagues, of which I am one, had previously proposed that the water jet we had observed was the result of a precessing jet and had a very short dynamical age. However, new results suggest that the H_2O structures seen are in fact the result of discontinuous ejection of a jet and its interaction with the walls of a previously excavated cavity; this changes the physical nature of the system.

Imai draws attention to the 28 talks and posters which refer in one way or another to observations of masers in evolved stars, a number too large for me on which to provide appropriate attention. The large number does, however, demonstrate that this is still an area of active and advancing research.

3. Conclusions

The Symposium was an excellent week of science, networking, food and drink in a beautiful part of Italy. The meeting was capped with a visit to the Sardinia Radio Telescope, which is now operational. On behalf of all the attendees, I thank the Local Organisers for all their hard work.

Maser research, our subject, is in rude health. The science progress since the Stellenbosch meeting in 2012 is clear and exciting. Although there is some concern about the long-term funding for one of the keystone facilities, the VLBA (or LBO), the loss of which would be catastrophic for maser and other high-resolution astronomy and astrometry, new or upgraded instruments such as ALMA and the JVLA are demonstrating their power. The future on that front is also truly exciting, with instruments such as MeerKAT, ASKAP and ultimately the SKA becoming available.

References

Beltran, M. T. 2018, *in this volume*

Booth, R. S., Vlemmings, W. H.T. & Humphreys, E. M.L. 2012 *Cosmic Masers - from OH to H_o (IAU S287)*

Braatz, J. A., *et al.* 2018, *in this volume*

Breen, S. L. 2018, *in this volume*

Gerhard, O. 2018, *in this volume*

Henkel, C., Greene, J. E., & Kamali, F. 2018, *in this volume*

Honma, M., *et al.* 2018, *in this volume*

Imai, H. 2018, *in this volume*

Panessa, F., *et al.* 2018, *in this volume*

Reid, M. J. 2018, *in this volume*

Sjouwerman, L. O., *et al.* 2018, *in this volume*

Zhang, J-S., *et al.* 2018, *in this volume*

Top row: Anna Bartkiewicz, Liz Humphreys, Ylva Pihlstrom, and Alison Peck in front of the Sardinia Radio Telescope (left); Symposium participants in a wood of ancient oak trees standing close to a menhir while listening about the prehistory of Sardinia (right) Mid row: Alberto Sanna and Anna Bartkiewicz (left); LOC members Tiziana Coiana and Paola Castangia, satisfied of the Symposium outcome (right) Bottom row: a group of participants (from left to right: Tomoya Hirota, Ciriaco Goddi, Jan Brand, and Gordon MacLeod) discussing at the end of a Symposium session (left); James Chibueze enjoying the view of the Sardinia Radio Telescope and the Gerrei area (right)

Photo credits (clockwise from top left): A. Peck, S. Poppi, S. Poppi; P. Soletta; S. Poppi, S. Poppi

Author index

Stecklum, B. – 37
Stevens, J. – 334
Strack, A. – 385
Stroh, M. C. – 399
Su, Y. – 187
Sugiyama, K. – 45, 162, 259, 267, 303, 305
Sun, Y. – 187
Sunada, K. – 162, 259, 307
Surcis, G. – 23, 27, 109, 129, 137, 215, 243, 285
Suyu, H. S. – 80
Suárez, O. – 377, 397
Szymczak, M. – 41, 211, 313, 319, 321

Tafoya, D. – 351, 355, 369
Takefuji, K. – 305
Tamura, Y. – 162
Tanaka, K. E. I. – 45
Tang, X.-D. – 449
Tarchi, A. – 96, 109, 129, 137
Tatematsu, K. – 247
Tobin, T. L. – 53
Tolmachev, A. – 422
Torrelles, J. M. – 351
Trinidad, M. A. – 315, 397
Trois, A. – 109

Ubertini, P. – 96
Uchiyama, M. – 45
Ueno, Y. – 162
Uscanga, L. – 397, 377

Val'tts, I. E. – 63
van den Heever, S. P. – 327
van der Avoird, A. – 23
van der Walt, D. J. – 13, 59, 225, 301, 327
van Langevelde, H. – 211, 411
van Langevelde, H. J. – 23, 27, 184

van Rooyen, R. – 13, 225
Varenius, E. – 285
Vasylenko, A. – 135, 23, 243, 347, 355
Vlemmings, W. H. T. – 27, 285, 369
Voronkov, M. A. – 105, 158

Wada, K. – 365
Walmsley, C. M. – 33
Walsh, A. J. – 267, 295, 397
West, M. – 389
Wilms, J. – 141
Winnberg, A. – 393
Wolak, P. – 41, 313, 319
Wu, G. – 291, 449
Wu, Y. – 259, 299
Wyrowski, F. – 331

Xu, Y. – 187

Yamaguchi, T. – 45
Yamamoto, S. – 391
Yamauchi, A. – 162, 307
Yang, J. – 187
Yonekura, Y. – 45, 267, 305
Yoon, D.-H. – 359
Youngjoo, Y. – 373
Yuan, J. – 299
Yuan, Y. – 449
Yun, Y. – 359
Yung, B. H. K. – 395
Yusef-Zadeh, F. – 172

Zhang, J. S. – 92
Zhang, Q. – 176, 235, 281, 301
Zhang, S.-B. – 187
Zhang, Y. – 395
Zhao, W. – 86
Zhdanov, V. I. – 135
Zheng, X. – 291
Zhou, J.-J. – 449
Zhou, X. – 187